Veterinary Drugs: Synonyms & Properties

Veterinary Drugs: Synonyms & Properties

Edited by

G W A Milne

Ashgate

Published by
Ashgate Publishing Limited
Gower House
Croft Road
Aldershot
Hampshire GU11 3HR
England

Ashgate Publishing Company
131 Main Street
Burlington, VT 05401-5600 USA

Ashgate website: http://www.ashgatechem.com

British Library Cataloguing in Publication Data
Veterinary drugs : names & synonyms
1. Veterinary drugs 2. Veterinary drugs - Terminology
I. Milne, G. W. A. (George William Anthony), 1937-
636 ' .089 ' 51

Library of Congress Cataloguing-in-Publication Data
Veterinary drugs : synonyms & properties / edited by G.W.A. Milne.
p. cm.
ISBN 0-566-08500-3
1. Veterinary drugs--Dictionaries. 2. Veterinary drugs--Indexes. I. Milne, George W. A., 1937-

SF917 .V483 2002
636.089'51--dc21

2001056481

ISBN 0 566 08500 3

Printed and bound in Great Britain by MPG Books Ltd, Bodmin, Cornwall

CONTENTS

PREFACE

The US FDA examines and approves all drugs that are to be marketed in the United States as veterinary medications. In addition to these approved agents, numerous drugs approved for human use are often employed in veterinary practice. Both FDA approved and "non-approved" drugs are described in this book.

Animals and humans have many of the same medical problems and, as a result, drugs that have been used in human medicine are often used in the veterinary context as well. Corticosteroids for example, are widely used in both domains as a treatment for arthritis-related problems, and many penicillin-based drugs are used in human and in veterinary medicine to combat infections.

Beyond this extensive common ground, veterinary medicine must deal with a number of problems that are rare or unknown in human medicine. The use of anthelmintics, for example, is rarely indicated in human medicine, particularly in the first world, but it represents one of the most prevalent treatments in animal medicine. Similarly, coccidiosis, a parasitic infection, is rare in humans, but many bird species, particularly chickens, are susceptible to it and its control has considerable economic benefit. The cost–benefit perceptions in animal medicine differ from those in human medicine. Contagious and life-threatening diseases are pursued aggressively in both areas, but treatment of disorders that affect quality of life is not perceived to be economically justified in, for example, farm animals. Pets or "companion animals" fall between these two extremes, leading to an increased demand for anxiolytics, as well as many other medications used to treat more benign conditions.

Veterinary drug development is a growing industry that, according to some estimates, already is involved in some 25% of pharmaceutical research and development.

This book contains summary information on over 750 chemicals used in approved veterinary drugs and in other preparations of use in animal medicine.

HOW TO USE THIS BOOK

Veterinary Drugs is divided into three parts. A brief description of each part is given below.

PART I: Main Entries

This book contains over 750 monographs on veterinary drugs. These monographs, or "entries", are organized by therapeutic category. There are 118 therapeutic or biological activity categories into which the veterinary drugs are classified (see page xxv). Each entry contains the FDA labeled use for the drug. A few of these drugs have multiple uses in the veterinary environment, and thus a single agent may have more than one entry. Each agent described is associated with its Chemical Abstracts Service (CAS) Registry Number (RN). Because different chemicals can have the same molecular formula or may share a trivial name synonym, the CAS RN is the only datum which uniquely identifies a chemical to the exclusion of all others. Each record also provides, as available, two other numeric identifiers for the chemical. These are the monograph number from the 12th Edition of the *Merck Index* and the European Inventory of Existing Commercial Chemical Substances (EINECS) number. The chemical name, molecular formula, and a list of trade names and synonyms are provided; whenever possible, the regulatory status and physical properties of each compound are described and the known biological activity and acute toxicity are recorded. Where available, a list of FDA-approved veterinary products is listed by New Animal Drug Approval (NADA) number along with the manufacturer or sponsor.

Record Structure

A typical monograph from the entries section of this book is shown on the following page. The first line contains, in bold face, the record number (**27**) and the name of the material (**Butorphanol Tartrate**). The second line gives the CAS Registry Number for the compound (58786-99-5), the corresponding *Merck Index* number (1565) and the EINECS number

(231-443-5). These numbers always appear in the same position (left, center or right) enabling the reader to determine which source they belong to. Whenever CAS Registry Numbers are used in the text, they are always enclosed in brackets, for example [58786-99-5]. The molecular formula and structure of the compound are provided. A list of synonyms follows, including proprietary and other trivial names.

A description of the material and its known uses then follows. Whenever possible, physical properties are presented. These include melting point, boiling point, and optical rotation, as well as density or specific gravity, uv absorption, solubility and acute toxicity, usually expressed as oral dosage in rodents. For agents approved as animal drugs, a list of veterinary products is provided along with their NADA numbers and manufacturers or sponsors.

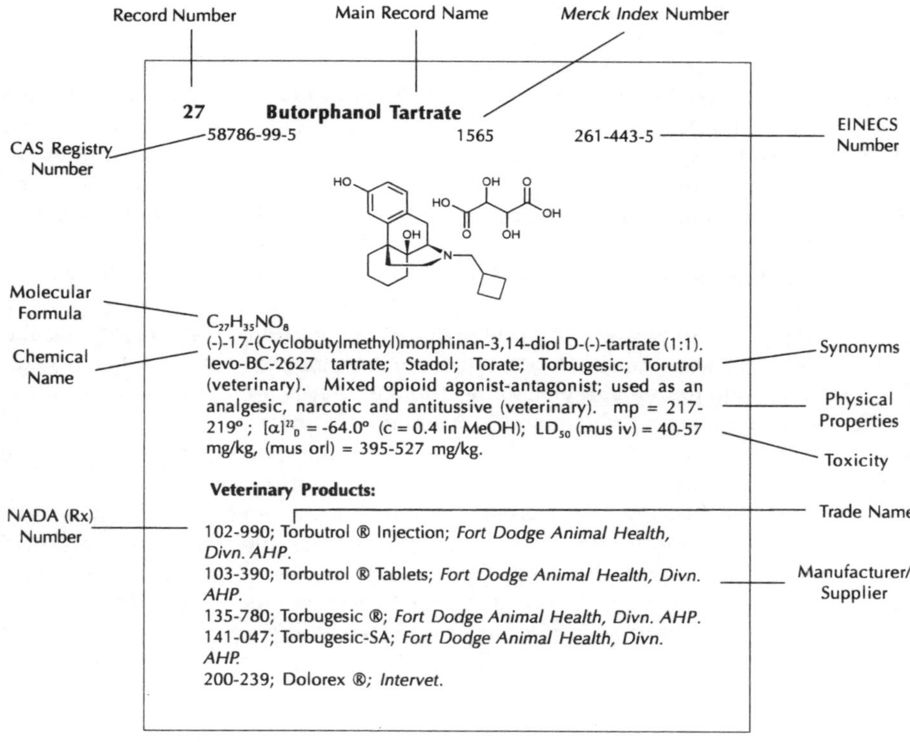

PART II: Indexes

This part contains four indexes. The purpose of each is described below:

- CAS Registry Number Index
  This index enables the reader to locate the record number, and thereby find the main entry for a veterinary drug, based on its CAS Registry Number.

- EINECS Number Index
  This index enables the reader to locate the record number, and thereby find the main entry for a veterinary drug, based on its EINECS number.

- NADA Number Index
  This index allows the reader to locate the record number, and thereby find the main entry for a veterinary drug, based on its NADA number, which the FDA assigns to a new animal drug when it approves the product for manufacture and marketing.

- Name and Synonym Index
  This is the master index containing all chemical and proprietary names found in Part I. It is the most convenient place for the reader to start if a name or synonym for a veterinary agent is known. This index enables the reader to locate the entry number in Part I that relates to the main entry for that chemical. In addition to listing the standard, commercial, and trivial names, the Name and Synonym Index contains the names of veterinary drug manufacturers and suppliers (in ***bold italics***). For each manufacturer or supplier, entry numbers for all the products produced by that company are listed. Therapeutic categories also are provided (in **boldface** type) with the range of entry numbers listed for each.

PART III: Manufacturer and Supplier Directory

This part contains a listing of companies that supply, manufacture, or sponsor veterinary medicinal agents. Each listing includes the company name, address, and contact information. Arranged alphabetically by company name, this directory provides information to help the reader contact the organization directly.

GLOSSARY OF TERMS

Term	**Definition**
aminoglycoside	An antibiotic family whose leading member is streptomycin.
amphilic, amphiphilic	A term describing materials that are attracted to acids and bases.
amphoteric	Capable of acting as an acid or a base.
anaphylaxis	A pronounced, unusual reaction of an organism due to hypersensitivity to a foreign substance, particularly a protein or drug.
anionic	Having the character of an anion, a negatively charged ion, which is normally attracted to an anode, or positive pole.
antioxidant	A substance that inhibits oxidation. Important in biological systems because it inhibits deterioration that is subsequent to oxidation.
antirachitic	Preventing or opposing the development of rickets, a disease most commonly caused by deficiency in vitamin D; an antirachitic agent.
antiscorbutic	Counteracting or preventing scurvy, a disease most commonly caused by deficiency in vitamin C; an antiscorbutic agent.
astringent	An agent that causes contraction of tissue and inhibits discharges of, for example, mucus or blood.
autoclavable	A material that will withstand autoclaving, the sterilization by steam at high temperatures and pressures.
bergamot	Small citrus tree (*citrus bergamia*) with mint-smelling fruit. The name is often applied (incorrectly) to members of the Mentha and Monarda families.
bioavailability	The degree and rate at which a substance, such as an administered drug, reaches the site of action and can be therapeutically effective.

buccal	Pertaining to or directed toward the cheek.
buffer	A substance or mixture that, in solution, resists changes in the pH of the solution.
carminative	Causing expulsion of gas from the stomach or bowel; a carminative agent.
cationic	Having the character of an cation, a positively charged ion, which is normally attracted to a cathode, or negative pole.
chelating agent	A chemical that can complex with other species, particularly metal ions, and so remove them from circulation in the system.
coagulant	An agent that promotes or accelerates the coagulation of blood.
colloid	A state of matter in which microscopic particles are dispersed and/or suspended within a gas, liquid, or solid.
conditioner	A material that, when added to a product, can modify the physical properties, such as viscosity, of the product.
contrast media	Materials that when introduced into a system can provide a background against which other parts of the system may be visualized. Thus barium sulfate, which reflects x-rays, is used to help discern soft tissue bodies in x-ray images.
demulcent	A material that soothes and protects an irritated mucous membrane.
denaturant	A chemical that can deprive a material of some or all of its natural character. Denaturing of a protein, achievable by addition of chemicals such as acids or by heating, follows from the destruction of the protein's tertiary structure, or folding modes.
dessicant	A drying agent.
diaphoretic	Able to induce perspiration; a diaphoretic agent.
emollient	Able to soften or soothe a target, such as the skin; an emollient agent.
emulsifier	An agent that can produce and stabilize an emulsion.
enteric-coated	A tablet or capsule that is coated with a special material such that the release of its contents is delayed until it reaches the enterin, or intestines.
excipient	A pharmacologically inert, adhesive substance (such as honey, syrup, gum arabic) used to bind the contents of a pill or tablet. Also, a relatively inert substance that is added to a drug to facilitate its administration.
exfoliation	The flaking and peeling of tissue cells, particularly of the skin. The process may be pathological or as the result of adminisatration of an exfoliant.
expectorant	An agent that promotes the ejection of fluids from the lungs and airways; having the activity of an expectorant.
gellant	A material that can convert a solution or a solid to a gel, a colloid with firm consistency.
glidant	An agent that promotes the flow characteristics of a substance.
GRAS	Generally Recognized as Safe – a regulatory designation for compounds that have no known toxicity.
hematinic	An agent that promotes an increase in the amount of circulatory iron-containing compounds such as hematin and hemoglobin.

hematopoietic	A drug that promotes blood formation in the body.
humectant	Promoting the retention of water; a humectant agent.
hydroscopic	Sensitive to the presence of water.
hydrotrope	A system or organism that responds in some way to the presence of water.
hygroscopic	Readily absorbing or attracting moisture from the surrounding atmosphere.
intraperitoneal	A route for drug administration involving injection of the drug through the peritoneum, the membrane that lines the abdominopelvic walls.
intravenous	A route for drug administration involving injection of the drug directly into a vein.
keratin	A proteinaceous material that is the major constituent of skin, hair, nails and similar tissues.
liposome	A particle of lipoidal material that is retained in emulsified form in tissue.
nebulizer	A chemical or a device that can create a spray from a liquid or solution.
oleaginous	Oily. Soluble in and a solvent for lipids.
OTC	Over the Counter. A medicinal agent that can be purchased without a medical prescription.
otic	Pertaining to the ear.
parenteral	Situated outside the intestine; an agent intended for parenteral administration. (Subcutaneous or intraparitoneal injections represent parenteral administrations.)
pectoral	Relating to the chest; an agent used to treat diseases of the respiratory tract.
pediculicide	An agent that kills lice.
PEG	Polyethyleneglycol. A polymer of variable molecular weight that is widely used as a carrier or vehicle for drugs.
percutaneous	Through the skin. A route of drug administration.
peritoneal	Pertaining to the peritoneum, the serous membrane lining the abdominopelvic walls. Intraperitoneal administration of a drug requires its injection through the serous membrane.
plaque	A term generally used to describe a conglomeration of material, as in a blood platelet or dental plaque – a deposit on the surface of a tooth.
plasticizer	A chemical that imparts mechanical flexibility to other chemical materials, especially polymers.
POE	Polyoxyethylene. A polymer of variable molecular weight that is widely used as a carrier or vehicle for drugs.
purgative	A material that causes evacuation of the bowels. Cathartic.
rheology	The science of the flow characteristics of materials.
scabicide	An agent that destroys the itch mite that causes scabies.
sequestrant	A chemical that can hold another species (for example, a metal ion) in solution to suppress chemical or biological activity. (See chelating agent.)
spermaticide	A compound that can kill sperm.

subcutaneous	Below the skin. A route for the administration of drugs.
superfatting	Describing an agent that can increase the effective amount of fat in a system.
surfactant	A compound that is surface active and can provide an interface between two phases. Most detergents, for example, are surfactants.
suture	A stitch or series of stitches used most commonly to join together two pieces of tissue such as flaps of skin bordering a wound.
sympathomimetic	Stimulating the sympathetic nervous system; a sympathomimetic agent.
synergist	An agent or drug that assists or cooperates with another drug, the two producing an effect that is greater than the sum of the individual effects.
teratogen	A material that when present in the body can lead to physical defects in a developing embryo.
thixotrope	A material with variable liquidity, for example a gel that becomes liquid upon shaking.
troche	A medicated tablet, disk or lozenge.
uremia	The presence in the blood of urinary chemicals. Caused by numerous conditions, including renal insufficiency.
urticaria	An allergic disorder of the skin characterized by edemous, elevated patches and usually accompanied by severe itching.
viscosity	The property of resistance to flow of a liquid or semifluid.
vulnerary	Promoting the healing of wounds; a vulnerary agent.

GLOSSARY OF UNITS

Name	Description
acute toxicity	Wherever possible the units of toxicity are LD_{50}, the dose which is lethal to 50% of the test animals. In most cases, acute toxicity is measured with the rat, orally administered, and the result is reported as LD_{50} (rat orl) = 50 mg/kg. Other species (for example, mus = mouse; rbt = rabbit; pgn = pigeon; gpg = guinea pig) and occasionally the animal's sex (m = male; f = female) may be cited as might other administration routes (sc = subcutaneous; ihl = inhalation; im = intramuscular; ip = intraperitoneal; iv = intravenous). Chronic toxicity data are not given.
boiling point	When measured at atmospheric pressure, boiling points are cited with no pressure, e.g. bp = 167°. At other pressures, the pressure is also cited, e.g. $bp_{0.01}$ = 167°.
density	The measurement temperature is given as a superscript; thus a density of 1.123 measured at 25° will appear as d^{25} = 1.123. If the measurement is explicitly referenced to the density of water at 4°, the citation will carry both a superscript and a subscript, as in d^{25}_{4} = 1.123. Specific gravities are denoted by the abbreviation sg.

mass	Unless otherwise specified, mass is expressed in a multiple of grams (g), such as micrograms (µg; 10^{-6} g), milligrams (mg; 10^{-3} g), kilograms (kg; 10^{3} g), etc.
melting point	Melting points are cited in degrees Celsius (°C) unless otherwise specified.
optical rotation	Optical rotations (α) are cited with the measurement temperature superscripted, and the measurement wavelength (often the sodium D line, at 549 nm) subscripted, as in $[\alpha]^{25}_{D}$ = 105°. When mutarotation can occur, the rotation given is an equilibrium value, measured after some time interval, which is cited, as in $[\alpha]^{25}_{D}$ = 105° (14 hr).
refractive index	Denoted by the letter n, refractive indexes are usually determined at a temperature which is cited as a superscript, as in n^{25} = 1.5432. The wavelength of the light used in the measurement is cited as a subscript, as in n^{25}_{546} = 1.5432. Most commonly, the sodium D line (wavelength 549 nm) is used, and in such cases the subscript is a D, as in n^{25}_{D} = 1.5432.
temperature	When no units are cited, the temperature given is in degrees Celsius (°C).
uv absorption	The ultraviolet absorption maxima given by the material are cited in nanometers (nm = 10^{-9} m) and the absorptivities (E, A, ε, or log ε, all of which are unitless) may also be given.
volume	Volume is expressed in liters (l) or milliliters (ml) unless otherwise specified.

ABBREVIATIONS AND SYMBOLS

Term	Definition
abs	absolute
abs config	absolute configuration
Ac–	acetyl (CH_3CO–)
ACE	angiotensin-converting enzyme
ACTH	adrenocorticotrophic hormone
AIDS	acquired immunodeficiency syndrome
alc	alcohol, alcoholic
amp.(s)	ampule(s)
AMP	adenosine 5′-monophosphate
aq	aqueous
atm	atmosphere, atmospheric
Bn–	benzyl (C_7H_7–)
BOC	tert-butoxycarbonyl ($C_5H_9O_2$–)
bp	boiling point
BP	British Pharmacopoeia
BPH	benign prostatic hypertrophy
Bu–	butyl (C_2H_5–)
Bz–	benzoyl (C_6H_5CO–)
c	concentration (usually in g/100 ml)
C	Celsius (temperature scale)
cAMP	cyclic AMP
CBZ	carbobenzyloxy ($C_8H_7O_2$–)
cc	cubic centimeters (milliliters)

CCK	cholecystokinin
CCL	*Candida* cylindrical lipase
CCl_4	carbon tetrachloride
$CHCl_3$	chloroform
CH_2Cl_2	methylene chloride
CH_3CN	acetonitrile
C_5H_5N	pyridine
C_6H_6	benzene
C_7H_8	toluene
cm	centimeter
CNS	central nervous system
CoA	coenzyme A
COD	cyclooctadiene
COMT	catechol-O-methyltransferase
CPMP	Commission on Proprietary Medicinal Products
d	density
d-	dextro(rotatory)
dec	decompose, decomposition
DIPT	diisopropyltartrate
dl-	racemic
DL-	racemic
DMA	dimethylacetamide
DMF	dimethylformamide
DMSO	dimethylsulfoxide
DNA	deoxyribonucleic acid
DOPA	dihydroxyphenylalanine
(E)-	*entgegen* (opposite)
EC	Enzyme Commission
ED	effective dose
EDTA	ethylenediamine tetraacetic acid
ee	enantiomeric equivalent
e.g.	for example
EINECS	European Inventory of Existing Commercial Chemical Substances
endo-	stereochemical descriptor
Et–	ethyl (C_2H_5–)
Et_2O	diethyl ether
EtOAc	ethyl acetate
EtOH	ethanol
exo-	stereochemical descriptor
F	Fahrenheit (temperature scale)
FDA	(United States) Food and Drug Administration
FEMA	Federal Emergency Management Agency
FMOC	fluoromethoxycarbonyl ($C_2F_3O_2$–)

g	gram(s)
g/l	grams/liter
gal	gallon(s)
GI	gastrointestinal
GLC	gas liquid chromatography
gpg	guinea pig
GRAS	Generally Recognized as Safe (a regulatory designation)
HCl	hydrochloric acid
HIV	human immunodeficiency virus
HKR	hydrolytic kinetic resolution
HMG-CoA	3-hydroxy-3-methylglutaryl coenzyme A
hmtr	hamster
H_2O	water
hr	hour
H_2SO_4	sulfuric acid
HT	hydroxytryptamine (serotonin)
ihl	inhalation
im	intramuscular
inj.	injection
ip	intraperitoneal
iPr–	isopropyl ($(CH_3)_2CH–$)
ir	infrared
iv	intravenous
JP	Japan
kcal	kilocalories
l	liter
l-	levo (rotatory)
λ (lambda)	wavelength
LC	lethal concentration
LC_{50}	median lethal concentration
LD	lethal dose
LD_{50}	median lethal dose
log	common logarithm
LSR	lanthnide shift reagent
MAO	monoamine oxidase
max	maximum, maxima
Me–	methyl (CH_3–)
Me_2CO	acetone
MeOH	methanol
mg	milligram
min	minimum, minima, minute
MLD	minimum lethal dose
mp	melting point
μg	microgram
mμ	millimicron (nanometer)

Ms–	mesyl (CH_3O_2S–)
mus	mouse
N	normal, normality
NADA	New Animal Drug Approval (by the US FDA)
NBD	norbornadiene
nm	nanometer (10^{-9} m)
NMO	N-methylmorpholine N-oxide
NMR	nuclear magnetic resonance
NSAID	non-steroidal anti-inflammatory drug
NSC	National Service Center (of the National Cancer Institute)
NTP	normal temperature, pressure
o-	ortho
OD	optical density
orl	oral
OTC	over the counter
p-	para
PEG	polyethylene glycol
pgn	pigeon
Ph–	phenyl (C_6H_5–)
pH	acid–base scale (log of reciprocal hydrogen ion concentration)
Ph. Eur.	European Pharmacopoeia
pK	log of the reciprocal of the dissociation constant
PLE	pig liver esterase
PMA	Pharmaceutical Manufacturing Association
POE	Polyoxyethylene
pOH	acid–base scale (log of reciprocal hydroxyl ion concentration)
ppb	parts-per-billion
PPL	porcine pancreatic lipase
ppm	parts-per-million
Pr–	propyl (C_3H_7–)
(R)	rectus (stereochemical descriptor)
rbt	rabbit
RNA	ribonucleic acid
(S)	sinister (stereochemical descriptor)
S-	symmetrical
sc	subcutaneous
SDM	site directed mutagenesis
sec	second
sec-	secondary
SG, sg	specific gravity
sp.	species (specific)
spp.	species (plural)
STP	standard temperature, pressure
tabl.	tablet

TBHP	tert-butyl hydroperoxide
temp	temperature
tert-	tertiary
THF	tetrahydrofuran
THP	tetrahydopyranyl (C_5H_7O–)
Ts–	tosyl ($C_7H_7O_7S$–)
TSCA	Toxic Substances Control Act
UK	United Kingdom
USA	United States of America
USAN	United States Adopted Names
USP	United States Pharmacopeia
USP/NF	United States Pharmacopeia/National Formulary
uv	ultraviolet
v/v	volume in volume
VIS	visible
viz.	namely
w/v	weight in volume
w/w	weight in weight
wt	weight
(Z)-	*zusammen* (on the same side)
>	greater than
<	less than
~	approximately
Å	Angstrom units (10^{-8} cm)

TABLE OF THERAPEUTIC CATEGORIES

PART I

MAIN ENTRIES

ACE Inhibitors

1 **Benazepril**
86541-75-5 1058

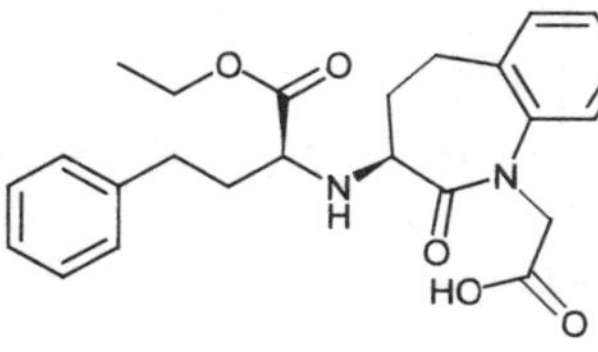

$C_{24}H_{28}N_2O_5$
(3S)-3-[[(1S)-1-Carboxy-3-phenylpropyl]amino]-2,3,4,5-tetrahydro-2-oxo-1H-1-benzazepine-1-acetic acid 3-ethyl ester.
CGS-14824A; Briem; Cibacen; Cibacène; Lotensin. Angiotensin-converting enzyme inhibitor. Used to treat hypertension in humans and dogs. mp= 148-149°; $[\alpha]_D$ = -159° (c = 1.2 EtOH).

2 **Benazepril Hydrochloride**
86541-74-4 1058

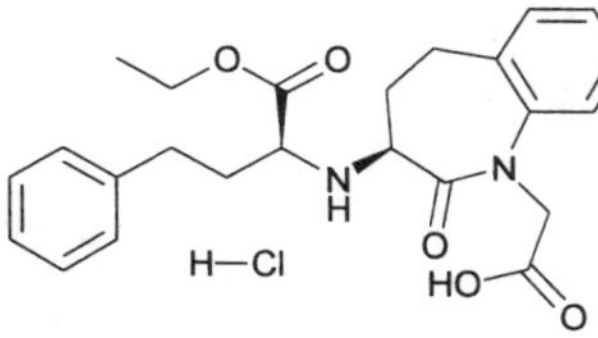

$C_{24}H_{29}ClN_2O_5$
(3S)-3-[[(1S)-1-Carboxy-3-phenylpropyl]amino]-2,3,4,5-tetrahydro-2-oxo-1H-1-benzazepine-1-acetic acid 3-ethyl ester hydrochloride.
Lotensin; CGS-14824A HCl; component of Lotensin-HCT, Lotrel, Lotrel capsules. Angiotensin-converting enzyme inhibitor. Used to treat hypertension in humans and dogs. mp = 188-190°; $[\alpha]_D$ = -141° (c = 0.9 EtOH).

3 **Captopril**
62571-86-2 1817 263-607-1

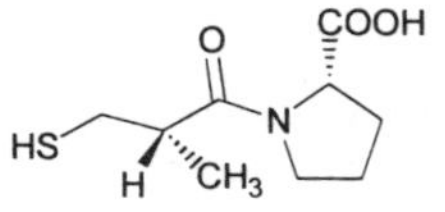

$C_9H_{15}NO_3S$
1-[(2S)-3-Mercapto-2-methylpropionyl]-L-proline.
Capoten; SQ-14225; Acediur; Acepril; Aceplus; Alopresin; Acepress; Capoten; Captolane; Captoril; Cesplon; Dilabar; Garranil; Hipertil; Lopirin; Lopril; Tensobon; Tensoprel; component of: Capozide, Acezide, Captea, Ecazide. Angiotensin-converting enzyme inhibitor. Orally active peptidyldipeptide hydrolase inhibitor. Used to treat hypertension and congestive heart failure. mp = 103-104°, 86°, 87-88°, 104-105°; $[\alpha]_D^{22}$ = -131.0° (c = 1.7 EtOH); freely soluble in H_2O, EtOH, $CHCl_3$, CH_2Cl_2; LD_{50} (mus iv) = 1040 mg/kg, (mus orl) = 6000 mg/kg.

4 **Enalapril Maleate**
76095-16-4 3605 278-375-7

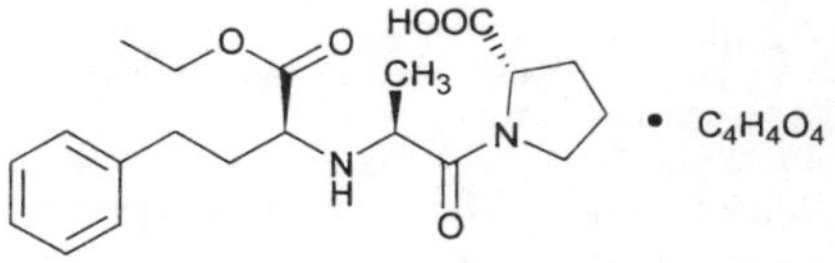

$C_{24}H_{32}N_2O_9$
1-[N-[(S)-1-Carboxy-3-phenylpropyl]-L-alanyl]-L-proline 1'-ethyl ester maleate (1:1).
Enacard; Renitec; Vasotec; MK-421; Amprace; Bitensil; Cardiovet; Enaloc; Enapren; Glioten; Hipoartel; Innovace; Lotrial; Olivin; Pres; Reniten; Renivace; Xanef; component of: Vaseretic, Acesistem, Co-Renitec, Innozide, Renacor, Xynertec. Angiotensin-converting enzyme inhibitor used in veterinary medicine as a vasodilator to treat heart failure and hypertension. mp = 143-144.5°; $[\alpha]_D^{25}$= -42.2° (c = 1 MeOH); soluble in H_2O (2.5 g/100 ml), EtOH (8 g/100 ml), MeOH (20 g/100 ml).

Veterinary Products:

141-015; Enacard ®; Enacard ® Tablets For Dogs; *Merial.*

alpha-Adrenergic Agonists

5 **Dopamine**
51-61-6 3479 200-110-0

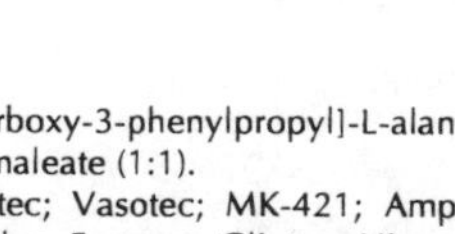

$C_8H_{11}NO_2$
4-(2-Aminoethyl)pyrocatechol.
3-hydroxytyramine. Adrenergic agent. Can also be used as an antihypotensive agent in treatment of, for example, shock.

6 **Dopamine Hydrochloride**
62-31-7 3479 200-527-8
$C_8H_{12}ClNO_2$
4-(2-Aminoethyl)pyrocatechol hydrochloride.
Dopastat; ASL-279; Cardiosteril; Dynatra; Inovan; Inotropin. Adrenergic; used as an antihypotensive agent in treatment of, for example, shock. Dec 241°; freely soluble in H_2O; soluble in EtOH, MeOH; insoluble in Et_2O, $CHCl_3$, C_6H_6, C_7H_8, petroleum ether; [hydrobromide]: mp = 210-214° (dec).

alpha-Adrenergic Antagonists

7 **Atipamezole Hydrochloride**
104075-48-1 894

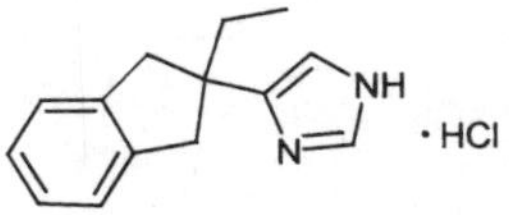

$C_{14}H_{17}ClN_2$
4-(2-Ethyl-2,3-dihydro-1H-inden-2-yl)-1H-imidazole hydrochloride.

MPV-1248; Antisedan. Antidote to α_2 adrenoceptor-induced sedation. Used to reverse sedative effects of medetomidine. Crystals; mp = 211-215°; LD_{50} (mrat sc) > 50 mg/kg, (frat sc) > 44 mg/kg.

Veterinary Products:

141-033; Antisedan®; *Orion-Farmos.*

8 **Yohimbine Hydrochloride**
65-19-0 10236 200-600-4

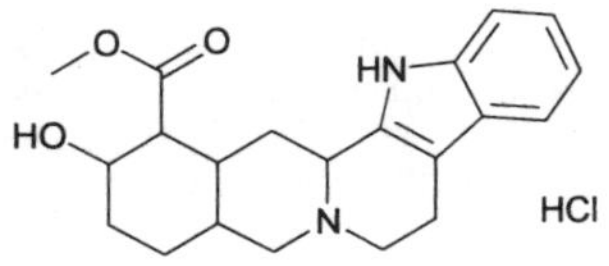

$C_{21}H_{27}ClN_2O_3$
(16α,17α)-17-Hydroxyyohimban-16-carboxylic acid methyl ester hydrochloride.
Aphrodyne; Yocon; Yohimex; Yohydrol. α-Adrenergic blocker. Has been used as an aphrodisiac in animals. Dec 302°; $[\alpha]_D^{22}$ = +105° (H_2O); soluble in H_2O (0.8 g/100 ml), EtOH (0.25 g/100 ml), pH of aqueous solution = 7.

Veterinary Products:

140-866; Yobine® Injectable Solution; Yobine® Injection; *Lloyd.*
140-874; Antagonil; *Wildlife.*

beta-Adrenergic Agonists

9 **Clenbuterol Hydrochloride**
21898-19-1 2407 244-643-7

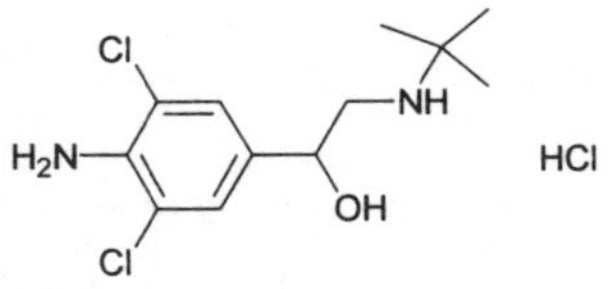

$C_{12}H_{19}ClN_2O$
4-Amino-α-[(tert-butylamino)methyl]-3,5-dichlorobenzyl alcohol monohydrochloride.

NAB-365Cl; Spiropent; Ventipulmin. Ephedrine derivative. β-Adrenergic agonist. Used as a bronchodilator, particularly in horses. mp = 174-175°; soluble in H_2O, MeOH, EtOH; slightly soluble in $CHCl_3$; insoluble in C_6H_6; LD_{50} (mus orl) = 176 mg/kg, (rat orl) = 315 mg/kg, (gpg orl) = 67.1, (mus iv) = 27.6 mg/kg.

Veterinary Products:

140-973; Vetipulmin® Syrup; *Boehringer Ingelheim Vetmedica.*

10 **Dobutamine**
34368-04-2 3456

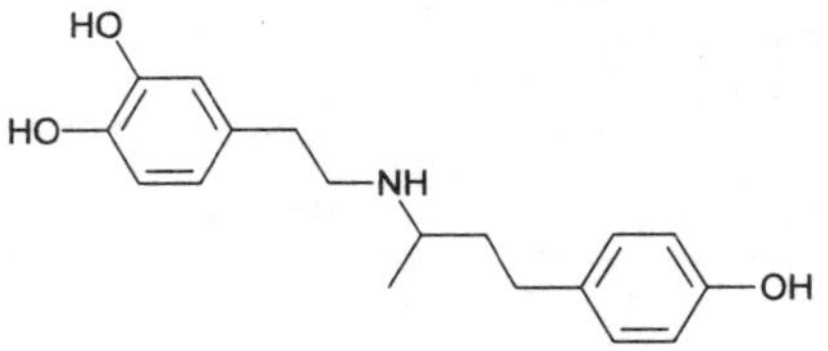

$C_{18}H_{23}NO_3$
(±)-4-[2-[[3-(p-Hydroxyphenyl)-1-methylpropyl]-amino]ethyl]pyrocatechol.
Compound 81929. Cardiotonic. Used in short-term treatment of heart failure.

11 **Dobutamine Hydrochloride**
49745-95-1 3456 256-464-1
$C_{18}H_{24}ClNO_3$
(±)-4-[2-[[3-(p-Hydroxyphenyl)-1-methylpropyl]-amino]ethyl]pyrocatechol hydrochloride.
Dobutrex; Inotrex; 46236. Cardiotonic. Used in short-term treatment of heart failure. mp = 184-186°; λ_m = 281, 223 nm (ε 4768, 14400, MeOH); LD_{50} (mus iv) = 73 mg/kg.

12 **Isoproterenol**
7683-59-2 5236 231-687-7

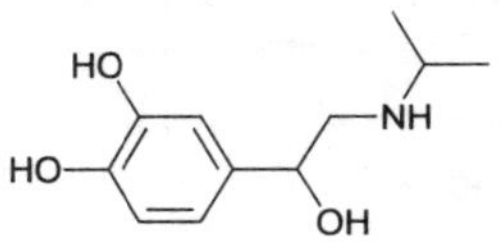

$C_{11}H_{17}NO_3$
3,4-Dihydroxy-α-[(isopropylamino)methyl]benzyl alcohol.
epinephrine isopropyl homolog; isoprenaline; dihydroxyphenylethanolisopropylamine; *N*-isopropyl-noradrenaline; A-21; Aludrine; Aleudrin; Isuprel; Norisodrine; Asiprenol; Asmalar; Neo-Epinine; Novodrin; Isupren; Neodrenal; Isopropydrin; Assiprenol; Respifral; Bellasthman; Saventrine; Proternol; Isorenin; Vapo-N-Iso; Isonorin. β-Adrenergic agonist; catecholamine derivative. Ephedrine derivative. Bronchodilator; anticholingergic. Used to treat cardiac arrhythmia and bronchial constriction. [(dl)-form]: mp = 155.5°; [(l)-form]: mp = 164-165°; $[\alpha]_D^{19}$ = -45.0° (c = 2 in 2N HCl); pKa = 8.64; LD_{50} (male rat orl) = 3675 mg/kg.

13 Isoproterenol Hydrochloride
51-30-9 5236 200-089-8
$C_{11}H_{18}ClNO_3$
3,4-Dihydroxy-α-[(isopropylamino)methyl]benzyl alcohol hydrochloride.
Aerolone; Aerotrol; Euspiran; Isomenyl; Isovon; Mistarel; Suscardia; component of: Aerolone Solution, Duo-Medihaler. β-Adrenergic agonist; catecholamine derivative. Ephedrine derivative. Bronchodilator; anticholingergic. Used to treat cardiac arrhythmia and bronchial constriction. [(dl)-form]: mp = 170-171° [(l)-form]: dec 162-164°; $[\alpha]_D^{19}$ = -50°; soluble in H_2O; less soluble in EtOH; nearly insoluble in many organic solvents; LD_{50} (rat orl) 2220 mg/kg.

14 Terbutaline
23031-25-6 9302 245-385-8

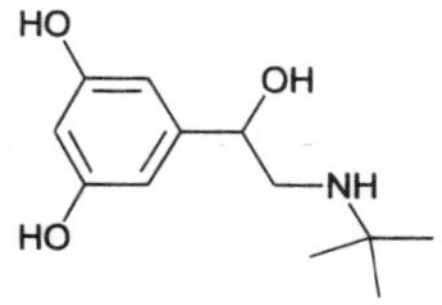

$C_{12}H_{19}NO_3$
5-[2-[(1,1-Dimethylethyl)amino]-1-hydroxyethyl]-1,3-benzenediol.
Bronchodilator; tocolytic. β-Adrenergic agonist. Ephedrine derivative. Used in small animals in the adjunctive treatment of cardiopulmonary diseases. Has been used as a brochodilator in horses, but has been superseded by clenbuterol in this application. mp = 119-120°.

15 Terbutaline Sulfate
23031-32-5 9302 245-386-3
$C_{24}H_{40}N_2O_{10}S$
5-[2-[(1,1-Dimethylethyl)amino]-1-hydroxyethyl]-1,3-benzenediol sulfate (2:1) (salt).
KWD-2019; Brethaire; Brethine; Bricanyl; Butaliret; Monovent; Terbasmin; Terbul. Bronchodilator; tocolytic. β-Adrenergic agonist. Ephedrine derivative. Used in small animals in the adjunctive treatment of cardiopulmonary diseases. Has been used as a brochodilator in horses, but has been superseded by clenbuterol in this application. mp = 246-248°; λ_m = 276 nm ($A^{1\%}_{1cm}$ 67.6 0.1 N HCl); pKa_1 = 8.8, pKa_2 = 10.1, pKa_3 = 11.2; soluble in H_2O (>2.0 g/100 ml), 0.1N HCl (>2.0 g/100 ml), 0.1N NaOH (>2.0 g/100 ml), EtOH (0.012 g/100 ml), 10% EtOH (>2.0 g/100 ml), MeOH (0.27 g/100 ml).

Adrenocorticotropic Hormones

16 Corticotropin
9002-60-2 136 232-659-7

Ser-Tyr-Ser-Met-Glu-His-Phe-Arg-
Trp-Gly-Lys-Pro-Val-Gly-Lys-Lys-
Arg-Arg-Pro-Val-Lys-Val-Tyr-Pro-NH_2

ACTH.
adrenocorticotropin; adrenocorticotrophin; Acethropan; Acortan; Acorto; Acthar; Acton; Actonar; Adrenomone; Alfatrofin; Cibacthen; Corstiline; Cortiphyson; Cortrophimn; Isactid; Reacthin; Solacthyl; Tubex; Ser-Tyr-Ser-Met-Glu-His-Phe-Arg-Trp-Gly-Lys-Pro-Val-Gly-Lys-Lys-Arg-Arg-Pro-Val-Lys-Val-Tyr-Pro-NH_2. Single chain peptide with 39 amino acid residues; isolated from the pituitary gland. Adrenocorticotropic hormone, used to stimulate glucocorticoid production. Used in dogs, cats, beef and diary cattle for diagnosis of hyper- or hypoadrenocorticism and to stimulate the adrenal cortex. Freely soluble in H_2O, appreciably soluble in 70% aqueous alcohol or acetone.

Veterinary Products:

008-760; Adrenomone; *Summit Hill.*
140-583; ACTH Gel; *Anthony Products.*

Anabolics

17 Boldenone 10-Undecenoate
13103-34-9 1354 236-024-5

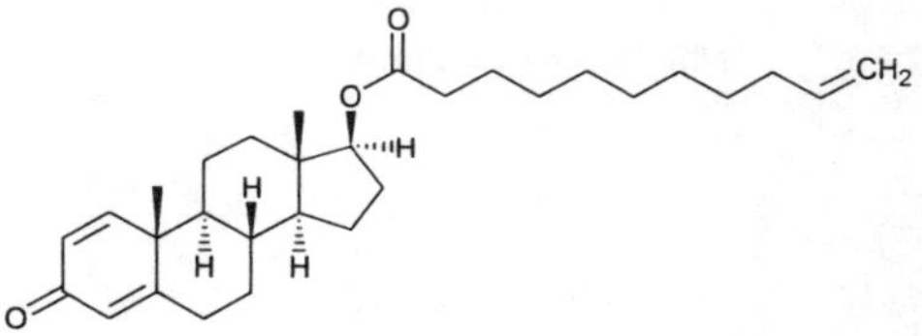

$C_{30}H_4O_3$
17β-Androsta-1,4-dien-3-one 10-undecenoate.
Equipoise; Ba-9038; Parenabol. Androgen with properties of an anabolic steroid.

Veterinary Products:

034-705; Equipoise®; *Fort Dodge Animal Health, Divn. AHP.*

18 Danazol
17230-88-5 2875 241-270-1

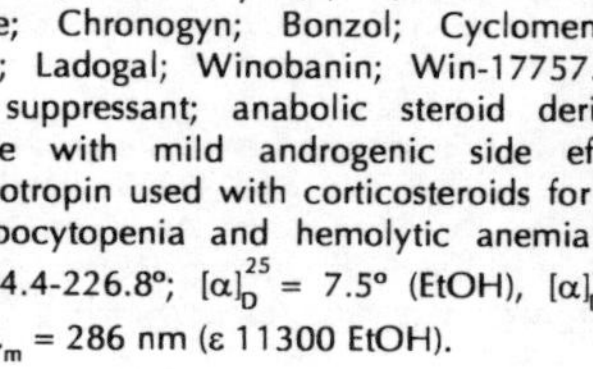

$C_{22}H_{27}NO_2$
17α-Pregna-2,4-dien-20-yno[2,3-d]isoxazol-17-ol.
Danocrine; Chronogyn; Bonzol; Cyclomen; Danol; Danovaol; Ladogal; Winobanin; Win-17757. Anterior pituitary suppressant; anabolic steroid derivative of ethisterone with mild androgenic side effects. An antigonadotropin used with corticosteroids for treatment of thrombocytopenia and hemolytic anemia in dogs. mp = 224.4-226.8°; $[\alpha]_D^{25}$ = 7.5° (EtOH), $[\alpha]_D^{25}$ = 21.9° ($CHCl_3$); λ_m = 286 nm (ε 11300 EtOH).

19 Mibolerone

3704-09-4 6262 223-046-5

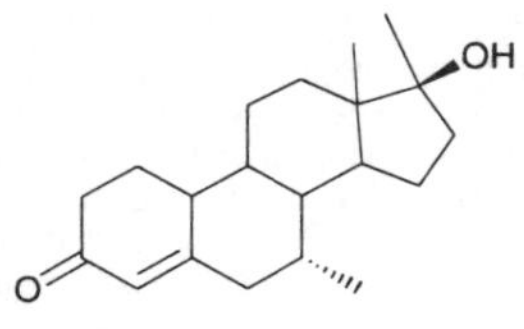

$C_{20}H_{30}O_2$

17β-Hydroxy-7α,17-dimethylestr-4-en-3-one.

U-10997; Cheque; Matenon. An anabolic steroid with androgenic properties. Soluble in H_2O (4.54 mg/100 ml at 37°).

Veterinary Products:

102-709; Cheque® Drops; *Pharmacia & Upjohn.*
107-347; Cheque® Medicated Dog Food; *Pharmacia & Upjohn.*

20 Nandrolone Decanoate

360-70-3 6452 206-639-3

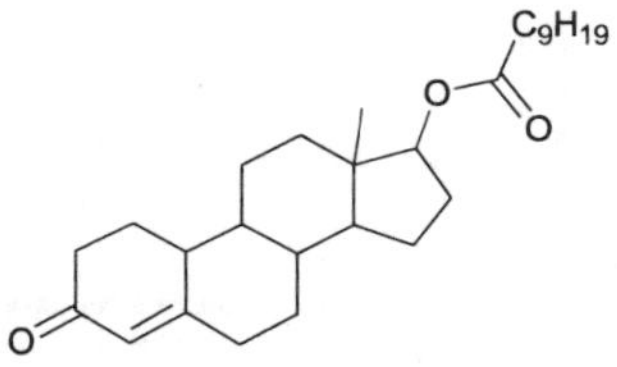

$C_{28}H_{44}O_3$

17β-Hydroxyestr-4-en-3-one decanoate.

19-nortestosterone decanoate; Deca-Durabol; Deca-Hybolin; Hybolin decanoate; Deca-Durabolin; Nandrobolic L.A.; Retabolil. Androgen. Used to stimulate erythropoiesis and possibly also appetite in anemic animals. mp= 32-35°; soluble in EtOH, Et_2O, Me_2CO, $CHCl_3$; insoluble in H_2O.

21 Stanozolol

10418-03-8 8951 233-894-8

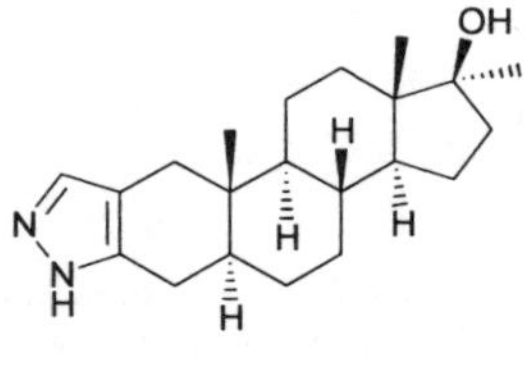

$C_{21}H_{32}N_2O$

17-Methyl-2'H-5α-androst-2-eno[3,2-c]pyrazol-17β-ol.

Winstrol; Win-14833; NSC-43193; androstanazole-stanazol; Stromba; Strombaject. Androgen. Used to treat anemia associated with chronic disease. mp = 229.8-242°; $[\alpha]_D$ = 35.7° ($CHCl_3$), 48.6° (EtOH); λ_m = 223 nm (ε 4740).

Veterinary Products:

015-506; Winstrol-V ® Tablets; *Pharmacia & Upjohn.*
030-844; Winstrol-V ® Sterile Suspension; *Pharmacia & Upjohn.*
135-544; Winstrol-V ® Chewable Tablets; *Pharmacia & Upjohn.*

22 Testosterone Propionate

57-85-2 9322 200-351-1

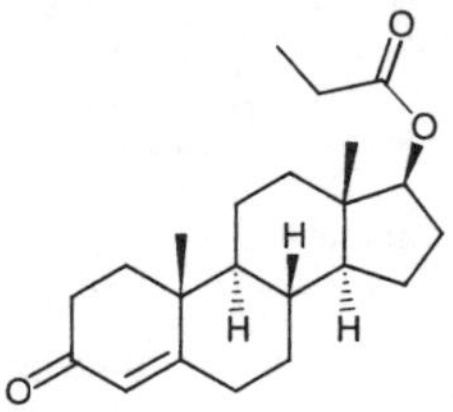

$C_{22}H_{32}O_3$

17-[(1-Oxopropanoyl)oxy]androst-4-en-3-one.

Anertan; Enarmon; Neo-Hombreol; Orchisterone; Perandren; Synandrol; Synerone; Testex; Testoviron; Virormone. Androgen. Used to treat testosterone-responsive urinary incontinence in neutered dogs. mp = 118-122°; $[\alpha]_D^{25}$ = 83° to 90° (c = 1 dioxane); soluble in EtOH, Et_2O, C_5H_5N, other organic solvents, insoluble in H_2O.

Veterinary Products:

011-427; (with Estradiol Benzoate) Synovex ®-H; *Fort Dodge Animal Health, Divn. AHP.*
135-906; (with Estradiol Benzoate, Tylosin Tartrate) Component™ T-H; Component™ E-H with Tylan®; Heifer-Oid; *Ivy.*

23 Trenbolone Acetate

10161-34-9 9716 233-432-5

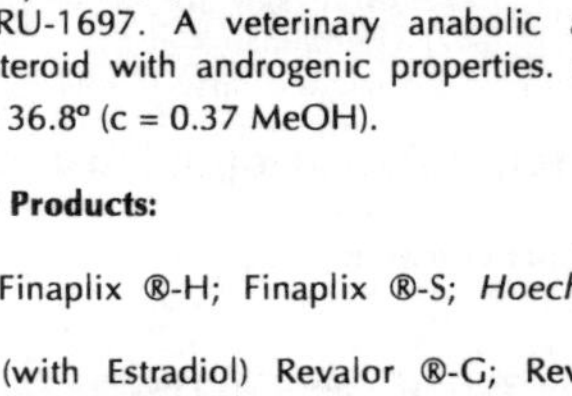

$C_{20}H_{24}O_3$

17β-Hydroxyestra-4,9,11-trien-3-one acetate.

Finaplix; RU-1697. A veterinary anabolic agent. An anabolic steroid with androgenic properties. mp = 96-97°; $[\alpha]_D^{20}$ = 36.8° (c = 0.37 MeOH).

Veterinary Products:

138-612; Finaplix ®-H; Finaplix ®-S; *Hoechst-Roussel Vet.*
140-897; (with Estradiol) Revalor ®-G; Revalor ®-S; *Hoechst-Roussel Vet.*
140-992; (with Estradiol) Revalor ®-H; *Hoechst-Roussel Vet.*
141-043; (with Estradiol Benzoate) Synovex ® Plus; *Fort Dodge Animal Health, Divn. AHP.*
200-221; (with Estradiol, Tylosin Tartrate) Component™

TE-S; Component™ TE-S with Tylan®; *Ivy*.
200-224; (with Tylosin Tartrate) Component™ T-H with Tylan®; Component™ T-S with Tylan®; *Ivy*.

24 Zeranol
26538-44-3 10251 247-769-0

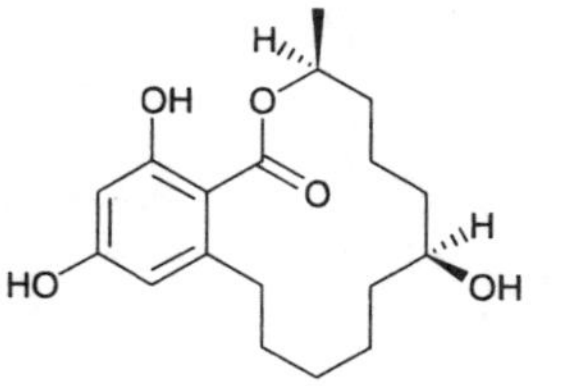

$C_{18}H_{26}O_5$
(3S,7X)-3,4,5,6,7,8,9,10,11,12-Decahydro-7,14,16-trihydroxy-3-methyl-1H-2-benzoxacyclotetradecin-1-one.
Ralabol; Ralgro; THFES(HM); P-1496; MK-188. An anabolic agent. Has two crystalline diastereoisomers; mp = 146-148°, $[\alpha]_D^{25}$ = +39° (MeOH) and mp = 178-180°; $[\alpha]_D^{25}$ = +46° (MeOH); λ_m = 218; 265; 284 nm (MeOH).

Veterinary Products:

038-233; Ralgro® Implants; Ralgro® Magnum; *Schering-Plough Animal Health*.

Analgesics, Narcotic

25 Buprenorphine
52485-79-7 1522 257-950-6

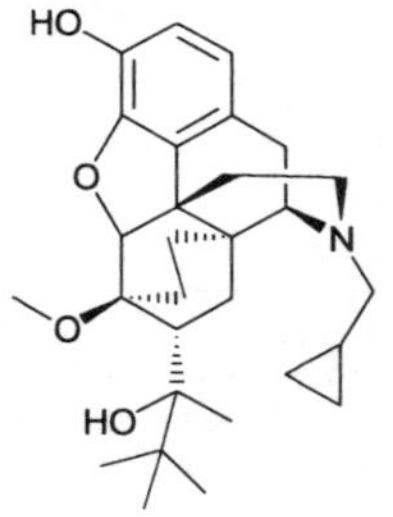

$C_{29}H_{41}NO_4$
[5α,7α(S)]-17-(Cyclopropylmethyl)-α-(1,1-dimethylethyl)-4,5-epoxy-18,19-dihydro-3-hydroxy-6-methoxy-α-methyl-6,14-ethenomorphinan-7-methanol.
RX-6029-M. An analgesic that demonstrates narcotic agonist-antagonist properties. Federally controlled substance (narcotic). Used in horses, in combination with acepromazine or xylazine, as a neuroleptanalgesic. mp = 209°.

26 Buprenorphine Hydrochloride
53152-21-9 1522 258-396-8
$C_{29}H_{42}ClNO_4$
[5α,7α(S)]-17-(Cyclopropylmethyl)-α-(1,1-dimethylethyl)-4,5-epoxy-18,19-dihydro-3-hydroxy-6-methoxy-α-methyl-6,14-ethenomorphinan-7-methanol hydrochloride.
CL-112302; NIH-8805; UM-952; Buprenex; Lepetan; Temgesic. An analgesic that demonstrates narcotic agonist-antagonist properties. Federally controlled substance (narcotic). Used in horses, in combination with acepromazine or xylazine, as a neuroleptanalgesic.

27 Butorphanol Tartrate
58786-99-5 1565 261-443-5

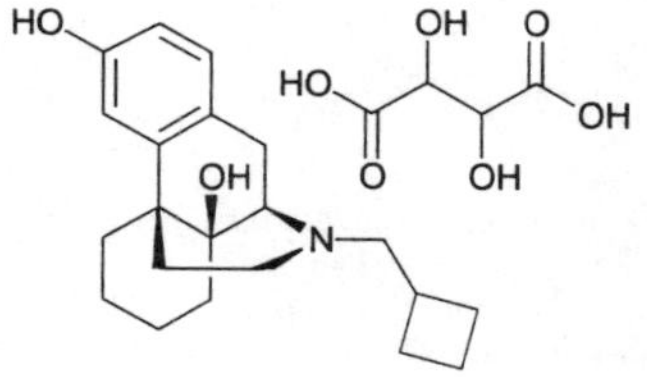

$C_{27}H_{35}NO_8$
(-)-17-(Cyclobutylmethyl)morphinan-3,14-diol D-(-)-tartrate (1:1).
levo-BC-2627 tartrate; Stadol; Torate; Torbugesic; Torutrol (veterinary). Mixed opioid agonist-antagonist; used as an analgesic, narcotic and antitussive (veterinary). mp = 217-219°; $[\alpha]_D^{22}$ = -64.0° (c = 0.4 in MeOH); LD_{50} (mus iv) = 40-57 mg/kg, (mus orl) = 395-527 mg/kg.

Veterinary Products:

102-990; Torbutrol ® Injection; *Fort Dodge Animal Health, Divn. AHP*.
103-390; Torbutrol ® Tablets; *Fort Dodge Animal Health, Divn. AHP*.
135-780; Torbugesic ®; *Fort Dodge Animal Health, Divn. AHP*.
141-047; Torbugesic-SA; *Fort Dodge Animal Health, Divn. AHP*.
200-239; Dolorex ®; *Intervet*.

28 Carfentanil Citrate
61380-27-6 262-748-6

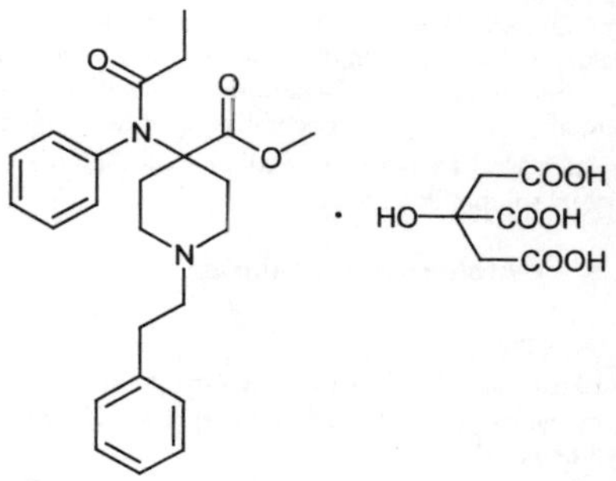

$C_{30}H_{38}N_2O_{10}$
Methyl 1-phenethyl-4-(N-phenylpropionamido)-isonipecotate citrate (1:1).
R-33799. Analgesic, narcotic.

Veterinary Products:

139-633; Wildnil®; *Wildlife*.

29 Codeine
76-57-3 2525 200-969-1

$C_{18}H_{23}NO_4$
7,8-Didehydro-4,5α-epoxy-3-methoxy-17-methyl morphinan-6α-ol monohydrate.
Analgesic, narcotic; antitussive. Federally controlled substance (opiate). Oral analgesic in small animals. mp = 154-156°; d_4^{20} = 1.32; $[\alpha]_D^{15}$ = -136° (c= 2, EtOH); pK = 6.05; soluble in H_2O (0.8 g/100 ml); more soluble in organic solvents.

30 Codeine Phosphate
52-28-8 2528 200-137-8

$C_{18}H_{24}NO_7P \cdot 0.5H_2O$
7,8-Dehydro-4,5α-epoxy-3-methoxy-17-methyl morphinan-6α-ol phosphate (1:1) (salt) hemihydrate.
Colrex Compound; Ambenyl Cough Syrup; component of: Actifed with Codeine Cough Syrup, ASA and Codeine Compound, Codis, Empracet Codeine Phosphate, Empirin with Codeine Tablets, Isoclor Expectorant C, Naldecon-CX, Nucofed, Percogesic with Codeine, Tussar-2, Tussar-SF, Tussi-Organidin, Tylenol. Analgesic, narcotic; antitussive. Federally controlled substance (opiate). Oral analgesic in small animals. Soluble in H_2O; slightly soluble in alcohol.

31 Etorphine Hydrochloride
3932

$C_{25}H_{34}ClNO_4$
[5α,7α(R)]-4,5-Epoxy-3-hydroxy-6-methoxy-α,17-dimethyl-α-propyl-6,14-ethenomorphinan-7-methanol hydrochloride.
M-99; component (with acepromazine) of Immobilon. Federally controlled substance (opiate derivative). CAUTION: very small quantities may cause respiratory paralysis and death. Narcotic analgesic. Used to immobilize large animals (veterinary). mp = 266-267°; 7β isomer hydrochloride dec 290°.

Veterinary Products:

047-870; (with Diprenorphine Hydrochloride) M50-50 Diprenorphine; M99 Etorphine; *Wildlife*.

32 Fentanyl Citrate
990-73-8 4043 213-588-0

$C_{28}H_{36}N_2O_8$
N-(1-Phenylethyl-4-piperidinyl)propionanilide citrate (1:1).
Fentanest; Leptanol; Pentanyl; Sublimaze. Analgesic, narcotic; tranquilizer (veterinary). Federally controlled substance (opiate). mp = 149-151°; soluble in H_2O (1 g/40 ml), less soluble in organic solvents; LD_{50} (mus iv) = 11.2 mg/kg, (mus sc) = 62 mg/kg.

Veterinary Products:

030-438; (with Droperidol) Innovar®-Vet Injection; *Schering-Plough Animal Health*.

33 Hydrocodone
125-29-1 4826 204-733-9

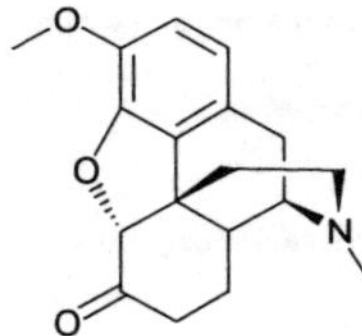

$C_{18}H_{21}NO_3$
4,5α-Epoxy-3-methoxy-17-methylmorphinan-6-one.
dihydrocodeinone; Bekadid; Dicodid. Analgesic, narcotic; used primarily as an antitussive. Federally controlled substance (opiate). CAUTION: may be habit forming. mp = 198°; soluble in alcohol, dilute acids; insoluble in H_2O; λ_m = 280 nm (ε 1310); LD_{50} (mus sc) = 85.7 mg/kg.

34 Hydrocodone Bitartrate
34195-34-1 4826

$C_{22}H_{27}NO_9 \cdot 2.5\ H_2O$
4,5α-Epoxy-3-methoxy-17-methylmorphinan-6-one tartrate (1:1) hydrate (2:5).
Calmodid; Codinovo; Dicodid; Duodin; Hydrokon; Kolikodal; Orthoxycol; Mercodinone; Norgan; Synkonin;

component of: Anexsia, Anodynos-DHC, Hycodan, Hycomine, Hycomine Compound, Hycomine Pediatric, Hycotuss, Hy-Phen, Kwelcof, PV Tussin Syrup, PV Tussin Tablet, Vicodin, Zydone. Analgesic, narcotic; used primarily as an antitussive. Federally controlled substance (opiate). CAUTION: may be habit forming. mp = 118-128°; soluble in H_2O (6/25 g/100 ml), 95% EtOH (0.67 g/100 g); nearly insoluble in Et_2O, $CHCl_3$; pH (2% aqueous solution) = 3.6.

35 Hydrocodone Bitartrate [anhydrous]

143-71-5 4826 205-608-1

$C_{22}H_{27}NO_9$

4,5α-Epoxy-3-methoxy-17-methylmorphinan-6-one tartrate (1:1).

Fortuss; DHC; Dico; component of: Synalgos-DC. Analgesic, narcotic; used primarily as an antitussive. Federally controlled substance (opiate). CAUTION: may be habit forming.

36 Meperidine

57-42-1 5894 200-329-1

$C_{15}H_{21}NO_2$

Ethyl 1-methyl-4-phenylisonipecotate.

isonipecaine; pethidine. Analgesic, narcotic. Federally controlled substance (opiate). Used as an analgesic in different animal species.

37 Meperidine Hydrochloride

50-13-5 5894 200-013-3

$C_{15}H_{22}ClNO_2$

Ethyl 1-methyl-4-phenylisonipecotate hydrochloride.

Algil; Alodan; Centralgin; Demerol hydrochloride; Dispadol; Dolantin; Dolestine; Dolosal; Mefedina. Analgesic, narcotic. Federally controlled substance (opiate). Used as an analgesic in different animal species. mp = 186-189°; soluble in H_2O, Me_2CO, EtOAc; slightly soluble in alcohol; insoluble in C_6H_6, Et_2O; LD_{50} (rat orl) = 170 mg/kg.

38 Morphine Sulfate

64-31-3 6361 200-582-8

$C_{34}H_{40}N_2O_{10}S$

7,8-Didehydro-4,5α-epoxy-17-methylmorphinan-3,6α-diol sulfate (2:1) (salt).

Kapanol; Moscontin; MS Contin; MST Continus; Oblioser; Oramorph; Relipain; Roxanol. Analgesic, narcotic. Used for treatment of acute pain. Federally controlled substance (opiate). CAUTION: may be habit forming.

39 Morphine Sulfate Pentahydrate

6211-15-0 6361

$C_{34}H_{50}N_2O_{15}S$

7,8-Didehydro-4,5α-epoxy-17-methylmorphinan-3,6α-diol sulfate (2:1) (salt) pentahydrate.

MST 10 Mundipharma; MST 30 Mundipharma. Analgesic, narcotic. Used for treatment of acute pain. Federally controlled substance (opiate). CAUTION: may be habit forming. mp = 250° (dec); $[\alpha]_D^{25}$ = -109° (c = 4 H_2O); soluble in H_2O (65 g/100 ml); less soluble in organic solvents.

40 Oxymorphone Hydrochloride

357-07-3 7103 206-610-5

$C_{17}H_{20}ClNO_4$

4,5α-Epoxy-3,14-dihydroxy-17-methylmorphinan-6-one hydrochloride.

Numorphan. Federally controlled substance (opiate). Analgesic, narcotic. Used in dogs and cats as a restraining agent. Crystalline solid.

Veterinary Products:

030-525; Numorphan; *Endo*.

41 Pentazocine Lactate

146-95-1 7261

$C_{22}H_{33}NO_4$

(2R*,6R*,11R*)-1,2,3,4,5,6-Hexahydro-6,11-dimethyl-3-(3-methyl-2-butenyl)-2,6-methano-3-benzazocin-8-ol lactate (salt).

Liticon; Pentalgina; Talwin Injection. Analgesic, narcotic. Mixed opioid agonist-antagonist. Federally controlled substance. Used for relief of colic pain in horses and of post-operative pain in dogs.

Veterinary Products:

044-611; Talwin®-V Sterile Aqueous Injection; *Pharmacia & Upjohn*.

Analgesics, Non-narcotic

42 Acetaminophen

103-90-2 45 203-157-5

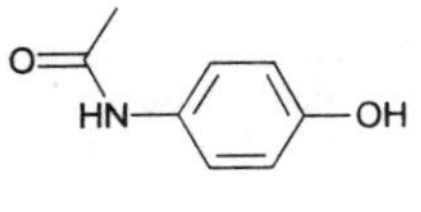

$C_8H_9NO_2$

4'-Hydroxyacetanilide.

p-hydroxyacetanilide; p-acetamidophenol; p-acetaminophenol; N-acetyl-p-aminophenol; paracetamol; Abensanil; Acamol; Acetalgin; Alpiny; Amadil; Anaflon; Anhiba; Apamide; APAP; Banesin; Ben-u-ron; Bickie-mol; Calpol; Captin; Claratal; Cetadol; Dafalgan; Datril; Dirox; Disprol; Doliprane; Dolprone; Dymadon; Enelfa; Eneril; Eu-Med; Exdol; Febrilex; Finimal; Gelocatil; Hedex; Homoolan; Korum; Momentum; Naprinol; Nebs; Nobedon; Ortensan; Pacemol; Paldesic; Panadol; Panaleve; Panasorb; Panets; Panodil; Parelan; Paraspen; Parmol; Pasolind N; Phenaphen; Salzone; Tabalgin; Tapar; Tempra; Tralgon; Tylenol; Valadol; component of: Actifed Plus, Allerest Sinus Pain Formula, Anexsia, Aspirin-Free Anacin, Children's Tylenol Cold Tablets, Contac Cough & Sore Throat Formula, Contac Jr Non-drowsy Formula, Contac Nighttime Cold Medicine, Contac Severe Cold Formula, Coricidin, Darvoset-N, Dristan Cold (Multisymptom), Dristan Cold (No Drowsiness), Empracet, Endecon, Gemnisyn, Headache Strength Allerest, Hycomine Compound, Hy-Phen, Intensin, Liquiprin, Maximum Strength Sine-Aid, Midol, Midol Maximum Strength, Midol PMS, Naldegesic, Naldetuss, Ornex, Percocet, Percogesic with Codeine, Propacet, Quiet World, Rhinex D-Lay Tablets, Sinarest, Sine-Off Maximum Strength Allergy/Sinus, Sine-Off Maximum Strength No Drowsiness Formula Caplets, Sinubid, St Joseph's Cold Tablets for Children, Sudafed Sinus, Supac, Teen Midol, TheraFlu, Tylenol Allergy Sinus, Tylenol Cold and Flu Multi-Symptom, Tylenol Cold Medication Caplets, Liquid and Tablets, Tylenol Cold Night TIme Liquid, Tylenol Cold No Drowsiness, Tylenol PM Tablets and Caplets, Tylenol with Codeine, Tylox, Vanquish, Vicodin, Wygesic, Zydone. Analgesic, antipyretic, anti-inflammatory. Used as an oral analgesic, sometimes with oral codeine phosphate, in dogs. mp = 169-170.5°; d_4^{21}= 1.293; λ_m = 250 nm (ε 13800 in EtOH); slightly soluble in cold H_2O, Et_2O; more soluble in hot H_2O; soluble in alcohol, dimethylformamide, ethylene dichloride, Me_2CO, EtOAc; nearly insoluble.

43 Acetylsalicylic Acid

50-78-2 102 200-064-1

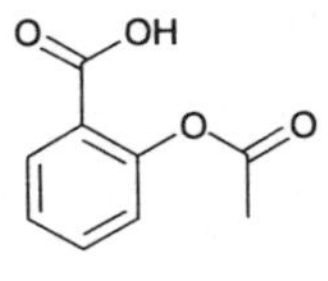

$C_9H_8O_4$

2-Acetoxybenzoic acid.

salycylacetylsalicylic acid; Aspirin; A.S.A; Acetyonyl; Bayer; Bufferin; Diplosal Acetate; Alka-Seltzer; Anacin; Ascriptin; Coricidin; Coricidin D; Darvon Compound; Easprin; Ecotrin; Empirin; Excedrin; Gelprin; Robaxisal; Vanquish; Ascoden-30; Measurin; Norgesic; Persistin; St Joseph's Aspirin for Adults; Supac; Triaminicin; ZORprin; Acidum Acetylsalicylicum; Acetilum Acidulatum; A.s.a.; Acenterine; Acetosal; Acetosalic Acid; Acetosalin; Acetylin; Acetylsal; Acylpyrin; Aspro; Asteric; Caprin; Colfarit; Duramax; Ecm; Endydol; Entrophen; Enterosarine; Helicon; Xaxa; Rhodine; Salacetin; Salcetogen; Saletin; Acesal; Acetisal; Acetonyl; Acimetten; Acisal; A.s.a. Empirin; Asagran; Asatard; Aspalon; Aspergum; Aspirdrops; AC 5230; Benaspir; Entericin; Extren; Bialpirinia; Contrheuma Retard; Crystar; Delgesic; Dolean Ph 8; Enterophen; Globoid; Idragin; Neuronika; Novid; Polopiryna; Solpyron; Claradin; Nu-seals; Solprin. Used in all species as an analgesic and antipyretic. mp = 159°; nearly insoluble in H_2O, C_6H_6; soluble in alkaline solutions.

Veterinary Products:

011-700; (with Methylprednisolone) Cortaba® Tablets; *Pharmacia & Upjohn.*

44 Dimethyl Sulfoxide

67-68-5 3308 200-664-3

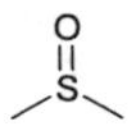

C_2H_6OS

Dimethyl sulfoxide.

Herpid; DMSO; SQ-9453; DMS-70; DMS-90; Demavet; Demeso; Dolicur; Domoso; Dromisol; Gamasol 90; Hyadur; Kemsol; Rimso-50; Sclerosol; Somipront; Syntexan. Anti-inflammatory agent. Use proposed in veterinary medicine as an analgesic, anti-inflammatory agent and penetrant carrier to enhance absorption. mp = 18.5°; bp = 189°; soluble in H_2O, organic solvents; LD_{50} (rat orl) = 19690 mg/kg.

Veterinary Products:

032-168; Domoso ® Solution; *Fort Dodge Animal Health, Divn. AHP.*

045-512; (with Fluocinolone Acetonide) Synotic ® Otic Solution; *Fort Dodge Animal Health, Divn. AHP.*

047-334; (with Fluocinolone Acetonide) Synsac Solution; *Fort Dodge Animal Health, Divn. AHP.*

047-925; Domoso ® Gel; *Fort Dodge Animal Health, Divn. AHP.*

45 Etodolac

41340-25-4 3920

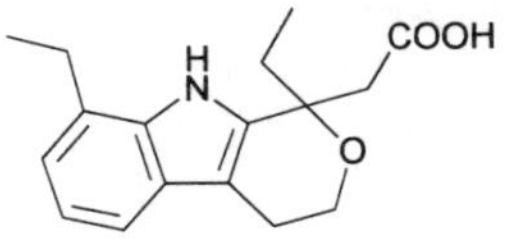

$C_{17}H_{21}NO_3$

1,8-Diethyl-1,3,4,9-tetrahydropyranol[3,4-b]indole-1-acetic acid.

etodolic acid; AY-24236; Edolan; Lodine; Ramodar; Tedolan; Ultradol; Zedolac. Arylacetic acid derivative. Anti-inflammatory; analgesic. Used to manage pain resulting from osteoarthritis in dogs. mp = 145-148°.

Veterinary Products:

141-108; EtoGesic™; *Fort Dodge Animal Health, Divn. AHP.*

46 Ketoprofen

22071-15-4 5316 244-759-8

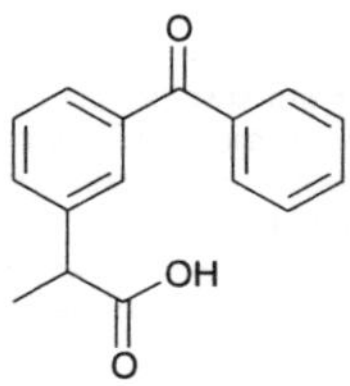

$C_{16}H_{14}O_3$
m-Benzoylhydratropic acid.
RP-19583; Alrheumat; Alrheumun; Capisten; Dexal; Epatec; Fastum; Iso-K; Kefenid; Ketopron; Lertus; Menamin; Meprofen; Orudis; Orugesic; Oruvail; Oscorel; Profenid; Toprec; Toprek. Anti-inflammatory; analgesic. Used as an anti-inflammatory in horses. mp = 94°; λ_m = 255 nm (log ε 4.33 in MeOH); soluble in Et_2O, alc, Me_2CO, $CHCl_3$, DMF, EtOAc; slightly soluble in H_2O; LD_{50} (rat orl) = 101 mg/kg.

Veterinary Products:

140-269; Ketofeb™; *Fort Dodge Animal Health, Divn. AHP.*

47 Medetomidine

86347-14-0 5830

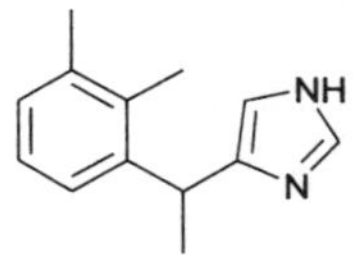

$C_{13}H_{16}N_2$
(±)-4-(α,2,3-Triethylbenzyl)imidazole.
Alpha-adrenergic agonist. Used in veterinary medicine as a sedative and analgesic.

Veterinary Products:

140-999; Domitor®; *Orion-Farmos.*

48 Naproxen

22204-53-1 6504 244-838-7

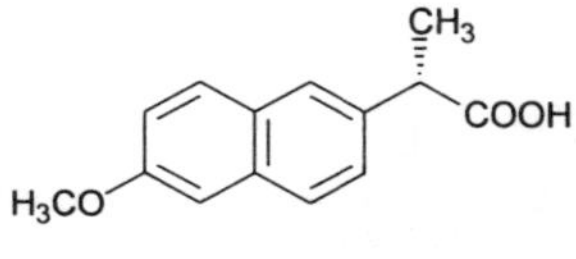

$C_{14}H_{14}O_3$
(+)-6-Methoxy-α-methyl-2-napthaleneacetic acid.
MNPA; RS-3540; Bonyl; Diocodal; Dysmenalgit; Equiproxen; Floginax; Laraflex; Laser; Naixan; Napren; Naprium; Naprius; Naprosyn; Naprosyne; Naprux; Naxen; Nycopren; Panoxen; Prexan; Proxen; Proxine; Reuxen; Veradol; Xenar. Arylpropionic acid derivative. Anti-inflammatory; analgesic; antipyretic. Used in horses to minimize pain associated with soft tissue diseases. mp = 152-154°; $[\alpha]_D$ = +66° (c = 1 in $CHCl_3$); nearly insoluble in H_2O; soluble in organic solvents; LD_{50} (mus orl) = 1234 mg/kg, (rat orl) = 534 mg/kg, (rat ip) = 575 mg/kg.

Veterinary Products:

096-674; Equiproxen Granules; *Fort Dodge Animal Health, Divn. AHP.*
096-675; Equiproxen 10% Solution; *Fort Dodge Animal Health, Divn. AHP.*

49 Phenylbutazone

50-33-9 7431 200-029-0

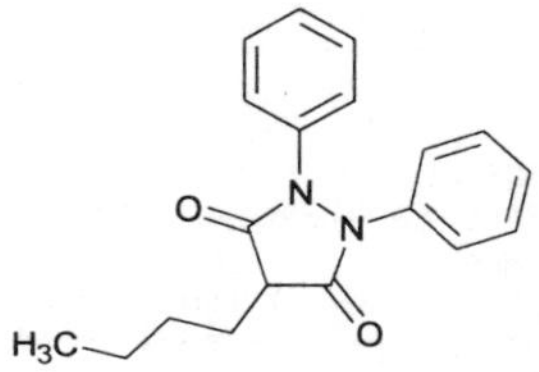

$C_{19}H_{20}N_2O_2$
4-Butyl-1,2-diphenyl-3,5-pyrazolidinedione.
flexazone; diphebuzol; fenibutazona; G-13871; R-3-ZON; Ambene; Artrizin; Azolid; Bizolin; Butacote; Butadion; Buitapirazol; Butadiona; Butatron; Butoz; Butazolidin; Buzon; Ecobutazone; Equipalazone; Exrheudon N; Fenibutol; Intrabutazone; Intrazone; Mepha-Butazon; Phenyzene; Robizone-V; Tevcodyne; Uzone. Analgesic and anti-inflammatory. Used to treat lameness in horses. Crystals; mp = 105°; soluble in H_2O (0.07 - 0.22 g/100 ml at 22.5°; λ_m = 239.5 nm (log ε 4.19, acidic MeOH).

Veterinary Products:

010-987; Butazolidin Bolus; Butazolidin Tablets; *Schering-Plough Animal Health.*
011-575; Butazolidin Injectable 20%; *Schering-Plough Animal Health.*
038-800; Butazolidin Granules; *Schering-Plough Animal Health.*
044-756; Tevcodyne; *Merial.*
045-416; Tevcodyne Injectable; *Merial.*
045-514; EquiBute Injection; *Fort Dodge Animal Health, Divn. AHP.*
045-515; EquiBute Tablets 100 mg; *Fort Dodge Animal Health, Divn. AHP.*
045-848; Phenylbutazone Injection; *Steris.*
046-780; Phen-Buta Vet Injection; *Anthony Products.*
047-712; Bizolin 200; *Boehringer Ingelheim Vetmedica.*
048-646; Therazone Injection; *Wendt.*
048-647; Therazone Tablets; *Wendt.*
049-187; Phen-Buta Vet Tablets; *Anthony Products.*
091-065; Robizone-V; *Robins.*
091-818; Phenylbutazone Tablets, USP 1 gram; *Phoenix.*
093-105; Robizone-V; *Robins.*
093-516; Bizolin Injection 20%; *Boehringer Ingelheim Vetmedica.*

094-170; Phenylbutazone Tablets, USP 100 mg; *Phoenix.*
096-671; Phen-Buta-Vet Injection; *Anthony Products.*
096-672; Phen-Buta-Vet Tablets; *Anthony Products.*
098-640; Robizone Injectable 20%; *Robins.*
099-618; Bizolin; *Boehringer Ingelheim Vetmedica.*
102-824; Phenylbutazone Tablets; *Pegasus.*
113-510; Equipalazone; *Merial.*
116-087; Butazolidin; Butazolidin Paste; Bute; Phenylzone Paste; *Schering-Plough Animal Health.*
118-979; Butatron Gel; *Merial.*
140-958; Equiphen ® Paste; *Luitpold.*
200-126; Phenylbutazone 20% Injection; *Phoenix.*

50 Xylazine Hydrochloride
23076-35-9 10213 245-417-0

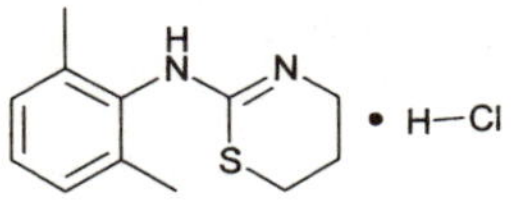

$C_{12}H_{17}ClN_2S$
5,6-Dihydro-2-(2,6-xylidino)-4H-1,3-thiazine hydrochloride.
Narcosyl; Rompun; Xylapan; Xylasol. Sedative (veterinary); analgesic (veterinary); muscle relaxant (veterinary).

Veterinary Products:

047-955; Rompun ® Injectable (20 mg); *Bayer.*
047-956; Rompun ® Injectable (100 mg); *Bayer.*
139-236; Anased ®; Anased ® Injectable; *Lloyd.*
140-442; Xylazine HCl Injection; *Boehringer Ingelheim Vetmedica.*
200-088; Sedazine ™; *Fort Dodge Animal Health, Divn. AHP.*
200-139; Chanazine ®; *Chanelle.*
200-184; Chanazine ®; *Chanelle.*

Anesthetics, Inhaled

51 Halothane
151-67-7 4634 205-796-5

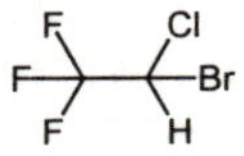

$C_2HBrClF_3$
2-Bromo-2-chloro-1,1,1-trifluoroethane.
Fluothane; Rhodialothan. Low cost inhalation anesthetic. bp = 50.2°, bp_{243} = 20°; slighlty soluble in H_2O (0.345 g/100 ml); soluble in organic solvents.

Veterinary Products:

014-170; Fluothane; Halothane, USP; *Fort Dodge Animal Health, Divn. AHP.*
200-200; Halothane, USP; *Halocarbon.*

52 Isoflurane
26675-46-7 5191 247-897-7

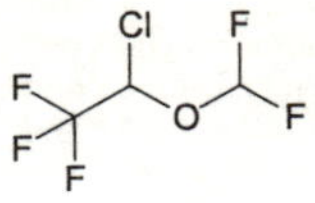

$C_3H_2ClF_5O$
1-Chloro-2,2,2-trifluoroethyl difluoromethyl ether.
Aerrane; Forane; Compound 469; Forene. Inhalation anesthetic. bp = 48.5°; sg = 1.45; non-flammable; soluble in most organic solvents.

Veterinary Products:

135-773; AErrane; *Baxter.*
200-070; IsoFlo ™; *Abbott.*
200-129; Isoflurane, USP; *Halocarbon.*
200-141; Isoflurane, USP; *Inhalon.*
200-187; Isoflurane, USP; *Marsam.*
200-237; Isoflurane, USP; *Rhodia Organique.*

53 Methoxyflurane
76-38-0 6072 200-956-0

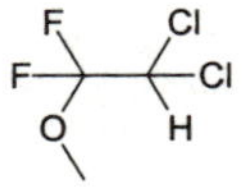

$C_3H_4Cl_2F_2O$
2,2-Dichloro-1,1-difluoroethyl methyl ether.
Penthrane; NSC-110432; DA-759; Metofane; Pentrane. Inhalation anesthetic. Reported to be nephrotoxic. bp = 105°, bp_{100} = 51°; mp = -35°; d_4^{20} = 1.4262, 1.4226.

Veterinary Products:

014-485; Metofane® Inhalation; *Schering-Plough Animal Health.*

Anesthetics, Intravenous

54 Midazolam
59467-70-8 6270 261-774-5

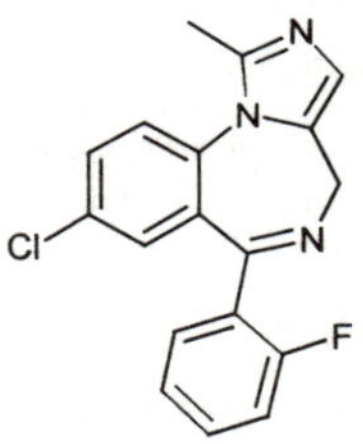

$C_{18}H_{13}ClFN_3$
8-Chloro-6-(o-fluorophenyl)-1-methyl-4H-imidazo[1,5-a]-[1,4]-benzodiazepine.
Benzodiazepine derivative. Intravenous anesthetic. When combined with ketamine or fentanyl, is a powerful sedative, suitable for use in small animals. mp = 158-160°; λ_m = 220 nm (ε 30000 iPrOH); soluble in H_2O.

55 **Midazolam Hydrochloride**

59467-96-8 6270 261-776-6

$C_{18}H_{14}Cl_2FN_3$

8-Chloro-6-(o-fluorophenyl)-1-methyl-4H-imidazo[1,5-a]-[1,4]-benzodiazepine hydrochloride.

Hypnovel; Rocam; Versed; Ro-21-3981/003. Benzodiazepine derivative. Intravenous anesthetic. When combined with ketamine or fentanyl, is a powerful sedative, suitable for use in small animals.

56 **Midazolam Maleate**

59467-94-6 6270 261-775-0

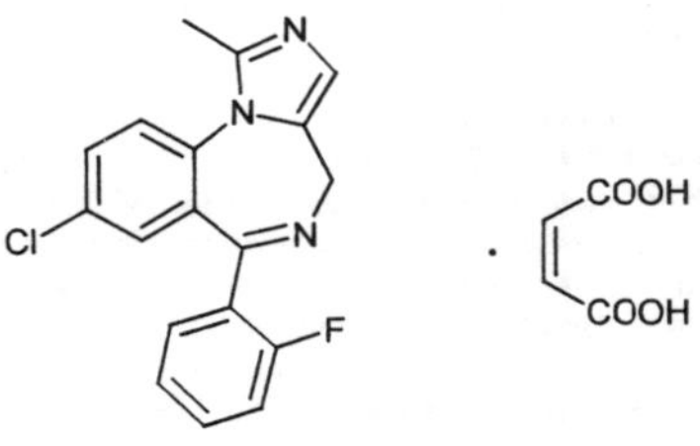

$C_{22}H_{17}ClFN_3O_4$

8-Chloro-6-(o-fluorophenyl)-1-methyl-4H-imidazo[1,5-a]-[1,4]-benzodiazepine maleate.

Ro-21-3981/001; Dormicum; Flormidal. Benzodiazepine derivative. Intravenous anesthetic. When combined with ketamine or fentanyl, is a powerful sedative, suitable for use in small animals. mp = 114-117°; LD_{50} (mus orl) = 760 mg/kg, (mus iv) = 86 mg/kg.

57 **Propofol**

2078-54-8 8020 218-206-6

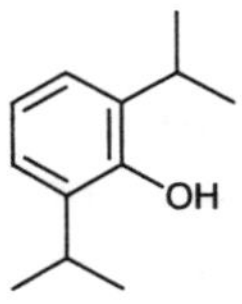

$C_{12}H_{18}O$

2,6-Diisopropylphenol.

Diprivan; Disoprivan; Rapinovet; ICI-35868. Rapidly acting intravenous anesthetic, useful in minor procedures. An alkylphenol. bp_{30} = 136°, bp_{17} = 126°; mp = 19°; d_{20} = 0.955.

Veterinary Products:

141-070; Rapinovet®; *Schering-Plough Animal Health.*
141-098; PropoFlo™; *Abbott.*

58 **Thialbarbitone Sodium**

9428

$C_{13}H_{15}N_2NaO_2S$

5-(2-cyclohexen-1-yl)dihydro-5-(2-propenyl)-2-thioxo-4,6(1H,5H)-pyrimidinedione sodium salt.

thialbarbital sodium; Kemithal sodium. Intravenous anesthetic. Pale yellow crystals; mp = 130-132°; very soluble in H_2O, EtOH, insoluble in C_6H_6, Et_2O; pH of 2.5% aqueous solution = 10.5.

Veterinary Products:

008-932; Kemthal L.A.; *Fort Dodge Animal Health, Divn. AHP.*

59 **Thiamylal Sodium**

337-47-3 9437 206-415-5

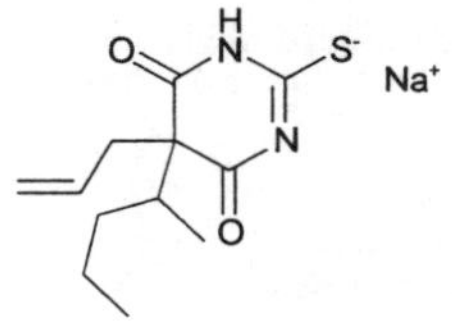

$C_{12}H_{17}N_2NaO_2S$

Sodium 5-allyl-5-(1-methylbutyl)-2-thiobarbiturate.

Surital. Intravenous anesthetic. Largely replaced by thiopental sodium.

Veterinary Products:

010-809; Surital; *Fort Dodge Animal Health, Divn. AHP.*
039-483; Bio-Tal; *Boehringer Ingelheim Vetmedica.*
132-028; Anestatal ™; *Boehringer Ingelheim Vetmedica.*

60 **Thiopental Sodium**

71-73-8 9487 200-763-1

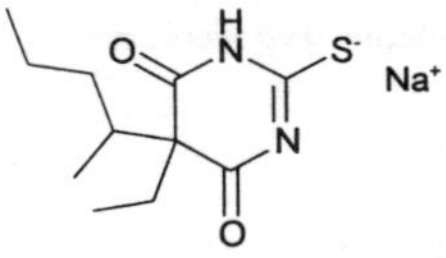

$C_{11}H_{17}N_2NaO_2S$

Sodium 5-ethyl-5-(1-methylbutyl)-2-thiobarbiturate.

Pentothal; thiomebumal sodium; penthiobarbital sodium; thiopentone sodium; thionembutal; Intraval Sodium; Nesdonal Sodium; Pentothal Sodium; Trapanal. Intravenous anesthetic. Rapid-acting and short-lasting, is used as an induction agent for general anesthesia. Soluble in H_2O, EtOH, insoluble in C_6H_6, Et_2O, petroleum ether; LD_{50} (mus ip) = 149 mg/kg, (mus iv) = 78 mg/kg.

Veterinary Products:

010-346; (with Pentobarbital Sodium) Combuthal Powder; *Merial.*
046-790; Sodium Thiopental; *Fort Dodge Animal Health, Divn. AHP.*

61 **Tricaine Methanesulfonate**

9751

Anesthetic for fish.

Veterinary Products:

042-427; Finquel; *Argent.*
200-226; Tricaine-S; *Western Chemical.*

Anesthetics, Local

62 Butacaine Sulfate

149-15-5 1531 205-733-1

$C_{36}H_{62}N_4O_8$
3-(Dibutyamino)-1-propanol-p-aminobenzoate (ester) sulfate (2:1).
Butyn Sulfate. Topical anesthetic. mp = 138.5-139.5°, 100-103°; soluble in H_2O (>100 g/100 ml), soluble in EtOH, Me_2CO, slightly soluble in $CHCl_3$, insoluble in Et_2O; LD_{50} (mus iv) = 12.4 mg/kg.

Veterinary Products:

130-872; (with Nitrofurazone) Nitrofurazone Anesthetic Dress; *Med-Pharmex.*

63 Dibucaine Hydrochloride

61-12-1 3081 200-498-1

$C_{20}H_{30}ClN_3O_2$
2-Butoxy-N-[2-(diethylamino)ethyl]cinchoninamide hydrochloride.
Nupercaine hydrochloride; Nupercainal; benzolin; Percaine; Cincaine; Sovcaine; Cinchocaine. Used in veterinary medicine as a local anesthetic. Topical anesthetic. Dec 90-98°; λ_m = 247, 320 nm (ε 24700, 8810 1N HCl); soluble in H_2O (200 g/100 ml), EtOH, Me_2CO, $CHCl_3$, slightly soluble in C_6H_6, C_7H_8, EtOAc, insoluble in Et_2O.

Veterinary Products:

128-967; (with Secobarbital Sodium) Repose Euthanasia Solution; *Fort Dodge Animal Health, Divn. AHP.*

64 Laureth-23

9002-92-0 7717 500-002-6

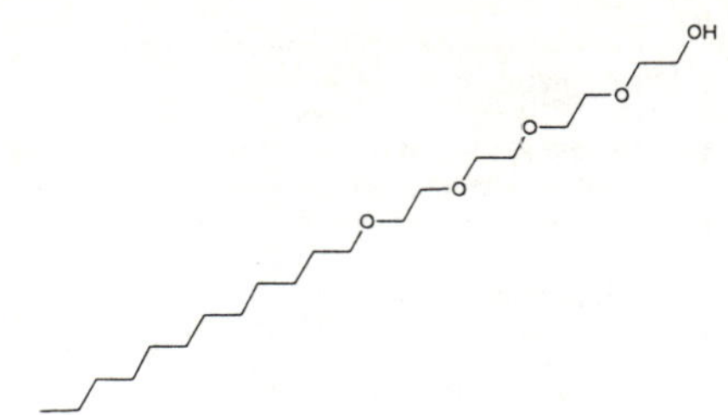

$C_{58}H_{118}O_{24}$
POE (23) lauryl ether.
component of: Cerasynt® 945. FDA approved for topicals, used as an emollient, emulsifier, thickener, and stabilizer. Used in topicals; anti-irritant in deodorant and anti-perspirants. Used in veterinary medicine as a local anesthetic and topical anti-pruritic.

65 Mepivacaine Hydrochloride

1722-62-9 5905 217-023-9

$C_{15}H_{23}ClN_2O$
1-Methyl-2',6'-pipecoloxylidide hydrochloride.
Carbocaine; Polocaine; Carbocaina; Chlorocain; Meaverin; Mepicaton; Mepident; Mepivastesin; Optocain; Scandicain. Topical anesthetic. mp = 262-264°; soluble in H_2O, LD_{50} (mus sc) = 280 mg/kg, (rat sc) = 500 mg/kg.

Veterinary Products:

100-703; Carbocaine+s0-V Sterile Aqueous Solution; *Pharmacia & Upjohn.*

66 Piperocaine Hydrochloride

24561-10-2 7627 246-315-9

$C_{16}H_{24}ClNO_2$
3-(2-Methylpiperidino)propyl benzoate hydrochloride.
Metycaine hydrochloride. Topical anesthetic. Used in veterinary medicine as a local anesthetic. mp = 172-175°; soluble in H_2O (100 g/100 ml), soluble in EtOH, $CHCl_3$, insoluble in Et_2O; LD_{50} (rat sc) = 1300 mg/kg, (rat iv) = 20 mg/kg.

Veterinary Products:

040-123; (with Cephalonium, Flumethasone, Iodochlorhydroxyquin, Polymyxin B Sulfate) Toptic Ointment (15 g); *Elanco.*

67 Proparacaine Hydrochloride
5875-06-9 7991 227-541-7

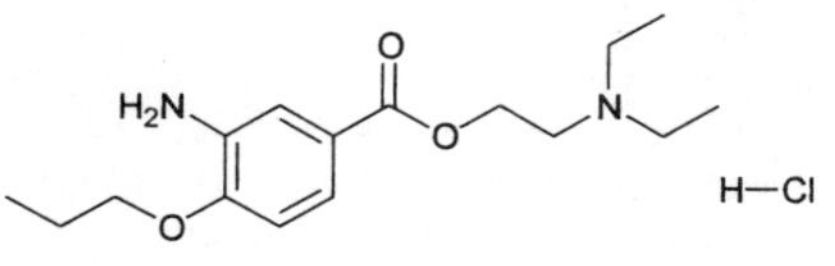

$C_{16}H_{27}ClN_2O_3$
2-(Diethylamino)ethyl 3-amino-4-propoxy benzoate hydrochloride.
Alcaine; Ophthaine; Ophthetic; Ak-Taine. Topical anesthetic used in human ophthalmology. Topical anesthetic, used in ophthalmology. mp = 182-183.3°; λ_m = 225, 270, 300 nm (MeOH); soluble in H_2O, EtOH, MeOH, insoluble in Et_2O, C_6H_6.

Veterinary Products:

009-035; Ophthaine® Solution; *Fort Dodge Animal Health, Divn. AHP.*

68 Tetracaine
94-24-6 9330 202-316-6

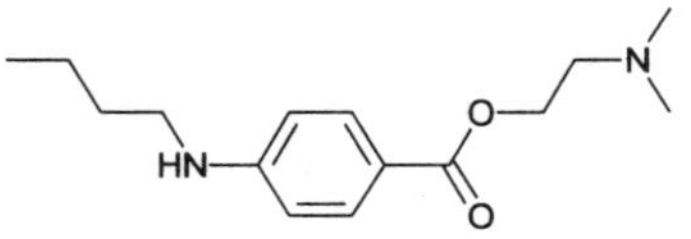

$C_{15}H_{24}N_2O_2$
2-(Dimethylamino)ethyl p-(butylamino)benzoate.
Medihaler-Tetracaine; Metraspray; Pontocaine. Topical anesthetic.

Veterinary Products:

032-322; (with Hexamethyltetracosane, Neomycin Sulfate, Prednisolone) Liquisone F with Cerumene; *Evsco.*
055-005; (with Chloramphenicol, Prednisolone, Squalene) Liquichlor with Cerumene; *Evsco.*

69 Tetracaine Hydrochloride
136-47-0 9330 205-248-5
$C_{15}H_{25}ClN_2O_2$
2-(Dimethylamino)ethyl p-(butylamino)benzoate hydrochloride.
Anethaine; Decicain, Pantocaine; Tonexol; Pontocaine Hydrochloride; component of: Fulvidex. Topical anesthetic. mp = 147-150°; λ_m 225, 310 nm (ε 14108, 26352 H_2O), 229, 281, 312 nm ($E^{1\%}_{1\ cm}$ 509, 55, 76 0.1N H_2SO_4), 226, 310 nm (ε 7586, 29512 MeOH), 308 nm (ε 27542 $CHCl_3$); soluble in H_2O (13.3 g/100 ml), EtOH (2.5 g/100 ml), $CHCl_3$ (3.3 g/100 ml), insoluble in Et_2O, C_6H_6, Me_2CO; LD_{50} (mus ip) = 70 mg/kg, (fmus iv) = 13 mg/kg, (fmus sc) = 35 mg/kg.

Veterinary Products:

010-524; (with Hydrocortisone Acetate, Neomycin Sulfate) Neo-Cortef® with Tetracaine; *Pharmacia & Upjohn.*
011-703; (with Neomycin Sulfate; Prednisolone Acetate) Neo-Delta-Cortef® with Tetracaine Ointment; *Pharmacia & Upjohn.*
015-433; (with Isoflupredone Acetate, Myristyl σ-Picolinium Chloride, Neomycin Sulfate) Neo-Predef® with Tetracaine Ointment; *Pharmacia & Upjohn.*
030-025; (with Isoflupredone Acetate, Neomycin Sulfate) Neo-Predef® with Tetracaine Topical Ointment; Tritop Topical Ointment; *Pharmacia & Upjohn.*

Antacids

70 Sodium Bicarbonate
144-55-8 8726 205-633-8

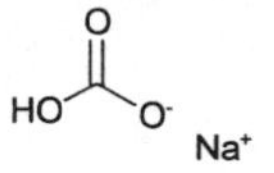

$CHNaO_3$
Sodium hydrogen carbonate.
sodium acid carbonate; baking soda. Antacid. Used in large and small animals against metabolic acidosis. Soluble in H_2O (10 g/100 ml at 25°, 8.3 g/100 ml at 18°), insoluble in EtOH; pH (aqueous solution) = 8.3.

Anthelmintics

71 Albendazole
54965-21-8 211 259-414-7

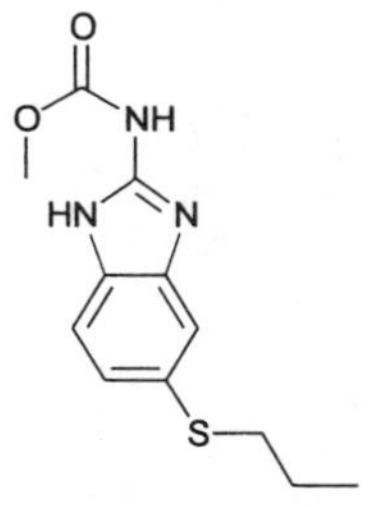

$C_{12}H_{15}N_3O_2S$
[5-(Propylthio)-1H-benzimidazol-2-yl]carbamic acid methyl ester.
SKF-62979; Bilutac; Eskazole; Proftril; 110-048 Valbazen®; 128-070 Valbazen®;140-934 Valbazen®; Zentel. Anthelmintic. Used to treat endoparasites (*Ostertagia ostertagi, Haemonchus* spp., *Trichostrongylus* spp., *Nematodius* spp., *Copperia* spp., *Bunostomum phlebotomum, Oesphagostomum* spp., *Dictacaulus* spp., *Fasciola hepatica* (adults) and *Moniezia* spp.) in cattle. Colorless crystals; mp = 208-210°.

Veterinary Products:

110-048; Valbazen®; *Pfizer.*
128-070; Valbazen®; *Pfizer.*
140-934; Valbazen®; *Pfizer.*

72 Albendazole Oxide
54029-12-8

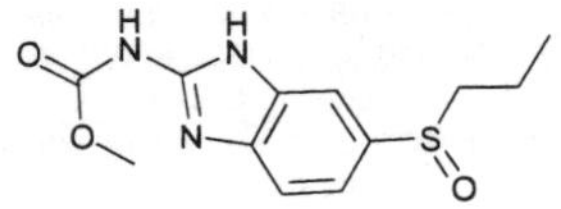

$C_{12}H_{15}N_3O_3S$
Methyl 5-(propylsulfinyl)-2-benzimidazolecarbamate.
Anthelmintic. Used to treat endoparasites *(Ostertagia ostertagi, Haemonchus* spp., *Trichostrongylus* spp., *Nematodius* spp., *Copperia* spp., *Bunostomum phlebotomum, Oesphagostomum* spp., *Dictacaulus* spp., *Fasciola hepatica* (adults) and *Moniezia* spp.) in cattle.

73 Arsenamide Sodium
831
$C_6H_6AsNNa_2O_5S_2$
[[(p-Carbamoylphenyl)arsylene]dithio]diacetic acid disodium salt.
Anthelmintic used for heartworm infection in dogs.

Veterinary Products:

006-623; Caparsolate Sodium; *Merial.*

74 Bunamidine Hydrochloride
1055-55-6 1510 213-890-2

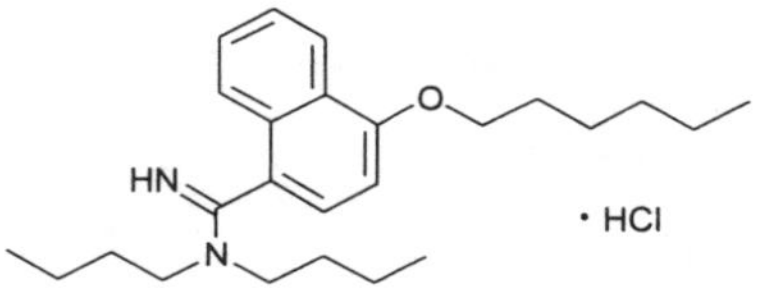

$C_{25}H_{39}ClN_2O$
N,N-Dibutyl-4-(hexyloxy)-1-naphthalenecarboximidamide hydrochloride.
Scolaban. Used as an anthelminitc for control of cestodes Crystals; mp = 214-214.8°; LD_{50} (mus orl) = 594 mg/kg.

Veterinary Products:

035-016; Scolaban 400; *Schering-Plough Animal Health.*

75 Butamisole Hydrochloride
54400-62-3 1540

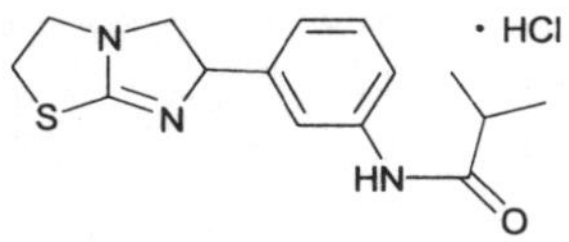

$C_{15}H_{20}ClN_3OS$
2-Methyl-N-[3-(2,3,5,6-tetrahydroimidazo[2,1-b]thiazol-6-yl)phenyl]propanamide hydrochloride.
CL-206214; Styquin. Anthelmintic.

Veterinary Products:

104-184; Styquin; *Cyanamid Agri., Puerto Rico.*
132-533; Styquin Parenteral 1.1%; *Bayer.*

76 n-Butyl Chloride
109-69-3 1595 203-696-6
C_4H_9Cl
1-Chlorobutane.
n-propyl carbinyl chloride; butyl chloride. Anthelmintic. Liquid; mp = -123.1°; bp = 78.5°; d_4^{15}= 0.89197, d_4^{20}= 0.88648, d_4^{25}= 0.88098; soluble in EtOH, Et_2O, poorly soluble in H_2O (0.066 g/100 ml at 12°); LD_{50} (rat orl) = 2670 mg/kg.

Veterinary Products:

046-746; Happy Jack Worm Capsules; *Happy Jack.*
046-922; Sergeants ® Sure Shot ® Capsules; *ConAgra.*
046-923; Sergeants ® Puppy Worm Capsules; *ConAgra.*
092-151; N-Butyl Chloride Canine Worming Caps; *Global.*
094-223; Canine Worm Caps; *K. C. Pharmacal.*
096-509; NBC Kaps Wormer; *Pfizer.*

77 Cambendazole
26097-80-3 1774 247-459-5

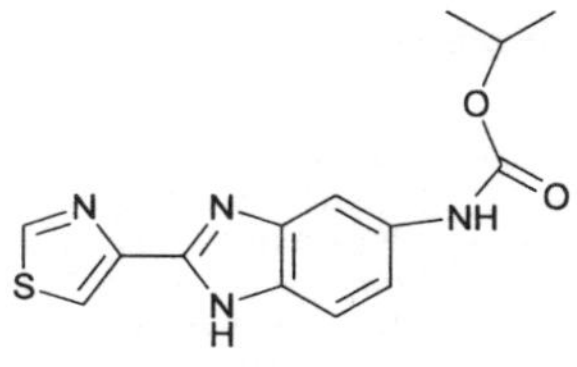

$C_{14}H_{14}N_4O_2S$
[2-(4-Thiazolyl)-1H-benzimidazol-5-yl]carbamic acid 1-methylethyl ester.
MK-905; Bonlam; Bovicam; Cambenzole; Cambet; Equiben; Novazole; Noviben. Anthelmintic. White crystals; mp = 238-240°; soluble in EtOH, DMF, sparingly soluble in Me_2CO, slightly soluble in C_6H_6, 0.1M HCl, H_2O (0.002 g/100 ml), insoluble in isooctane; λ_m = 232, 319 nm ($A_{1\,cm}^{1\%}$ = 670, 740 0.1N HCl).

Veterinary Products:

094-642; Camvet Suspension Horse Wormer; *Merial.*
096-506; Camvet Horse Wormer Pellets; *Merial.*
096-731; Camvet Horse Wormer Paste 45%; *Merial.*

78 Clorsulon
60200-06-8 2471 262-100-2

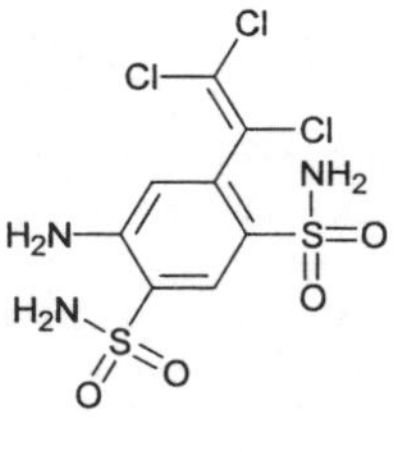

$C_8H_8Cl_3N_3O_4S_2$
4-Amino-6-(trichloroethenyl)-1,3-benzenesulfonamide.
MK-401; Curatrem. Anthelmintic, used to control trematodes and liver fluke (*Fasciola hepatica*) in cattle.

Crystals; mp = 194-203°, 203-205°; λ_m = 227, 267, 325 nm (ε 36310, 17395, 4530, MeOH); LD_{50} (mus ip) = 761 mg/kg, (mus orl) > 10000 mg/kg.

Veterinary Products:

136-742; Curatrem ® Drench; Curatrem ® Drench For Cattle; *Merial.*
140-833; (with Ivermectin) Ivomec ® F Injection For Cattle; Ivomec ® Plus Injection For Cattle; *Merial.*

79 Coumaphos

56-72-4 2626 200-285-3

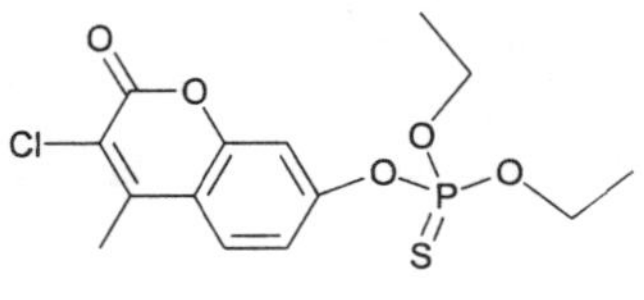

$C_{14}H_{16}ClO_5PS$
O-(3-Chloro-4-methyl-2-oxo-2H-1-benzopyran-7-yl) phosphorothioate.
Agridip; Asantol; Asunthol; Asuntol; Bay 21/199; Baymix; Bazmix; 3-chloro-7-hydroxy-4-methyl-coumarin O,O-diethyl phosphorothioate; 3-chloro-7-hydroxy-4-methyl-, O-ester with O,O-diethylpyrophosphorothioate; O-(3-chloro-4-methyl-7-coumarinyl) phosphorothioate; 3-chloro-4-methyl-7-hydroxycoumarin diethyl thiophosphoric acid ester; O-(3-chloro-4-methyl-2-oxo-2H-1-benzopyran-7-yl) phosphorothioate; O-(3-chloro-4-methyl-2-oxo-2H-1-benzopyran-7-yl) O,O-diethyl phosphorothioate; O-(3-chloro-4-methyl-2-oxo-2H-benzopyran-7-yl)phosphorothioate; 3-chloro-4-methylumbelliferone, O-ester with O,O-diethyl phosphorothioate; Co-Ral; Coumafos; Coumaphos; Diolice; Meldane; Muscatox; Resistox; Resitox; Umbethion. Anthelmintic, used against nematodes Crystals; mp = 91-92°; d = 1.474, very poorly soluble in H_2O (< 0.1 g/100 ml at 22°).

Veterinary Products:

040-001; Meldane 2; *Bayer.*
042-116; Purina ® 6 Day Worm-Kill Feed; *Purina Mills.*
042-117; Purina ® 6 Day Worm-Kill Concentrate; *Purina Mills.*

80 Dichlorophene

97-23-4 3120 202-567-1

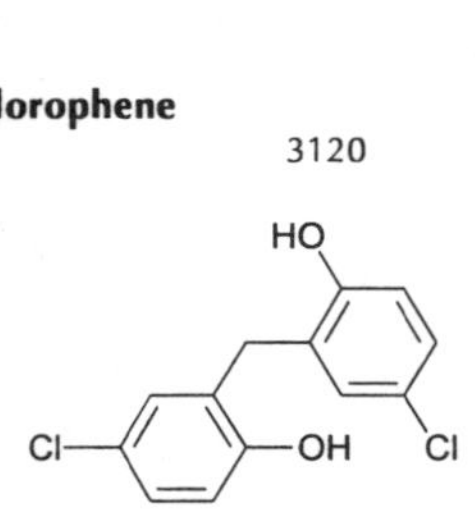

$C_{13}H_{10}Cl_2O_2$
2,2'-Methylenebis(4-chlorophenol).
dichlorophene; G-4; Anthipohen; Dicestal; Didroxane; Di-phenthane-70; Parabis; Plath-Lyse; Preventol G-D; Teniathane; Teniatol; Wespuril. Anthelmintic. Targets Cestodes. Used in veterinary medicine as an antiprotozoal, anthelmintic and antifungal agent. Inhibits 2-propanol-supported respiration of the parasite. Crystals; mp = 177-178°; insoluble in H_2O, sparingly soluble in C_7H_8, soluble in 95% EtOH (100 g/100 ml), Et_2O (< 100 g/100 ml), MeOH, isopropyl ether, petroleum ether; LD_{50} (mrat orl) = 1506 mg/kg, (frat orl) = 1683 mg/kg.

Veterinary Products:

006-691; (with Toluene) Vermiplex; *Schering-Plough Animal Health.*
007-829; Happy Jack Tapeworm Tablets; *Happy Jack.*
048-010; (with Toluene) Anaplex Caps; Canine and Feline Wormer Caps; *Boehringer Ingelheim Vetmedica.*
095-333; (with Toluene) Difolin No. 1 Capsules; Difolin No. 2.5 Capsules; Difolin No. 5 Capsules; Difolin No. 25 Capsules; Difolin No. 40 Capsules; *Fort Dodge Animal Health, Divn. AHP.*
101-497; (with Toluene) Tiny Tiger ™ Worming Capsules For Cats; Worming Capsules For Puppies, Small Dogs And Cats; *Lambert-Kay.*
101-498; (with Toluene) L.K. Worming Capsules For Dogs; *Lambert-Kay.*
102-020; (with Toluene) Tri-Plex Worm Capsules; *Schering-Plough Animal Health.*
102-942; (with Toluene) Pulvex Multipurpose Worm Caps; *Zema.*
103-953; (with Toluene) Trivermicide Worm Capsules; *Happy Jack.*
107-189; (with Toluene) Worm Capsules; *Farnam.*
117-688; (with Toluene) Dichlorophene & Toluene Caps; *Texas Vitamin.*
120-671; (with Toluene) Pet-Worm-Caps; *K. C. Pharmacal.*
121-557; (with Toluene) THR Worm Capsules; *Natchez Animal Supply.*
126-676; (with Toluene) D & T Worm Capsules; *Boehringer Ingelheim Vetmedica.*
138-900; (with Toluene) Dichlorophene / Toluene; *Global.*
140-850; (with Toluene) Elite Dog & Cat Wormer; *RSR Labs.*

81 Dichlorvos

62-73-7 3129 200-547-7

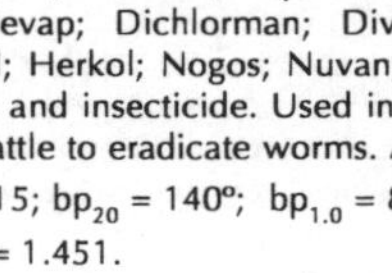

$C_4H_7Cl_2O_4P$
Dimethyl 2,2-dichloroethenyl phosphate.
phosphoric acid 2,2-dichlorovinyl dimethyl ester; O,O-dimethyl O-(2,2,-dichlorovinyl) phosphate; dichlorophos; dichlorovos; DDVP; SD 1750; Vapona; Astrobot; Atgard; Canogard; Dedevap; Dichlorman; Divipan; Equigard; Equigel; Estrosol; Herkol; Nogos; Nuvan; Task; Verdisol. An anthelmintic and insecticide. Used internally in dogs, cats, pigs and cattle to eradicate worms. Also used in flea collars. d_4^{25} = 1.415; bp_{20} = 140°; $bp_{1.0}$ = 84°; $bp_{0.5}$ = 72°; $bp_{0.01}$ = 30°; n_D^{25} = 1.451.

Veterinary Products:

031-512; Atgard ® Swine Wormer 27g; Atgard ® Swine Wormer 5.4g; Atgard ® V Swine Wormer 27g; Atgard ® V Swine Wormer 5.4; *Boehringer Ingelheim Vetmedica.*
033-803; Task Dog Anthelmintic; *Boehringer Ingelheim Vetmedica.*

035-918; Equigard; Verdisol; *Boehringer Ingelheim Vetmedica.*
040-848; Atgard ® C Swine Wormer; *Boehringer Ingelheim Vetmedica.*
043-606; Atgard ® Swine Wormer; Atgard ® V Swine Wormer; *Boehringer Ingelheim Vetmedica.*
048-237; Equigel; *Boehringer Ingelheim Vetmedica.*
048-271; Task ® Tabs; *Boehringer Ingelheim Vetmedica.*
049-032; Atgard ® C Production Efficiency Improver; *Boehringer Ingelheim Vetmedica.*

82 Diethylcarbamazine Citrate

1642-54-2 3165 216-696-6

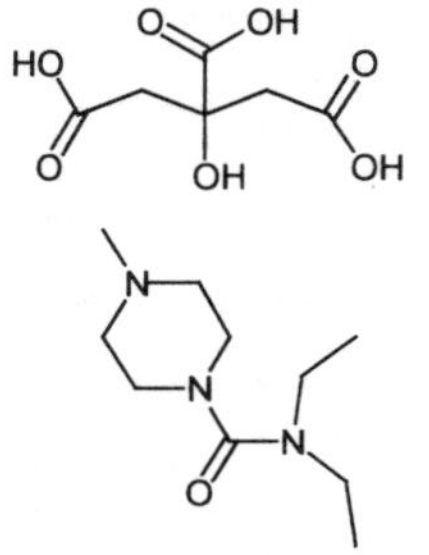

$C_{16}H_{29}N_3O_8$
N,N-Diethyl-4-methyl-1-piperazinecarboxamide citrate.
Banocide; Dec; Dirocide; Filaribits; Filazine; Franocide; Hetrazan; Loxuran; Longicid. Anthelmintic. Targets nematodes. mp = 141-143°; soluble in H_2O (> 75 g/100 ml), soluble in EtOH, insoluble in C_6H_6, Me_2CO, Et_2O, $CHCl_3$; LD_{50} (rat orl) = 1.38 g/kg.

Veterinary Products:

092-836; Diethylcarbamazine Citrate Tab.; Diro-Form; *Lloyd.*
092-837; D.E.C.- Sol. Syrup; Nemacide Oral Syrup; *Boehringer Ingelheim Vetmedica.*
093-512; Dirocide Tablets; *Fort Dodge Animal Health, Divn. AHP.*
100-689; Difil Syrup; *Evsco.*
100-690; Difil Tablets; *Evsco.*
104-493; Filaribits ® Chewable Tablets; *Pfizer.*
107-506; Carbam ® Tablets; *Merial.*
108-487; DEC Tabs 400 mg; *Wendt.*
108-863; DEC Chewable Tabs; *Wendt.*
118-032; Carbam ® Palatabs ®; *Merial.*
120-326; Filban Chewable Wafers; *Schering-Plough Animal Health.*
123-116; Diethylcarbamazine Citrate Cap.; *R. P. Scherer.*
124-842; Filban Tablets; *Schering-Plough Animal Health.*
125-137; Filaricide Capsules; *Happy Jack.*
127-627; Nemacide; Nemacide-C; *Boehringer Ingelheim Vetmedica.*
128-069; Nemacide Chewable Tablets; *Boehringer Ingelheim Vetmedica.*
128-517; Pet-Dec; *Pfizer.*
131-808; Dirocide Syrup; *Fort Dodge Animal Health, Divn. AHP.*
136-483; (with Oxibendazole) Filaribits ® Plus Chewable Tablets; *Pfizer.*

83 Dipiperazine Sulfate

71607-28-8 275-672-3

$C_8H_{22}N_4O_4S$
Di(hexahydropyridazine) sulfate.
Anthelmintic. Targets nematodes.

Veterinary Products:

010-005; Pig Wormer; Wazine®Wazine-34; *Fleming.*

84 Dithiazanine Iodide

514-73-8 3434 208-186-7

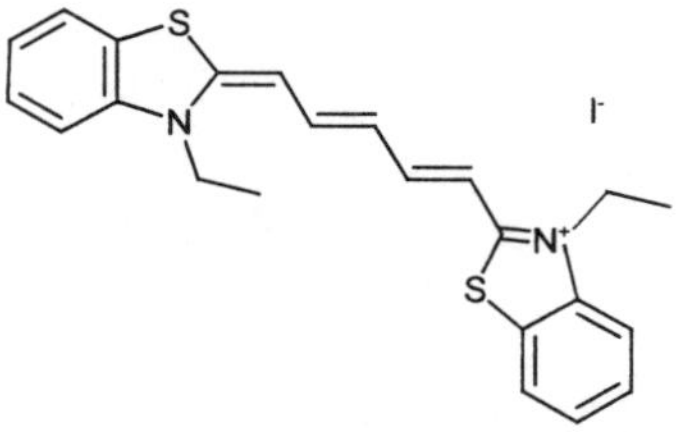

$C_{23}H_{23}IN_2S_2$
3-Ethyl-2-[5-(3-ethyl-2-benzothiazolinylidene)-1,3-pentadienyl]benzothiazolium iodide.
Developed as a photographic sensitizer. Used as an anthelmintic targeting nematodes. mp = 248° (dec); insoluble in H_2O.

Veterinary Products:

011-531; Dizan Tablets; *Boehringer Ingelheim Vetmedica.*
011-674; Dizan Powder; *Boehringer Ingelheim Vetmedica.*
012-469; (with Piperazine Citrate) Dizan Suspension With Piperazine; *Boehringer Ingelheim Vetmedica.*

85 Doramectin

117704-25-3 3483

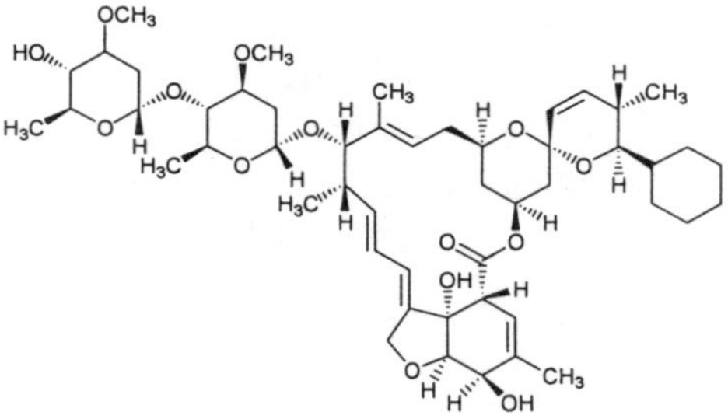

$C_{50}H_{74}O_{14}$
(2aE,4E,8E)-(5'S,6S,6'R,7S,11R,13S,15S,17aR,20R, 20aR,20bS)-6'-Cyclohexyl-5',6,6',7,10,11,14,15,17a,20, 20a,20b-dodecahydro-20,20b-dihydroxy-5',6,8,19-tetramethyl-17-oxospiro[11,15-methano-2H,13H,17H-furo[4,3,2-pq]-[2,6]-benzodioxacyclooctadecin-13,2'-[2H]pyran]-7-yl 2,6-dideoxy-4-O-(2,6-dideoxy3-O-methyl-α-L-arabinohexopyranosyl)-3-O-α-L-arabinohexapyranoside.
25-cyclohexyl-5-O-demethyl-25-de(1-methylpropyl)avermectin A_{1a}; 25-cyclohexylavermectin B_1; UK-67994.

Endectocide. Biosynthetic antiparacitic used to treat sheep scabies. Structurally related to avermectins. mp = 116-119°.

Veterinary Products:

141-061; Dectomax ® Injectable Solution for Cattle; Dectomax ® Injectable Solution for Cattle and Swine; *Pfizer.*
141-095; Dectomax ® Pour-on Solution; *Pfizer.*

86 Epsiprantel

98123-83-2 3671

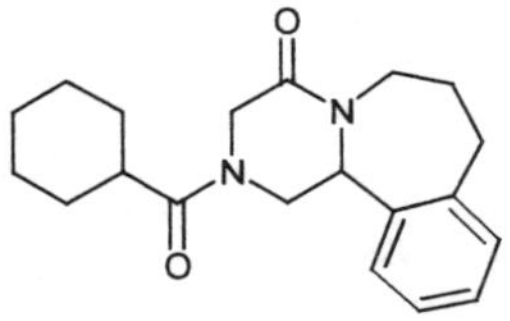

$C_{20}H_{26}N_2O_2$
2-(Cyclohexylcarbonyl)-2,3,6,7,8,12b-hexahydro-pyrazino[2,1-a][2]benzazepin-4(1H)-one.
BRL-38705; Cestex. Anthelmintic used in digs and cats to eradicate cestodes. Structurally related to praziquantel. White crystals; mp = 189-190°.

Veterinary Products:

045-578; Cestex®; *Steris.*

87 Febantel

58306-30-2 3982 261-205-0

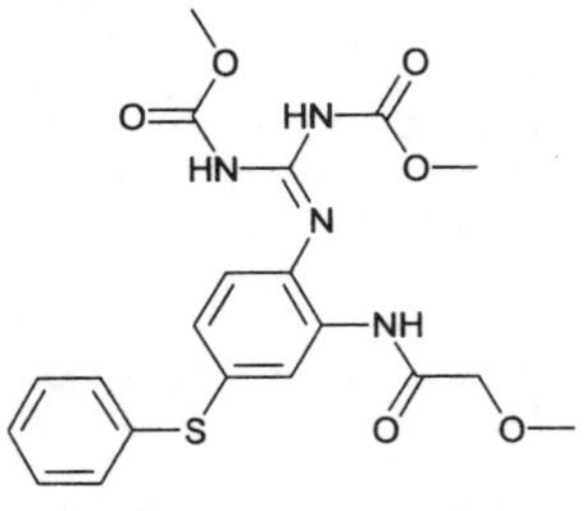

$C_{20}H_{22}N_4O_6S$
[[2-[(Methoxyacetl)amino]-4-(phenylthio)phenyl]-carbonimidoyl]biscarbamic acid dimethyl ester.
Bay Vh 5757; Bay h 5757; Rintal. Anthelmintic. Used to control strongyles, ascarids and pinworms in horses and intestinal parasites in dogs. Crystals; mp = 129-130°.

88 Fenbendazole

43210-67-9 4000 256-145-7

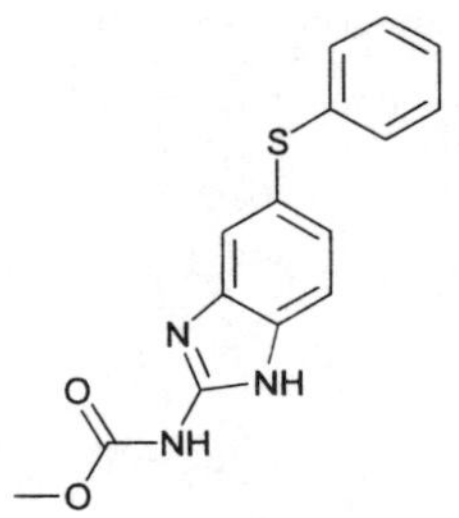

$C_{15}H_{13}N_3O_2S$
[5-(Phenylthio)-1H-benzimidazol-2-yl]carbamic acid methyl ester.
HOE-881v; Panacur; Safe-Guard. Anthelmintic targeting nematodes Used in dogs, cattle, horses and pigs. Has also been used in cats, sheep, goats, birds and llamas. Odorless, tasteless, light brown crystals; mp = 233° (dec); insoluble in H_2O, most organic solvents, soluble in DMSO.

Veterinary Products:

107-345; Cutter Paste; Rintal ® Paste; *Bayer.*
107-346; Rintal ® Suspension; *Bayer.*
133-953; (with Praziquantel) Vercom Paste Anthelmintic; *Bayer.*
140-912; Rintal ® Tabs Anthelmintic Tablets; *Bayer.*
141-007; (with Praziquantel, Pyrantel Pamoate) Drontal ™ Plus Broad Spectrum Anthelmintic Tablets; *Bayer.*

89 Haloxon

321-55-1 4636 206-289-1

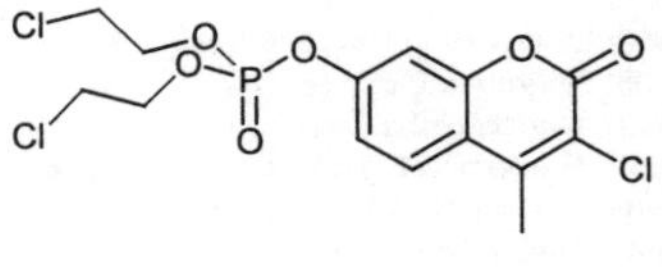

$C_{14}H_{14}Cl_3O_6P$
7-[[Bis(2-chloroethoxy)phosphinyl]oxy]-3-chloro-4-methyl-2H-1-benzopyran-2-one.
3-chloro-4-methylumbelliferone di(2-chloroethyl) phosphate; galoxone; helmirone; 96-H-60; Galloxon; Loxon; Luxon. Anthelmintic. Crystals; mp = 91°; LD_{50} (rat orl) = 900 mg/kg.

Veterinary Products:

048-913; Halox Wormer Drench; *Schering-Plough Animal Health.*
092-483; Halox Bolus; *Schering-Plough Animal Health.*

90 Hygromycin B
31282-04-9 4901 250-545-5

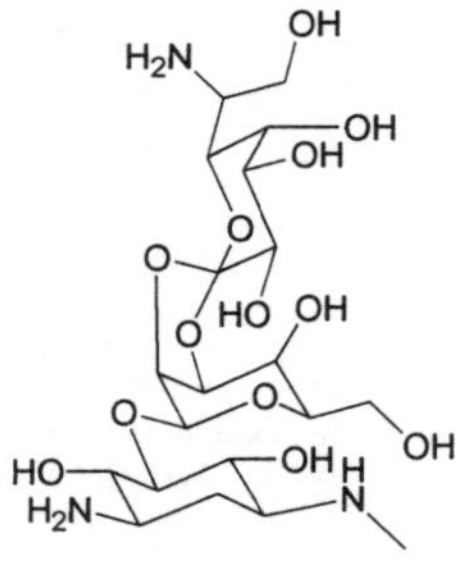

$C_{20}H_{37}N_3O_{13}$
O-6-Amino-6-deoxy-L-glycero-D-galacto-heptopyranosylidene-(1→2-3)-O-β-D-talopyranosyl-(1→5)-2-deoxy-N^3-methyl-D-streptamine.
Hygromix. Antibiotic produced by *Streptomyces hygroscopicus*. Used as an anthelmintic in pigs and chickens Amorphous powder; dec 160-180°; $[\alpha]_D^{20} = 20.2°$ (c = 1, H_2O); soluble in H_2O, MeOH, EtOH, insoluble in less polar solvents; weakly basic, pKa = 7.1, 8.8.

Veterinary Products:

010-918; Hygromix 2.4; *Elanco.*
011-948; Hygromix-2.4; *Elanco.*
012-548; (with Tylosin Phosphate)Tylosin plus Hygromix; *Elanco.*
013-214; Purina ® Hygromix-Swine; *Purina Mills.*
013-388; (with Tylosan) Hygromix-Tylan ® Premix; *Elanco.*
109-688; Hygromix 2.4 Premix; I.N. Hygromycin-2.4 Premix; *Intl. Nutrition.*
110-439; HFA Hygromix 0.48 Medicated Premix; HFA Hygromix 2.4 Medicated Premix; *ADM Animal Health & Nutrition.*
110-440; Doboy Hygromix 2.4 Medicated; *Quali-Tech.*
127-508; Hygromix 0.6; *Akey*r0.*
129-395; Hygromix 0.6 Premix; *I.M.S.*
140-443; Hygromix 0.6 Premix; Hygromix 1.6 Premix; *Alpharma.*
140-842; Hygromix 2.4 Premix (Type A Medicated Article); *ADM Animal Health & Nutrition.*

91 Imidocarb Dipropionate
4951

$C_{25}H_{32}N_6O_7$
N,N'-Bis[3-(4,5-dihydro-1H-imidazol-2-yl)phenyl]urea dipropionate.
imizol; Imizad Equine Injection. Antiprotozoal (Babesia).

Veterinary Products:

097-288; Imizol ® Equine Injection; *Schering-Plough Animal Health.*
141-071; Imizol ®; *Schering-Plough Animal Health.*

92 Ivermectin
70288-86-7 5264 274-536-0

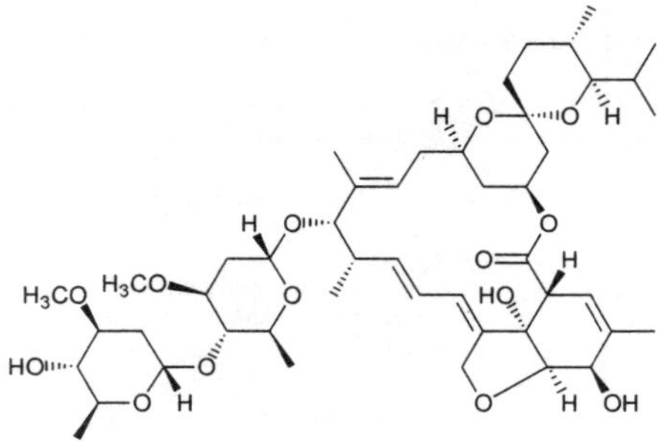

$C_{48}H_{74}O_{14}$
(2aE,4E,8E)-(5'S,6S,6'R,7S,11R,13R,15S,17aR,20R,20aR,20bS)-6'-(S)-sec-Butyl-3',4',5',6,6',7,10,11,14,15,17a,20,20a,20b)-tetradecahydro-20,20b-dihydroxy-5',6,8,19-tetramethyl-17-oxospiro[11,15-methano-2H,13H,17H]-furo[4,3,2-pq][2,6]benzodioxacyclooctadecin-13,2'-[2H]pyran]-7-yl 2,6-dideoxy-4-O-(2,6-dideoxy-3-O-methyl-α-L-arabino-heopyranosyl)-3-O-methyl-α-L-arabino-hexopyranoside.
22,23-dihydro C-076B; MK-933; Cardomec; Cardotek 30; Eqvalan; Heartgard 30; Ivomec; Mectizan; Zimecterin. Primarily 22,23-dihydroavermectin B_{1a}. Anthelmintic. Targets onchocerca (filarial worms). Used in dogs and large animals to control a variety of worm parasites. $[\alpha]_D = 71.5° \pm 3°$; λ_m = 238, 245 nm (ε = 27100, 30100 MeOH); slightly soluble in H_2O (0.4 mg/100 ml), insoluble in hydrocarbons, very soluble in MEK, prolylene glycol, polyethylene glycol.

Veterinary Products:

127-443; Eqvalan ® Injection; *Merial.*
128-409; Ivomec ® .27% Injection Grower And Feeder Pigs; Ivomec ® 1% Injection; Ivomec ® 1% Injection Cattle And Swine; Ivomec ® Injection; *Merial.*
131-392; Ivomec ® Liquid; *Merial.*
134-314; Eqvalan ®; Eqvalan ® Paste For Horses; *Merial.*
137-006; Ivomec ® Cattle Paste 0.153%; *Merial.*
138-412; Heartgard ® Tablets; Heartgard 30 ®; *Merial.*
140-439; Eqvalan ® Oral Liquid; Eqvalan ® Oral Liquid For Horses; *Merial.*
140-833; Ivomec ® F Injection For Cattle; Ivomec ® Plus Injection For Cattle; *Merial.*
140-841; Ivomec ® Pour-On For Cattle; *Merial.*
140-886; Heartgard ® Chewables For Dogs; Heartgard 30 ® Chewables; *Merial.*
140-971; Heartgard ® Plus; Heartgard ® Plus For Dogs; *Merial.*
140-974; Ivomec ® Premix for Swine; *Merial.*
140-988; Ivomec ® Sustained-Release Bolus for Cattle; *Merial.*
141-054; Ivomec ® Premix for Swine / Lincomix ® Premix; *Merial.*
141-078; Heartgard ™ for Cats; *Merial.*
141-097; Ivomec ® Premix + BMD ® Ivermectin; *Merial.*
200-202; Ivermectin Liquid for Horses; Phoenectin ™ Liquid for Horses; *Phoenix.*
200-219; Ivermectin Pour-On for Cattle; Phoenectin ™; Phoenectin ™ Pour-On for Cattle; *Phoenix.*

93 Levamisole

14769-73-4 5486 238-836-5

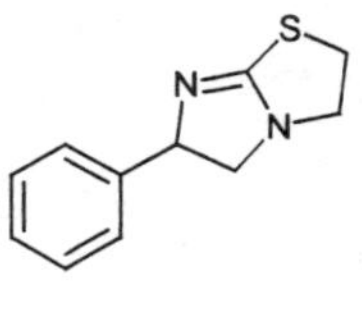

$C_{11}H_{12}N_2S$
(-)-2,3,5,6-Tetrahydro-6-phenylimidazo[2,1-b]thiazole.
Levovermax; Totalon. Immunomodulator with anthelmintic activity against nematodes. mp = 60-61.5°; $[\alpha]_D^{25}$ = -85.1° (c = 10 $CHCl_3$); [DL-form (tetramisole, Tetramizole)]: mp = 87-89°; [D-(+)-form (dexamisole)]: mp = 60-61.5°; $[\alpha]_D^{25}$ = 85.1° (c = 10 $CHCl_3$).

Veterinary Products:

139-877; Totalon ™ Topical Cattle Anthelmintic; *Schering-Plough Animal Health.*
140-844; Tramisol ® Pour-On; *American Cyanamid Divn., AHP Corp.*

94 Levamisole Hydrochloride

16595-80-5 5486 240-654-6

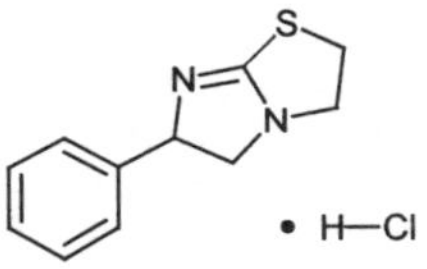

$C_{11}H_{13}ClN_2S$
(-)-2,3,5,6-Tetrahydro-6-phenylimidazo[2,1-b]thiazole monohydrochloride.
Ergamisole; R-12654; Ascaridil; Decaris; Ergamisol; Levacide; Levadin; Levasole; Meglum; Nemicide; Nilverm; Ripercol; Solaskil; Spartakon; Tramisol. Immunomodulator with anthelmintic activity against nematodes. mp = 227-229°; $[\alpha]_D^{20}$ = -124° ± 2° (c = 0.9 H_2O); soluble in H_2O; [DL-form (Bayer 9051, McN-JR-8299, R-8299)]: mp = 264-265°; soluble in H_2O (21 g/100 ml), MeOH, propylene glycol; sparingly soluble in EtOH; slightly soluble in $CHCl_3$, Me_2CO, C_6H_{14}; LD_{50} (mus iv) = 22 mg/kg, (mus sc) = 84 mg/kg, (mus orl) = 210 mg/kg, (rat iv) = 24 mg/kg, (rat sc) = 130 mg/kg, (rat orl) = 480 mg/kg; [D-(+)-form]: mp = 227-227.5°; $[\alpha]_D^{20}$ = 125° (c = 0.7 H_2O).

Veterinary Products:

039-356; Tramisol ® Cattle Wormer Bolus; Ripercol L Bolus; *Cyanamid Agri., Puerto Rico.*
039-357; Ripercol L Soluble Drench Powder; *Cyanamid Agri., Puerto Rico.*
042-740; Tramisol ® Soluble Drench Powder for Sheep; Ripercol L; *Cyanamid Agri., Puerto Rico.*
042-837; Tramisol ® Sheep Wormer Oblets; Ripercol L Wormer Oblets; *Cyanamid Agri., Puerto Rico.*
044-015; Tramisol ® Type A Medicated Article; *Cyanamid Agri., Puerto Rico.*
045-455; Tramisol ® Type A Medicated Article; *Cyanamid Agri., Puerto Rico.*
045-513; Ripercol L; *Cyanamid Agri., Puerto Rico.*
091-826; Levasole ® Cattle Wormer Bolus; Tramisol Cattle Wormer Oblets Tablets; *Schering-Plough Animal Health.*
092-237; (with Piperazine Dihydrochloride) Ripercol L-Piperazine Soluble Dihydrochloride; *Cyanamid Agri., Puerto Rico.*
093-688; (with Piperazine Dihydrochloride) Ripercol L-Piperazine Dihydrochloride; *Cyanamid Agri., Puerto Rico.*
107-085; Tramisol ®; *Cyanamid Agri., Puerto Rico.*
112-049; Levasole ® Soluble Pig Wormer; *Schering-Plough Animal Health.*
112-050; Levasole ® Soluble Drench Powder; Tramisol Soluble Drench Powder; *Schering-Plough Animal Health.*
112-051; Levamisole Soluble Drench Powder; Levasole Soluble Drench Powder; Tramisol ® Soluble Drench Powder; *Schering-Plough Animal Health.*
112-052; Levasole ® Sheep Wormer Bolus; Tramisole Sheep Wormer Oblets Tablets; *Schering-Plough Animal Health.*
126-237; Tramisol ® Gel; *Cyanamid Agri., Puerto Rico.*
200-225; Prohibit ™ Soluble Drench Powder; *Agri Labs.*

95 Mebendazole

31431-39-7 5807 250-635-4

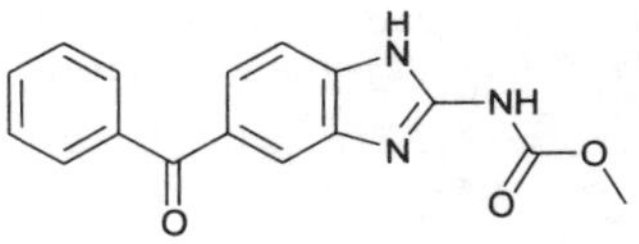

$C_{16}H_{13}N_3O_3$
Methyl 5-benzoyl-2-benzimidazolecarbamate.
Vermox; R-17635; Bantenol; Equivurm Plus; Lomper; Mebenvet; Noverme; Ovitelmin; Pantelmin; Telmin; Vermicidin; Vermirax. Anthelmintic. Targets nematodes. mp = 288.5°; insoluble in H_2O, EtOH, Et_2O, $CHCl_3$, soluble in formic acid; LD_{50} (sheep orl) > 80 mg/kg, (mus, rat, chk) > 40 mg/kg.

Veterinary Products:

091-736; Telmin ™ Equine Wormer; *Schering-Plough Animal Health.*
097-221; Telmin ™ Syringe Formula; *Schering-Plough Animal Health.*
100-402; Telmin ™ B Equine Wormer; *Schering-Plough Animal Health.*
102-987; Telmintic Powder; *Schering-Plough Animal Health.*
118-394; Telmin ™; *Schering-Plough Animal Health.*
138-954; Telmin ™ B Paste; *Schering-Plough Animal Health.*

96 Melarsomine Dihydrochloride

$C_{13}H_{23}AsCl_2N_8S_2$
Bis(2-aminoethyl) p-[(4,6-diamino-3-triazin-2-yl)amino]-dithiobenzenearsonite dihydrochloride.
Used to treat heartworm disease in dogs.

Veterinary Products:

141-042; Immiticide Sterile Powder; *Merial.*

97 Milbemycin Oxime

129496-10-2 6277

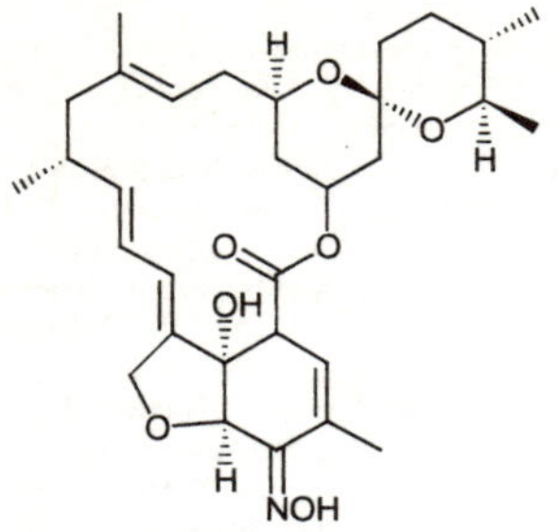

CGA-179246; Interceptor; Milbemax. Mixture of 4 parts milbemycin A_4 5-oxime and 1 part milbemycin A_3 5-oxime. Anthelmintic used to treat infection in dogs. Yellow-white powder; mp = 169.6-177.4°; soluble in H_2O, pH of aqueous solution = 6.3; LD_{50} (mus orl) = 1000-1500 mg/kg.

Veterinary Products:

140-915; Interceptor®; Interceptor® Flavor Tabs; Interceptor® Flavor Tabs® for Cats; SafeHeart™; *Novartis Animal Health.*
141-084; (with Lufenuron) Sentinel®; Sentinel® Flavor Tabs®; *Novartis Animal Health.*

98 Moxidectin

113507-06-5 6373

$C_{37}H_{53}NO_8$
[6R,23E,25S(E)]-5-O-Demethyl-28-deoxy-25-(1,3-dimethyl-1-butenyl)-6,28-epoxy-23-(methoxyimino)milbemycin B.
CL-301423; Cydectin; Moxidec; Vetdectin. Anthelmintic and antiparasitic. Used for prevention of heartworm in dogs.

Veterinary Products:

141-051; ProHeart ™; *Fort Dodge Animal Health, Divn. AHP.*
141-087; Quest ™ (moxidectin) 2% Equine Oral Gel; *Fort Dodge Animal Health, Divn. AHP.*
141-099; Cydectin ® (moxidectin) 0.5% Pour-On for Cattle; *Fort Dodge Animal Health, Divn. AHP.*

99 Oxfendazole

53716-50-0 7069 258-714-5

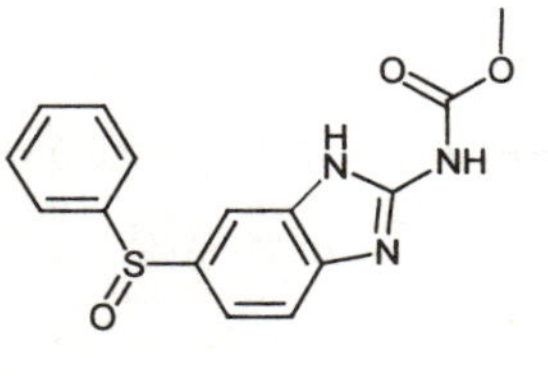

$C_{15}H_{13}N_3O_3S$
[5-(Phenylsulfinyl)-1H-benzimidazol-2-yl]carbamic acid methyl ester.
5-phenylsulfinyl-2-carbomethoxyaminobenzimidazole; RS-8858; Autoworm; Benzelmin; Repidose; Synanthic; Systamex. Anthelmintic. Used to control strongyles and roundworms in horses and other farm animals. Crystals; mp = 253° (dec); LD_{50} (dog) > 1600 mg/kg, (rat) >6400 mg/kg, (mus) >6400 mg/kg.

Veterinary Products:

110-776; Benzelmin ® Powder For Suspension; *Fort Dodge Animal Health, Divn. AHP.*
110-777; Benzelmin ® Top Dress.; *Fort Dodge Animal Health, Divn. AHP.*
132-105; Benzelmin ® Equine Anthelmic Paste; Synanthic ® Oral Paste; *Fort Dodge Animal Health, Divn. AHP.*
133-841; Benzelmin ® Equine Anthelmintic Suspension; Synanthic ® Suspension; *Fort Dodge Animal Health, Divn. AHP.*
136-740; (with Trichlorfon) Benzelmin ® Plus Paste; *Fort Dodge Animal Health, Divn. AHP.*
140-854; Synanthic ® Bovine Dewormer Suspension 22.5%; Synanthic ® Bovine Dewormer Suspension 9.06%; *Fort Dodge Animal Health, Divn. AHP.*
140-892; Synanthic ® Bovine Dewormer Paste, 18.5%; *Fort Dodge Animal Health, Divn. AHP.*

100 Oxibendazole

20559-55-1 7070 243-877-7

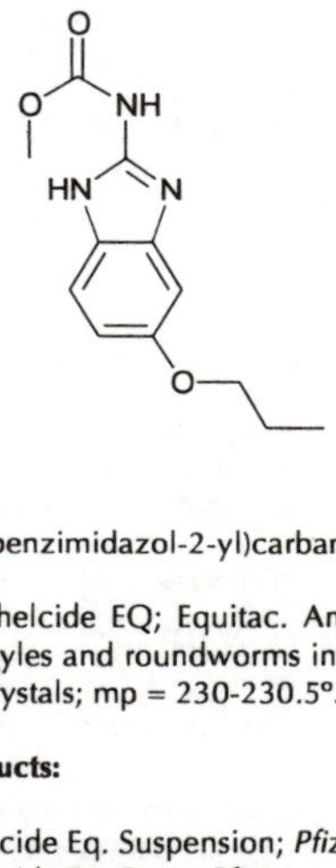

$C_{12}H_{15}N_3O_3$
(5-Propoxy-1H-benzimidazol-2-yl)carbamic acid methyl ester.
SKF-30310; Anthelcide EQ; Equitac. Anthelmintic. Used to control strongyles and roundworms in horses and other farm animals. Crystals; mp = 230-230.5°.

Veterinary Products:

109-722; Anthelcide Eq. Suspension; *Pfizer.*
121-042; Anthelcide Eq. Paste; *Pfizer.*
136-483; (with Diethylcarbamazine Citrate) Filaribits® Plus Chewable Tablets; *Pfizer.*

101 Phenothiazine

92-84-2 7404 202-196-5

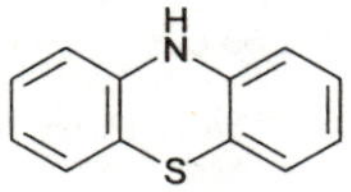

$C_{12}H_9NS$
10H-Phenothiazine.
dibenzothiazine; thiodiphenylamine; AFI-Tiazin; Antiverm; Fentiazin; Helmetina; Lethelmin; Nemazine;

Orimon; Phénégic; Phenoverm; Phenovis; Phenoxur; Reconox; Souframine; Vermitin. Anthelmintic, also used as a pesticide. Yellow crystals; mp = 185.1°; bp = 371°, bp_{40} = 290°; soluble in C_6H_6, Et_2O, hot AcOH, slightly soluble in EtOH, mineral oils, insoluble in petroleum ether, $CHCl_3$, H_2O.

Veterinary Products:

015-154; (with Piperazine Dichloride, Trichlorfon); *Fort Dodge Animal Health, Divn. AHP.*
033-149; (with Piperazine-Carbon Disulfide Complex)Parvex Plus ® Suspension Complex; *Pharmacia & Upjohn.*

102 Piperazine Citrate

144-29-6 7617 205-622-8

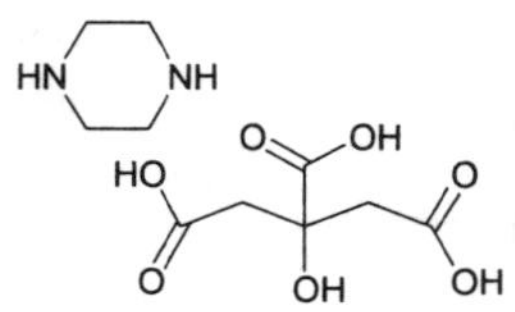

$C_{24}H_{46}N_6O_{14}$
Tripiperazine dicitrate.
Helmezine; Oxucide; Patazine; Pinozan; Pipizan Citrate; Pipracid (syrup); Rhomex; Ta-Verm; Worm Away. Ascaricide and Anthelmintic. Targets nematodes. Used as an ascaricide in dogs, cats, horses, pigs and poultry. mp = 182-187° (dec); insoluble in EtOH, Et_2O, $CHCl_3$.

Veterinary Products:

012-469; (with Dithiazanine Iodide) Dizan Suspension with Piperazine; *Boehringer Ingelheim Vetmedica.*
047-333; (with Thiabendazole) Equizole ® A Liquid; *Merial.*
092-709; Sergeants ® Worm Away For Dogs; *ConAgra.*
092-710; Sergeants ® Worm Away For Cats; *ConAgra.*

103 Piperazine Complex with Carbon Disulfide

Ascaricide and Anthelmintic. Targets nematodes. Used as an ascaricide in dogs, cats, horses, pigs and poultry.

Veterinary Products:

011-299; Parvex ® Suspension; *Pharmacia & Upjohn.*
011-590; Parvex ® Bolus; *Pharmacia & Upjohn.*
033-149; (with Phenothiazine) Parvex Plus ® Suspension Complex; *Pharmacia & Upjohn.*

104 Piperazine Dihydrochloride

142-64-3 7617 205-551-2

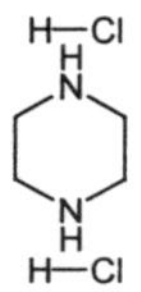

$C_4H_{12}Cl_2N_2$
Hexahydropyrazine dihydrochloride.
piperazidine hydrochloride; diethylenediamine hydrochloride. Ascaricide and Anthelmintic. Targets nematodes. Used as an ascaricide in dogs, cats, horses, pigs and poultry.

Veterinary Products:

015-154; (with Phenothiazine, Trichlorfon) Dyrex T.F - 1000; Dyrex T.F - 500; Dyrex T.F. 200 Trichlorfon; *Fort Dodge Animal Health, Divn. AHP.*
092-237; (with Levamisole Hydrochloride) Ripercol L-Piperazine Soluble Dihydrochloride; *Cyanamid Agri., Puerto Rico.*
093-688; (with Levamisole Hydrochloride) Ripercol L-Piperazine Dihydrochloride; *Cyanamid Agri., Puerto Rico.*

105 Piperazine Hydrochloride

6094-40-2 7617 228-042-7

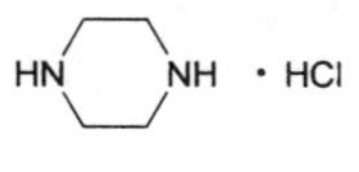

$C_4H_{11}ClN_2$
Hexahydropyrazine monohydrochloride.
Ascaricide and Anthelmintic. Targets nematodes. Used as an ascaricide in dogs, cats, horses, pigs and poultry.

Veterinary Products:

010-005; (with Dipiperazine Sulfate) Pig Wormer; Wazine®; Wazine-34; *Fleming.*

106 Piperazine Phosphate

1951-97-9 7617 217-775-8

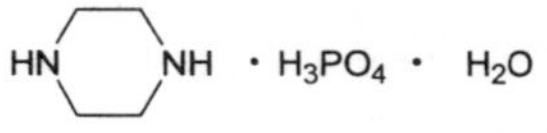

$C_4H_{13}N_2O_4P$
Hexahydropyrazine monophosphate.
Ascaricide and Anthelmintic. Targets nematodes. Used as an ascaricide in dogs, cats, horses, pigs and poultry. Slightly soluble in H_2O, pH of aqueous solution = 6.5.

Veterinary Products:

037-410; (with Thiabendazole) Equizole ® A; *Merial.*
091-260; Pulvex Worm Caps; *Zema.*
101-161; (with Thenium Closylate) Thenatol PW Tablets; *Schering-Plough Animal Health.*

107 Praziquantel

55268-74-1 7896 259-559-6

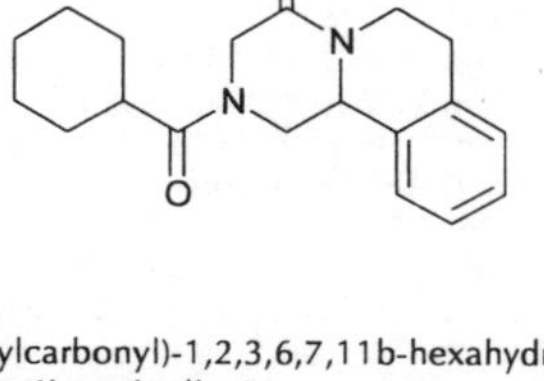

$C_{19}H_{24}N_2O_2$
2-(Cyclohexylcarbonyl)-1,2,3,6,7,11b-hexahydro-4H-pyrazino[2,1-a]isoquinolin-4-one.
EMBAY 8440; Biltricide; Cesol; Droncit. Anthelmintic.

Targets schistosoma. mp = 136-138°; soluble in H_2O (0.04 g/100 ml), EtOH (9.7 g/100 ml), $CHCl_3$ (56.7 g/100 ml); LD_{50} (mus orl) = 2000-3000 mg/kg, (mus sc) > 3000 mg/kg, (rat orl) = 2000-3000 mg/kg, (rat sc) > 3000 mg/kg.

Veterinary Products:

111-607; Droncit ® 5.68% Injectible Solution; *Bayer.*
111-798; Cutter Tape-Tabs (OTC); Droncit ® Canine Cestode Tablets (Rx); *Bayer.*
133-953; (with Febantel) Vercom Paste *Anthelmintic; Bayer.*
141-007; (with Febantel, Pyrantel Pamoate) Drontal ™ Plus Broad Spectrum Anthelmintic Tablets; *Bayer.*
141-008; (with Pyrantel Pamoate) Drontal ™ Tablets; *Bayer.*

108 Pyrantel Pamoate

22204-24-6 8139 244-837-1

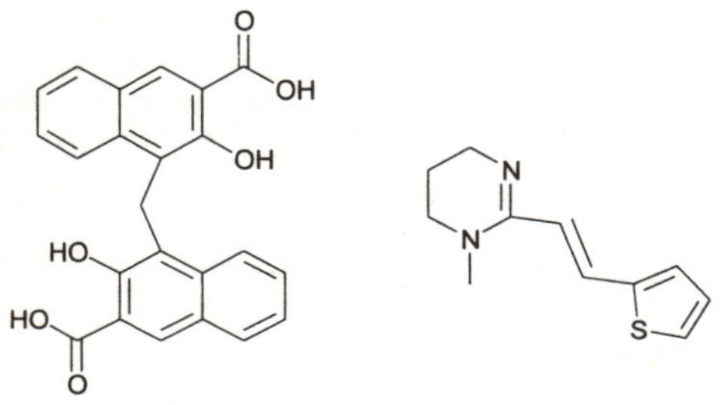

$C_{34}H_{30}N_2O_6S$
(E)-1,4,5,6-Tetrahydro-1-methyl-2-[2-(2-thienyl)vinyl]pyrimidine compound with 4,4'methylenebis[3-hydroxy-2-naphthoate].
Antiminth; Combantrin; Cobantril; Early Bird; Helmex; Helmintox; Piranver; CP-10423-16; component of: Drontal, Drontal Plus, HeartGard Plus. Anthelmintic. Targets nematodes. Used to control parasites in small and large animals. Insoluble in H_2O.

Veterinary Products:

091-739; Strongid ® T; *Pfizer.*
100-237; Nemex; Nemex-2; RFD Liquid Wormer; *Pfizer.*
101-331; Dog Wormer Tablets; *Farnam.*
129-831; Banminth ®-P Paste; *Pfizer.*
139-191; Dog Wormer Chewable Tablets; *Farnam.*
140-971; (with Ivermectin) Heartgard ® Plus; Heartgard ® Plus For Dogs; *Merial.*
141-007; (with Febantel, Praziquantel) Drontal ™ Plus Broad Spectrum Anthelmintic Tablets; *Bayer.*
141-008; (with Praziquantel) Drontal ™ Tablets; *Bayer.*
200-007; Liqui-Vict 2X ™; *Happy Jack.*
200-028; Evict ®; Evict ® DS ™ Wormer For Puppies and Dogs; Evict ® Liquid Wormer For Puppies and Dogs; Lassie ® DS Liquid Wormer For Puppies and Dogs; Vet's-Own ™ Canine Anthelmintic Suspension; Vet's-Own ™ DS Canine Anthelmintic Suspension; *Lambert-Kay.*
200-246; Anthelban V; Pyrantel Pamoate Oral Suspension; *Phoenix.*
200-248; Pyrantel Pamoate Suspension; Pyrantel Pamoate Suspension 2.27mg; Pyrantel Pamoate Suspension 4.54mg; *Phoenix.*

109 Pyrantel Tartrate

33401-94-4 8139 251-501-8

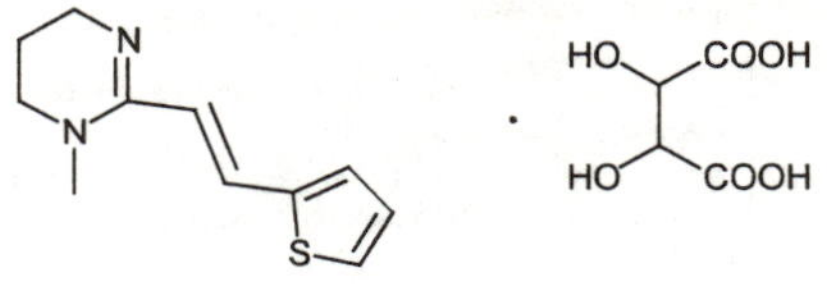

$C_{15}H_{20}N_2O_6S$
(E)-1,4,5,6-Tetrahydro-1-methyl-2-[2-(2-thienyl)-vinyl]pyrimidine tartrate (1:1).
Banminth; Strongid; CP-10423-18. Anthelmintic. Targets nematodes. Used to control parasites in small and large animals. mp = 148-150°; λ_m = 312 nm (log ε 4.27 H_2O).

Veterinary Products:

042-888; Banminth ® (5.66 g); Strongid ® (5.66 g); *Pfizer.*
043-290; Banminth ® Premix 80; *Pfizer.*
092-150; Purina ® Horse & Colt Wormer; *PM Resources.*
092-955; (with Carbadox) Mecadox ® / Banminth ®; *Pfizer.*
097-258; Purina ® Ban Worm For Pigs; *Purina Mills.*
110-047; (with Tylosin Phosphate, Lincomycin Hydrochloride Monohydrate) Banminth ® Tylan ®; *Pfizer.*
116-044; Banminth ® /Lincomix ®; *Pfizer.*
118-814; Worm-Ban 10; Worm-Ban 5; *Akey.*
118-815; Q.T. Ban-Tech; *Quali-Tech.*
118-877; Ban-A-Worm Pyrantel Tartrate; Ban-A-Worm Pyrantel Tartrate Ton Pack; *ADM Animal Health & Nutrition.*
119-062; Cadco-BN-10 Banminth ® Premix; Cadco-BN-5 Banminth ® Premix; *Triple F.*
119-063; Pyrantel Tartrate Ton Pack; *Bioproducts.*
121-337; INI Swine Ban-Wormer B-9 19.2 Banminth.; INI Swine Ban-Wormer B-9.6 Banminth; *Intl. Nutrition.*
123-000; Super Swine Wormer B-9 19.2 Banminth; Super Swine Wormer B-9.6 Banminth; *Seeco.*
129-415; Custom Ban Wormer 19.2 Banminth ®; Custom Ban Wormer 9.6 Banminth ®; *Custom Feed Blenders.*
132-136; Ban-A-Worm II; *Wayne Feed Divn.*
133-490; Ban-D-Wormer II Banminth ®; *ADM Animal Health & Nutrition.*
135-243; Swine Guard-BN Banminth ® Premix; *Southern Micro-Blenders.*
135-941; Check-E-Ton BM; *Purina Mills.*
136-369; Custom Ban Wormer 9.6; *S. St. Paul Feeds.*
136-384; Swine Wormer-BN Banminth ®; *Nutra-Blend.*
136-601; Swine Guard-BN; *I.M.S.*
137-484; Swine Guard-BN; *Custom Feed Services.*
138-656; BN Wormer-19.2 Banminth ® Premix; *Farmland.*
138-941; (with Lincomycin Hydrochloride Monohydrate) Lincomix ® / Banminth ®; Lincomix ® 20/ Banminth ®; *Pharmacia & Upjohn.*
140-819; Strongid ® 48; *Pfizer.*
140-825; Banminth ® Intermediate Premix; Swine Wormer-B; *Furst-McNess.*
200-168; CW 48 (Type A Medicated Article); *Equi Aid.*

110 Sodium Iodide

7681-82-5 8777 231-679-3

INa

Sodium iodide.

Ioduril; Anayodin. Iodine supplement; expectorant. Used in treatment of actinobacillosis and actinomycosis in cattle. Soluble in H_2O (200 g/100 ml), EtOH (50 g/100 ml), glycerol (100 g/100 ml); Me_2CO; pH = 8-9.5; MLD (rat iv) = 1300 mg/kg.

111 Thenium Closylate

4304-40-9 9414 224-318-6

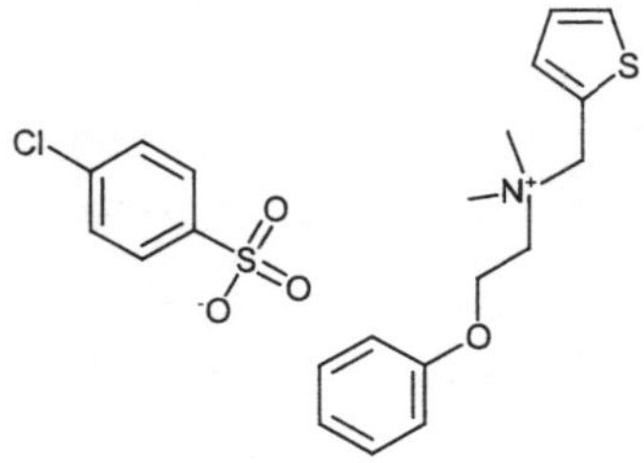

$C_{21}H_{24}ClNO_4S_2$

N,N-Dimethyl-N-(2-phenoxyethyl)-2-thiophenemethanaminium salt with 4-chlorobenzenesulfonic acid (1:1). dimethyl(2-phenoxyethyl)-2-thenylammonium p-chlorobenzensulfonate; 611C55; Bancaris; Canopar. Anthelmintic. Crystals; mp = 159-160°; soluble in H_2O (0.6 g/100 ml at 20°.

Veterinary Products:

015-182; Canopar Tablets; *Schering-Plough Animal Health.*
101-161; (with Piperazine Phosphate)Thenatol PW Tablets; *Schering-Plough Animal Health.*

112 Thiabendazole

148-79-8 9426 205-725-8

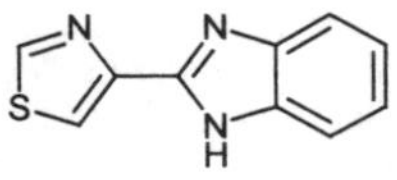

$C_{10}H_7N_3S$

2-(4-Thiazolyl)-1H-benzimidazole.

MK-360; Omnizole; Thiaben; Thibenzole; Bovizole; Eprofil; Equizole; Mintezol; Top Form Wormer; Mertect; Lombristop; Minzolum; Nemapan; Polival; TBZ; Tecto. Anthelmintic. Targets nematodes. mp = 304-305°; λ_m = 298 nm (ε 23330 MeOH); slightly soluble in H_2O (3.84 g/100 ml at pH 2.2), soluble in DMF, DMSO, slightly soluble in alcohols, esters and chlorocarbons; LD_{50} (mus orl) = 3600 mg/kg, (rat orl) = 3100 mg/kg, (rbt orl) > 3800 mg/kg.

Veterinary Products:

013-022; Thibenzole Sheep & Goat Wormer; *Merial.*
013-407; Equizole ® Horse Wormer; *Merial.*
013-954; Thibenzole 20% Swine Premix; *Merial.*
014-350; Omnizole; *Merial.*
015-123; TBZ ® Cattle Wormer (Drench); *Merial.*
015-875; TBZ 200 Medicated Feed Premix; Thibenzole - 100; Thibenzole - 200; Thibenzole - 45; *Merial.*
030-103; Thibenzole (18 mg); *Merial.*
030-578; E-Z-EX Wormer Pellets; *Moorman.*
034-114; Equizole ®; *Merial.*
035-631; Thibenzole Pig Wormer; *Merial.*
037-410; (with Piperazine Phosphate) Equizole ® A; *Merial.*
040-205; Purina ® Horse Wormer Medicated; *Purina Mills.*
042-633; (with Dexamethasone, Neomycin Sulfate) Tresaderm Dermatologic Solution; *Merial.*
042-910; E-Z-EX Wormer Mintrate Block; *Moorman.*
043-141; Thibenzole 300 Medicated; *Merial.*
044-654; Equizole ® Horse Wormer Pellets; *Merial.*
047-333; (with Piperazine Citrate) Equizole ® A Liquid; *Merial.*
048-487; TBZ ® Wormer Paste 50%; *Merial.*
049-461; TBZ ® Wormer Paste 43%; *Merial.*
091-067; (with Trichlorfon) Equivet-14; Equizole ®-B; *Farnam.*
098-689; Equizole ® 50% Wormer Paste; Equizole ® 50% Wormer Paste For Horses; *Merial.*

113 Ticarbodine

31932-09-9 9567

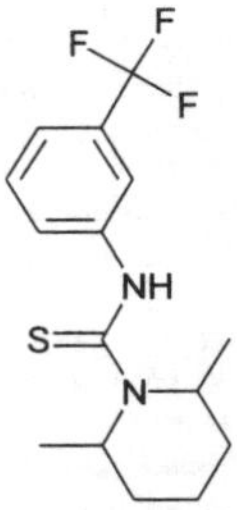

$C_{15}H_{19}F_3N_2S$

2,6-Dimethyl-N-[3-(trifluoromethyl)phenyl]-1-piperidinecarbothioamide.

EL-974; Tribodine; α,α,α-trifluoro-2,6-dimethylthio-1-piperidinecarboxy-m-toluidide. Anthelmintic. Effective dogs. Light yellow crystals; mp = 123-126°.

114 Tioxidazole

61570-90-9 9599 262-854-2

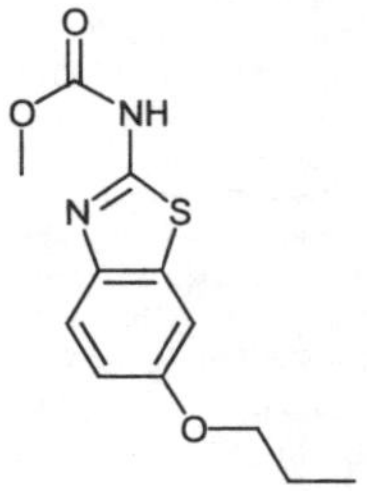

$C_{12}H_{14}N_2O_3S$

(6-Propoxy-2-benzothiazolyl)carbamic acid methyl ester.

Sch-21480; Tiox. Anthelmintic used in horses. White

crystals; mp = 178-180°; insoluble in H_2O, slightly soluble in organic solvents.

Veterinary Products:

047-092; Tribodine 900 mg; *Elanco.*
096-678; Tribodine Capsules 900 mg; *Elanco.*

115 Trichlorfon

52-68-6 9753 200-149-3

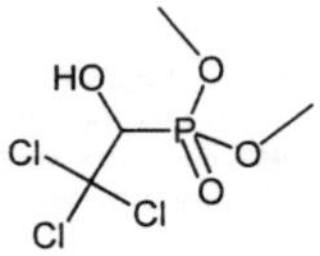

Dimethyl 2,2,2-trichloro-1-hydroxyethylphosphonate.
An ectoparasiticide and anthelmintic targeting nematodes.

Veterinary Products:

012-956; Dyrex Bolus; Dyrex Capsules; Dyrex Granules; Dyrex Tablets; *Fort Dodge Animal Health, Divn. AHP.*
013-248; (with Atropine) Freed No. 25; Freed No.10; *Fort Dodge Animal Health, Divn. AHP.*
015-154; (with Phenothiazine, Piperazine Dihydrochloride) Dyrex T.F - 1000; Dyrex T.F - 500; Dyrex T.F. 200; *Fort Dodge Animal Health, Divn. AHP.*
035-650; (with Atropine Sulfate) Dyrex Powder; *Fort Dodge Animal Health, Divn. AHP.*
048-915; Purina ® Bot Control; *Purina Mills.*
091-067; (with Thiabendazole) Equivet-14; Equizole ®-B; *Farnam.*
100-402; (with Mebendazole) Telmin ™ B Equine Wormer; *Schering-Plough Animal Health.*
136-740; (with Oxfendazole) Benzelmin ® Plus Paste; *Fort Dodge Animal Health, Divn. AHP.*
138-954; (with Mebendazole) Telmin ™ B Paste; *Schering-Plough Animal Health.*

Antiamebics

116 Carbarsone

121-59-5 1830 204-484-6

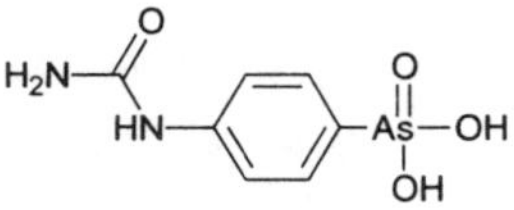

$C_7H_9AsN_2O_4$
N-Carbamoylarsanilic acid.
p-arsonophenylurea; Amabevan; Ameban; Amibiarson; Arsambide; Carb-O-Sep; Histocarb; Fenarsone; Leucarsone; Aminarsone; Amebarsone. Antiamebic; used as an antihistomonad in turkeys. mp = 174°; slightly soluble in H_2O, EtOH; insoluble in Et_2O, $CHCl_3$; LD_{50} (rat orl) = 510 mg/kg.

Veterinary Products:

010-285; Carb-O-Sep ® Type A Medicated Article; *Alpharma.*
038-879; (with Zoalene) Carb-O-Sep ® / Zoamix®; *Alpharma.*
039-646; Carb-O-Gain; *Alpharma.*
118-507; (with Amprolium) Amprol / Carb-O-Sep ®; *Merial.*
130-661; (witn Bambermycins) Flavomycin ® / Carb-O-Sep; *Hoechst-Roussel Vet.*
136-484; (with Bacitracin Zinc) Carb-O-Sep ® / Baciferm ®; *Roche Vitamins.*
200-203; (with Bacitracin Zinc) Carb-O-Sep ® / Albac ®; *Alpharma.*

117 Chlortetracycline

57-62-5 2245 200-341-7

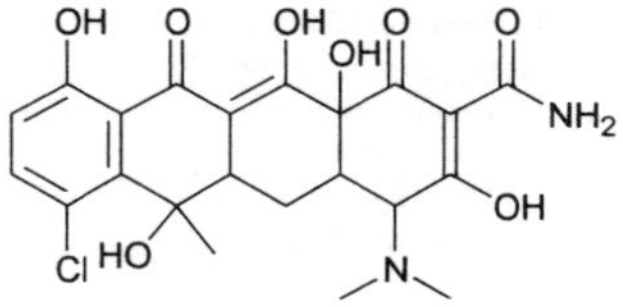

$C_{22}H_{23}ClN_2O_8$
7-Chloro-4-(dimethylamino)-1,4,4a,5,5a,6,11,12a-octahydro-3,6,10,12,12a-pentahydroxy-6-methyl-1,11-dioxo-2-naphthacenecarboxamide.
7-chlorotetracycline; Acronize; Aureocina; Aureomycin; Biomitsin; CentrAureo; Chrusomykine; Orospray. Antibacterial, antiamebic and antiprotozoal. Used in veterinary medicine as an antimicrobial agent. mp = 168-169°; $[\alpha]_D^{23}$=-275.0° (MeOH); λ_m= 230, 262.5, 367.5 nm (0.1N HCl), 255, 285, 345 nm (0.1N NaOH); soluble in H_2O (0.5-0.6 mg/ml), cellosolves, dioxane, carbitol, poorly soluble in other organic solvents.

Veterinary Products:

048-761; Aureomycin ® Type A Medicated Article; *Roche Vitamins.*
091-647; (with Amprolium, Ethopabate) Rainbow Broiler Base Concentrate; *Stockton Hay & Grain.*
138-935; Chlortetracycline - 50; Chlortetracycline - 60; Chlortetracycline - 70; Chlortetracycline - 80; Chlortetracycline - 100; Chlortetracycline - 100MR; Pennchlor 100MR; Pennchlor 50; Pennchlor 50-G; Pennchlor 70; Pennchlor 80; Pennchlor 90; *Pennfield Oil.*
141-059; (with Bacitracin Methylene Disalicylate) BMD ® - 10, 25, 30, 40, 50, 60, 75 and ChlorMax™ -50, 65, 70/ Micro-CTC ®100; *Alpharma.*
200-259; (with Roxarsone, Salinomycin Sodium) ChlorMax ™ / Sacox ® / 3-Nitro ®; ChlorMax ™ / Bio-Cox ® / 3-Nitro ®; *Alpharma.*
200-260; (with Roxarsone, Salinomycin Sodium) ChlorMax ™ / Sacox ® / 3-Nitro ®; ChlorMax ™ / Bio-Cox ® / 3-Nitro ®; *Alpharma.*
200-261; (with Salinomycin Sodium) ChlorMax ™ / Sacox ®; ChlorMax ™ / Bio-Cox ®; *Alpharma.*
200-262; (with Salinomycin Sodium) ChlorMax ™ / Sacox ®; *Hoechst-Roussel Vet.*
200-263; (with Monensin Sodium) ChlorMax™ / Coban ®; *Alpharma.*

118 Iodochlorhydroxyquin

130-26-7 5052 204-984-4

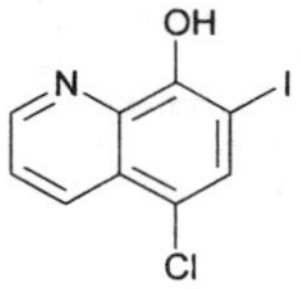

$C_9H_{15}ClINO$
5-Chloro-7-iodo-8-quinolinol.
Domeform-HC; Quin-O-Crème; Rheaform Boluses; Vioform; Clioquinol; Cort-Quin; Formtone-HC; Lidaform-HC; Nystaform; Nystaform-HC; Racet; Vioform-Hydrocortisone; Amebil; Alchloquin; Amoenol; Bactol; Barquinol; Budoform; Chinoform; Clioquinol; Cliquinol; Eczecidin; Enteroquinol; Entero-Septol; Entero-Vioform; Enterozol; Entrokin; Hi-Eneterol; Iodoenterol; Nioform; Quinambicide; Rometin; Vioform; Vioformio. Antiamebic. Also has been used as a topical and intestinal anti-infective. mp = 178-179° (dec); λ_m = 266 nm ($A^{1\%}_{1\ cm}$ = 990 H_2O/HCl), 269 nm ($A^{1\%}_{1\ cm}$ = 1120 (MeOH/KOH), 255 nm ($A^{1\%}_{1\ cm}$ = 1570 EtOH); soluble in EtOH (7.8 mg/ml), EtOAC (58.8 mg/ml), AcOH (5.9 mg/ml), insoluble in H_2O, EtOH, Et_2O; LD_{50} (cat orl) = 400 mg/kg.

Veterinary Products:

031-448; Rheaform Bolus; *Fort Dodge Animal Health, Divn. AHP.*
040-123; (with Cephalonium, Flumethasone, Piperocaine Hydrochloride; Polymyxin B Sulfate) Toptic Ointment (15 g); *Elanco.*

119 Tetracycline

60-54-8 9337 200-481-9

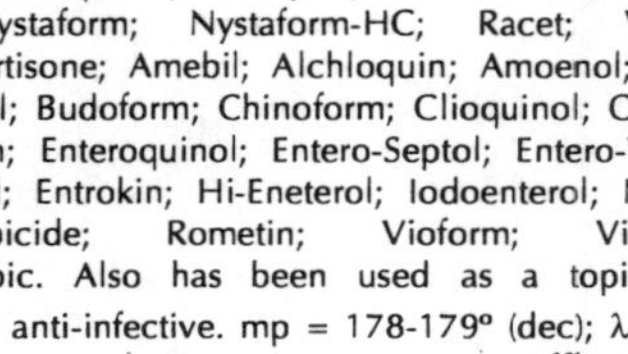

$C_{22}H_{24}N_2O_8$
(4S,4aS,5aS,6S,12aS)-4-(Dimethylamino)-1,4,4a,5,5a,6,11,12a-octahydro-3,6,10,12,12a-pentahydroxy-6-methyl-1,11-dioxo-2-naphthacenecarboxamide.
Liquamycin; Mysteclin-F; Talsutin; tsiklomitsin; Abricycline; Ambramycin; Bio-Tetra; Cyclomycin; Dumocyclin; Tetradecin. Antiamebic, antibacterial and antirickettsial. mp = 170-175° (dec); $[\alpha]_D^{25}$ = -257.9° (0.1 N HCl), -239° (MeOH); λ_m = 220, 268, 355 nm (ε 13000, 18040, 13320 0.1N HCl); soluble in H_2O (1.7 mg/ml), MeOH (> 20 mg/ml); LD_{50} (rat orl) = 707mg/kg, (mus orl) = 808 mg/kg.

Veterinary Products:

065-124; Tetracycline; *Pfizer.*

120 Tetracycline Hydrochloride

64-75-5 9337 200-593-8

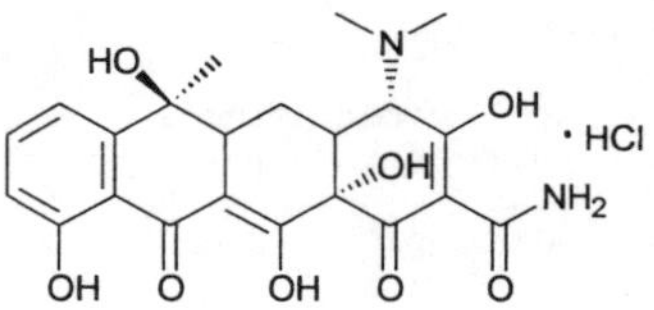

$C_{22}H_{26}Cl_2N_2O_8$
(4S,4aS,5aS,6S,12aS)-4-(Dimethylamino)-1,4,4a,5,5a,6,11,12a-octahydro-3,6,10,12,12a-pentahydroxy-6-methyl-1,11-dioxo-2-naphthacene-carboxamide monohydrochloride.
Achro; Achromycin; Ala-Tet; Cyclopar; Panmycin; Robitet; Steclin; SumycinTetraSURE; Ambracyn; Ambramicina; Bristaciclina; Cefracycline; Criseociclina; Cyclopar; Diocyclin; Helvecyclin; Hostacyclin; Imex; Mediletten; Mephacyclin; Panmycin; Partrex; Polycycline; Purociclina; Quadracyclin; Remicyclin; Riocyclin; Ro-cycline; Sanclomycine; St; Supramycin; Sustamycin; Tefilin; Tetrabakat; Tetrabid; Tetrablet; Tetrabon; Tetrachel; Tetracompren; Tetracyn; Tetrakap; Tetralution; Tetramavan; Tetramycin; Tetrosol; Topicycline; Totomycin; Triphacyclin; Unicin; Vetquamycin-324. Antiamebic, antibacterial, antirickettsial. mp = 214°; $[\alpha]_D^{25}$ = -257.9° (c = 0.5 on 0.1N HCl); soluble in H_2O, MeOH, EtOH, insoluble in Et_2O, hydrocarbons; LD_{50} (rat orl) = 6443 mg/kg.

Veterinary Products:

055-073; Panmycin ® Tablets; *Pharmacia & Upjohn.*
055-076; (with Novobiocin Sodium) Albaplex ® 3x Tablets; Albaplex ® Tablets; *Pharmacia & Upjohn.*
065-004; Panmycin ® 500 Bolus; Panmycin ® Hydrochloride; *Pharmacia & Upjohn.*
065-060; Panmycin ® Aquadrops Liquid; *Pharmacia & Upjohn.*
065-061; Tetrachel-Vet Drops; Tetrachel-Vet Syrup; *Pfizer.*
065-063; Tetracycline Capsules; *Eon Labs.*
065-065; Tetracycline HCl Caps; *Global.*
065-066; Tetrachel-Vet Tablets 100; *Pfizer.*
065-067; Tetracycline HCl Tablets; *Premo.*
065-069; Tetrachel-Vet Capsules 500; *Pfizer.*
065-090; (with Novobiocin Sodium, Prednisolone) Delta Albaplex ® 3x Tablets; Delta Albaplex ® Tablets; *Pharmacia & Upjohn.*
065-121; Tetracycline-Vet Capsules 250; *Pfizer.*
065-122; Tetracyn Ointment; *Pfizer.*
065-123; Tetracycline Soluble Powder; *Pfizer.*
065-140; Tet-Sol 10; Tet-Sol 324™; *Alpharma.*
065-269; Polyotic ® Soluble Powder; American Cyanamid Divn., AHP Corp.*.
065-270; Polyotic ® Oblets ®; American Cyanamid Divn., AHP Corp.*.
065-409; Panmycin ® Capsules; *Pharmacia & Upjohn.*
065-410; Tetra-Sal; *Fort Dodge Animal Health, Divn. Am. Cyanamid.*
065-441; Polyotic ® Soluble Powder Concentrate; American Cyanamid Divn., AHP Corp.*.
065-496; Tetracycline Soluble Powder; *Boehringer Ingelheim Vetmedica.*
140-578; Solu-Tet ® 324; *Alpharma.*
200-049; Tetra-Bac 324 Soluble Powder; Tetracycline Hydrochloride Soluble Powder-324; *Agri Labs.*

200-136; Tetracycline HCL Powder; Tetracycline Hydrochloride Soluble Powder-324; *Phoenix*.
200-234; Tetrasol Soluble Powder; *Med-Pharmex*.

Antianginals

121 Atenolol
29122-68-7 892 249-451-7

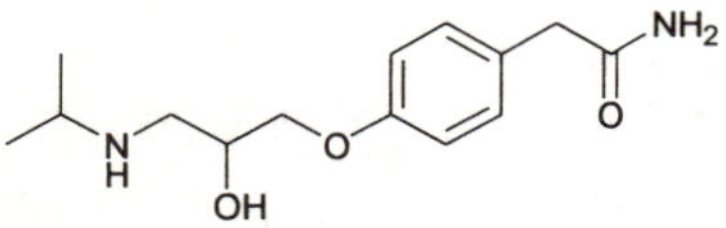

$C_{14}H_{22}N_2O_3$
2-[p-[2-Hydroxy-3-(isopropylamino)propoxy]-phenyl]acetamide.
ICI-66082; AteHexal; Atenol; Cuxanorm; Ibinolo; Myocord; Prenormine; Seles Beta; Selobloc; Teno-basan; Tenoblock; Tenormin; Uniloc. Cardioselective β-adrenergic blocking agent. Has anti-anginal, anti-arrhythmic and antihypertensive properties. Used in preference to propranolol for treatment of bronchospastic disease in small animals. mp = 146-148°, 150-152°; λ_m = 225, 275, 283 nm (MeOH); very soluble in MeOH; soluble in AcOH, DMSO; less soluble in Me_2CO, dioxane; insoluble in CH_3CN, EtOAc, $CHCl_3$; LD_{50} (mus orl) = 2000 mg/kg, (mus iv) = 98.7 mg/kg, (rat orl) = 3000 mg/kg, (rat iv) = 59.24 mg/kg.

122 Propranolol
525-66-6 8025 208-378-0

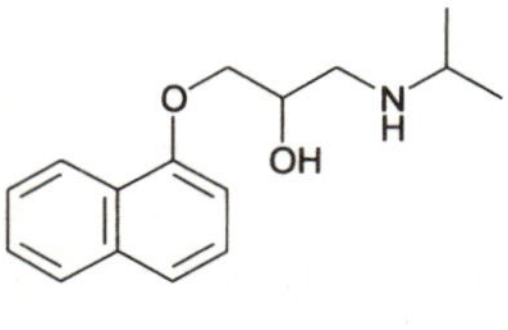

$C_{16}H_{21}NO_2$
1-(Isopropylamino)-3-(1-naphthyloxy)-2-propanol.
Avlocardyl; Euprovasin. A β-adrenergic blocker. Anti-anginal and antiarrhythmic (class II) agent. Used to treat hypertension, migraine headache and angina. mp = 96°.

123 Propranolol Hydrochloride
318-98-9 8025 206-268-7

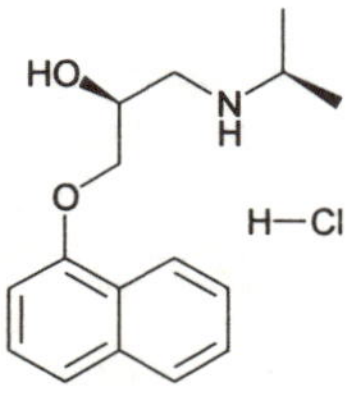

$C_{16}H_{22}ClNO_2$
1-(Isopropylamino)-3-(1-naphthyloxy)-2-propanol hydrochloride.
AY-64043; ICI-45520; NSC-91523; Angilol; Apsolol; Bedranol; Beprane; Berkolol; Beta-Neg; Beta-Tablinen; Beta-Timelets; Cardinol; Caridolol; Deralin; Dociton; Duranol; Efektolol; Elbol; Frekven; Inderal; Indobloc; Intermigran; Kemi S; Oposim; Prano-Puren; Prophylux; Propranur; Pylapron; Rapynogen; Sagittol; Servanolol; Sloprolol; Sumial; Tesnol. A β-adrenergic blocker. Antianginal and antiarrhythmic (class II) agent. Used to treat hypertension, migraine headache and angina. mp = 163-164°; soluble in H_2O, alcohol; insoluble in Et_2O, C_6H_6, EtOAc; LD_{50} (mus orl) = 565 mg/kg, (mus iv) = 22 mg/kg, (mus ip) = 107 mg/kg.

124 Verapamil
52-53-9 10083 200-145-1

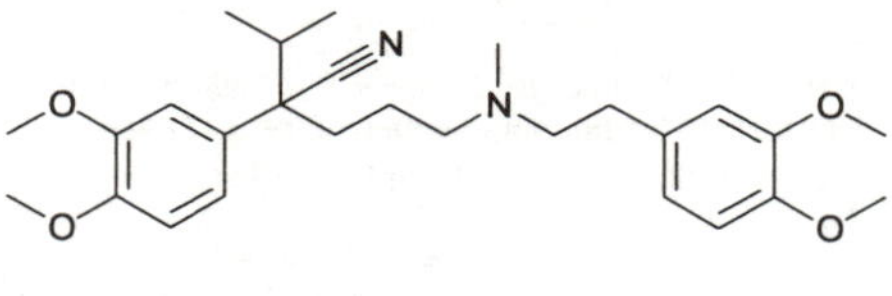

$C_{27}H_{38}N_2O_4$
5-[(3,4-Dimethoxyphenethyl)methylamino]-2-(3,4-dimethoxyphenyl)-2-isopropylvaleronitrile.
D-365; CP-16533-1; iproveratril. Antianginal and Class IV antiarrhythmic agent. Coronary vasodilator with calcium channel blocking activity. Used to treat tachycardias in dogs. $bp_{0.001}$ = 243-246°; insoluble in H_2O; slightly soluble in C_6H_6, C_6H_{16}, Et_2O; soluble in EtOH, MeOH, Me_2CO, EtOAc, $CHCl_3$.

125 Verapamil Hydrochloride
152-11-4 10083 205-800-5
$C_{27}H_{39}ClN_2O_4$
5-[(3,4-Dimethoxyphenethyl)methylamino]-2-(3,4-dimethoxyphenyl)-2-isopropylvaleronitrile hydrochloride.
Calan; isoptin; Verelan. Antianginal and Class IV antiarrhythmic agent. Coronary vasodilator with calcium channel blocking activity. Used to treat tachycardias in dogs. mp = 138.5-140.5°; soluble in H_2O (7 g/100 g, 83 mg/ml), EtOH (26 mg/ml), propylene glycol (93 mg/ml), MeOH (> 100 mg/ml), iPrOH (4.6 mg/ml), EtOAc (1.0 mg/ml), DMF (> 100 mg/ml), CH_2Cl_2 (> 100 mg/ml), C_6H_{14} (0.001 mg/ml); LD_{50} (rat iv) = 16 mg/kg, (mus iv) = 8 mg/kg.

Antiarrhythmics

126 Amiodarone
1951-25-3 504 217-772-1

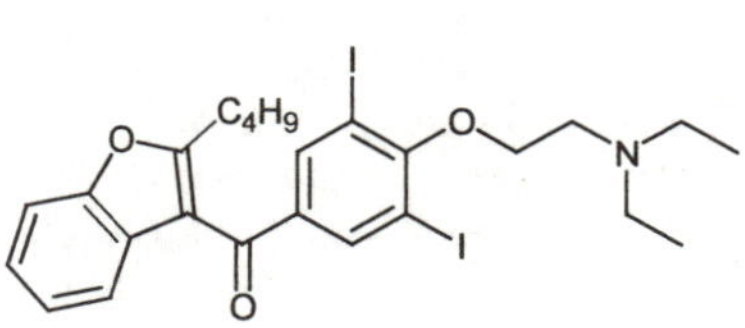

$C_{25}H_{29}I_2NO_3$
2-Butyl-3-benzofuranyl 4-[2-(diethylamino)ethoxy]-3,5-diiodophenyl ketone.
Cordarone; L-3428; SKF-33134-A. A benzofuran derivative. Antianginal and Class III antiarrhythmic.

Blocks both α- and β-receptors and is used to treat recurrent ventircular tachycardia, but is highly toxic and muist be used with care.

127 Amiodarone Hydrochloride

19774-82-4 504 243-293-2

$C_{25}H_{30}ClI_2NO_3$

2-Butyl-3-benzofuranyl 4-[2-(diethylamino)ethoxy]-3,5-diiodophenyl ketone hydrochloride.

L-3428; Amiodar; Ancoron; Angiodarona; Atlansil; Cordarex; Cordarone; Cordarone X; Miocard; Miodaron; Ortacrone; Ritmocardyl; Rythmarone; Trangorex. A benzofuran derivative. Antianginal and Class III antiarrhythmic. Blocks both α- and β-receptors and is used to treat recurrent ventircular tachycardia, but is highly toxic and muist be used with care. mp = 156°, 159 ± 2°; λ_m = 208, 242 nm ($E^{1\%}_{1\,cm}$ 662 ± 8, 623 ± 10 MeOH); soluble in EtOH (1.28 g/100 ml), MeOH (9.98 g/100 ml), $CHCl_3$ (44.51 g/100 ml), n-PrOH (0.13 g/100 ml), Et_2O (0.17 g/100 ml), THF (0.60 g/100 ml), C_6H_6 (0.65 g/100 ml), CH_2Cl_2 (19.20 g/100 ml), CH_3CN (0.32 g/100 ml), 1-octanol (0.30 g/100 ml), H_2O (0.07 g/100 ml), C_6H_{14} (0.03 g/100 ml), petroleum ether (0.001 g/100 ml).

128 Disopyramide

3737-09-5 3424 223-110-2

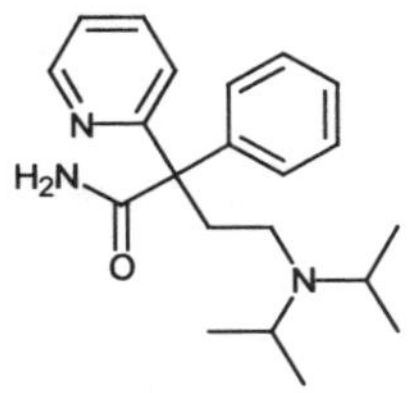

$C_{21}H_{29}N_3O$

α-[2-(Diisopropylamino)ethyl]-α-phenyl-2-pyridineacetamide.

SC-7031; H-3292; Dicorantil; Isorythm; Lispine; Ritmodan; Rythmodan. Class IA antiarrhythmic agent and cardiac depressant. Used to manage arrhythmia in dogs. mp = 94.5-95°; LD_{50} (mus ip) = 175 mg/kg.

129 Disopyramide Phosphate

22059-60-5 3424 244-756-1

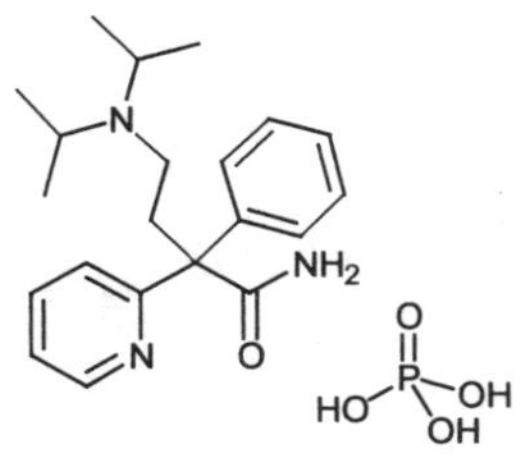

$C_{21}H_{32}N_3O_5P$

α-[2-(Diisopropylamino)ethyl]-α-phenyl-2-pyridineacetamide phosphate.

SC-13957; Norpace; Dirythmin SA; Diso-Duriles; Rythmodul. Class IA antiarrhythmic agent and cardiac depressant. Used to manage arrhythmia in dogs.

130 Esmolol

103598-03-4 3741

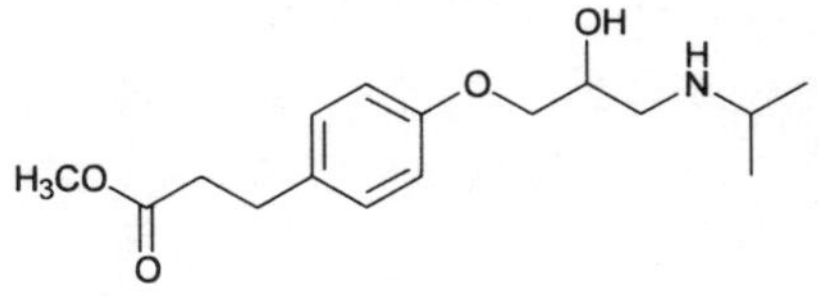

$C_{16}H_{25}NO_4$

(±)-Methyl p-[2-hydroxy-3-(isopropylamino)propoxy]-hydrocinnamate.

Class II antiarrhythmic. Ultra-short-acting β-adrenergic blocker. mp = 48-50°.

131 Esmolol Hydrochloride

81161-17-3 3741

$C_{16}H_{26}ClNO_4$

(±)-Methyl p-[2-hydroxy-3-(isopropylamino)propoxy]-hydrocinnamate hydrochloride.

ASL-8052; Brevibloc. Class II antiarrhythmic. Ultra-short-acting β-adrenergic blocker. mp = 85-86°.

132 Lidocaine

137-58-6 5505 205-302-8

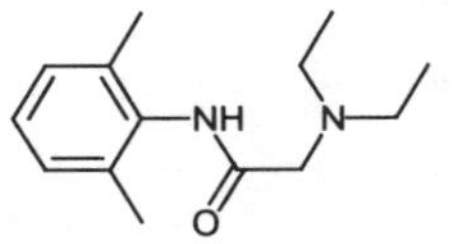

$C_{14}H_{22}N_2O$

2-(Diethylamino)-2',6'-acetoxylidide.

Dalcaine; Solarcaine; Xylocaine; lignocaine; Cuivasil; Duncaine; Leostesin; Lidothesin; Rucaina; Xylocaine; Xylocitin; Xylotox. Class IB antiarrhythmic agent. Not recommended for use in cats. mp = 68-69°; bp_4 = 180-182°, bp_2 = 159-160°; soluble in organic solvents, insoluble in H_2O.

Veterinary Products:

013-146; (with Oxytetracycline Hydrochloride) Liquamycin ® Intramuscular; *Pfizer.*

049-948; (with Oxytetracycline Hydrochloride) Aquachel 100 mg; *Pfizer.*

133 Lidocaine Hydrochloride

73-78-9 5505 200-803-8

$C_{14}H_{23}ClN_2O.H_2O$

2-(Diethylamino)-2',6'-acetoxylidide hydrochloride monohydrate.

Lidesthesin; Lignavet; Odontalg; Sedagul; Xylocard; Xyloneural. Class IB antiarrhythmic agent. Not recommended for use in cats. mp = 77-78°; [anhydrous form]: mp = 127-129°; very soluble in H_2O, EtOH, less soluble in $CHCl_3$, insoluble in Et_2O; LD_{50} (mus orl) = 292 mg/kg, (mus ip) = 105 mg/kg, (mus iv) = 19.5 mg/kg.

Veterinary Products:

045-578; (with Epinephrine Acetate) Lidocaine HCl with Epinephrine; *Steris*.

134 Mexiletine

31828-71-4 6257 250-825-7

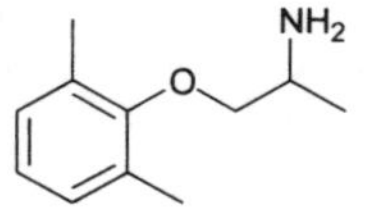

$C_{11}H_{17}NO$

1-Methyl-2-(2,6-xylyloxy)ethylamine.

Class IB antiarrhythmic agent. Has been used to treat ventricular arrhythmias.

135 Mexiletine Hydrochloride

5370-01-4 6257 226-362-1

$C_{11}H_{18}ClNO$

1-Methyl-2-(2,6-xylyloxy)ethylamine hydrochloride.

Mexitil; Ko-1173Cl; Katen; ritalmex. Class IB antiarrhythmic agent. Has been used to treat ventricular arrhythmias. mp = 203-205°; LD_{50} (mrat orl) = 350 mg/kg, (mrat iv) = 27 mg/kg, (frat orl) = 400 mg/kg, (frat iv) = 30 mg/kg, (mmus orl) = 310 mg/kg, (mmus iv) = 43 mg/kg), (fmus orl) = 400 mg/kg, (fmus iv) = 50 mg/kg, (mrbt orl) = 180 mg/kg, (frbt orl) = 160 mg/kg.

136 Tocainide

41708-72-9 9629 255-505-0

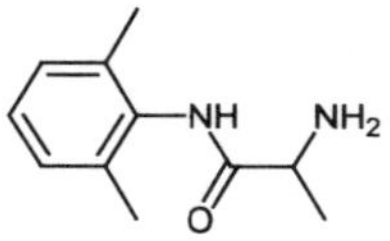

$C_{11}H_{16}N_2O$

2-Amino-2',6'-propionoxylidide.

W-36095. Class IB antiarrhythmic agent with cardiodepressant properties. Used in dogs in oral therapy of ventricular arrhythmias.

137 Tocainide Hydrochloride

71395-14-7 9629 275-361-2

$C_{11}H_{17}ClN_2O$

2-Amino-2',6'-propionoxylidide hydrochloride.

Tonocard; Taquidil; Xylotocan. Class IB antiarrhythmic agent with cardiodepressant properties. Used in dogs in oral therapy of ventricular arrhythmias. mp = 246-247°.

138 Tocainide, R-(-)-form Hydrochloride

53984-74-0 9629

$C_{11}H_{17}ClN_2O$

(R)-(-)-2-Amino-2',6'-propionoxylidide.

Class IB antiarrhythmic with cardiodepressant properties. Used in dogs in oral therapy of ventricular arrhythmias. mp = 265-266°; $[\alpha]_D$ = -42.16° (c = 2.63 MeOH).

139 Tocainide, S-(+)-form Hydrochloride

53984-76-2 9629

$C_{11}H_{17}ClN_2O$

(S)-(+)-2-Amino-2',6'-propionoxylidide.

Class IB antiarrhythmic agent with cardiodepressant properties. Used in dogs in oral therapy of ventricular arrhythmias. mp = 264.5°; $[\alpha]_D$ = +42.35° (c = 2.63 MeOH).

Antiarthritics

140 Auranofin

34031-32-8 911 251-801-9

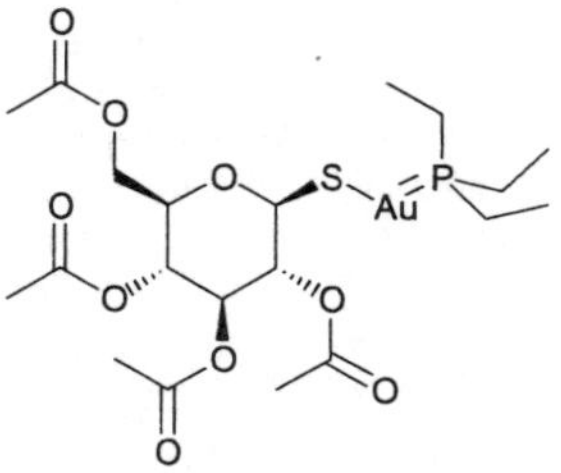

$C_{20}H_{34}AuO_9PS$

(1-Thio-β-D-glucopyranosato)(triethylphosphine)gold 2,3,4,6-tetraacetate.

Ridaura; Ridauran; SKF-39162; Aktil; Crisinor; Crisofin. Antirheumatic agent. Used to treat arthritis in dogs. mp = 110-111°; LD_{50} (rat orl) = 265 mg/kg, (mus orl) = 310 mg/kg.

141 Aurothioglucose

12192-57-3 917 235-365-7

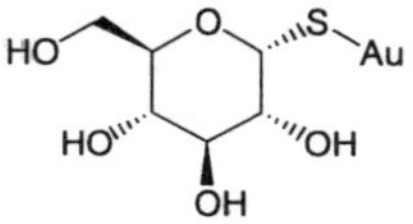

$C_6H_{11}AuO_5S$

(1-Thio-D-glucopyranosato)gold.

Solganal; Solganal B; Aureotan; Aurumine; Oronol. Antirheumatic agent. Used to treat arthritis in dogs and has been used to produce obesity in animals. Soluble in H_2O, slightly soluble in propylene glycol, insoluble in other organic solvents.

142 Azathioprine

446-86-6 935 207-175-4

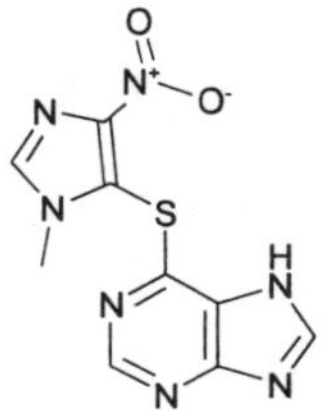

$C_9H_7N_7O_2S$

6-[(1-Methyl-4-nitroimidazol-5-yl)thio]purine.

Imuran; BW-57-322; NSC-39084; Azamune; Azanin; Azoran; Imurel. Immunosuppressant and antirheumatic agent. Used in dogs as a second or third line immunosuppressive agent. mp = 243-244° (dec); λ_m = 276 nm (ε 18200 MeOH); slightly soluble in H_2O, $CHCl_3$, EtOH.

143 Methotrexate
59-05-2 6065 200-413-8

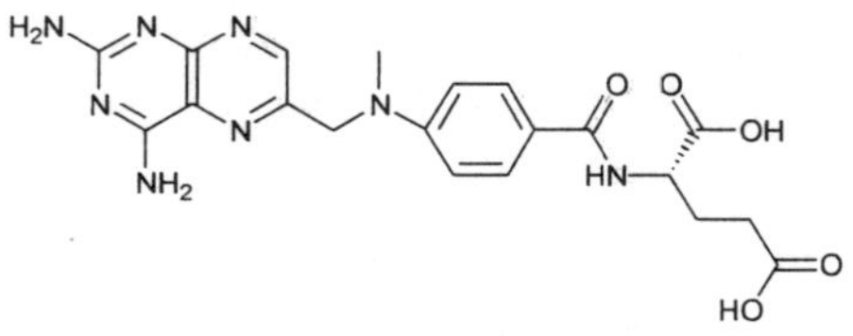

$C_{20}H_{22}N_8O_5$
L-(+)-N-[p-[[(2,4-Diamino-6-pteridinyl)methyl]-methylamino]benzoyl]glutamic acid.
Amethopterin; Mexate; CL-14377; NSC-740; MTX; Emtexate; Ledertrexate; A-Methopterin; Rheumatrex. Antineoplastic and antirheumatic agent. Used to treat solid tumors, rheumatoid arthritis and psoriasis.

144 Methotrexate Disodium Salt
7413-34-5 6065 231-022-0

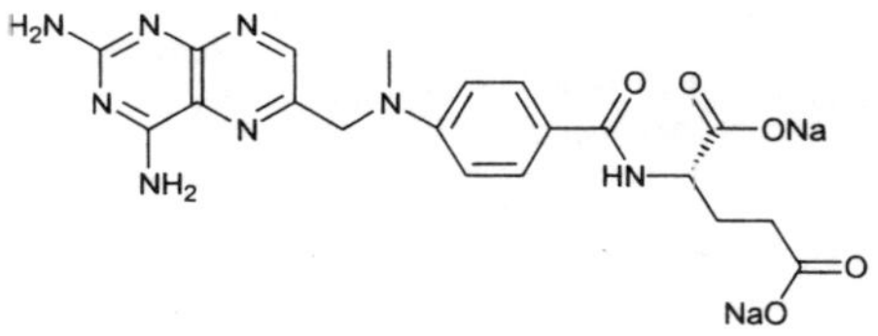

$C_{20}H_{20}N_8Na_2O_5$
L-(+)-N-[p-[[(2,4-Diamino-6-pteridinyl)methyl]-methylamino]benzoyl]glutamic acid disodium salt.
Folex; Mexate. Antineoplastic and antirheumatic agent. Used to treat solid tumors, rheumatoid arthritis and psoriasis.

145 Methotrexate Monohydrate
6745-93-3 6065
$C_{20}H_{22}N_8O_5 \cdot H_2O$
L-(+)-N-[p-[[(2,4-Diamino-6-pteridinyl)methyl]-methylamino]benzoyl]glutamic acid monohydrate.
Antineoplastic and antirheumatic agent. Used to treat solid tumors, rheumatoid arthritis and psoriasis. mp = 185-204° (dec); λ_m = 244, 307 nm (0.1N HCl), 257, 302, 370 nm (0.1N NaOH); LD_{50} (rat iv) = 14 mg/kg.

146 Penicillamine
52-67-5 7214 200-148-8

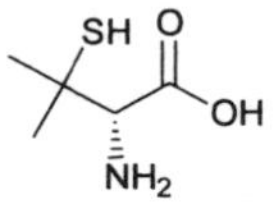

$C_5H_{11}NO_2S$
D-3-Mercaptovaline.
Depen; Cuprimine; β-Thiovaline; Cuprenil; Depamine; Mercaptyl; Pendramine; Perdolat; Sufirtan; Trolovol. Antirheumatic agent. Also used as a copper chelator in treatment of Wilson's disease and, in dogs, for problems associated with copper storage. mp = 202-206°; $[\alpha]_D^{25}$ = -63° (c = 1 C_5H_5N); LD_{50} (rat orl) > 10000 mg/kg, (rat ip) > 660 mg/kg.

147 Penicillamine, DL-form
52-66-4 7214 200-147-2

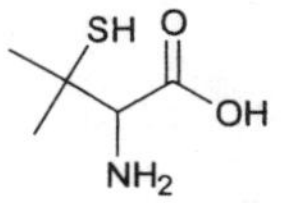

$C_5H_{11}NO_2S$
DL-3-Mercaptovaline.
Antirheumatic agent. Also used as a copper chelator in treatment of Wilson's disease and, in dogs, for problems associated with copper storage. mp = 201° (dec); LD_{50} (rat orl) = 365 mg/kg.

148 Penicillamine, DL-form Hydrochloride
22572-05-0 7214 245-093-0
$C_5H_{12}ClNO_2S$
DL-3-Mercaptovaline hydrochloride.
Antirheumatic agent. Also used as a copper chelator in treatment of Wilson's disease and, in dogs, for problems associated with copper storage. mp = 145-148°.

149 Penicillamine Hydrochloride
2219-30-9 7214 218-727-9

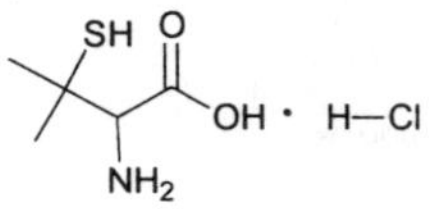

$C_5H_{12}ClNO_2S$
3-Mercaptovaline hydrochloride.
Distamine; Metalcaptase. Antirheumatic agent. Also used as a copper chelator in treatment of Wilson's disease and, in dogs, for problems associated with copper storage. mp = 177.5° (dec); $[\alpha]_D^{25}$ = -63° (1N NaOH); soluble in H_2O, EtOH; LD_{50} (mus iv) = 2289 mg/kg.

150 Penicillamine, L-form
1113-41-3 7214 214-203-9

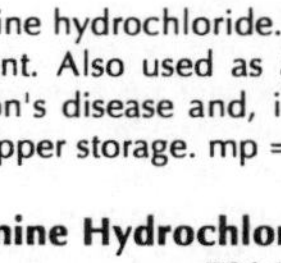

$C_5H_{11}NO_2S$
L-3-Mercaptovaline.
Antirheumatic agent. Also used as a copper chelator in treatment of Wilson's disease and, in dogs, for problems associated with copper storage. mp = 190-194°; $[\alpha]_D^{25}$ = 63° (1N NaOH); LD_{50} (rat ip) = 350 mg/kg.

Antiasthmatic Agents

151 Dexamethasone

50-02-2 2986 200-003-9

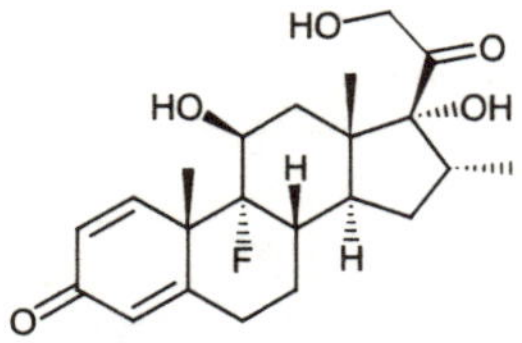

$C_{22}H_{29}FO_5$

9-Fluoro-11β,17,21-trihydroxy-16α-methylpregna-1,4-diene-3,20-dione.

1-dehydro-16α-methyl-9α-fluorohydrocortisone; 16α-methyl-9α-fluoro-Δ^1-hydrocortisone; 16α-methyl-9α-fluoroprednisolone; hexadecadrol; Aeroseb-Dex; Corson; Cortisumman; Decacort; Decaderm; Decadron; Decalix; Decasone; Dekacort; Deltafluorene; Deronil; Deseronil; Dexacortal; Dexacortin; Dexafarma; Dexa-Mamallet; Dexameth; Dexamonozon; Dexapos; Dexa-sine; Dexasone; Dexinoral; Dinormon; Fluormone; Isopto-Dex; Lokalison F; Loverine; Luxazone; Maxidex; Millicorten; Pet-Derm III; component of: Azimycin, Azium, Deronil, Dexacidin, Fulvidex, Maxitrol, Naquasone, Tobradex, Tresaderm. An antiasthmatic glucocorticoid used in diagnosis of Cushing's syndrome and depression and used as an anti-inflammatory in many animal species. mp = 262-264°, 278-281°; $[\alpha]_D^{25}$ = 77.5° (dioxane); slightly soluble in H_2O (0.010 g/100 ml), soluble in Me_2CO, EtOH, $CHCl_3$.

Veterinary Products:

011-885; Azium ® Tablets 0.25 mg; *Schering-Plough Animal Health.*
011-901; Azium ® Aqueous Suspension Veterinary; *Schering-Plough Animal Health.*
012-559; Azium ® Solution 2 mg; *Schering-Plough Animal Health.*
030-136; (with Trichlormethiazide) Naquasone ® Bolus; *Schering-Plough Animal Health.*
030-434; Azium ® Powder 10 mg; *Schering-Plough Animal Health.*
030-435; Azium ® Boluses 10 mg; *Schering-Plough Animal Health.*
042-633; (with Neomycin Sulfate, Thiabendazole) Tresaderm Dermatologic Solution; *Merial.*
092-835; Azium ® Oral Solution 2 mg; *Schering-Plough Animal Health.*
095-218; Dexamethasone Tablets 0.25 mg; *Merial.*
099-606; Dexameth-A-Vet Injection; *Anthony Products.*
099-607; Dexameth-A-Vet; *Pfizer.*
108-687; Pet Derm ® III Tablets; *Pfizer.*
110-350; Dexamethasone; *Steris.*
111-369; Dexamethasone Sterile Solution; *Pfizer.*
128-089; Zonometh Solution; *Merial.*
130-660; Dexachel; *Pfizer.*
200-108; Dexamethasone Solution; *Phoenix.*

152 Dexamethasone 21-Phosphate Disodium Salt

2392-39-4 2986 219-243-0

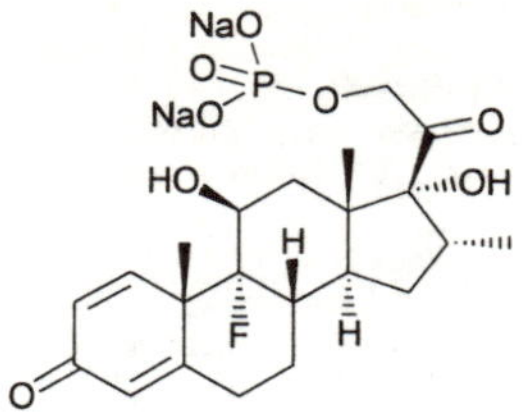

$C_{22}H_{28}FNa_2O_8P$

9-Fluoro-11β,17,21-trihydroxy-16α-methylpregna-1,4-diene-3,20-dione 21-phosphate disodium salt.

Dalalone; Decadron Inhalation, Injection, Ophthalmic Solution and Ointment, and Topical Cream; Hexadrol Injectable; Maxidex Ointment; Ak-Dex; Baldex; Colvasone; Dexabene; Dezone; Fortecortin; Oradexon; Orgadrone; Solu-Decadron; Soldesam. An antiasthmatic glucocorticoid used in diagnosis of Cushing's syndrome and depression and in many animal species. mp = 233-235°; $[\alpha]_D$ = 257° (H_2O); $[\alpha]_D^{25}$ = 74° ± 4° (c = 1); λ_m = 238-239 nm (ε 14000); soluble in H_2O.

Veterinary Products:

099-604; Dex-A-Vet Injection; *Anthony Products.*
099-605; Dex-A-Vet Injection; *Anthony Products.*
104-606; Dexamethasone Injection; *Steris.*
110-349; Dexamethasone Injection; *Steris.*
123-815; Dexamethasone Sodium Phosphate Inj.; *Merial.*

153 Triamcinolone Acetonide

76-25-5 9728 200-948-7

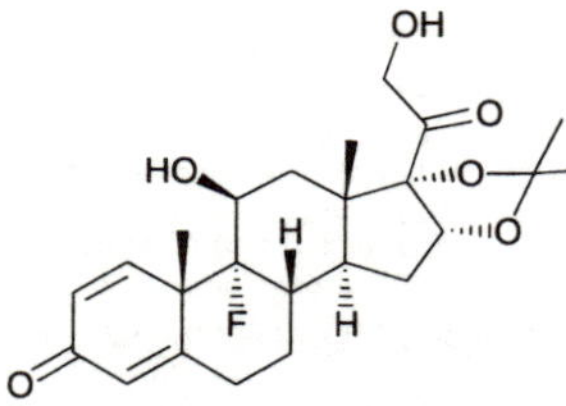

$C_{24}H_{31}FO_6$

9-Fluoro-11β,16α,17,21-tetrahydroxypregna-1,4-diene-3,20-dione cyclic 16,17-acetal with acetone.

Aristocort; Aristocort A; Azmacort; Kenalog; Nasacort; Nasacort AQ; Panolog Ointment; TAC-3; Triacet; Triamonide 40; Trymex; Vetalog; Adcortyl; Delphicort; Extracort; Ftorocort; Kenacort A; Kenalog; Ledercort Cream; Nasacort; Respicort; Rineton; Solodelf; Tramasin; Triam; Tricinolon; Volon A; Volonimat; component of: Mycolog II, MycoTriacet II, Mytrex. Glucocorticoid; anti-asthmatic (inhaled) and nasal antiallergic. mp = 292-294°; $[\alpha]_D^{23}$ = 109° (c = 0.75 $CHCl_3$); λ_m = 238 nm (ε 14600 EtOH); sparingly soluble in MeOH, Me_2CO, EtOAc.

Veterinary Products:

012-198; Vetalog Parenteral Veterinary; *Fort Dodge Animal Health, Divn. Am. Cyanamid.*

012-258; (with Neomycin Sulfate, Nystatin, Thiostrepton) Panolog ® Ointment; *Fort Dodge Animal Health, Divn. Am. Cyanamid.*
013-624; Vetalog Tablets; *Fort Dodge Animal Health, Divn. Am. Cyanamid.*
046-146; Vetalog Cream; *Fort Dodge Animal Health, Divn. Am. Cyanamid.*
096-676; (with Neomycin Sulfate, Nystatin, Thiostrepton) Panolog Cream; *Fort Dodge Animal Health, Divn. Am. Cyanamid.*
099-388; Vetalog Oral Powder; *Fort Dodge Animal Health, Divn. Am. Cyanamid.*
137-694; Triamcinolone Acetonide Tablet; *Boehringer Ingelheim Vetmedica.*
138-869; Triamcinolone Acetonide Suspension; *Boehringer Ingelheim Vetmedica.*
140-810; (with Neomycin Sulfate, Nystatin, Thiostrepton) Panavet Ointment; *Med-Pharmex.*
140-847; (with Neomycin Sulfate, Nystatin, Thiostrepton) Animax; *Altana.*
140-879; (with Neomycin Sulfate, Nystatin, Thiostrepton) Derma 4 Ointment; *Pfizer.*
140-889; (with Neomycin Sulfate, Nystatin, Thiostrepton) Derm-Otic Ointment; *Biocraft.*
141-003; (with Neomycin Sulfate, Nystatin, Thiostrepton) Derm-Otic Ointment; *Robins.*
200-245; (with Neomycin Sulfate, Nystatin, Thiostrepton) Derma-Vet Cream; *Med-Pharmex.*

Antibacterial Adjuncts

154 Cilastatin

82009-34-5 2331

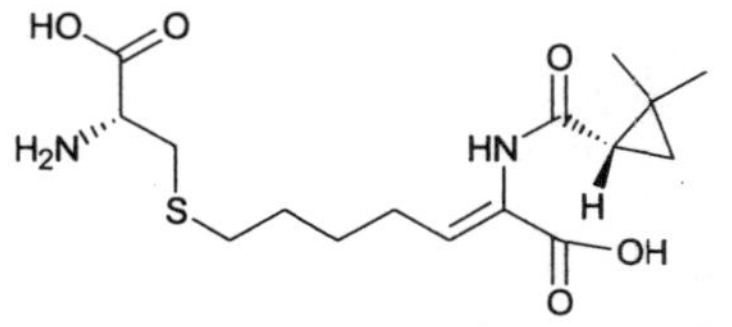

$C_{16}H_{26}N_2O_5S$
[R-[(R*,S*(Z)]-7-[(2-Amino-2-carboxyethyl)thio]-2-[[(2,2-dimethylcyclopropyl)carbonyl]amino]-2-heptenoic acid.
MK-791. Used in combination with Imipenem, an antibacterial. A dipeptidase I inhibitor which prevents renal metabolism of penem and carbapenem antibiotics.

155 Cilastatin Sodium salt

81129-83-1 2331 24390-14-5

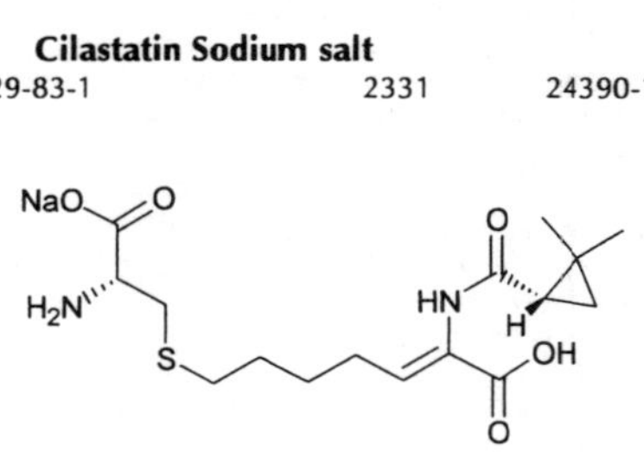

$C_{16}H_{25}N_2NaO_5S$
[R-[(R*,S*(Z)]-7-[(2-Amino-2-carboxyethyl)thio]-2-[[(2,2-dimethylcyclopropyl)carbonyl]amino]-2-heptenoic acid acid sodium salt (1:1).
Cilastatin Sodium. Used in combination with Imipenem, an antibacterial. A dipeptidase I inhibitor which prevents renal metabolism of penem and carbapenem antibiotics.
Very soluble in H_2O, MeOH.

156 Clavulanate Potassium

61177-45-5 2402 262-640-9

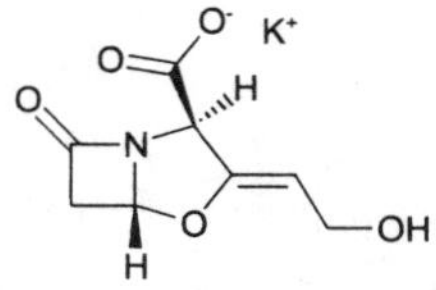

$C_8H_8KNO_5$
Potassium 3-(2-hydroxyethylidene)-7-oxo-4-oxa-1-azabicyclo[3.2.0]heptane-2-carboxylate.
Weak antibacterial agent. A powerful β-lactamase inhibitor, it is often used in conjunction with penicillins.

Veterinary Products:

055-099; (with Amoxicillin Trihydrate) Clavamox ® Tablets; *Pfizer.*
055-101; (with Amoxicillin Trihydrate) Clavamox ® Drops; *Pfizer.*

Antibiotics, 2,4-Diaminopyrimidine

157 Ormetoprim

6981-18-6 230-246-6

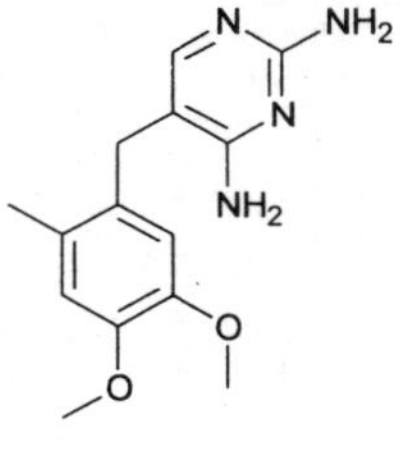

$C_{14}H_{18}N_4O_2$
5-[(4,5-Dimethoxy-2-methylphenyl)methyl]-2,4-pyrimidinediamine.
Ro-5-9754; NSC-95072. A 2,4-diaminopyrimidine antibiotic.

Veterinary Products:

040-209; (with Sulfadimethoxine) Rofenaid ® 40; *Roche Vitamins.*
041-984; (with Sulfadimethoxine, Roxarsone) Rofenaid ® Plus Roxarsone; *Roche Vitamins.*
100-929; (with Sulfadimethoxine) Primor ® Tablets; *Pfizer.*
125-933; (with Sulfadimethoxine) Romet ® -30; *Roche Vitamins.*

158 Trimethoprim
738-70-5 9840 212-006-2

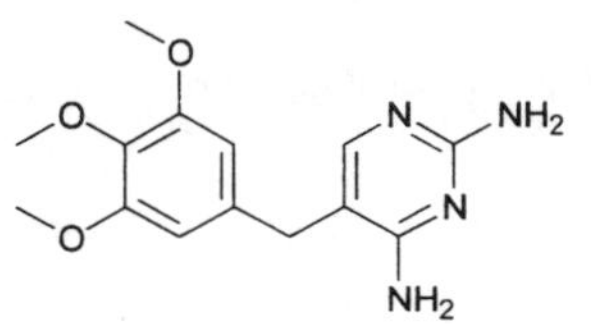

$C_{14}H_{18}N_4O_3$
5-[(3,4,5-Trimethoxyphenyl)methyl]-2,4-pyrimidinediamine.
Instalac; Monotrim; Proloprim; Syraprim; Tiempe; Trimanyl; Trimogal; Trimopan; Trimpex; Uretrim; Wellcoprim. A 2,4-diaminopyrimidine antibiotic. Often used in conjunction with various sulfa drugs. mp = 199-203°; soluble in dimethylacetamide (13.86 g/100 ml), benzyl alcohol (7.29 g/100 ml), propylene glycol (2.57 g/100 ml), $CHCl_3$ (1.82 g/100 ml), MeOH (1.21 g/100 ml), H_20 (0.04 g/100 ml), Et_2O (0.003 g/100 ml), C_6H_6 (0.002 g/100 ml); LD_{50} (mus orl) = 7000 mg/kg.

Veterinary Products:

095-614; (with Sulfadiazine) Tribrissen ® 20 mg Tablets; Tribrissen ® 30 mg Tablets; Tribrissen ® 480 mg Tablets; Tribrissen ® 960 mg Tablets; *Schering-Plough Animal Health.*
105-093; (with Sulfadiazine Sodium) Tribrissen ® 24% Injection; *Schering-Plough Animal Health.*
106-965; (with Sulfadiazine) Tribrissen ® 48% Injection; *Schering-Plough Animal Health.*
115-578; (with Sulfadiazine) Di-Trim ® Tablets; *Fort Dodge Animal Health, Divn. AHP.*
131-918; (with Sulfadiazine) Tribrissen ® 400 Oral Paste; *Schering-Plough Animal Health.*
132-486; (with Sulfadiazine) Di-Trim ® 24%; *Fort Dodge Animal Health, Divn. AHP.*
134-778; (with Sulfadiazine) Di-Trim ® 48% Injection; *Fort Dodge Animal Health, Divn. AHP.*
136-342; (with Sulfadiazine) Di-Trim ® 400 Paste; *Fort Dodge Animal Health, Divn. AHP.*
136-741; (with Sulfadiazine) Tribrissen ® 60 Oral Suspension; *Schering-Plough Animal Health.*
200-033; (with Sulfadiazine) Uniprim ™ Powder; *Macleod.*

Antibiotics, Aminoglycoside

159 Apramycin Sulfate
65710-07-8 792 265-890-7
$C_{42}H_{84}N_{10}O_{26}S$
O-4-Amino-4-deoxy-α-D-glucopyranosyl-(1→8)-(8R)-2-amino-2,3,7-trideoxy-7-(methylamino)-D-glycero-α-D-allo-octadialdo-1,5:8,4-dipyranosyl-(1→4)-2-deoxy-D-streptamine sulfate.
Aminoglycoside antibiotic used to treat infections in pigs and cattle.

Veterinary Products:

106-964; Apralan® Soluble Powder; *Elanco.*
126-050; Apralan® 75; *Elanco.*

160 Bambermycins
11015-37-5 979 234-246-7

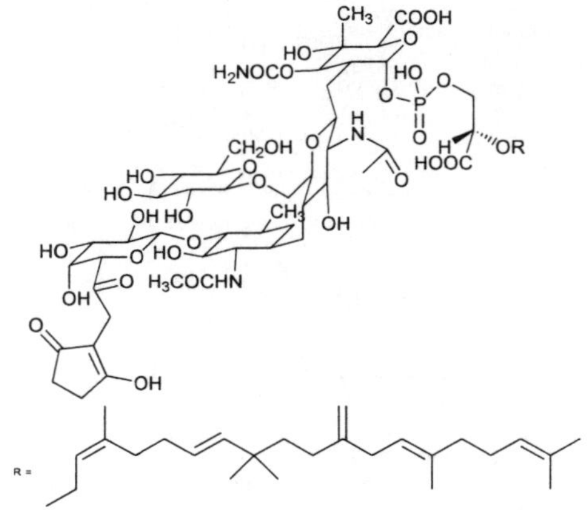

Bambermycins.
Moenomycin; flavophospholipol; Flavomycin. Aminoglycoside antibiotic complex containing moenomycins A, B_1, B_2 and C. Dec ≅ 200°; λ_m = 258 nm ($E^{1\%}_{1cm}$ 60 H_2O pH 7); soluble in H_2O, MeOH, DMF; less soluble in EtOH, PrOH, slightly soluble in Et_2O, EtOAc, insol in C_6H_6, $CHCl_3$; LD_{50} (mus orl, sc, ip) > 2000mg/kg, (mus iv) = 1400 mg/kg.

Veterinary Products:

044-759; Flavomycin ® Type A Medicated Article; *Hoechst-Roussel Vet.*
095-543; (with Amprolium, Ethopabate) Amprol HI-E ® + Flavomycin ®; *Hoechst-Roussel Vet.*
095-547; (with Amprolium, Ethopabate, Roxarsone) Amprol HI-E ® + Flavomycin ® + 3-Nitro ®; *Hoechst-Roussel Vet.*
095-548; (with Amprolium, Roxarsone) Amp +3-Nitro ® + Flavomycin ®; *Hoechst-Roussel Vet.*
095-549; (with Amprolium, Ethopabate, Roxarsone) Amprol + 3-Nitro + Flavomycin ®; *Hoechst-Roussel Vet.*
098-340; (with Monensin Sodium) Flavomycin ® + Monensin; *Hoechst-Roussel Vet.*
098-341; (with Monensin, Roxarsone) Flavomycin ® + 3-Nitro ® + Coban ®; *Hoechst-Roussel Vet.*
101-628; (with Roxarsone, Zoalene) Flavomycin ® + 3-Nitro + Zoalene; *Hoechst-Roussel Vet.*
101-629; (with Zoalene) Flavomycin ® + Zoalene; *Hoechst-Roussel Vet.*
112-687; (with Lasalocid Sodium, Roxarsone) Avatec ® / 3-Nitro ® / Flavomycin ®; *Roche Vitamins.*
130-185; (with Amprolium) Flavomycin ® + Amprolium; *Hoechst-Roussel Vet.*
130-661; (with Carbarsone) Flavomycin ® / Carb-O-Sep; *Hoechst-Roussel Vet.*
131-146; Flavomycin ® 0.4; Flavomycin ® 2; *Triple F.*
131-413; Flavomycin ® 2; Flavomycin ® 0.4; *Akey.*
132-448; Flavomycin ®; *ADM Animal Health & Nutrition.*
132-705; Flavomycin ®; *Quali-Tech.*
134-185; (with Roxarsone, Salinomycin Sodium) Biocox ® / 3-Nitro ® / Flavomycin ®; *Roche Vitamins.*
134-284; (with Salinomycin Sodium) Biocox ® /

Flavomycin ®; *Roche Vitamins.*
140-339; (with Nicarbazin) Flavomycin ® / Nicarb ®; *Hoechst-Roussel Vet.*
140-843; (with Narasin, Roxarsone) Monteban ® + Flavomycin ® + 3-Nitro ®; *Hoechst-Roussel Vet.*
140-845; (with Narasin) Flavomycin ® + Monteban ®; *Hoechst-Roussel Vet.*
140-918; (with Halofuginone Hydrobromide) Stenorol ® + Flavomycin ®; *Hoechst-Roussel Vet.*
141-034; Gainpro ® Type A Medicated Article; *Hoechst-Roussel Vet.*
141-129; (with Lasalocid Sodium) Avatec ® / Flavomycin ®; *Hoechst-Roussel Vet.*
200-080; (with Roxarsone, Salinomycin Sodium) Sacox ® + 3-Nitro ® + Flavomycin ®; *Hoechst-Roussel Vet.*
200-083; (with Salinomycin Sodium) Sacox ® + Flavomycin ®; *Hoechst-Roussel Vet.*

161 Dihydrostreptomycin Sesquisulfate

5490-27-7 3222 226-823-7

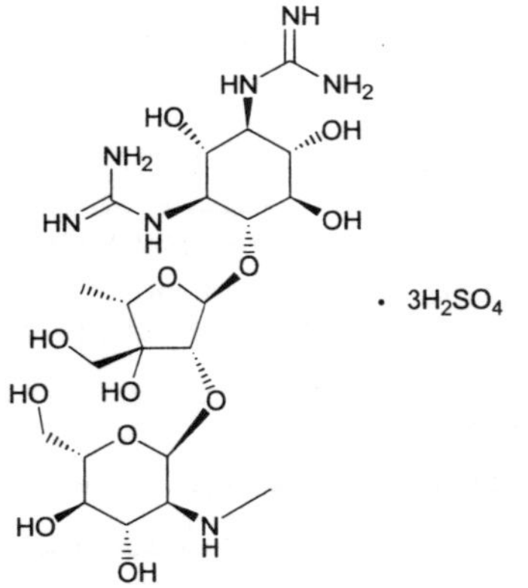

· $3H_2SO_4$

$C_{42}H_{88}N_{14}O_{36}S_3$
O-2-Deoxy-2-(methylamino)-α-L-glucopyranosyl-(1→2)-O-5-deoxy-3-C-(hydroxymethyl)-α-L-lyxofuranosyl-(1→4)-N,N'-bis(aminoiminomethyl)-D-streptamine sulfate (2:3) (salt).
Didromycine; Double-Mycin; Sol-Mycin; Strepto-Magma. Antibacterial (tuberculostatic in humans). Used in veterinary medicine as a general antibacterial. Dec 255-265°, 250°; $[\alpha]_D^{25}$ = -88.5° (c = 1); very soluble in H_2O; soluble in 50% MeOH/H_2O (crystals: 0.08 g/100 ml; powder: 10 g/100 ml); at 28° soluble in H_2O (> 2 g/100 ml), MeOH (0.035 g/100 ml), EtOH (0.010 g/100 ml).

Veterinary Products:

055-028; (with Penicillin G Procaine) Quartermaster ® Dry Cow Treatment; *West Agro.*
055-097; (with Penicillin G Procaine) Dry-Mast; *Boehringer Ingelheim Vetmedica.*
065-013; Dihydrostreptomycin; *Norbrook.*
065-291; (with Streptomycin Sulfate) Dihydrostreptomycin Sulfate; *Pfizer.*
065-483; Pfizer-Strep; *Pfizer.*

162 Gentamicin Sulfate

1405-41-0 4398 215-778-9

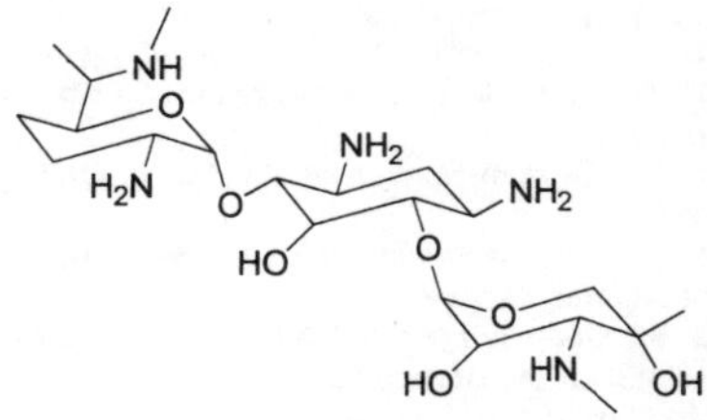

Alcomicin; Bristagen; Cidomycin; Duragentam; Garamycin; Garasol; Genoptic; Gentacin; Gentak; Gentalline; Gentalyn; Gentibioptal; Genticin; Gentocin; Gentogram; Gent-Ophtal; Gentrasul; Lugacin; Nichogencin; Ophtagram; Pangram; Refobacin; Septopal; Sulmycin; U-gencin. Aminoglycoside antibiotic. Used parenterally in dogs, cats, chickens, turkeys and pigs. mp = 218-237°; $[\alpha]_D^{25}$ = 102°; soluble in ethylene glycol, formamide; LD_{50} (mus ip) = 430 mg/kg, (mus sc) = 485 mg/kg, (mus orl) > 9050 mg/kg.

Veterinary Products:

034-267; (with Betamethasone Acetate) Gentocin ® Durafilm Ophthalmic Solution; *Schering-Plough Animal Health.*
038-292; Gentocin ® Solution; *Schering-Plough Animal Health.*
046-724; Gentocin ® Solution; *Schering-Plough Animal Health.*
046-821; (with Betamethasone Valerate) Gentocin ® Otic Solution; *Schering-Plough Animal Health.*
047-486; Garasol ® Injection; *Schering-Plough Animal Health.*
091-191; Garacin Oral Solution; Gentocin ® Oral Solution; *Schering-Plough Animal Health.*
092-523; Garasol ® Solution; *Schering-Plough Animal Health.*
098-989; Gentocin ® Ophtalmic Ointment; *Schering-Plough Animal Health.*
099-008; Gentocin ® Ophthalmic Solution; *Schering-Plough Animal Health.*
101-862; Garasol ® Injection; *Schering-Plough Animal Health.*
103-037; Garacin ® Piglet Injection; *Schering-Plough Animal Health.*
113-231; (with Betamethasone Valerate) Topagen ® Ointment; *Schering-Plough Animal Health.*
130-464; Gentocin ® (Garacin) Pig Pump Oral Solution; *Schering-Plough Animal Health.*
130-952; Gentocin ® Pink Eye Spray; *Schering-Plough Animal Health.*
132-338; (with Betamethasone Valerate) Gentocin ® Topical Spray; *Schering-Plough Animal Health.*
133-836; Gentocin ® (Garacin) Soluble Powder; *Schering-Plough Animal Health.*
137-310; Gentamicin Sulfate Inj. Sol.; *Boehringer Ingelheim Vetmedica.*
140-896; (with Betamethasone Valerate, Clotrimazole) Otomax ®; *Schering-Plough Animal Health.*
200-023; Gentamicin Sulfate Solution 100 mg/ml; *Boehringer Ingelheim Vetmedica.*

200-037; Gentamicin Sulfate Solution; Legacy Sterile Solution; *Agri Labs.*
200-102; Gentaglyde ™ Solution; *Fort Dodge Animal Health, Divn. AHP.*
200-115; Gentamex ™ 100; *Anthony Products.*
200-137; Gentamicin Sulfate Solution; *Phoenix.*
200-147; Genta-Ject ®; Gentamicin Sulfate Injection; *Merial.*
200-174; Gentamicin Sulfate Pig Pump Oral Solution; *Phoenix.*
200-183; (with Betamethasone Valerate) Gentavet ® Otic Solution; *Med-Pharmex.*
200-185; Gen-Gard ™ Soluble Powder; *Agri Labs.*
200-188; (with Betamethasone Valerate) Betagen ™ Topical Spray; *Med-Pharmex.*
200-190; Gentoral; *Med-Pharmex.*
200-191; Gentasol; *Med-Pharmex.*
200-229; (with Betamethasone Valerate, Clotrimazole) Tri-Otic Ointment; *Med-Pharmex.*

163 Kanamycin A Sulfate

25389-94-0 5293 246-933-9

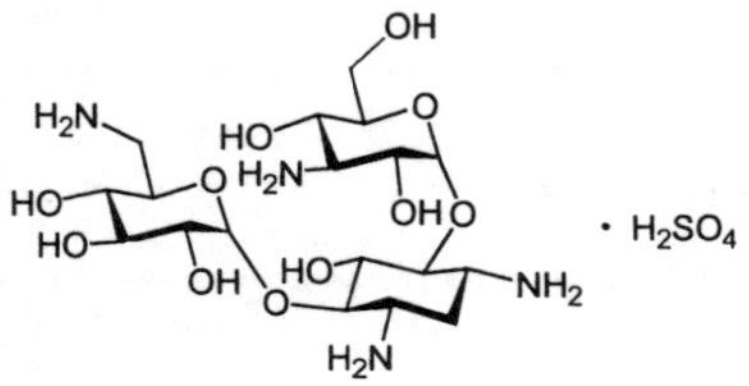

$C_{18}H_{38}N_4O_{15}S$
O-3-Amino-3-deoxy-α-D-glucopyranosyl-(1→6)-O-[6-amino-6-deoxy-α-D-glucopyranosyl-(1→4)]-2-deoxy-D-streptamine sulfate.
Cantrex; Cristalomicina; Enterokanacin; Kamycin; Kasmynex; Kanabristol; Kanacedin; Kanamytrex; Kanasig; Kanatrol; Kanicin; Kannasyn; Kantrex; Kantrox; Klebcil; Otokalixin; Resistomycin; Ophtalmokalixan; Kantrexil; Kano; Kanescin; Kanaqua; component of: Amforol. Aminoglycoside antibiotic. Dec > 250°; freely soluble in H_2O; insoluble in most organic solvents; LD_{50} (mus orl) = 20700 mg/kg, (mus ip) = 1450 mg/kg.

Veterinary Products:

041-836; Kantrim ® 200; *Fort Dodge Animal Health, Divn. AHP.*
042-548; (with Aminopentamide Hydrogen Sulfate, Attapulgite, Bismuth Subcarbonate, Pectin) Amforol ® Suspension; *Fort Dodge Animal Health, Divn. AHP.*
042-661; Kantrim ® ophthalmic Ointment; *Fort Dodge Animal Health, Divn. AHP.*
042-841; (with Attapulgite, Bismuth Subcarbonate, Pectin) Amforol ® Veterinary Oral Tablets; *Fort Dodge Animal Health, Divn. AHP.*
042-883; Kantrim ® ophthalmic Solution; *Fort Dodge Animal Health, Divn. AHP.*
043-784; (with Amphomycin Calcium, Hydrocortisone Acetate) Kanfosone Ointment; *Fort Dodge Animal Health, Divn. AHP.*
047-997; (with Amphomycin Calcium, Hydrocortisone Acetate) Amphoderm Ointment; *Fort Dodge Animal Health, Divn. AHP.*

164 Neomycin B Sulfate

1405-10-3 6542 248-770-9

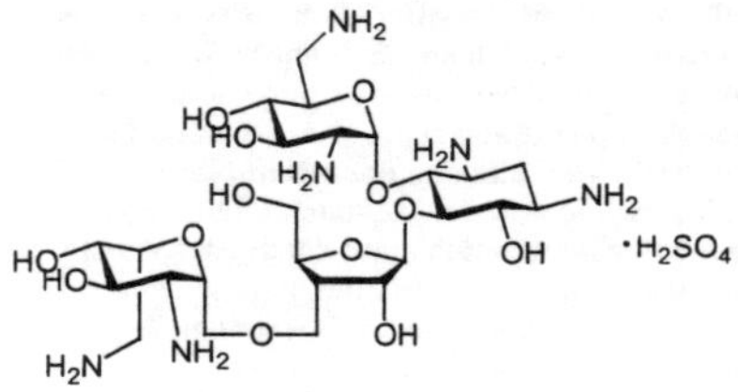

$C_{23}H_{48}N_6O_{17}S$
Neomycin B sulfate.
Biosol; Bykomycin; Endomixin; Fraquinol; Myacine; Neosulf; Neomix; Neobrettin; Nivemycin; Tuttomycin. Aminoglycoside antibiotic. Used for oral treatment of enteric infections. $[\alpha]_D^{20}$ = 54° (c = 2 H_2O); soluble in H_2O (0.63 g/100 ml), MeOH (0.0225 g/100 ml), EtOH (0.95 g/100 ml), iPrOH (0.008 g/100 ml), isoamyl alcohol (0.0247 g/100 ml), cyclohexane (0.008 g/100 ml), C_6H_6 (0.005 g/100 ml), insoluble in Et_2O, Me_2CO, $CHCl_3$.

Veterinary Products:

010-524; (with Hydrocortisone Acetate, Tetracaine Hydrochloride) Neo-Cortef ® with Tetracaine; *Pharmacia & Upjohn.*
011-315; Neomix ® 325 Soluble Powder; Neomix ® AG 325 Soluble Powder; Neomycin 325 Soluble Powder; *Pharmacia & Upjohn.*
011-437; (with Prednisolone Sodium Phosphate) Hydeltrone Ointment; *Merial.*
011-703; (with Prednisole Acetate, Tetracaine Hydrochloride) Neo-Delta-Cortef ® With Tetracaine Ointment; *Pharmacia & Upjohn.*
011-953; Biosol ® Sterile Solution; Biosol ® Sterile Solution 50 mg; *Pharmacia & Upjohn.*
012-258; (with Nystatin, Thiostreptone, Triamcinolone Acetonide) Panolog ® Ointment; *Fort Dodge Animal Health, Divn. Am. Cyanamid.*
013-181; (with Aminopropazine Fumarate) Jenomycin Tablets; *Schering-Plough Animal Health.*
015-151; (with Fluocinolone Acetonide) Neo-synalar Cream; *Medicis Dermatologics.*
015-433; (with Isoflupredone Acetate, Myristyl σ-Picolinium Chloride, Tetracaine Hydrochloride) Neo-Predef ® With Tetracaine Ointment; *Pharmacia & Upjohn.*
030-025; (with Isoflupredone Acetate, Tetracaine Hydrochloride) Neo-Predef ® with Tetracaine Top. Ointment; Tritop ® Topical Ointment; *Pharmacia & Upjohn.*
031-914; (with Isopropamide Iodide, Prochlorperazine Dimaleate) Neo-Darbazine Spansule Capsule No.1; Neo-Darbazine Spansule Capsule No.3; *Pfizer.*
031-962; (with Tylosin) Tylan ®Plus Neomycin Eye Powder; *Elanco.*
032-322; (with Hexamehyltetracosane, Prednisolone, Tetracaine) Liquisone F With Cerumene; *Evsco.*
034-872; (with Isoflupredone Acetate) Neo Predef ® Sterile Ointment; *Pharmacia & Upjohn.*
038-801; (with Flumethasone, Polymyxin B Sulfate) Anaprime ® Ophthalmic Solution; *Fort Dodge Animal Health, Divn. AHP.*
042-633; (with Dexamethasone, Thiabendazole) Tresaderm Dermatologic Solution; *Merial.*
044-655; Neomycane Ophthalmic Ointment; *Evsco.*
045-288; (with Prednisolone Acetate)Optisone; *Evsco.*

049-725; (with Flumethasone, Polymyxin B Sulfate) Anaprime ® Opthakote Ophthalmic; *Fort Dodge Animal Health, Divn. AHP.*
049-726; (with Polymyxin B Sulfate) Optiprime ® Opthakote Ophthalmic Solution; *Fort Dodge Animal Health, Divn. AHP.*
065-015; (with Bacitracin Zinc, Hydrocortisone Acetate, Polymyxin B Sulfate) Bacitracin-Neomycin-Polymyxin With Hydrocortisone Acetate Ophthalmic Ointment; Vetropolycin HC ophthalmic Ointment; *Altana.*
065-016; (with Bacitracin Zinc, Polymyxin B Sulfate) Bac-Neo-Poly Ophthalmic Ointment; Vetropolycin Ophthalmic Ointment; *Altana.*
065-114; (with Bacitracin Zinc, Polymyxin B Sulfate) Mycitracin ® Sterile ophthalmic Ointment; *Pharmacia & Upjohn.*
065-119; (with Hydrocortisone Acetate, Penicillin G Procaine, Polymyxin B Sulfate) Forte Topical ® Ointment; *Pharmacia & Upjohn.*
065-476; (with Bacitracin Zinc, Hydrocortisone Acetate, Polymyxin B Sulfate) Cortisporin Veterinary Ophthalmic Ointment; *Schering-Plough Animal Health.*
065-485; (with Bacitracin Zinc, Polymyxin B Sulfate) Neosporin Ophthalmic Ointment; *Schering-Plough Animal Health.*
091-534; (with Prednisolone Acetate) Neo-Delta Cortef ® Sterile Solution; *Pharmacia & Upjohn.*
093-514; (with Hydrocortisone Acetate) Neo-Cortef ® Sterile Ointment; *Pharmacia & Upjohn.*
096-676; (with Nystatin, Thiostreptone, Tramcinolone Acetonide) Panolog Cream; *Fort Dodge Animal Health, Divn. Am. Cyanamid.*
140-810; (with Nystatin, Thiostreptone, Tramcinolone Acetonide) Panavet Ointment; *Med-Pharmex.*
140-847; (with Nystatin, Thiostreptone, Tramcinolone Acetonide) Animax; *Altana.*
140-879; (with Nystatin, Thiostreptone, Tramcinolone Acetonide) Derma 4 Ointment; *Pfizer.*
140-889; (with Nystatin, Thiostreptone, Tramcinolone Acetonide) Derm-Otic Ointment; *Biocraft.*
141-003; (with Nystatin, Thiostreptone, Tramcinolone Acetonide) Derm-Otic Ointment; *Robins.*
200-046; Neomycin Soluble Powder 325g/lb; Neomycin Sulfate Soluble Powder, 325 g/lb; *Pfizer.*
200-050; Neomycin 325 Soluble Powder; *Merial.*
200-113; Biosol ® Liquid; *Pharmacia & Upjohn.*
200-118; Neomycin Oral Solution; *Phoenix.*
200-130; Neo-Sol 50 ®; *Alpharma.*
200-153; Neo 200 Oral Solution; *Merial.*
200-235; Neosol Soluble Powder; *Med-Pharmex.*
200-245; (with Nystatin, Thiostreptone, Tramcinolone Acetonide) Derma-Vet Cream; *Med-Pharmex.*

165 Neomycin Palmitate

1405-12-5 6542 259-582-1

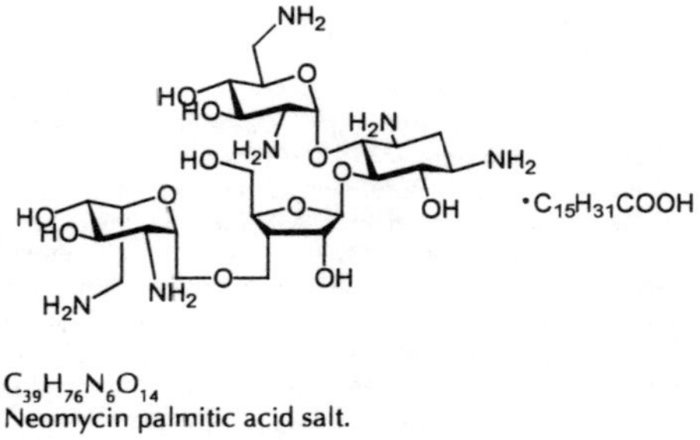

$C_{39}H_{76}N_6O_{14}$
Neomycin palmitic acid salt.

Aminoglycoside antibiotic. Used for oral treatment of enteric infections.

Veterinary Products:

040-322; (with Chymotrypsin, Hydrocortisone Acetate, Trypsin) Kymar Ointment Improved; *Schering-Plough Animal Health.*

166 Spectinomycin Dihydrochloride Pentahydrate

22189-32-8 8890

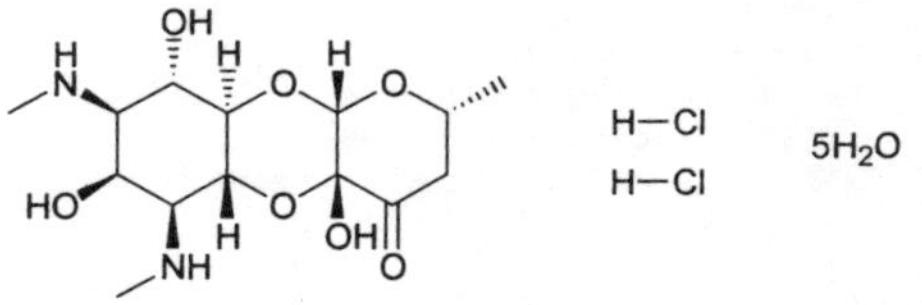

$C_{14}H_{26}Cl_2N_2O_7{\cdot}H_2O$
Decahydro-4a,7,9-trihydroxy-2-methyl-6,8-bis(methyl-amino)-4H-pyrano[2,3-b][1,4]benzodioxin-4-one dihydrochloride pentahydrate.
Spectam; Spectogard; Stanilo; Togamycin; Trantan; Trobicin. Aminoglycoside antibiotic. Used as an antibacterial in chickens, turkeys and pigs. Colorless needles; mp = 205-207° (dec); $[\alpha]_D$ = +14.8° (c = 0.42 H_2O); soluble in $CHCl_3$ (0.0042 g/100 ml), Me_2CO (0.0015 g/100 ml), Et_2O (0.001 g/100 ml); soluble in H_2O, MeOH, propylene glycol, formamide, DMSO, 0.1N HCl, 0.1N NaOH (> 2 g/100 ml).

Veterinary Products:

015-126; Spectinomycin Tablet & Injection; *Fort Dodge Animal Health, Divn. AHP.*
033-157; Spectam ® Scour Halt; *Merial.*
038-661; Spectam ® Water Soluble Concentrate; *Merial.*
040-040; Spectam ® Injectable; *Merial.*
041-629; Spectinomycin Oral Liquid; *Fort Dodge Animal Health, Divn. AHP.*
093-483; Spectam ® Injectable; *Merial.*
093-515; Spectam ® Tablets; *Merial.*
200-127; Prospec ® Injectable; Spectinomycin Injectable; *Pharmacia & Upjohn.*

167 Spectinomycin Sulfate Tetrahydrate

64058-48-6 8890

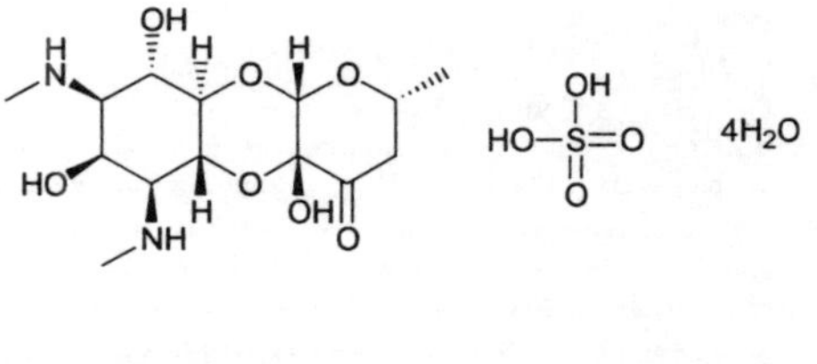

$C_{14}H_{26}N_2O_{11}S{\cdot}4H_2O$
Decahydro-4a,7,9-trihydroxy-2-methyl-6,8-bis(methyl-amino)-4H-pyrano[2,3-b][1,4]benzodioxin-4-one.
Aminoglycoside antibiotic. Used as an antibacterial in chickens, turkeys and pigs. White crystals; mp = 185° (dec); $[\alpha]_D^{25}$ = +17.0° (c = 1, H_2O); soluble in H_2O (22.5

g/100 ml), DMSO (0.5 - 1.0 g/100 ml), insoluble in C_5H_5N, EtOH, EtOAc, cyclohexane, C_6H_6, Me_2CO, DMF, dioxane, CH_3CN.

Veterinary Products:

046-109; (with Lincomycin Hydrochloride Monohydrate) L-S 50 Water Soluble ® Powder; *Pharmacia & Upjohn.*
141-077; Adspec ® Sterile Solution; *Pharmacia & Upjohn.*

168 Streptomycin
57-92-1 8983 200-355-3

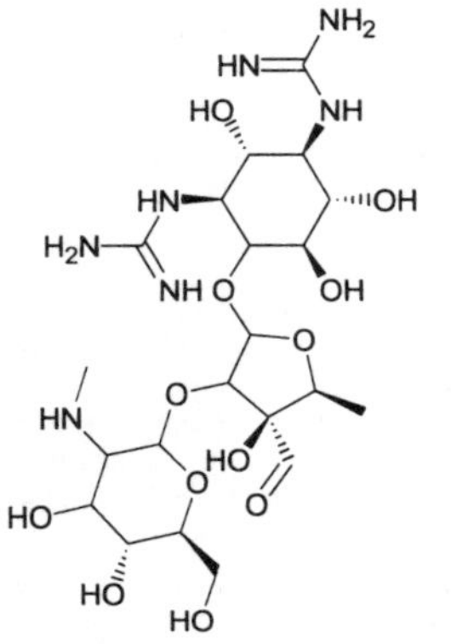

$C_{21}H_{39}N_7O_{12}$
O-2-Deoxy-2-(methylamino)-α-L-glucopyranosyl-(1→2)-O-5-deoxy-3-C-formyl-α-L-lyxofuranosyl-(1→4)-N,N'-bis(aminoiminomethyl)-D-streptamine.
streptomycin A. Aminoglycoside antibiotic. [tri-hydrochloride ($C_{21}H_{42}Cl_3N_7O_{12}$)]: $[\alpha]_D^{26}$ = -84°; soluble in H_2O (> 2g/100 ml); MeOH (> 2 g/100 ml), EtOH (0.09 g/100 ml), iPrOH (0.012 g/100 ml), isoamyl alcohol (0.012 g/100 ml), petroleum ether (0.002 g/100 ml), CCl_4 (0.004 g/100 ml), Et_2O (0.001 g/100 ml).

Veterinary Products:

049-462 (with Amprolium, Arsanilic Acid, Ethopabate, Penicillin G Procaine) Rainbow Broiler Premix No. 1; *Stockton Hay & Grain.*

169 Streptomycin Sesquisulfate
3810-74-0 8983 223-286-0

$C_{42}H_{84}N_{14}O_{36}S_3$
O-2-Deoxy-2-(methylamino)-α-L-glucopyranosyl-(1→2)-O-5-deoxy-3-C-formyl-α-L-lyxofuranosyl-(1→4)-N,N'-bis(aminoiminomethyl)-D-streptamine sulfate (2:3) salt.
streptomycin sulfate; AgriStrep; Streptobrettin; Vetstrep. An aminoglycoside antibiotic. Soluble in H_2O (> 2 g/100 ml), MeOH (0.085 g/100 ml), EtOH (0.030 g/100 ml), iPrOH (0.001 g/100 ml), petroleum ether (0.0015 g/100 ml), CCl_4 (0.0035 g/100 ml), Et_2O (0.0035 g/100 ml).

Veterinary Products:

065-107; (with Bacitracin Methylene Disalicylate) Entromycin Powder; *Veterinary Specialties.*
065-252; Strep-Sol; *Veterinary Services.*
065-291; (with Dihydrostreptomycin Sulfate) Dihydrostreptomycin Sulfate; *Pfizer.*
200-197; Streptomycin Oral Solution; *Contemporary Products.*

170 Tobramycin
32986-56-4 9628 251-322-5

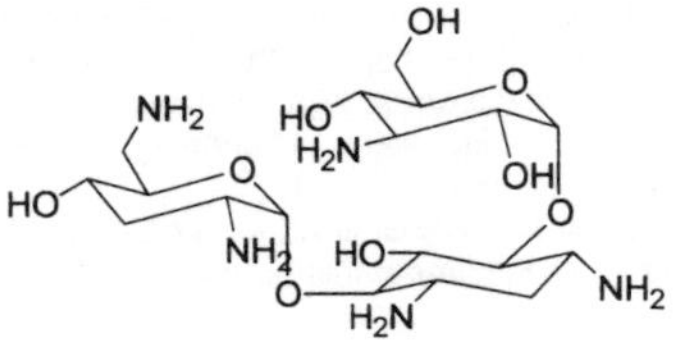

$C_{18}H_{37}N_5O_9$
O-3-Amino-3-deoxy-α-D-glucopyranosyl-(1→6)-O-[2,6-diamino-2,3,6-trideoxy-α-D-ribo-hexopyranosyl-(1→4)-]-2-deoxy-D-streptamine.
nebramycin factor 6; NF 6; Gernebcin; Tobracin; Tobradistin; Tobralex; Tobramaxin; Tobrex. Aminoglycoside antibiotic. May be used to treat serious Gram-negative infections, particularly where gentamicin-resistant bacteria are involved. Soluble in H_2O; $[\alpha]_D$ = 128°; LD_{50} (mus sc) = 441 mg/kg, (rat sc) = 969 mg/kg.

171 Tobramycin Sulfate
79645-27-5 9628

$C_{36}H_{84}N_{10}O_{38}S_5$
O-3-Amino-3-deoxy-α-D-glucopyranosyl-(1→6)-O-[2,6-diamino-2,3,6-trideoxy-α-D-ribo-hexopyranosyl-(1→4)-]-2-deoxy-D-streptamine slufate.
Nebcin; Obracin; Tobra. Aminoglycoside antibiotic. May be used to treat serious Gram-negative infections, particularly where gentamicin-resistant bacteria are involved.

Antibiotics, Amphenicol

172 Chloramphenicol
56-75-7 2120 200-287-4

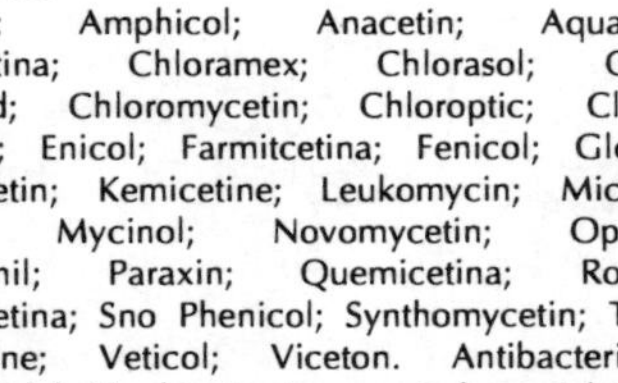

$C_{11}H_{12}Cl_2N_2O_5$
D-Threo-N-dichloroacetyl-1-p-nitrophenyl-2-amino-1,3-propanediol.
Ak-Chlor; Amphicol; Anacetin; Aquamycetin; Chemicetina; Chloramex; Chlorasol; Chloricol; Chlorocid; Chloromycetin; Chloroptic; Cloramfen; Clorocyn; Enicol; Farmitcetina; Fenicol; Globenicol; Intramycetin; Kemicetine; Leukomycin; Micoclorina; Mychel; Mycinol; Novomycetin; Ophthoclor; Pantovernil; Paraxin; Quemicetina; Rompheni; Sintomicetina; Sno Phenicol; Synthomycetin; Tevcocin; Tyfomycine; Veticol; Viceton. Antibacterial and antirickettsial. Used to treat anaeroic bacterial infections in small animals and horses. mp = 150.5-151.5°; $[\alpha]_D^{27}$ =

18.6° (c = 4.86 EtOH), $[\alpha]_D^{25}$ = -25.5° (EtOAc); λ_m 278 nm ($E_D^{1\%}$ = 298); soluble in H_2O (0.25 g/100 ml at 25°), propylene glycol (15.1 g/100 ml at 25°); very soluble in MeOH, EtOH, BuOH, EtOAc; Me_2CO, fairly soluble in Et_2O, insoluble in C_6H_6, petroleum ether.

Veterinary Products:

055-002; Tevcocin; *Merial.*
055-005; (with Prednisolone, Squalene, Tetracaine) Liquichlor With Cerumene; *Evsco.*
055-034; Chlorasol; *Evsco.*
055-051; Chloromycetin Tablets 100 mg; Chloromycetin Tablets 250 mg; Chloromycetin Tablets 500 mg; *Fort Dodge Animal Health, Divn. AHP.*
055-052; Chlora-Tabs 100; *Evsco.*
055-059; Tevcocin Tablets; *Merial.*
065-137; Amphicol-V; *Ferrante.*
065-149; Chloromycetin Ophthalmic Ointment; *Fort Dodge Animal Health, Divn. AHP.*
065-150; Chloramphenicol Capsules; *Nylos.*
065-158; Chloricol; *Evsco.*
065-241; Mychel-Vet Capsules (50 mg); *Pfizer.*
065-259; (with Prednisolone Acetate) Chlorasone Ophthalmic Ointment; *Evsco.*
065-345; Chloramphenicol Capsules; *Eon Labs.*
065-460; Chloramphenicol 1% Ophthalmic; Vetrocloricin ophthalmic Ointment; *Altana.*
065-461; Anacetin Tablets; *Boehringer Ingelheim Vetmedica.*
065-463; Mychel-Vet Injection; *Pfizer.*
065-489; Mychel-Vet Tabs; *Pfizer.*
065-491; Medichol Tablets; *Boehringer Ingelheim Vetmedica.*

173 Chloramphenicol Palmitate
530-43-8 2120 208-477-9

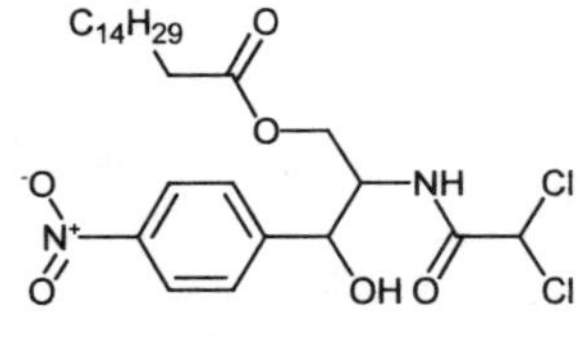

$C_{27}H_{42}Cl_2N_2O_6$
D-Threo-(-)-2,2-dichloro-N[β-hydroxy-α-(hydroxymethyl)-p-nitrophenyl]acetamide α-palmitate.
Chlorambon; Chloropal; Clorolifarina. Antibacterial; anti-rickettsial. For treating anaerobic bacterial infections in small animals, horses. mp = 90°; $[\alpha]_D^{26}$ = 24.6° (c = 5 EtOH); λ_m 271 nm ($E_{1\,cm}^{1\%}$ 179 EtOH); slightly soluble in H_2O (0.105 g/100 ml at 28°), petroleum ether (0.0225 g/100 ml), freely soluble in MeOH, EtOH, $CHCl_3$, Et_2O, C_6H_6.

Veterinary Products:

055-047; Chloromycetin Palmitate Oral Suspension; *Fort Dodge Animal Health, Divn. AHP.*

174 Florfenicol
73231-34-2 [76639-94-6] 4145
$C_{12}H_{14}Cl_2FNO_4S$
2,2-Dichloro-N-[1-fluoromethyl)-2-hydroxy-2-[4-(methylsulfonyl)phenyl]ethyl]acetamide.
Nuflor; Sch-25298; Aquafen. Antibacterial and antirickettsial. Use to treat respiratory disease in cattle. mp = 153-154°; soluble in H_2O.

Veterinary Products:

141-063; Nuflor® Injectable Solution; *Schering-Plough Animal Health.*

Antibiotics, Ansamycin

175 Rifampin
13292-46-1 8382 236-312-0

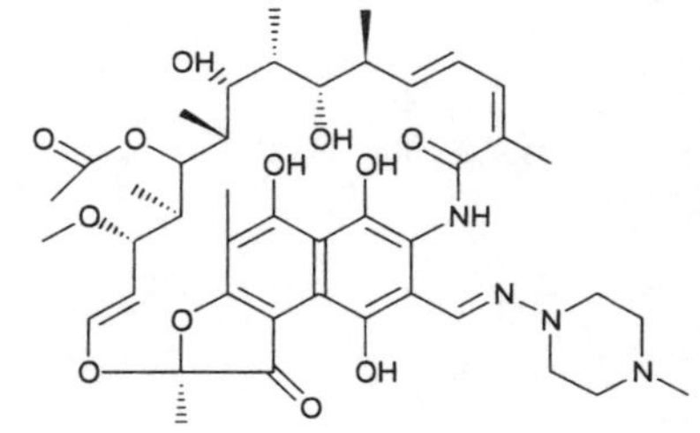

$C_{43}H_{58}N_4O_{12}$
5,6,9,17,19,21-Hexahydroxy-23-methoxy-2,4,12,16,18,20,22-heptamethyl-8-[N-(4-methyl-1-piperazinyl)formimidoyl]-2,7-(epoxypentadeca[1,11,13]-trienimino)naphtho[2,1-b]furan-1,11(2H)-dione 21-acetate.
Rifadin; Rimactane; L-5103 Lepetit; Ba 41166/E; NSC-113926; component of: Rifater. Antibacterial of the ansamycin family. Used to treat *Corynebacterium* infections in young horses. Dec 183-188°; λ_m 237, 255,334, 475 nm (ε 33200, 32100, 27000, 15400 pH 7.38); freely soluble in CH_3Cl, DMSO; soluble in EtOAc, MeOH, THF; slightly soluble in H_2O, Me_2CO, CCl_4; LD_{50} (mus orl) = 885 mg/kg, (mus iv) = 260 mg/kg, (mus ip) = 640 mg/kg, (rat orl) = 1720 mg/kg, (rat iv) = 330 mg/kg, (rat ip) = 550 mg/kg.

176 Rifampin SV
6998-60-3 8384 230-273-3

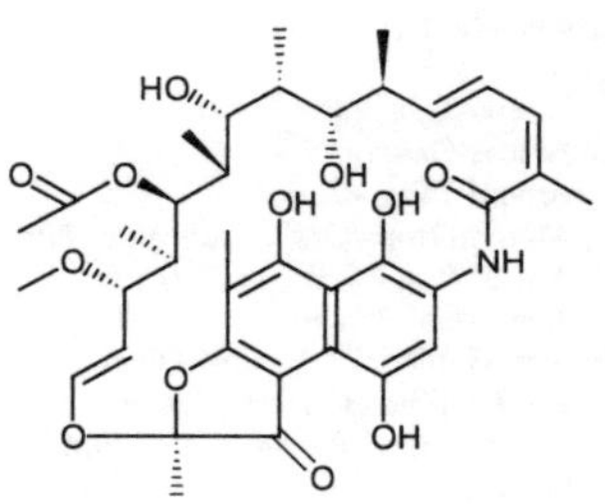

$C_{37}H_{47}NO_{12}$
5,6,9,17,19,21-Hexahydroxy-23-methoxy-2,4,12,16,18,20,22-heptamethyl-2,7-(epoxypentadeca[1,11,13]trienimino)naphtho[2,1-b]furan-1,11(2H)-dione 21-acetate.
rifomycin SV; rifamicine SV. Antibacterial of the ansamycin family. Used to treat *Corynebacterium*

infections in young horses. mp 300° (dec > 140°); $[\alpha]_D^{20}$ = -4° (MeOH); λ_m = 223, 314, 445 nm ($E^{1\%}_{1\ cm}$ 586, 322, 204 phosphate buffer pH 7.3); slightly soluble in H_2O, petroleum ether; soluble in MeOH, EtOH, Me_2CO, EtOAc; LD_{50} (mus iv) = 550 mg/gk, (mus ip) = 625 mg/kg, (mus orl) = 2120 mg/kg.

Antibiotics, beta-Lactam

177 Amoxicillin Trihydrate

61336-70-7 617

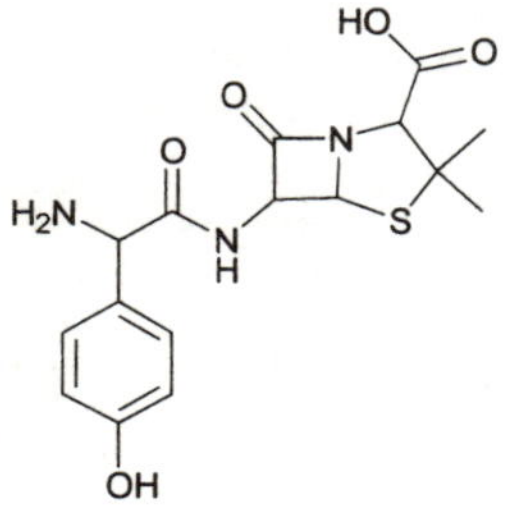

$C_{16}H_{19}N_3O_5S.3H_2O$

[2S-[2α,5α,6β(S*)]]-6-[[Amino(4-hydroxyphenyl)acetyl]-amino]-3,3-dimethyl-7-oxo-4-thia-1-azabicyclo[3.2.0]-heptane-2-carboxylic acid trihydrate.

BRL-2333; Agram; Alfamox; Almodan; Amodex; Amoxi; Amoxidal; Amoxidin; Amoxil; Amoxillat; Amoxi-Wolff; Amoxypen; Ardine; AX 250; Clamoxyl; Cuxacillin; Dura AX; Flemoxin; Hiconcil; Ibiamox; Larocin (obsolete); Larotid; Moxal; Moxaline; Neamoxyl; Polymox; Raylina; Robamox; Sigamopen; Silamox; Trimox; Uro-Clamoxyl; Utimox; Zamocillin. Antibacterial, often used in conjunction with clavulanate potassium; a β-lactamase inhibitor. Used to deal with urinary tract infections in dogs and cats. Has been used to treat canine periodontal dieease. $[\alpha]_D^{20}$ = +246° (c = 0.1); λ_m 230, 274 nm (ε 10850, 1400 EtOH); λ_m 229, 272 nm (ε 9500, 1080 0.1N HCl); λ_m = 248, 291 nm (ε 2200, 3000 KOH); soluble in H_2O (0.4 g/100 ml), MeOH (0.75 g/100 ml), EtOH (0.34 g/100 ml), insoluble in C_6H_6, EtOAc, CH_3CN, C_6H_{14}.

Veterinary Products:

055-078; Amoxi-Tabs; *Pfizer.*
055-080; Amoxi-Doser; *Pfizer.*
055-081; Amoxi-Tabs; *Pfizer.*
055-085; Amoxi-Drops Oral Suspension; *Pfizer.*
055-087; Amoxi-Bol; *Pfizer.*
055-088; Amoxi-Sol; *Pfizer.*
055-089; Amoxi-Inject 25 Grams; *Pfizer.*
055-091; Amoxi-Inject 3 Grams; *Pfizer.*
055-099; (with Clavulanate Potassium) Clavamox® Tablets; *Pfizer.*
055-100; Amoxi-Mast; *Pfizer.*
055-101; (with Clavulanate Potassium) Clavamox® Drops; *Pfizer.*
065-492; Robamox®-V Tablets; *Teva.*
065-495; Robamox®-V; *Teva.*
141-004; Robamox®-V; *Robins.*
141-005; Robamox®-V Tablets; *Robins.*

178 Ampicillin

69-53-4 628 200-709-7

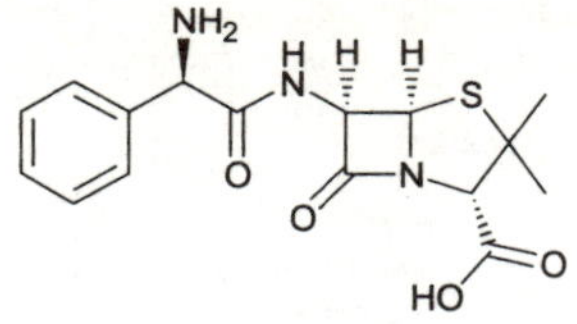

$C_{16}H_{19}N_3O_4S$

[2S-[2α,5α,6β(S*)]]-6-[(Aminophenylacetyl)amino]-3,3-dimethyl-7-oxo-4-thia-1-azabicyclo[3.2.0]heptane-2-carboxylic acid.

(2S,5R,6R)-6-[(R)-2-Amino-2-phenylacetamido]-3,3-dimethyl-7-oxo-4-thia-1-azabicyclo[3.2.0]-heptane-2-carboxylic acid; 6-[(Aminophenylacetyl)amino]-3,3-dimethyl-7-oxo-4-thia-1-azabicyclo[3.2.0]heptane-2-carboxylic acid; 6-[D(-)-α-aminophenylacetamido]-penicillanic acid; Adobacillin; Albipen; Alpen; Amblosin; Amfipen; Aminobenzylpenicillin; Amipenix; Amipenix S; Ampi-Bol; Ampi-Tablinen; Ampicillin A; Ampicillin Sulbactam; Ampicin; Ampicina; Ampicyn; Ampilar; Ampimed; Ampipenin; Ampitab; Amplisom; Amplital; Ampy-Penyl; Austrapen; AY-6108; Binotal; Bonapicillin; Britacil; BRL-1341; Copharcilin; D(-)-α-amino-benzylpenicillin; Doktacillin; Domicillin; Dumopen; Grampenil; Guicitrina; Marisilan; Nuvapen; Omnipen; P-50; Pen-Bristol; Penbritin; Penbrock; Penicline; Pénicline; Penstabil; Pentrex; Pentrexyl; Polycillin; Ponecil; QI Damp; Rosampline; Synpenin; Tokiocillin; Totacillin; Totalciclina; Totapen; Ultrabion; Unasyn; Viccillin; Wymox. Antibacterial. Not well absorbed orally and has been supplanted by amoxicillin. Crystals; dec 199-202°; $[\alpha]_D^{23}$ = +287.9° (c = 1 H_2O); soluble in H_2O (1.3 g/100 ml at 42°); pH of saturated aqueous solution = 5 - 7.5.

Veterinary Products:

055-013; Omnipen 250 mg.; *Fort Dodge Animal Health, Divn. AHP.*

179 Ampicillin Sodium

69-52-3 628 200-708-1

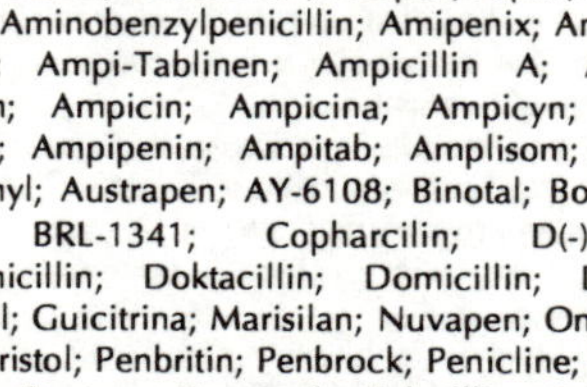

$C_{16}H_{18}NaN_3O_4S$

[2S-[2α,5α,6β(S*)]]-6-[(Aminophenylacetyl)amino]-3,3-dimethyl-7-oxo-4-thia-1-azabicyclo[3.2.0]heptane-2-carboxylic acid sodium salt.

Alpen-N; Amcill-S; Ampicin; Cilleral; Omnipen-N; Pen A/N; Penbritin-S; Pentrex; Polycillin-N; Principen/N; Synpenin; Viccillin. Antibacterial. Not well absorbed orally and has been supplanted by amoxicillin. Odorless, off-white, crystalline hygroscopic powder; dec = 205°; $[\alpha]_D^{23}$ = +209° (c = 0.2 H_2O); soluble in H_2O, pH of aqueous solution = 8 - 10.

Veterinary Products:

055-084; Amp-Equine; *Pfizer.*

180 Ampicillin Trihydrate

7177-48-2 628

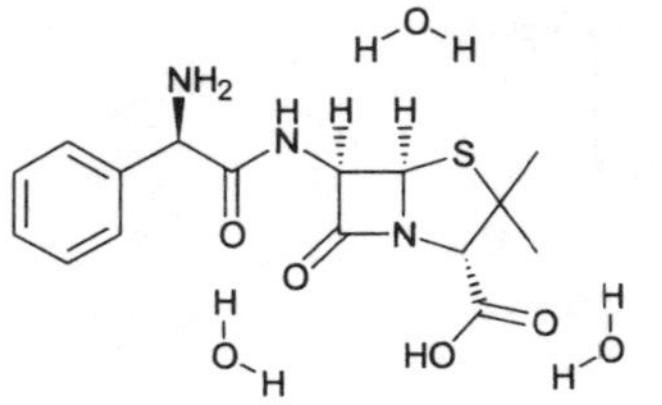

$C_{16}H_{19}N_3O_4S.3H_2O$
[2S-[2α,5α,6β(S*)]]-6-[(Aminophenylacetyl)amino]-3,3-dimethyl-7-oxo-4-thia-1-azabicyclo[3.2.0]heptane-2-carboxylic acid trihydrate.
Alpen; Amblosin; Amcill; Ampilag; Ampilar; Ampi-Tablinen; Amplital; Austrapen; Binotal; Cetampin; Cymbi; Pen A; Penbristol; Penbritin; Penbrock; Pensyn; Pentrexyl; Polycillin; Princillin; Principen; Rosampline; Totacillin; Totalciclina; Totapen; Ukapen; Ultrabion; Vidopen. Antibacterial. Not well absorbed orally and has been supplanted by amoxicillin.

Veterinary Products:

055-030; Polyflex®; *Fort Dodge Animal Health, Divn. AHP.*
055-036; Princillin Capsules125 mg; Princillin Capsules250 mg; Princillin Capsules500 mg; *Norbrook.*
055-042; Ampi-Tab; *Pfizer.*
É055-050; Princillin Soluble Powder; Norbrook.
055-056; Princillin Bolus; *Norbrook.*
055-061; Princillin 125 For Oral Suspension; *Norbrook.*
055-064; Princillin Injection; *Fort Dodge Animal Health Divn., Am. Cyanamid.*
055-066; Princillin Injection; *Fort Dodge Animal Health Divn., Am. Cyanamid.*
055-071; Princillin Injection 200 mg; *Fort Dodge Animal Health Divn., Am. Cyanamid.*
055-074; Ampi-Bol; *Pfizer.*
055-079; Ampi-Ject; *Pfizer.*
200-180; Ampicillin Trihydrate; *Hanford.*

181 Carbenicillin

4697-36-3 1838 225-171-0

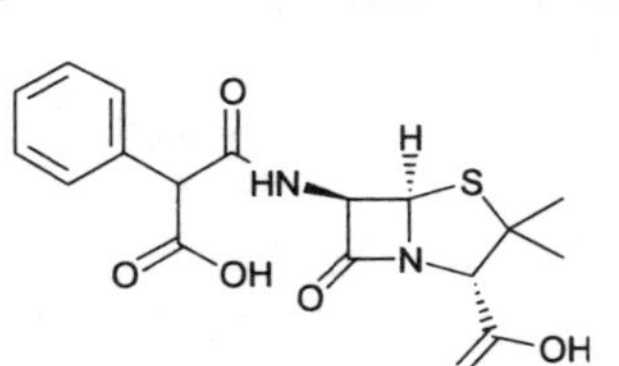

$C_{17}H_{18}N_2O_6S$
[2S-(2α,5α,6β)]-6-[(Carboxyphenylacetyl)amino]-3,3-dimethyl-7-oxo-4-thia-1-azabicyclo[3.2.0]heptane-2-carboxylic acid.
Antibacterial. Used to treat urinary tract infections.

182 Carbenicillin Disodium

4800-94-6 1838 225-360-8

$C_{17}H_{16}N_2Na_2O_6S$
[2S-(2α,5α,6β)]-6-[(Carboxyphenylacetyl)amino]-3,3-dimethyl-7-oxo-4-thia-1-azabicyclo[3.2.0]heptane-2-carboxylic acid disodium salt.
BRL-2064; CP-15639-2; Anabactyl; Carbapen; Carbecin; Geocillin; Geopen; Hyoper; Microcillin; Pyocianil; Pyopen. Antibacterial. Used to treat urinary tract infections.

183 Carbenicillin Indanyl Sodium

26605-69-6 1838 247-845-3

$C_{26}H_{25}N_2NaO_6S$
[2S-(2α,5α,6β)]-6-[[3-[(2,3-Dihydro-1H-inden-5-yl)oxy]-1,3-dioxo-2-phenylproply]amino]-3,3-dimethyl-7-oxa-4-thia-1-azabicyclo[3.2.0]heptane-2-carboxylic acid sodium salt.
Carindacillin Sodium; CP-15464-2; Carindapen; Geocillin; G.U.-Pen. Antibacterial. Used to treat urinary tract infections.

184 Carbenicillin Phenyl

27025-49-6 1838 248-171-2

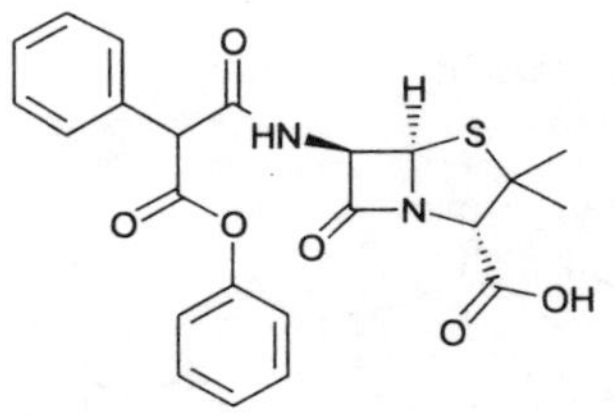

$C_{23}H_{22}N_2O_6S$
[2S-(2α,5α,6β)]-6-[(Carboxyphenylacetyl)amino]-3,3-dimethyl-7-oxo-4-thia-1azabicyclo[3.2.0]heptane-2-carboxylic acid phenyl.
Carfecillin. Antibacterial. Used to treat urinary tract infections.

185 Carbenicillin Phenyl Sodium

21649-57-0 1838 244-496-9

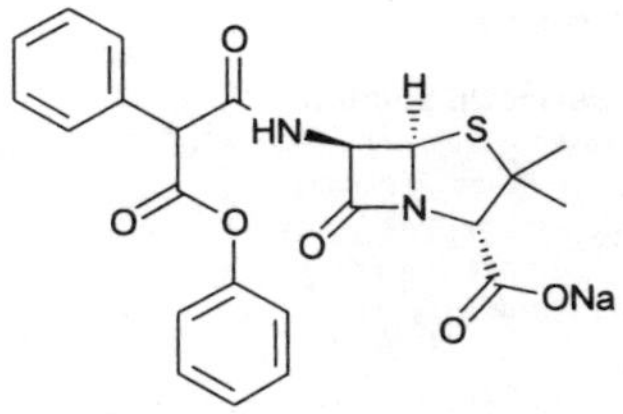

$C_{23}H_{21}N_2NaO_6S$
[2S-(2α,5α,6β)]-6-[(Carboxyphenylacetyl)amino]-3,3-dimethyl-7-oxo-4-thia-1-azabicyclo[3.2.0]heptane-2-carboxylic acid phenyl sodium.
carfecillin sodium; BRL-3475; Gripenin-O; Urocarf; Uticillin; carbenicillin phenyl sodium. Antibacterial. Used to treat urinary tract infections. $[\alpha]_D^{20} = 216.2°$ (H_2O).

186 Cefadroxil

66592-87-8 1963 256-555-6

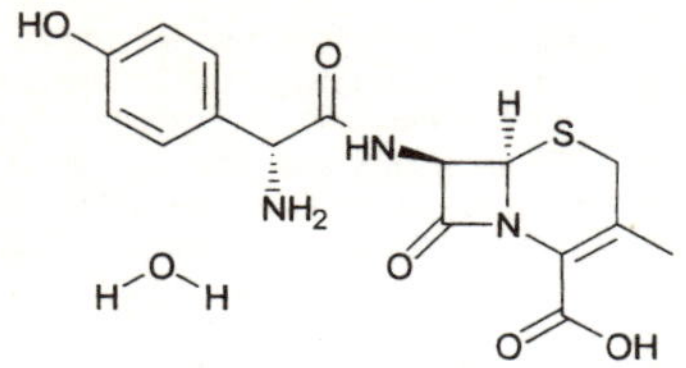

$C_{16}H_{17}N_3O_5S.H_2O$

[6R-[6α,7β(R*)]]-7-[[Amino-(4-hydroxyphenyl)acetyl]-amino]-3-methyl-8-oxo-5-thia-1-azabicyclo[4.2.0]oct-2-ene 2-carboxylic acid .

BL-S578; MJF-11567-3; Baxan; Bidocef; Cefa-Drops; Cefamox; Ceforal; Cephos; Duracef; Duricef; Kefroxil; Oracéfal; Sedral; Ultracef. Antibacterial. mp = 197° (dec).

Veterinary Products:

119-688; Cefa-Tabs ®; *Fort Dodge Animal Health, Divn. AHP.*

140-684; Cefa-Drops ®; *Fort Dodge Animal Health, Divn. AHP.*

187 Cefazolin

25953-19-9 1967 247-362-8

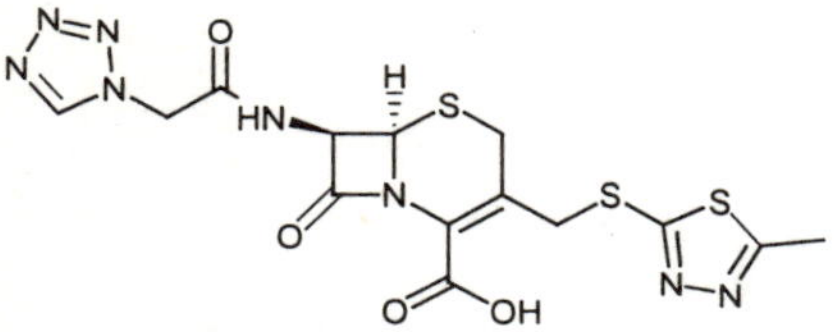

$C_{14}H_{14}N_8O_4S_3$

(6R-trans)-3-[[(5-Methyl-1,3,4-thiadiazol-2-yl)thio-methyl]-8-oxo-7-[(1H-tetrazol-1-ylacetyl)amino]-5-thia-1-azabicyclo[4.2.0]oct-2-ene 2-carboxylic acid.

CEZ. A cephalosporin antibacterial. mp = 198-200° (dec); λ_m = 272 nm (ε 13150 pH 6.4); easily soluble in DMF, C_5H_5N; soluble in Me_2CO, aqueous dioxane, aqueous EtOH, slightly soluble in MeOH; practically insoluble in $CHCl_3$, C_6H_6, Et_2O.

188 Cefazolin Sodium

27164-46-1 1967 248-278-4

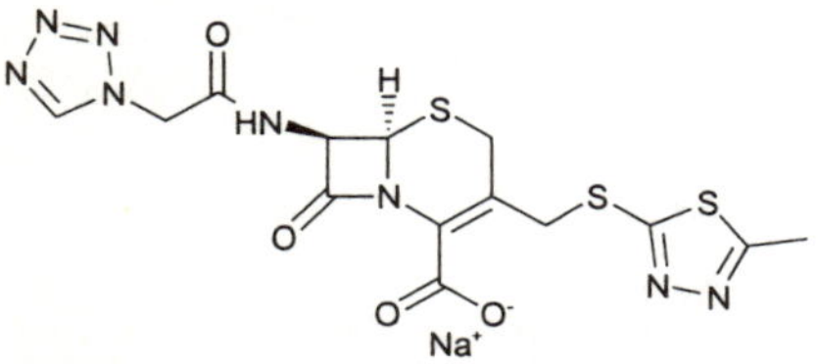

$C_{14}H_{13}N_8NaO_4S_3$

(6R-trans)-3[[(5-Methyl-1,3,4-thiadiazol-2-yl)thio]methyl]-8-oxo-7-[(1H-tetrazol-1-ylacetyl)amino]-5-thia-1-aza-bicyclo[4.2.0]oct-2-ene-2-carboxylic acid sodium salt.

sodium CEZ; SKF-41558; Acef; Ancef; Atirin; Biazolina; Bor-Cefazol; Cefacidal; Cefamedin; Cefamezin; Cefazil; Cefazina; Elzogram; Firmacef; Gramaxin; Kefzol; Lampocef; Liviclina; Totacef; Zolicef. A cephalosporin antibacterial. Crystallizes in three forms, readily soluble in H_2O, less soluble in organic solvents; LD_{50} (rat ip) = 7.4 mg/kg.

189 Cefoperazone

62893-19-0 1981 263-749-4

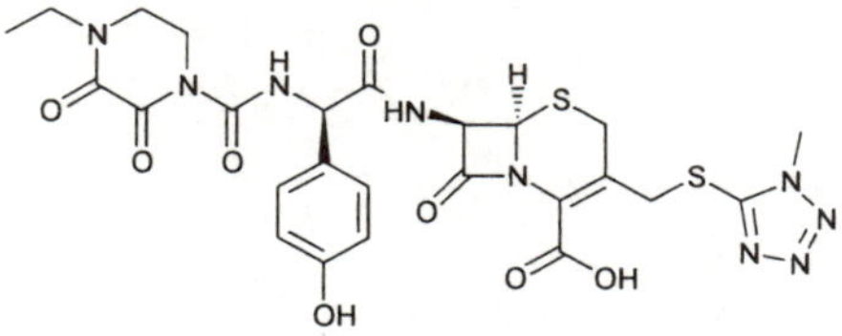

$C_{25}H_{27}N_9O_8S_2$

[6R-[6α,7β(R*))]]-7-[[[[(4-Ethyl-2,3-dioxo-1-piperazinyl)carbonyl]amino](4-hydroxyphenyl)-acetyl]amino]-3-[[(1-methyl-1H-tetrazol-5-yl)thio]methyl]-8-oxo-5-thia-1-azabicyclo[4.2.0]oct-2-ene-2-carboxylic acid .

A cephalosporin antibacterial. Used to treat enterobacterial infections. mp = 169-171° (hydrated).

190 Cefoperazone Sodium

62893-20-3 1981 263-751-5

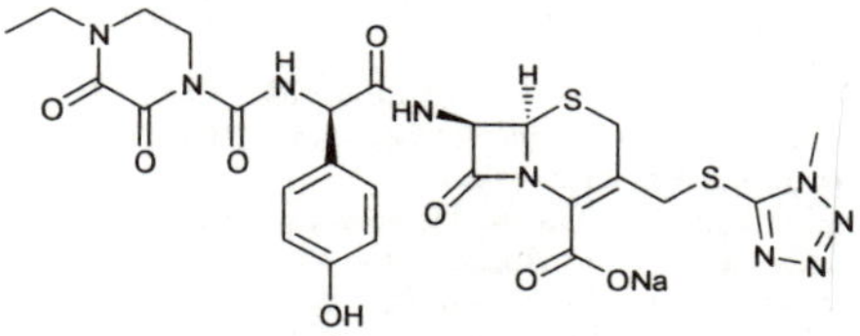

$C_{25}H_{26}N_9NaO_8S_2$

[6R-[6α,7β(R*))]]-7-[[[[(4-Ethyl-2,3-dioxo-1-piperazinyl)carbonyl]amino](4-hydroxyphenyl)-acetyl]amino]-3-[[(1-methyl-1H-tetrazol-5-yl)thio]methyl]-8-oxo-5-thia-1-azabicyclo[4.2.0]oct-2-ene-2-carboxylic acid sodium salt.

CP-52640-2; T-1551; Bioperazone; Cefazone; Cefobid; Cefobine; Cefobis; Cefogram; Cefoneg; Cefosint; Dardum; Farecef; Kefazon; Novobiocyl; Pathozone; Peracef; Perocef; Tomabef. A cephalosporin antibacterial. Used to treat enterobacterial infections.

191 Cefotaxime

63527-52-6 1983 264-299-1

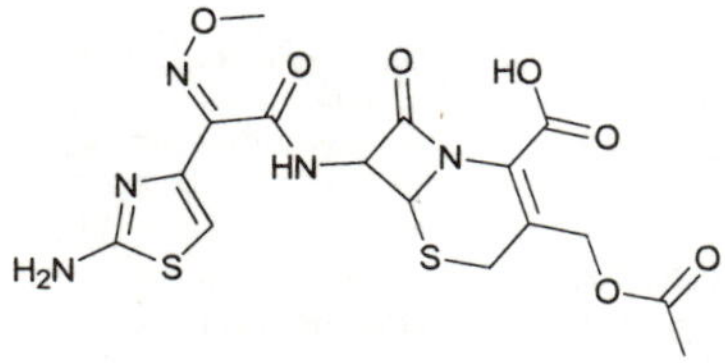

$C_{16}H_{17}N_5O_7S_2$

[6R-[6α,7β(Z)]]-3-[(Acetyloxy)methyl]-7-[[(2-amino-4-thiazolyl)(methoxyimino)acetyl]amino]-8-oxo-5-thia-1-

azabicyclo[4.2.0]oct-2-ene-2-carboxylic acid.
A cephalosporin antibacterial.

192 Cefotaxime Sodium

64485-93-4 1983 264-915-9

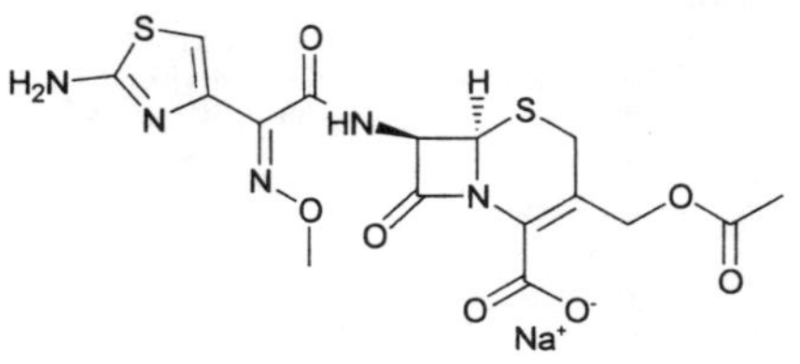

$C_{16}H_{16}N_5NaO_7S_2$

[6R-[6α,7β(Z)]]-3-[(Acetyloxy)methyl]-7-[[(2-amino-4-thiazolyl)(methoxyimino)acetyl]amino]-8-oxo-5-thia-1-azabicyclo[4.2.0]oct-2-ene-2-carboxylic acid sodium salt.
HR-756; RU-24756; Cefotax; Chemcef; Claforan; Pretor; Tolycar. Antibiotic. $[\alpha]_D^{20}$ = +55 ± 2° (c = 0.8 in H_2O).

193 Cefoxitin

35607-66-0 1986 252-641-2

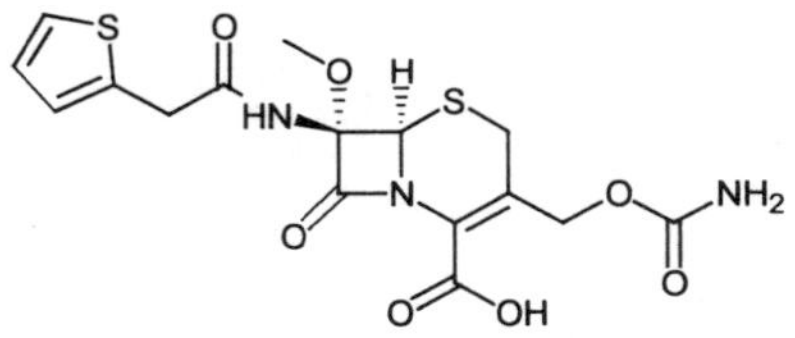

$C_{16}H_{17}N_3O_7S_2$

(6R-cis)-3-[[(Aminocarbonyl)oxy]methyl]-7-methoxy-8-oxo-7-[(2-thienylacetyl)amino]-5-thia-1-azabicyclo[4.2.0]oct-2-ene-2-carboxylic acid.
Mefoxin. Antibacterial. mp = 149-150°; poorly soluble in H_2O, soluble in organics; LD_{50} (rat orl) = 8980 mg/kg.

194 Cefoxitin Sodium

33564-30-6 1986 251-574-6

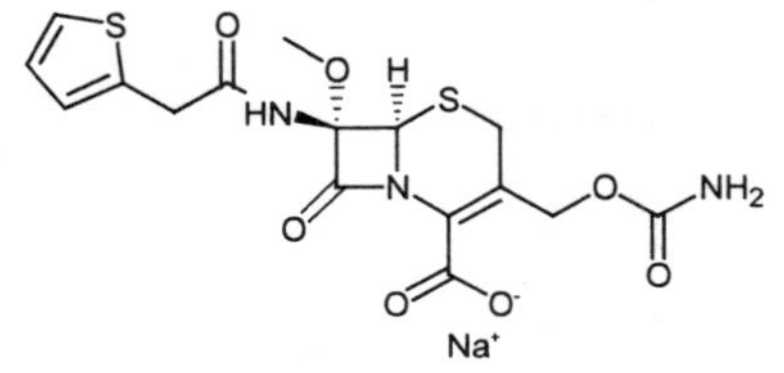

$C_{16}H_{16}N_3NaO_7S_2$

(6R-cis)-3-[[(Aminocarbonyl)oxy]methyl]-7-methoxy-8-oxo-7-[(2-thienylacetyl)amino]-5-thia-1-azabicyclo-[4.2.0]oct-2-ene-2-carboxylic acid sodium salt.
Cefoxitin sodium salt; MK-306; Betacef; Farmoxin; Mefoxin; Mefoxitin; Merxin; Cenomycin. Cephalosporin antibiotic. $[\alpha]_{589nm}^{25}$ = +210° (c = 1 in MeOH); very soluble in H_2O; soluble in MeOH; sparingly soluble in EtOH, Me_2CO; insoluble in aromatic and aliphatic hydrocarbons; LD_{50} (mus iv) = 5.10 mg/kg, (rat iv) = 8.98 mg/kg, (dog iv) > 10 mg/kg.

195 Ceftiofur Hydrochloride

103980-44-5 1999

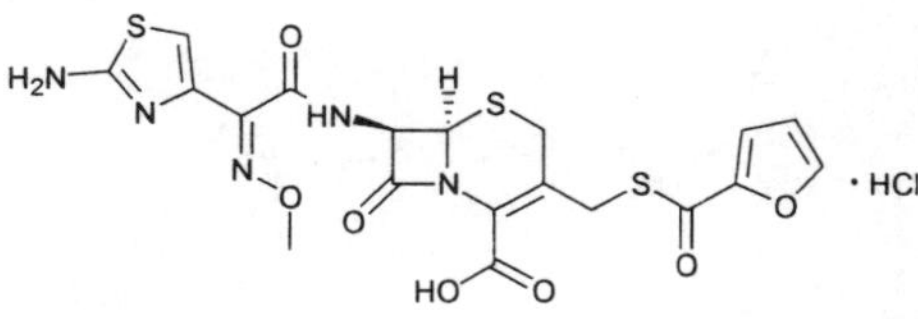

$C_{19}H_{17}N_5O_7S_3$.HCl

[6R-[6α,7β(Z)]]-7-[[(2-Amino-4-thiazolyl)(methoxyimino)-acetyl]amino]-3-[[(2-furanylcarbonyl)-thio]methyl]-8-oxo-5-thia-1-azabicyclo[4.2.0]oct-2-ene-2-carboxylic acid monohydrocholoride.
U-67279A. Antibacterial used to treat bovine respiratory disease.

Veterinary Products:

140-890; Excenel Sterile Suspension; *Pharmacia & Upjohn.*

196 Ceftiofur Sodium

104010-37-9 1999

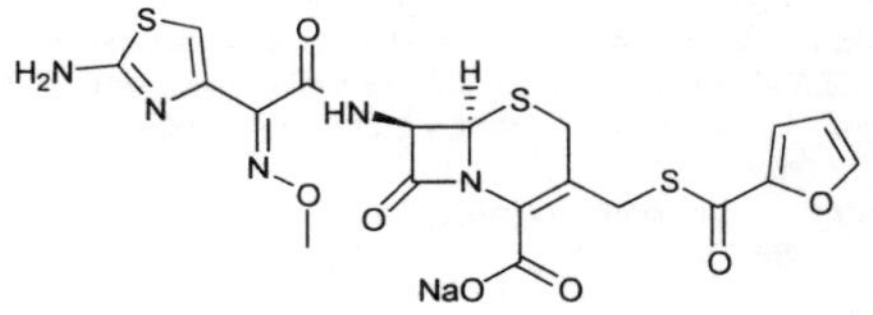

$C_{19}H_{16}N_5NaO_7S_3$

[6R-[6α,7β(Z)]]-7-[[(2-Amino-4-thiazolyl)-(methoxyimino)acetyl]amino]-3-[[(2-furanylcarbonyl)-thio]methyl]-8-oxo-5-thia-1-azabicyclo[4.2.0]oct-2-ene-2-carboxylic acid monosodium salt.
CM-31916; U-64279E; Excenel; Naxcel. Antibacterial used to treat bovine respiratory disease .

Veterinary Products:

140-388; Naxcel® Sterile Powder; *Pharmacia & Upjohn.*

197 Ceftriaxone

73384-59-5 2001 277-405-6

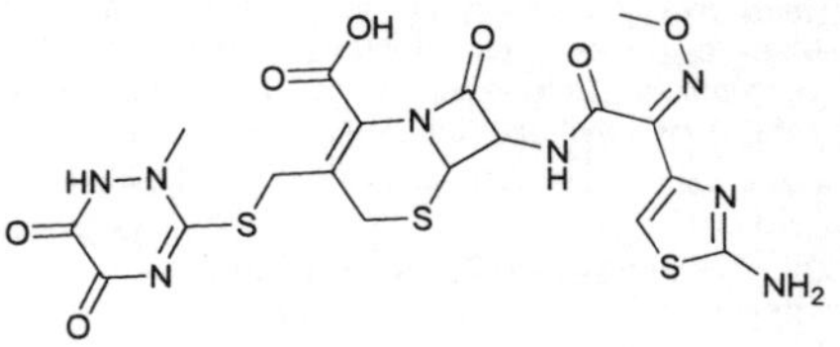

$C_{18}H_{18}N_8O_7S_3$

[6R-[6α,7β(Z)]]-7-[[(2-Amino-4-thiazolyl)(methoxyimino)-acetyl]amino]-8-oxo-3-[[(1,2,5,6-tetrahydro-2-methyl-5,6-dioxo-1,2,4-triazin-3-yl)thio]methyl]-5-thia-1-azabicyclo[4.2.0]oct-2-en-2-carboxylic acid.
ceftriaxone. A cephalosporin antibacterial.

198 Ceftriaxone Sodium
104376-79-6 2001

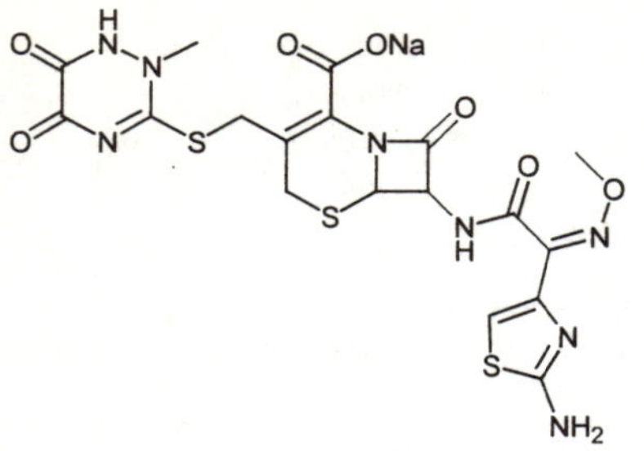

$C_{18}H_{16}N_8Na_2O_7S_3 \cdot 3.5H_2O$
[6R-[6α,7β(Z)]]-7-[[(2-Amino-4-thiazolyl)(methoxyimino)-acetyl]amino]-8-oxo-3-[[(1,2,5,6-tetrahydro-2-methyl-5,6-dioxo-1,2,4-triazin-3-yl)thio]methyl]-5-thia-1-azabicyclo-[4.2.0]oct-2-ene-2-carboxylic acid disodium salt hemiheptahydrate.
Ro-13-9904/001; Rocefin; Rocephin(e). A cephalosporin antibacterial. mp >155° (dec); $[\alpha]_D^{25}$ = -165° (c = 1 in H_2O); λ_m (H_2O) = 242, 272 nm (ε 32300, 29530); soluble in H_2O (40 g/100 ml); LD_{50} (mmus iv) = 3000 mg/kg, (mmus orl) > 10000 mg/kg, (fmus iv) = 2800 mg/kg, (fmus orl) > 10000 mg/kg, (mrat iv) = 2175 mg/kg, (mrat orl) > 10000 mg/kg, (frat iv) = 2175 mg/kg, (frat orl) > 10000 mg/kg.

199 Cephalexin [anhydrous]
15686-71-2 2021 239-773-6

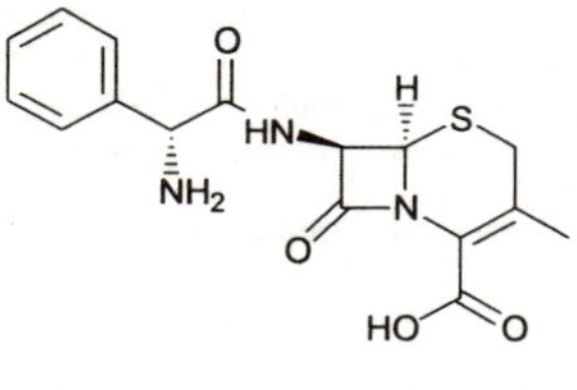

$C_{16}H_{17}N_3O_4S$
[6R-[6α,7β(R*)]]-7-[(Aminophenylacetyl)amino]-3-methyl-8-oxo-5-thia-1-azabicyclo[4.2.0]oct-2-ene-2-carboxylic acid.
Cefadros; Cefaloto; Cefanex; Cefaseptin; Ceporexine; Cex; Derantel; Efalexin; Farexin; Fergon 500; Garasin; Ibilex; Iwalexin; larixin; Lexibiotico; Llonexina; Madlexin; Mamalexin; Mecilex; Ohlexin; Oracocin; Rinesal; Sencephalin; Sintolexyn; Syncl; Taicelexin; Tokiolexin; Xahl; Alfaspoven [as sodium salt]. A semi-synthetic oral cephalosporin antibacterial agent. λ_m = 260 nm (ε 7750).

200 Cephalexin Hydrochloride
105879-42-3 2021
$C_{16}H_{20}ClN_3O_5S$
[6R-[6α,7β(R*)]]-7-[(Aminophenylacetyl)amino]-3-methyl-8-oxo-5-thia-1-azabicyclo[4.2.0]oct-2-ene-2-carboxylic acid hydrochloride monohydrate.
LY-061188; Keftab. A semi-synthetic oral cephalosporin antibacterial agent. LD_{50} (mus orl) = 1600-4500 mg/kg, (mus ip) = 400-1300 mg/kg, (rat orl) > 5000 mg/kg, (rat ip) > 3700 mg/kg.

201 Cephalexin Monohydrate
23325-78-2 2021 239-773-6
$C_{16}H_{17}N_3O_4S \cdot H_2O$
[6R-[6α,7β(R*)]]-7-[(Aminophenylacetyl)amino]-3-methyl-8-oxo-5-thia-1-azabicyclo[4.2.0]oct-2-ene-2-carboxylic acid monohydrate.
Cefa-Iskia; Cefibacter; Ceporex; Ceporexin; Keforal; Keflet; Keflex; Oracef; Ortisporina; Sartosona; Servispor. A semi-synthetic oral cephalosporin antibacterial agent. λ_m = 260 nm (ε = 7750); pKa = 5.2, 7.3; [monohydrate]: LD_{50} (rat orl) > 5000 mg/kg.

202 Cephalonium
5575-21-3 226-948-7

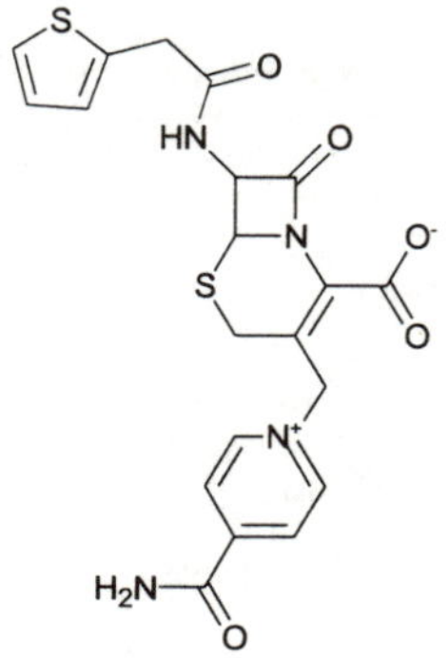

$C_{20}H_{18}N_4O_5S_2$
3-(4-Carbamoylpyridylmethyl)-8-oxo-7-(phenyl-acetamido)-5-thia-1-azabicyclo[4.2.0]oct-2-ene-2-carboxylic acid.
Cepravin D.C. Cephalosporin antibiotic; used to treat mastitis in cows.

Veterinary Products:

040-123; (with Flumethasone, Iodochlorhydroxyquin, Piperocaine Hydrochloride, Polymyxin B sulfate) Toptic Ointment (15 g); *Elanco.*

203 Cephalothin
153-61-7 2028 205-815-7

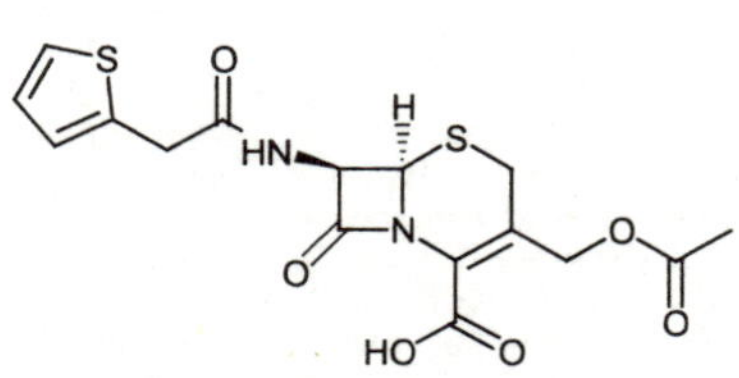

$C_{16}H_{16}N_2O_6S_2$
(6R-trans)-3-[(Acetyloxy)methyl]-8-oxo-7-[(2-thienyl-acetyl)amino]-5-thia-1-azabicyclo[4.2.0]oct-2-carboxylic acid.
7-(2-Thienylacetamido)cephalosporanic acid; 7-(thio-phene-2-acetamido)cephalosporanic acid. Short-acting, injectable antibiotic. mp = 160-160.5°; $[\alpha]_D^{20}$ = 50° (CH_3CN, c = 1.03).

Veterinary Products:

092-602; Cephalothin Discs; *Elanco.*

204 Cephapirin Benzathine

97468-37-6

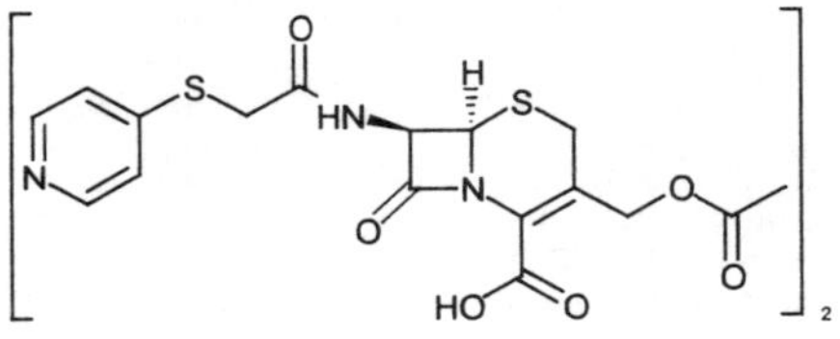

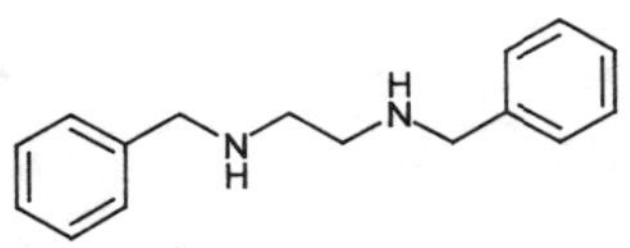

$C_{50}H_{54}N_8O_{12}S_4$

(6R-trans)-3-[(Acetyloxy)methyl]-8-oxo-7-[[(4-pyridinylthio)acetyl]amino]-5-thia-1-azabicyclo[4.2.0]oct-2-ene 2-carboxylic acid compound with N,N-bis(phenylmethyl))-1,2-ethanediamine (2:1).

Short-acting, injectable antibiotic. Used to treat mastitis in dairy cows.

Veterinary Products:

108-114; Cefa-Dry®; Tomorrow Infusion; *Fort Dodge Animal Health, Divn. AHP.*

205 Cephapirin Sodium

24356-60-3 2030 246-194-2

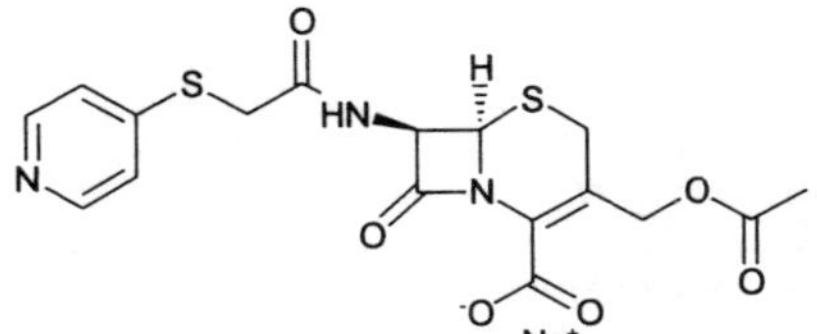

$C_{17}H_{16}N_3NaO_6S_2$

(6R-trans)-3-[(Acetyloxy)methyl]-8-oxo-7-[[(4-pyridinylthio)acetyl]amino]-5-thia-1-azabicyclo[4.2.0]oct-2-ene 2-carboxylic acid monosodium salt.

7-[α-(4-pyridylthio)-acetamido]cephalosporanic acid sodium salt; sodium 7-(pyrid-4-ylthioacetamido)-cephalosporanate; Cephapirin sodium; BL-P-1322; Ambrocef; Brisfirina; Bristocef; Cefadyl; Cef-Lak; Cefatrexyl; Piricef; ToDay. Short-acting, injectable antibiotic. Used to treat mastitis in dairy cows. Soluble in H_2O.

Veterinary Products:

097-222; Cefa-Lak®; Today Intramammary Infusion; *Fort Dodge Animal Health, Divn. AHP.*

206 Cloxacillin Benzathine

23736-58-5 2480 245-855-2

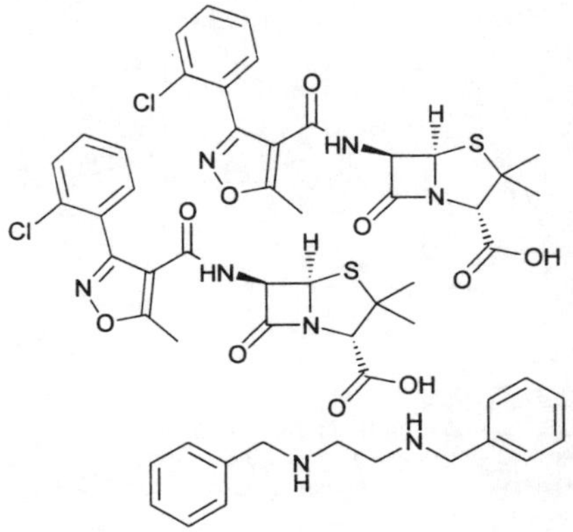

$C_{54}H_{56}Cl_2N_8O_{10}S_2$

[2S-(2α,5α,6β)]-6-[[[3-(2-Chlorophenyl)-5-methyl-4-isoxazolyl]carbonyl]amino]-3,3-dimethyl-7-oxo-4-thia-1-azabicyclo[3.2.0]heptane-2-carboxylic acid benzathine salt.

Boviclox; Dry-Clox; Noroclox DC; Opticlox; Orbenin Dry Cow; Triclox. Antibacterial used in treatment of bone, skin and tissue infections in small animals and mastitis in cows.

Veterinary Products:

055-058; Dry-Clox ®; Dry-Clox ® Intramammary Infusion; *Fort Dodge Animal Health, Divn. AHP.*
055-068; Boviclox; *Norbrook.*
055-069; Orbenin DC; *Pfizer.*

207 Cloxacillin Sodium Monohydrate

7081-44-9 2480

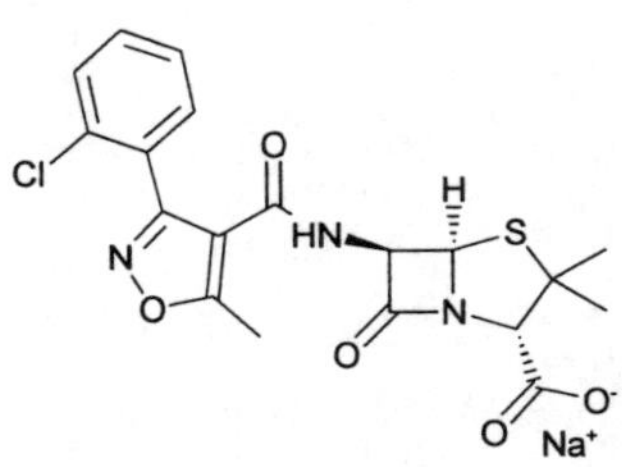

$C_{19}H_{17}ClN_3NaO_5S \cdot H_2O$

[2S-(2α,5α,6β)]-6-[[[3-(2-Chlorophenyl)-5-methyl-4-isoxazolyl]carbonyl]amino]-3,3-dimethyl-7-oxo-4-thia-1-azabicyclo[3.2.0]heptane-2-carboxylic acid sodium salt monohydrate.

sodium cloxacillin; BRL-1621; Bactopen; Cloxapen; Cloxypen; Ekvacillin; Gelstaph; Orbenin; Methoxillin-S; Prostaphlin-A; Staphobristol-250; Staphybiotic; Tegopen; Tepogen. Antibacterial used in treatment of bone, skin and tissue infections in small animals and mastitis in cows. mp = 170° (dec); $[\alpha]_D^{20}$ = 163°; soluble in H_2O, polar organic solvents; LD_{50} (rat ip) = 1630 ± 112 mg/kg, (mus ip) = 1280 ± 50 mg/kg.

208 Dicloxacillin Sodium Monohydrate
13412-64-1 3134

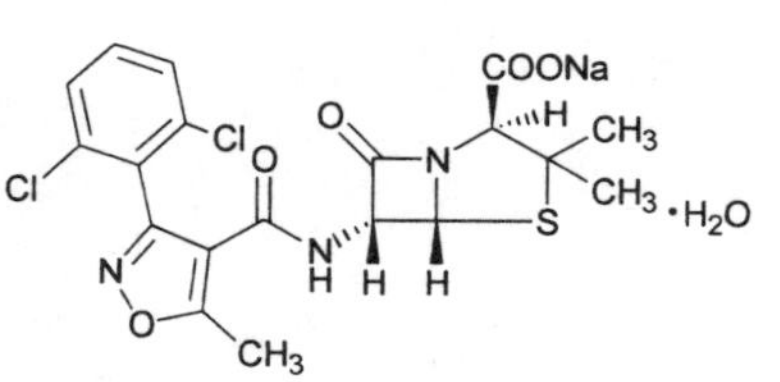

$C_{19}H_{16}Cl_2N_3NaO_5 \cdot H_2O$
[2S-(2α,5α,6β)]-6-[[[3-(2,6-Dichlorophenyl)-5-methyl-4-isoxazolyl]carbonyl]amino]-3,3-dimethtyl-7-oxo-4-thia-1-azabicyclo[3.2.0]heptane-2-carboxylic acid sodium salt monohydrate.
sodium dicloxacillin monohydrate; P-1011; Brispen; Constphyl; Dichlor-Stapenor; Diclocil; Dycill; Dynapen; Noxaben; Pathocil; Pen-Sint; Stampen; Syntarpen; Veracillin. Antibacterial used in treatment of bone, skin and tissue infections in small animals and mastitis in cows. Dec 222-225°; $[\alpha]_D^{20}$ = 127.2° (H_2O); soluble in H_2O, less soluble in BuOH, slightly soluble in Me_2CO and the usual organic solvents; LD_{50} (mus iv) = 900 mg/kg, (rat ip) = 630 mg/kg, (rat orl) > 5000 mg/kg.

Veterinary Products:

055-070; Dariclox®; *Pfizer.*

209 Hetacillin Potassium
5321-32-4 4706 226-182-3

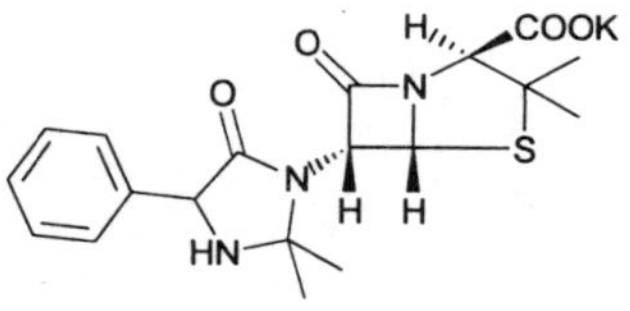

$C_{19}H_{22}KN_3O_4S$
[2S-[2α,5α,6β(S*)]]-6-(2,2-Dimethyl-5-oxo-4-phenyl-1-imidazolidinyl)-3,3-dimethyl-7-oxo-4-thia-1-azabicyclo-[3.2.0]heptane-2-carboxylic acid potassium salt.
Uropen; Versapen K; HetacinK; Natacillin. Antibacterial. Used orally in dogs and cats.

Veterinary Products:

055-021; Hetacin ® K Capsules Vet; *Fort Dodge Animal Health, Divn. AHP.*
055-022; Hetacin ® K Tablets; *Fort Dodge Animal Health, Divn. AHP.*
055-048; Hetacin ® K Oral Liquid; *Fort Dodge Animal Health, Divn. AHP.*
055-054; Hetacin ® K; Hetacin ® K Intramammary Infusion; *Fort Dodge Animal Health, Divn. AHP.*

210 Imipenem [anhydrous]
64221-86-9 4954 264-734-5

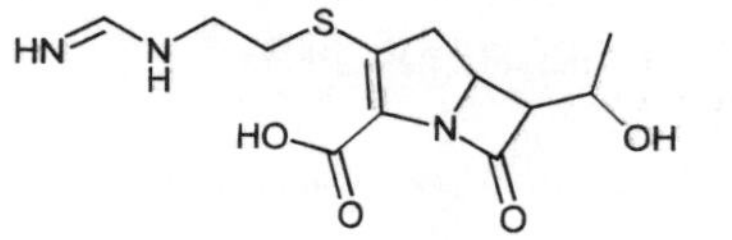

$C_{12}H_{17}N_3O_4S$
[5R-[5α,6α(R*)]]-6-(1-Hydroxyethyl)-3-[[2-[(imino-methyl)amino]ethyl]thio]-7-oxo-1-azabicyclo[3.2.0]hept-2-ene-2-carboxylic acid.
Broad spectrum semi-synthetic antibiotic used in combination with Cilastatin in horses and small animals.

211 Imipenem Monohydrate
74431-23-5 4954
$C_{12}H_{17}N_3O_4S.H_2O$
[5R-[5α,6α(R*)]]-6-(1-Hydroxyethyl)-3-[[2-[(imino-methyl)amino]ethyl]thio]-7-oxo-1-azabicyclo[3.2.0]hept-2-ene-2-carboxylic acid monohydrate.
imipemide; MK-787; MK-0787; component of [in combination with cilastatin sodium]: Primaxin, Imipem, Tenacid, Tienam, Tracix, Zienam. Broad spectrum semi-synthetic antibiotic used in combination with Cilastatin in horses and small animals. $[\alpha]_D^{25}$ = +86.8° (c = 0.05 in 0.1M phosphate, pH 7); λ_m 299 nm (ε 9670, 98% H_2O, NH_2OH ext); soluble in H_2O (1 g/100 ml), MeOH (0.5 g/100 ml), EtOH (0.02 g/100 ml), Me_2CO (< 0.01 g/100 ml), DMF (< 0.01 g/100 ml), DMSO (0.03 g/100 ml).

212 Oxacillin
66-79-5 7036 200-635-5

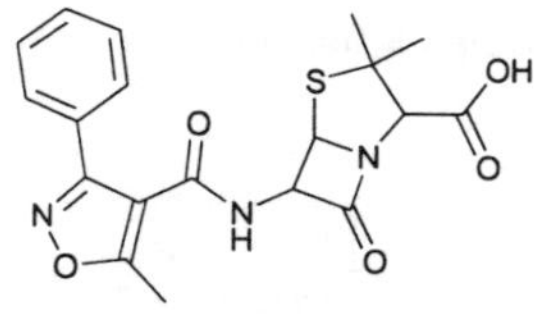

$C_{19}H_{19}N_3O_5S$
[2S-(2α,5α,6β)]-3,3-Dimethyl-6-[[(5-methyl-3-phenyl-4-isoxazolyl)carbonyl]amino]-7-oxo-4-thia-1-azabicyclo[3.2.0]heptane-2-carboxylic acid.
oxazocilline. Antibacterial. Used to treat bacterial infections in veterinary medicine.

213 Oxacillin Sodium
1173-88-2 7036 214-636-3

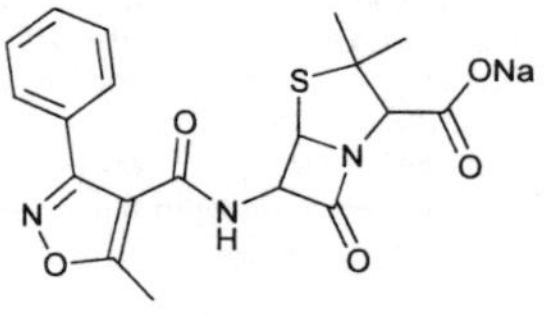

$C_{19}H_{18}N_3NaO_5S$
[2S-(2α,5α,6β)]-3,3-Dimethyl-6-[[(5-methyl-3-phenyl-4-isoxazolyl)carbonyl]amino]-7-oxo-4-thia-1-azabicyclo-[3.2.0]heptane-2-carboxylic acid sodium salt.

penicillin P-12; sodium oxacillin; BRL-1400; Bactocill; Bristopen; Cryptocillin; Micropenin; Oxabel; Penstapho; Penstaphocid; Prostaphlin; Resistopen; Stapenor; Bactocil sodium. Antibacterial. Used to treat bacterial infections in veterinary medicine. mp = 188° (dec) $[\alpha]_D^{20}$ = +201° (c = 1 in H_2O); LD_{50} (rat orl) > 8000 mg/kg.

214 Oxacillin Sodium Monohydrate

7240-38-2 7036

$C_{19}H_{18}N_3NaO_5S{\cdot}H_2O$

[2S-(2α,5α,6β)]-3,3-Dimethyl-6-[[(5-methyl-3-phenyl-4-isoxazolyl)carbonyl]amino]-7-oxo-4-thia-1-azabicyclo[3.2.0]heptane-2-carboxylic acid sodium salt monohydrate.

oxazocilline; Stapenor; Stapenor Retard; penicillin P-12; sodium oxacillin; BRL-1400; Bactocill; Bristopen; Cryptocillin; Micropenin; Oxabel; Penstapho; Penstaphocid; Prostaphlin; Resistopen. Antibacterial. Used to treat bacterial infections in veterinary medicine. mp = 188° (dec); $[\alpha]_D^{20}$ = 201° (c = 1 H_2O); LD_{50} (rat orl) > 8000 mg/kg.

215 Penicillin G Benzathine

1538-09-6 7221 216-260-5

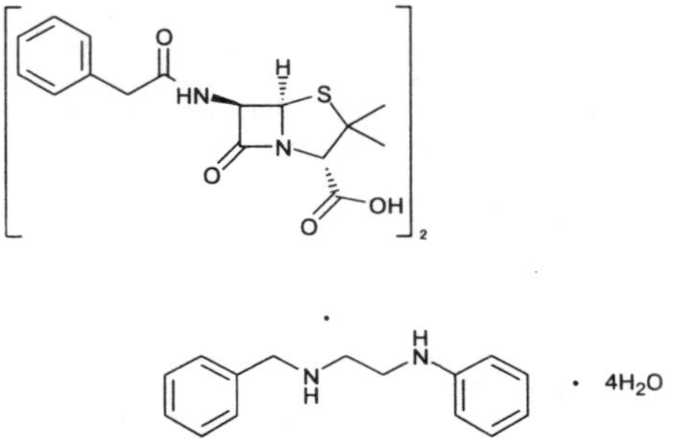

$C_{48}H_{56}N_6O_8S_2$

[2S-(2α,5α,6β)]-3,3-Dimethyl-7-oxo-6-[(phenylacetyl)-amino]-4-thia-1-azabicyclo[3.2.0]heptane-2-carboxylic acid compound with N,N'-dibenzylethylenediamine (1:1).

Penicillin G Benzathine; Benethamine penicillin; Benepen; Betapen; Benetolin; Potassium penicillin G; Potassium benzylpenicillinate; Benzylpenicillinic acid potassium salt; Notaral; benzathine penicillin G; diamine penicillin; Beacillin; Megacillin suspension; Bicillin; Cillenta; Permapen; Duropenin; Dibencillin; DBED Penicillin; Penidural; Tardocillin; Dibencil; Lentopenil; Vicin Neolin; Pen-Di-Ben; Penidure; Moldamin; Extencilline; Longacilina; Longicil; Penadur; Penditan; Cepacilina. Antibacterial. mp = 123-124°; slightly soluble in H_2O; $[\alpha]_D^{25}$ = +206° (c = 0.105 in formamide).

Veterinary Products:

055-008; (with Penicillin G Procaine) Bicillin Fortified; *Wyeth.*
065-087; (with Penicillin G Procaine) Longicil Fortified; *Fort Dodge Animal Health, Divn. AHP.*
065-169; (with Penicillin G Procaine) Dura-biotic; Flo-cillin ®; *Fort Dodge Animal Health, Divn. AHP.*
065-488; (with Penicillin G Procaine) Benza-Pen; *Walco.*
065-498; (with Penicillin G Procaine) Pen BP-48; *Pfizer.*
065-500; (with Penicillin G Procaine) Penicillin G Procaine Aqueous; Sterile Penicillin G Benzathine; *Hanford.*
065-506; (with Penicillin G Procaine) Combicillin ®-AG or Combicillin ®; *Anthony Products.*

216 Penicillin G Potassium

113-98-4 7225 204-038-0

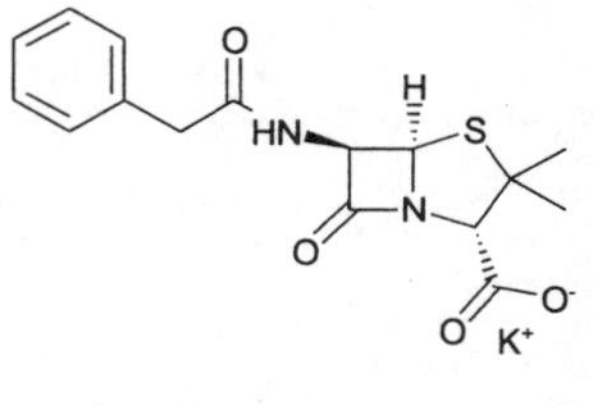

$C_{16}H_{17}KN_2O_4S$

[2S-(2α,5α,6β)]-3,3-Dimethyl-7-oxo-6-[(phenylacetyl)amino]-4-thia-1-azabicyclo[3.2.0]heptane-2-carboxylic acid monopotassium salt.

potassium penicillin G; potassium benzylpenicillinate; Notaral; Crytapen; Hipercilina; Pentid; Tabilin; Eskacillin; Forpen; Hylenta; Cosmopen; Falapen; Hyasorb; Cristapen; M-Cillin; Monopen; Megacillin Tablets. Antibacterial. mp = 214-217° (dec); $[\alpha]_D^{22}$ = 285-310° (c = 0.7); freely soluble in H_2O.

Veterinary Products:

055-060; Penicillin G Potassium, USP; *Fort Dodge Animal Health, Divn. AHP.*
200-103; Penicillin G Potassium, USP; *Merial.*
200-106; R-Pen; *Russell.*
200-122; Solu-Pen; *Alpharma.*

217 Penicillin G Procaine

54-35-3 7226 200-205-7

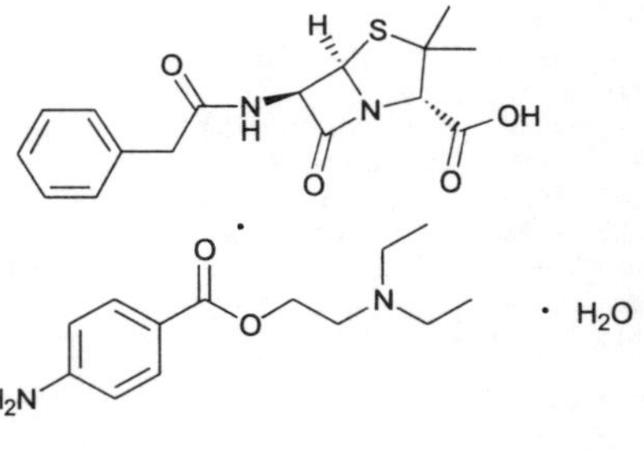

$C_{29}H_{38}N_4O_6S$

[2S-(2α,5α,6β)]-3,3-Dimethyl-7-oxo-6-[(phenylacetyl)-amino]-4-thia-1-azabicyclo[3.2.0]heptane-2-carboxylic acid compound with 2-(diethylamino)ethyl 4-aminobenzoate (1:1) .

benzylpenicillin procaine; Abbocillin-DC; Afsillin; Ampinpenicillin; Aquacillin; Aquasuspen; Avloprocil; Cilicaine; Crysticillin; Despacilina; Depocillin; Distaquaine; Dorsallin A.R.; Duracillin; Flo-cillin Aqueous; Hydracillin; Ilcocillin P; Kabipenin; Megapen; Mylipen; Ledercillin; Lenticillin; Mammacillin; Neoproc; Pentaquacaine G; Pen-50; Premocillin; Procanodia; Pro-Pen; Wycillin. Antibacterial. Monoclinic crystals; mp = 106-110° (dec); d = 1.255 - 1.256; dextrorotatory in aqueous solutions; soluble in H_2O (0.68 g/100 ml), MeOH (> 2 g/100 ml), iPrOH (0.65 g/100 ml), C_6H_6 (0.0075 g/100 ml), C_7H_8 (0.105 g/100 ml), petroleum ether (0.012 g/100 ml), CCl_4 (0.012 g/100 ml), EtOAc (0.335 g/100 ml), insoluble in isooctane; LD_{50} (mus sc) = 2300 mg/kg.

Veterinary Products:

035-688; (with Chlortetracycline Calcium Complex, Sulfamethazine) Aureo SP-250; Aureomix 500; *Roche Vitamins.*
039-077; (with Chlortetracycline Hydrochloride, Sulfathiazole) CSP ™ 250; CSP ™ 500; *Boehringer Ingelheim Vetmedica.*
046-666; Penicillin 100 Type A Medicated Article; Penicillin G Procaine 50 Type A Medicated Article; *Alpharma.*
046-668; Penicillin G Procaine 50% Type A Medicated Article; *Pfizer.*
049-462; (with Amprolium, Arsanilic Acid, Ethopabate, Streptomycin) Rainbrook Broiler Premix No.1; *Stockton Hay & Grain.*
055-008; (with Penicillin G Benzathine) Bicillin Fortified; *Wyeth.*
055-028; (with Dihydrostreptomycin Sulfate) Quartermaster ® Dry Cow Treatment; *West Agro.*
055-072; (with Novobiocin Sodium) Albacillin ® Suspension; Special Formula 17900-Forte ® Suspension; *Pharmacia & Upjohn.*
055-097; (with Dihydrostreptomycin Sulfate) Dry-Mast; *Boehringer Ingelheim Vetmedica.*
055-098; (with Novobiocin Sodium) Albadry Plus ® Suspension; *Pharmacia & Upjohn.*
065-010; Agricillin Pen Aqueous; Aqua-Cillin; Pen Aqueous; Penicillin G Co-op; *Norbrook.*
065-081; Go-Dry; Hanfords Penicillin; Masti-Clear; *Hanford.*
065-087; (with Penicillin G Benzathine) Longicil Fortified; *Fort Dodge Animal Health, Divn. AHP.*
065-110; Pro-Pen G In Aqueous Suspension; *Pfizer.*
065-119; (with Hydrocortione Acetate, Neomycin Sulfate, Polymyxin B Sulfate) Forte Topical ® Ointment; *Pharmacia & Upjohn.*
065-130; Crystalline Pro Penicillin G; *Fort Dodge Animal Health Divn., Am. Cyanamid.*
065-169; (with Penicillin G Benzathine) Dura-biotic; Flo-cillin ®; *Fort Dodge Animal Health, Divn. AHP.*
065-174; Crysticillin; *Fort Dodge Animal Health Divn., Am. Cyanamid.*
065-276; Veesyn Granules for Oral Soln; *Merial.*
065-383; Formula A-34; Uni Biotic 4 Dose; *Merial.*
065-465; Aqua-Mast; Hanfords Four-Pen; *Hanford.*
065-488; (with Penicillin G Benzathine) Benza-Pen; *Walco.*
065-493; Penicillin G Procaine Aqueous Suspension; *Hanford.*
065-498; (with Penicillin G Benzathine) Pen BP-48; *Pfizer.*
065-500; (with Penicillin G Benzathine) Penicillin G Procaine Aqueous; Sterile Penicillin G Benzathine; *Hanford.*
065-505; Microcillin-AG ®; *Anthony Products.*
065-506; (with Penicillin G Benzathine) Combicillin ®-AG or Combicillin ®; *Anthony Products.*
091-668; (with Chlortetracycline Calcium Complex, Sulfamethazine) ChlorMax™-SP 250; ChlorMax™-SP 500; ChlorMax™-SP 1000; Chlorachel 250 Swine; Pficlor 250; *Alpharma.*
200-140; (with Chlortetracycline Hydrochloride, Sulfathiazole) Aureozol ®; *Roche Vitamins.*
200-167; (with Chlortetracycline Calcium Complex, Sulfathiazole) Aureozol ® 500 Granular; *Roche Vitamins.*

218 Penicillin V Potassium

132-98-9 7230 205-086-5

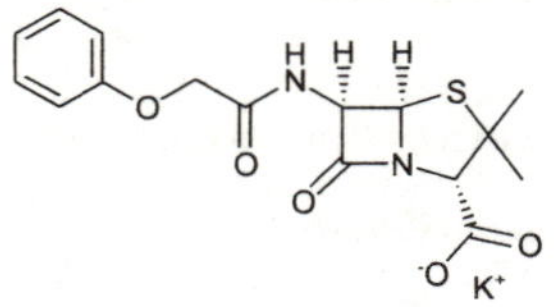

$C_{16}H_{17}KN_2O_5S$
[2S-(2α,5α,6β)]-3,3-Dimethyl-7-oxo-6-[(phenoxyacetyl)-amino]-4-thia-1-azabicyclo[3.2.0]heptane-2-carboxylic acid monopotassium salt.
D-α-phenoxymethylpencillinate K salt; phenoxymethylpenicillin Potassium; V-cillin-k; Compocillin-VK; Penvikal; Apsin VK; Arcacil; Beromycin 400; Cliacil; Distakaps V-k; Dowpen V-k; Fenoxypen; Oracil-VK; PVK; Suspen; Betapen V; phenoxymethylpenicillin potassium; V-Cillin-K; Compocillin-VK; Penvikal; Apsin VK; Arcacil; Beromycin 400; Cliacil; Distakaps V-K; Dowpen V-K; Fenoxypen; Oracil-VK; PVK; Suspen; Uticillin VK; Beromycin; Antibiocin; Arcasin; Betapen-VK; Calciopen K; Distaquaine V-K; DQV-K; Isocillin; Distaquaine V-K; Icipen; Ispenoral; Ledercillin VK; Megacillin Oral; Orapen; Ospeneff; Pedipen; Penagen; Pencompren; penicillin potassium phenoxymethyl; Pen-Vee-K; Pen-V-K; Qidpen VK; Robicillin VK; ocillin-VK; Roscopenin; Fenocin; Fenocin Forte; Penapar VK; Robicillin Vk; SK-Penicillin VK; Stabillin VK syrup 125; Stabillin VK syrup 62.5; SumapenVK; V-CIL-K; Veetids; Vepen; Pfizerpen VK; Uticillin VK; Beromycin; Antibiocin; Arcasin; Beromycin (Penicillin); Betapen-VK; Calciopen K; Distaquaine V-K; DqV-K; Isocillin; Icipen; Ispenoral; Ledercillin VK; Megacillin Oral; Orapen; Ospeneff; Pedipen; Penagen; Pencompren; Penicillin Potassium Phenoxymethyl; Pen-vee-k; Pen-V-K; Qidpen VK; Robicillin VK; Rocillin-VK; Roscopenin; Sk-penicillin VK; Stabillin VK Syrup 125; Stabillin VK Syrup 62.5; Sumapen VK; V-CIL-K; Veetids; Vepen; Pfizerpen VK. Antibacterial. mp = 120-128° (dec); λ_m = 268, 274 nm (ε 1330 1100); soluble in H_2O (0.025 g/100 ml), polar organics.

Veterinary Products:

065-275; Penicillin VK Filmtab 250 mg; *Merial.*

219 Ticarcillin Disodium

4697-14-7 9568 249-642-5

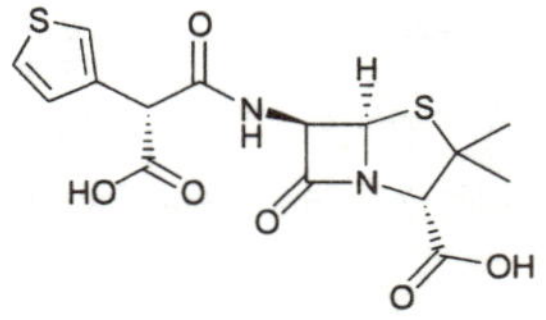

$C_{15}H_{14}N_2Na_2O_6S_2$
[2S-[2α,5α,6β(S*)]]-6-[(Carboxy-3-thienylacetyl)amino]-3,3-dimethyl-7-oxo-4-thia-1-azabicyclo[3.2.0]heptane-2-carboxylic acid disodium salt.

BRL-2288; Aerugipen; Monapen; Ticarpen; Ticillin. Antibacterial. Used to treat endometriosis in horses and to combat *Pseudomonas* infections in most animals. Soluble in H_2O (>100 g/100 ml).

Veterinary Products:

055-095; Ticillin®; *Pfizer*.

Antibiotics, Lincosamide

220 Clindamycin Hydrochloride Monohydrate

58207-19-5 2414

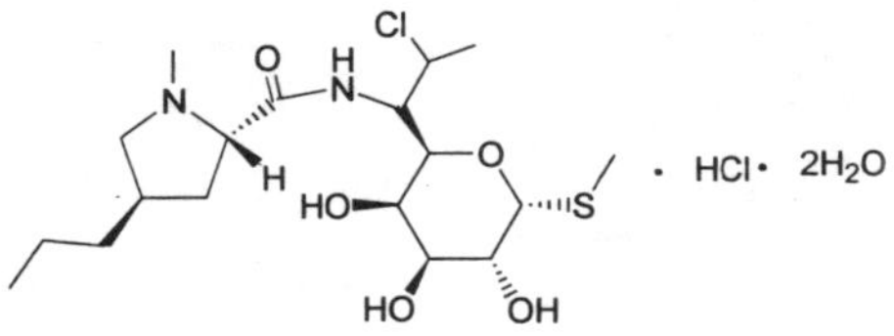

$C_{18}H_{34}Cl_2N_2O_5S.H_2O$
Methyl 7-chloro-6,7,8-trideoxy-6-(1-methyl-trans-4-propyl-L-2-pyrrolidinecarboxamido)-1-thio-L-threo-α-D-galactooctopyranoside hydrochloride monohydrate.
Cleocin HCl; Dalacine. Antibacterial. Used to treat wounds and abscesses in dogs and to counter most pathogenic anaerobic organisms. mp = 141-143°; $[\alpha]_D$ = 144° (H_2O); pKa = 7.6; soluble in H_2O, C_5H_5N, EtOH, DMF; LD_{50} (mus iv) = 245 mg/kg, (mus ip) = 361 mg/kg, (mus orl) = 2618 mg/kg.

Veterinary Products:

120-161; Antirobe ® Capsules; *Pharmacia & Upjohn*.
135-940; Antirobe ® Aquadrops Liquid; *Pharmacia & Upjohn*.
200-193; Clindamycin Hydrochloride Oral Liquid; *Phoenix*.

221 Lincomycin

154-21-2 5525 205-824-6

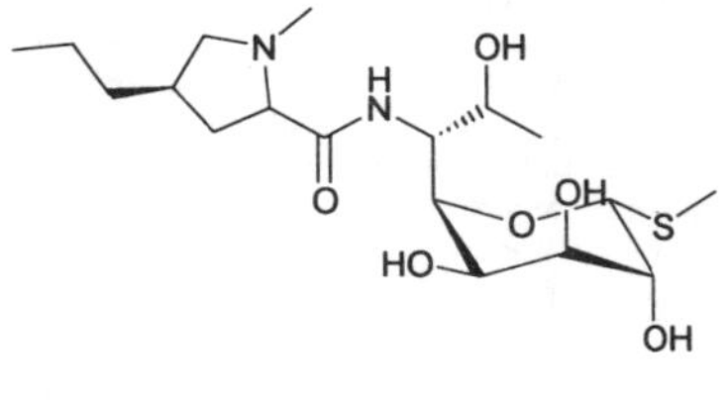

$C_{18}H_{34}N_2O_6S$
Methyl 6,8-dideoxy-6-trans-(1-methyl-4-propyl-L-2-pyrrolidinecarboxamido)-1-thio-D-erythro-α-D-galactooctopyranoside.
U-10149; NSC-70731; lincolnensin; Lincolcina. Antibacterial. Has been supplanted by Clindamycin. Slightly soluble in H_2O, soluble in MeOH, EtOH, EtOAc, Me_2CO, $CHCl_3$.

Veterinary Products:

047-262; (with Decoquinate)Deccox®/ Lincomycin; *Alpharma*.
137-537; (with Salinomycin Sodium) Bio-Cox®/ Lincomix ®; *Roche Vitamins*.
140-947; (with Narasin, Nicarbazin) Maxiban®/ Lincomix ®; *Elanco*.
140-954; (with Fenbendazole) Lincomix ® Type A Medicated Article / Safe-Guard® Type A Medicated Article; *Hoechst-Roussel Vet.*
141-054; (with Ivermectin) Ivomec ® Premix for Swine / Lincomix ® Premix; *Merial*.
200-090; (with Roxarsone, Salinomycin Sodium) Sacox ® + Lincomix ® + 3-Nitro ®; *Hoechst-Roussel Vet.*
200-093; (with Salinomycin Sodium) Sacox ® + Lincomix ®; *Hoechst-Roussel Vet.*
200-170; (with Nicarbazin, Roxarsone) Nicarmix 25 / 3-Nitro ® / Lincomix ®; *Planalquimica*.
200-171; (with Nicarbazin) Nicarmix 25® / Lincomix®; *Planalquimica*.

222 Lincomycin Hydrochloride Hemihydrate

859-18-7 5525 212-726-7

$C_{18}H_{35}ClN_2O_6S \cdot 1/2H_2O$
Methyl 6,8-dideoxy-6-trans-(1-methyl-4-propyl-L-2-pyrrolidinecarboxamido)-1-thio-D-erythro-α-D-galactooctopyranoside.
Frademicina; Lincocin; Mycivin; Waynecomycin. Antibacterial. Has been supplanted by Clindamycin. Crystals; mp = 145-147°; $[\alpha]_D^{25}$ = 137° (H_2O); freely soluble in H_2O, MeOH, EtOH, sparingly soluble in most organic solvents; LD_{50} (mus, rat orl) = 4000 mg/kg, (mus, rat ip) = 1000 mg/kg.

Veterinary Products:

092-482; (with Monensin Sodium) Coban ® / Lincomix ®; Lincomix ® / Coban ®; *Pharmacia & Upjohn*.
092-522; (with Monensin) Lincomix ® + Coban ® with or without Roxarsone; Lincomix ® / Coban ® / Roxarsone; *Pharmacia & Upjohn*.
093-106; (with Robenidine Hydrochloride) Lincomix ® + Robenz ®; *Pharmacia & Upjohn*.
101-689; (with Lasalocid Sodium) Lincomix ® / Avatec ®; *Pharmacia & Upjohn*.
107-997; (with Nicarbazin, Roxarsone) Medicated Broiler Feed; *Koffolk*.
108-116; (with Nicarbazin) Nicarbazin Lincomycin Premix; *Koffolk*.
112-661; (with Lasalocid Sodium, Roxarsone) Avatec ® / Lincomycin / 3-Nitro ®; *Roche Vitamins*.
132-923; Linco 8; Linco 20; *Intl. Nutrition*.
140-340; (with Halofuginone Hydrobromide) Stenorol ® + Lincomix ®; *Hoechst-Roussel Vet.*
140-581; (with Roxarsone, Salinomycin Sodium) Biocox ® / 3-Nitro ® / Lincomix ®; *Roche Vitamins*.
200-189; Lincomycin Soluble; *Russell*.
200-233; Linco Soluble; *Alpharma*.
200-241; Lincosol Soluble Powder; *Med-Pharmex*.

223 Lincomycin Hydrochloride Monohydrate

7179-49-9 5525

$C_{18}H_{35}ClN_2O_6S.H_2O$
Methyl 6,8-dideoxy-6-trans-(1-methyl-4-propyl-L-2-pyrrolidinecarboxamido)-1-thio-D-erythro-α-D-galactooctopyranoside hydrochloride monohydrate.
Albiotic; Cillimycin; Lincomix; Lincocin [as hydrochloride hemihydrate]. Antibacterial. Has been

supplanted by Clindamycin.

Veterinary Products:

033-887; Lincocin ® Tablets; *Pharmacia & Upjohn.*
034-025; Lincocin ® Sterile Solution (Rx); Lincomix ® Injectable 25; Lincomix ® Injectable 50; Lincomix ® Injectable 100; Lincomix ® Injectable 300; *Pharmacia & Upjohn.*
034-085; Lincomix ® Feed Medication; *Pharmacia & Upjohn.*
040-587; Lincocin ® Aquadrops; *Pharmacia & Upjohn.*
041-178; (with Amprolium, Ethopabate, Roxarsone) Lincomix ® / Amprol Plus / Roxarsone; *Pharmacia & Upjohn.*
044-820; (with Amprolium, Ethopabate) Lincomix ® & Amprol Plus; *Pharmacia & Upjohn.*
044-972; (with Clopidol) Lincomix ® / Coyden; *Pharmacia & Upjohn.*
045-738; (with Buquinolate) Lincomix ® / Bonaid; Lincomycin & Buquinolate; *Steris.*
046-109; (with Spectinomycin Sulfate Tetrahydrate) L-S 50 Water Soluble ® Powder.
047-261; (with Decoquinate) Lincomix ® / Deccox ®; *Pharmacia & Upjohn.*
048-954; (with Zoalene) Lincomix ® / Zoamix; *Pharmacia & Upjohn.*
097-505; Lincomix ® 10 Feed Medications; Lincomix ® 20 Feed Medications; Lincomix ® 50 Feed Medications; *Pharmacia & Upjohn.*
111-636; Lincomix ® Soluble Powder; *Pharmacia & Upjohn.*
116-044; (with Pyrantel Tartrate) Banminth ® / Lincomix ®; *Pfizer.*
132-574; Purina ® Check-R-Ton Li; *Purina Mills.*
138-941; (with Pyrantel Tartrate) Lincomix ® / Banminth ®; Lincomix ® 20/ Banminth ®; *Pharmacia & Upjohn.*

Antibiotics, Macrolide

224 Carbomycin A

4564-87-8 1854

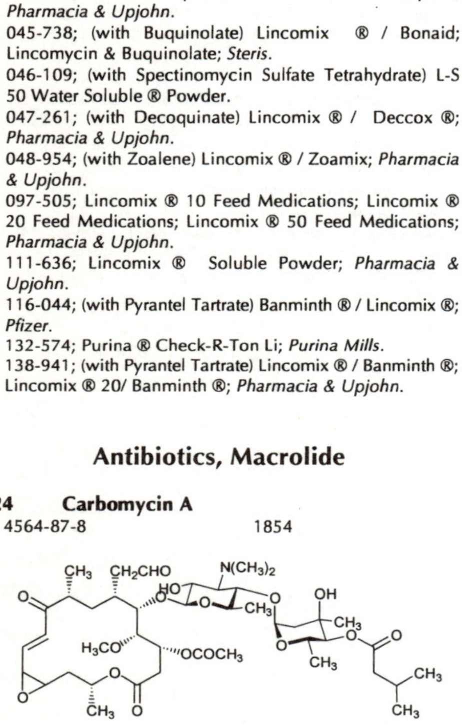

$C_{42}H_{67}NO_{16}$
(12S,13S)-9-Deoxy12,13-epoxy-12,13-dihydro-9-oxoleucomycin V 3-acetate 4^B-(3-methylbutanoate).
magnamycin A; deltamycin A_4; M-4209; NSC-51001. Macrolide antibiotic. mp = 214°; $[\alpha]_D^{25}$ = -58.6° ($CHCl_3$); λ_m 238, 327 nm ($E_{1cm}^{1\%}$ 185, 0.9, EtOH); pK_b = 7.2; soluble in H_2O (0.029 g/100 ml), MeOH (> 2 g/100 ml), EtOH (> 2 g/100 ml); LD_{50} (mus iv) = 550 mg/kg.

Veterinary Products:

032-946; (with Oxytetracycline Hydrochloride) Magna Terramycin Soluble Powder; *Pfizer.*

225 Carbomycin B

21238-30-2 1854

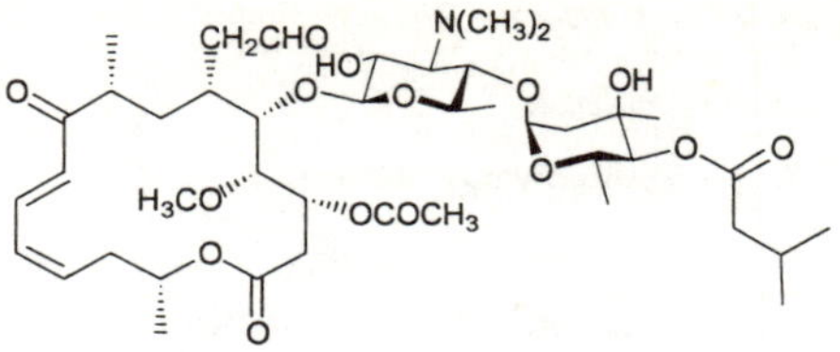

$C_{42}H_{67}NO_{15}$
9-Deoxy-9-oxo-leucomycinV 3-acetate 4^b-(3-methylbutanoate).
magnamycin B. Macrolide antibiotic. mp = 141-144° (dec); $[\alpha]_D^{25}$ = -35° (c = 1 $CHCl_3$); λ_m = 278 nm ($E_{1\,cm}^{1\%}$ 276, EtOH); soluble in EtOH (45 g/100 ml), H_2O (0.01g/100 ml).

226 Erythromycin

114-07-8 3720 204-040-1

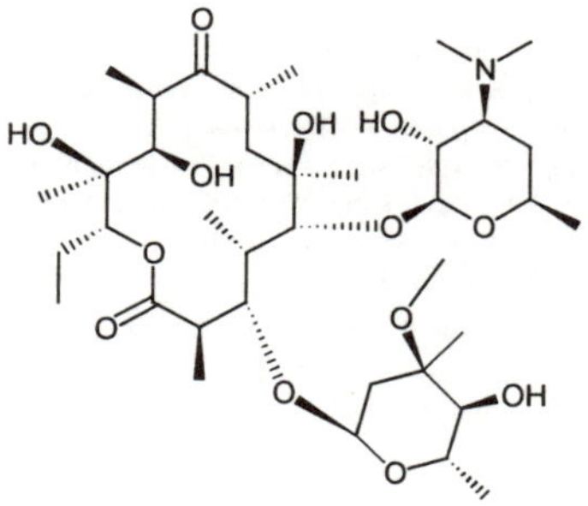

$C_{37}H_{67}NO_{13}$
(3S*,4S*,5S*,6R*,7R*,9R*,11R*,12R*,13S*,14R*)-4-[(2,6-Dideoxy-3-C-methyl-3-O-methyl-α-L-ribohexo-pyranosyl)oxy]-14-ethyl-7,12,13-trihydroxy-3,5,7,9,11,13-hexamethyl-6-[[3,4,6-trideoxy-3-(dimethylamino)-β-D-xylohexopyranosyl]oxy]-oxacyclotetradecane-2,10-dione.
A/T/S; Emgel; ERYC; Erycette; EryDerm; Erygel; Erymax; Ery-Tab; Erythro; Ilotycin; PCE; Erythromycin A; Abomacetin; Ak-Mycin; Aknin; E-Base; EMU; E-Mycin; Eritrocina; Erythromast 36; Erythromid; Erycen; Erycin; Erycinum; Ermycin; Inderm; Retcin; Staticin; Stiemycin; Torlamicina; component of: Benzamycin, Staticin, T-Stat. Macrolide antibiotic. Used in dogs, cats, pigs, sheep and cattle, particularly if the animal is thought to be penicillin-sensitive. mp = 135-140°, 190-193°; $[\alpha]_D^{25}$ = -78° (c = 1.99, EtOH); λ_m 280 nm (pH 6.3); poorly soluble in H_2O (0.020 g/100 ml); freely soluble in MeOH, EtOH, Me_2CO; $CHCl_3$, EtOAc, CH_3CN, moderately soluble in Et_2O, ethylene dichloride.

Veterinary Products:

012-123; Erythro ® - 100, 200; Gallimycin ® Injectable; *Merial.*

035-455; Erythro ®-36 Dry; Gallimycin ®-36 Dry; *Merial.*
035-456; Gallimycin ®-36 Sterile; *Merial.*
037-586; Erythromast 36; *Fort Dodge Animal Health, Divn. AHP.*
101-690; Erythro ®-100 Injection; *Merial.*

227 Erythromycin Phosphate

4501-00-2 3720 224-812-1
Macrolide antibiotic. Used in dogs, cats, pigs, sheep and cattle, particularly if the animal is thought to be penicillin-sensitive.

Veterinary Products:

035-157; Gallimycin® 100; Gallimycin® 500; *Merial.*

228 Erythromycin Thiocyanate

7704-67-8 3720 231-723-1
Macrolide antibiotic. Used in dogs, cats, pigs, sheep and cattle, particularly if the animal is thought to be penicillin-sensitive.

Veterinary Products:

010-092; Gallimycin ® 50; *Merial.*
038-241; (with Arsanilic Acid, Zoalene) Erythro ® (High Lev) / Zoalene + Arsanilic Acid; *Merial.*
038-242; (with Amprolium, Arsanilic Acid, Ethopabate) Erythro ® (Low Lev) / Amp + Etho; *Merial.*
038-624; (with Arsanilic Acid) Pro-Gallimycin-10; *Merial.*
041-955; Erythromycin Medicated Premix; *Merial.*

229 Oleandomycin

3922-90-5 6962 223-495-7

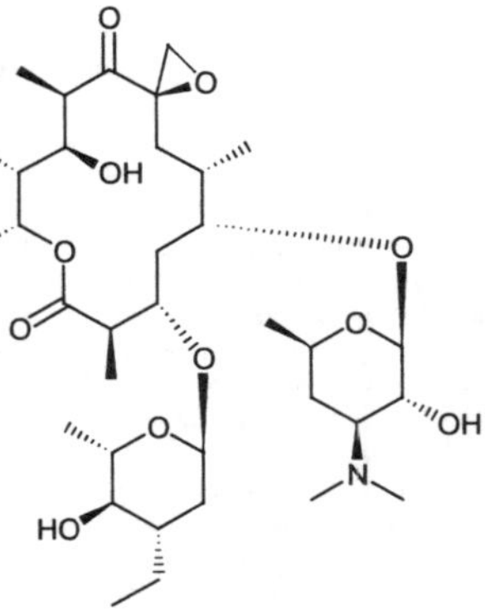

$C_{35}H_{61}NO_{12}$
Oleandomycin.
PA-105; Amimycin; Landomycin; Romicil. Macrolide antibiotic. Moderately soluble in H_2O, freely soluble in MeOH, EtOH, BuOH, Me_2CO, insoluble in C_6H_{14}, C_6H_6, CCl_4; λ_m = 286-289 nm (MeOH).

Veterinary Products:

035-287; OM-5 Premix; *Pfizer.*

230 Tylosin

1401-69-0 9963 215-754-8

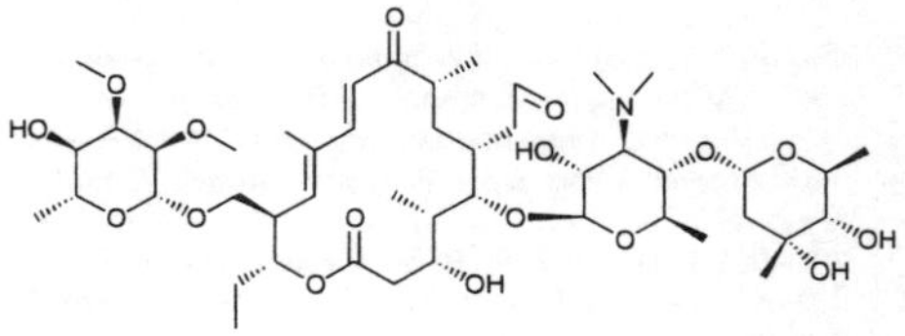

$C_{46}H_{77}NO_{17}$
Tylan.
Tylon. Macrolide antibiotic. Used to treat infections in pigs and cattle. Crystals; mp = 128-132°; $[\alpha]_D^{25}$ = -46° (c = 2, MeOH); λ_m = 282 nm ($E_{1\,cm}^{1\%}$ 245); soluble in H_2O (0.5 g/100 ml at 25°), soluble in lower alcohols, esters, ketones, chlornatdhydrocarbons, C_6H_6, Et_2O, solutions are stable between pH 4 and pH 9.

Veterinary Products:

012-965; Tylan ® Injection 50 mg; Tylan ® Injection 200 mg; *Elanco.*
013-388; (with Hygromycin B) Hygromix-Tylan ® Premix; *Elanco.*
030-330; (with Sulfadiazine, Sulfamerazine, Sulfamethazine) Tylocine Sulfa Tablets 50; *Elanco.*
031-962; (with Neomycin Sulfate) Tylan ®Plus Neomycin Eye Powder; *Elanco.*
131-956; (with Sulfamethazine) Tylan ® Sulfa-G; *ADM Animal Health & Nutrition.*
138-955; Tylosin Injection; *Boehringer Ingelheim Vetmedica.*

231 Tylosin Phosphate

1405-53-4 9963 215-779-4
Macrolide antibiotic. Used to treat infections in pigs and cattle.

Veterinary Products:

012-491; Tylan ® 100 CAL Type A Medicated Article; Tylan ® 100 CAL Type A Medicated Article; Tylan ® 40 CAL Type A Medicated Article; *Elanco.*
012-548; (with Hygromycin B) Tylosin plus Hygromix; *Elanco.*
013-162; Tylan ® Premix No.10; *Elanco.*
015-166; Tylan ® 100 Premix; *Elanco.*
041-275; (with Sulfamethazine) Tylan ® 10 Sulfa-G Premix; Tylan ® 40 Sulfa-G Premix; *Elanco.*
042-660; (with Sulfamethazine) Purina ® Pork-Plus Medicated; *Purina Mills.*
043-387; Purina ® Hog Plus II; *Purina Mills.*
046-415; Tylosin ® 5 Type A Medicated Article; Tylosin ® 10 Type A Medicated Article; Tylosin ® 20 Type A Medicated Article; Tylosin ® 40 Type A Medicated Article; CO-OP Tylan ® 10 Mix; *Alpharma.*
049-890; Norco T-2 Pre-Pak; *Norco Mills.*
091-582; Gilt Edge Tylan ® Mix; *ADM Animal Health & Nutrition.*
091-749; (with Sulfamethazine) Tylan ® 40 Plus Sulfa-G; CO-OP Tylosin 40 Plus Sulfamethazine; *Alpharma.*
093-518; Tylan ®-10 Plus; *Bioproducts.*
095-551; Tylan ®-4-Premix; *Intl. Nutrition.*
095-628; Tylosin Antibiotic Premix; *Heinold Feeds.*
095-953; Moormaboost TY 4000 Medicated; *Moorman.*

096-161; Hy-Con Tylan ® Premix; *Yoder Feed.*
096-162; Hog Grow-R-Mix-4000; Hog-Grow-R-Mix-800; *Young's.*
096-780; Tylan ® 40; Tylan ® 10; *J&R Specialty.*
096-837; M & M Tylosin Premix; *M&M Livestock.*
097-567; Tylan ® 10 Premix; *Golden Sun Feeds.*
097-615; (with Sulfamethazine) Swine Med-A-Mix TS 8000 Premix; Tylan ® 5, 10, 20, 40 Sulfa-G; *Golden Sun Feeds.*
097-980; Quali-Tech Tylan ®-10 Premix; *Quali-Tech.*
097-981; (with Sulfamethazine) Quali-Tech Tylan ®-Sulfa Premix 10 -10; Tylan ® 5, 10, 20, 40 Sulfa-G; *Quali-Tech.*
098-429; Medic-Meal-T Premix; *J. C. Feed Mills.*
098-431; Tylan ® 10 Premix; *Pfizer.*
098-639; (with Sulfamethazine) Tylan ® 10 Sulfa-G; Tylan ® 20 Sulfa-G; Tylan ® 40 Sulfa-G; Tylan ® 5 Sulfa-G; *Bioproducts.*
098-687; Hy-Test Hy-Boost TY 5 Medicated; *Kerber.*
099-468; Waynextra For Swine; Wayne Feed Divn.*.
099-767; (with Sulfamethazine) Purina ® Tylan ®-40 Plus Sulfamethazine; *Purina Mills.*
100-128; Supersweet Medipak Tylan ® 10; *Merial.*
100-352; Seeco Inc T-10 Premix; *Nutribasics.*
100-556; Vigorena Feeds Hy-Ty Premix; *Springfield Milling.*
100-991; McNess Custom Premix L200; *Furst-McNess.*
101-905; Mill Co-Medicator TY-10; Mill Co-Medicator TY-4; *Waterloo Mills.*
101-906; (with Sulfamethazine) Mill Co Medicator TS-40 Premix; *Waterloo Mills.*
103-089; Tylan ® 5; Tylan ® 10; Tylan ® 20; Tylan ® 40; *Akey.*
104-646; (with Monensin Sodium) Tylan ® / Rumensin ®; *Elanco.*
106-507; Tylan ® 5; Tylan ® 10; Tylan ® 20; Tylan ® 40; *Custom Feed Blenders.*
107-002; (with Sulfamethazine) Seeco Tylan ®-Sulfa 10 Premix Med.; *Seeco.*
107-957; (with Sulfamethazine) Tylan ® 20 Sulfa-G; Tylan ® 40 Sulfa-G; *ADM Animal Health & Nutrition.*
108-484; (with Sulfamethazine) HFA Tylosin-10 Plus Sulfa; *ADM Animal Health & Nutrition.*
109-816; (with Sulfamethazine) Tylan ® 5 Sulfa-G; Tylan ® 10 Sulfa-G; Tylan ® 20 Sulfa-G; Tylan ® 40 Sulfa-G; *Intl. Nutrition.*
110-044; Peavy Pro-Tone Plus Pak GF T-1; *Peavy.*
110-045; Good-Life Tylan ® 10 Premix; *ADM Animal Health & Nutrition.*
110-047; (with Pyrantel Tartrate) Banminth ® Tylan ®; *Pfizer.*
111-069; (with Sulfamethazine) Tylan ® 5 Sulfa-G; Tylan ® 10 Sulfa-G; Tylan ® 20 Sulfa-G; Tylan ® 40 Sulfa-G; *Moorman.*
111-637; Tylosin ® 10 Type A Medicated Article; Tylosin ® 20 Type A Medicated Article; Tylosin ® 40 Type A Medicated Article; Tylosin ® 5 Type A Medicated Article; *Alpharma.*
116-196; Webel Tylan ® Premix; *Webel Feeds.*
120-614; (with Sulfamethazine) Tylan ® 5 Sulfa-G; Tylan ® 10 Sulfa-G; Tylan ® 20; Tylan ® 40 Sulfa-G; *Webel Feeds.*
121-147; Nutra-Mix Tylan ® 5; Nutra-Mix Tylan ® 10; Nutra-Mix Tylan ® 20; Nutra-Mix Tylan ® 40; *Ag-Mark.*
121-200; Tylosin 10 Premix; *Custom Feed Services.*
122-522; (with Sulfamethazine) Tylan ® 5 Sulfa-G; Tylan ® 10 Sulfa-G; Tylan ® 20 Sulfa-G; Tylan ® 40 Sulfa-G; *Custom Feed Blenders.*
124-391; (with Sulfamethazine) Nutra-Mix Tylan ® - Sulfa Premixes; *Ag-Mark.*
127-195; Tylan ® 10; Tylan ® 40; *I.M.S.*
127-506; (with Sulfamethazine) Heinold Tylan ® 5 Sulfa Premix; Tylan ® 5 Sulfa-G; Tylan ® 10 Sulfa-G; Tylan ® 20 Sulfa-G; Tylan ® 40 Sulfa-G; *Heinold Feeds.*
127-507; (with Sulfamethazine) Tylan ® 5 Sulfa-G; Tylan ® 10 Sulfa-G; Tylan ® 20 Sulfa-G; Tylan ® 40 Sulfa-G; *Akey.*
128-411; (with Sulfamethazine) Tylan ® 5 Sulfa Premix; *ADM Animal Health & Nutrition.*
129-159; (with Sulfamethazine) Tylan ® 5 Sulfa-G; Tylan ® 10 Sulfa-G; Tylan ® 20 Sulfa-G; Tylan ® 40 Sulfa-G; *Custom Feed Services.*
129-161; (with Sulfamethazine) Nutra-Blend Tylan ® 5 Sulfa Premix; Tylan ® 10 Sulfa Premix; Tylan ® 20 Sulfa Premix; Tylan ® 40 Sulfa Premix; Tylan ® 5 Sulfa Premix; *Nutra-Blend.*
129-646; (with Sulfamethazine) Tylan ® 5 Sulfa-G; Tylan ® 10 Sulfa-G; Tylan ® 20 Sulfa-G; Tylan ® 40 Sulfa-G; Tylan ® 5 Sulfa Premix; *I.M.S.*
131-537; Baby Pig Premix; Gossett G-F Swine Premixes; *Gossett Nutrition.*
131-957; Tylan ® 5; Tylan ® 10; Tylan ® 20; Tylan ® 40; *ADM Animal Health & Nutrition.*
131-958; (with Sulfamethazine) Tylan ® Sulfa-G; *Wayne Feed Divn.*
133-833; Tylan ® 10 Premix; *Southern Micro-Blenders.*
138-187; Tylan ® 10 Premix; Tylan ® 100 Premix; Tylan ® 40 Premix; *Micro Chemical.*
138-342; (with Sulfamethazine) Tylan ® 5 Sulfa-G Premix; Tylan ® 10 Sulfa-G Premix; Tylan ® 20 Sulfa-G Premix; Tylan ® 40 Sulfa-G Premix; *Feed Service Co.*
138-792; (with Melengestrol Acetate, Monensin Sodium) MGA ® 100-200 / Rumensin / Tylan ®; *Pharmacia & Upjohn.*
138-870; (with Melengestrol Acetate, Monensin Sodium) MGA ® (liquid) / Rumensin / Tylan; MGA ® 100-200 Premix / MGA 500 Liquid Premix / Rumensin ® / Tylan ®; *Pharmacia & Upjohn.*
138-904; (with Lasalocid Sodium, Melengestrol Acetate) MGA ® 100-200 Premix / MGA ® 500 Liquid Premix / Bovatec ® / Tylan ®; MGA ® 100 / Bovatec / Tylan; *Pharmacia & Upjohn.*
138-992; (with Lasalocid Sodium, Melengestrol Acetate) MGA ® (liquid) / Bovatec / Tylan; *Pharmacia & Upjohn.*
138-995; (with Melengestrol Acetate) MGA ® 100 / Tylan; MGA ® 200 / Tylan; *Pharmacia & Upjohn.*
139-192; (with Melengestrol Acetate) MGA ® 500 (liquid) / Tylan ®; *Pharmacia & Upjohn.*
140-680; Tylan ® 5 Premix; Tylan ® 10 Premix; Tylan ® 20 Premix; Tylan ® 40 Premix; *Pennfield Oil.*
140-681; (with Sulfamethazine) Tylan ® 5 Sulfa-G Premix; Tylan ® 10 Sulfa-G Premix; Tylan ® 20 Sulfa-G Premix; Tylan ® 40 Sulfa-G Premix; *Pennfield Oil.*
140-820; (with Sulfamethazine) Tylan ® 5 Sulfa-G Premix; Tylan ® 10 Sulfa-G Premix; Tylan ® 20 Sulfa-G Premix; Tylan ® 40 Sulfa-G Premix; *Furst-McNess.*
140-939; (with Monensin Sodium) Liquid Type B Medicated Cattle Feed R/T 150; Liquid Type B Medicated Cattle Feed R/T 400; *Elanco.*

232 Tylosin Tartrate

1405-54-5 9963 215-781-5
Antibacterial. Used to treat infections in pigs and cattle.

Veterinary Products:

012-585; Tylan ® Injectable; *Elanco.*
013-076; Tylan ® Soluble; *Elanco.*
110-315; (with Estradiol Benzoate, Progesterone) Calf-Oid; Component ™ E-C with Tylan ®; Component ™ E-S;

Component ™ E-S with Tylan ®; Component™ E-C; Implus-C ®; Steer-Oid; *Ivy*.
135-906; (with Estradiol Benzoate, Testosterone Propionate) Component ™ E-H; Component ™ E-H with Tylan ®; Heifer-Oid; *Ivy*.
200-221; (with Estradiol, Trenbolone Acetate) Component ™ TE-S; Component ™ TE-S with Tylan ®; *Ivy*.
200-224; (with Trenbolone Acetate) Component ™ T-H with Tylan ®; Component ™ T-S with Tylan ®; *Ivy*.

Antibiotics, Nitrofuran

233 Nifurpirinol

13411-16-0 6628 236-503-9

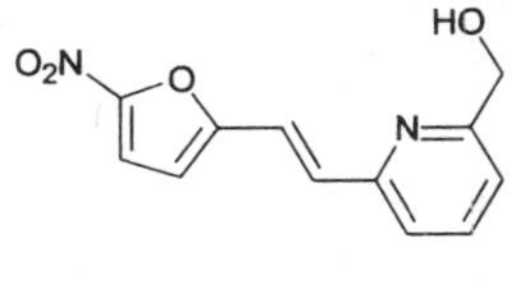

$C_{12}H_{10}N_2O_4$
6-[2-(5-Nitro-2-furyl)vinyl]-2-pyridinemethanol.
Furanace; P-7138; furpyrinol; furpirinol. Antibacterial. Used to treat bacterial diseases in fish. mp = 170-171°; LD_{50} (eel orl) = 1780 mg/kg.

Veterinary Products:

099-568; Furanace Caps; *Abbott*.

234 Nitrofurantoin

67-20-9 6696 200-646-5

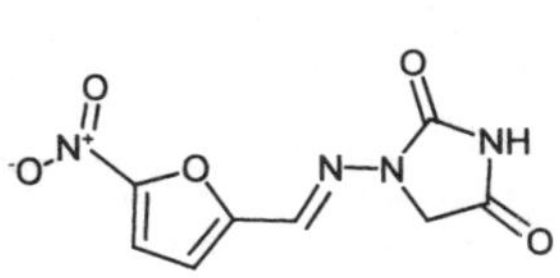

$C_8H_6N_4O_5$
1-[(5-Nitrofurfurylidene)amino]hydantoin.
Macrobid; Macrodantin; Parfuran; Berkfurin; Chemiofuran; Cyantin; Cystit; FuaMed; Furachel; Furalan; Furadantin; Furadantine MC; Furadoin; Furantoina; Furobactina; Furophen T-Caps; Ituran; Trantoin; Urizept; Urodin; Urolong; Uro-Tabvlinen; Welfurin. Antibacterial. Used to treat mastitis in cows. Dec 270-272°; λ_m = 370 nm ($E^{1\%}_{1\ cm}$ 776); soluble in H_2O (0.019 g/100 ml at pH 7), EtOH 0.051 g/1oo ml), Me_2CO (0.51o g/100 ml), DMF (8 g/100 ml), glycerol (0.06 g/100 ml), polyethylene glycol (1.5 g/100 ml).

235 Nitrofurazone

59-87-0 6697 200-443-1

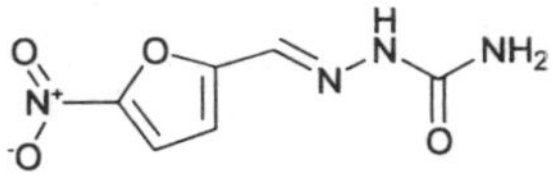

$C_6H_6N_4O_4$
5-Nitro-2-furaldehyde semicarbazone.
Amifur; Furacin; Chemofuran; Furesol; Nifuzon; Nitrofural; Nitrozone; Furacinetten; Furacoccid; Furazol W; Mammex; Furaplast; Coxistat; Aldomycin; Nefco; Vabrocid. Antibacterial. Used as a topical anti-infective. Pale yellow needles; dec 236-240°; λ_m = 260 375 nm; slightly soluble in H_2O (0.023 g/100 ml), EtOH (0.17 g/100 ml), proylene glycol (0.29 g/100 ml), insoluble in Et_2O; LD_{50} (rat orl) = 590 mg/kg, (rat sc) = 3000 mg/kg.

Veterinary Products:

011-154; NFZ ® Puffer; *Hess & Clark*.
100-854; Fura Ointment; *Farnam*.
121-723; Nitrofurazone Dressing; *Happy Jack*.
122-447; Fura-Septin Soluble Dressing; *Anthony Products*.
125-797; Nitrofurazone Dressing; *Boehringer Ingelheim Vetmedica*.
126-236; Nitrofurazone Soluble Powder; *Boehringer Ingelheim Vetmedica*.
126-504; Nitrofurazone Ointment; *Merial*.
130-872; (with Butacaine Sulfate) Nitrofurazone Anesthetic Dress.; *Med-Pharmex*.
132-427; Fura-Zone; *Squire*.
138-657; Nitrofurazone Ointment; *Veterinary Labs*.
140-851; NFZ ® Wound Dressing; *Hess & Clark*.
140-881; Nitrofurazone Soluble Dressing; *Med-Pharmex*.
140-910; NFZ ® Wound Powder; *Hess & Clark*.

Antibiotics, Other

236 Carbadox

6804-07-5 1825 229-879-0

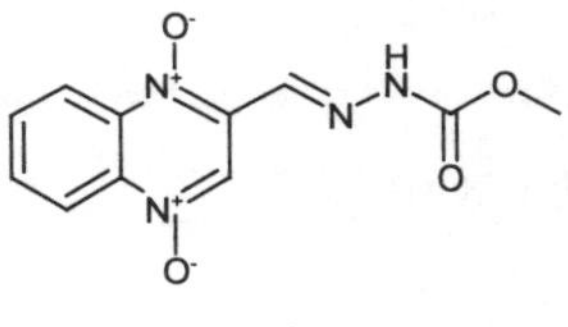

$C_{11}H_{10}N_4O_4$
Methyl 3-(2-quinoxalinylmethylene)carbazate N^1,N^4-dioxide.
Mecadox; GS-6244. Antibacterial. Yellow crystals; mp = 239.5-240°; insoluble in H_2O; λ_m = 236, 251, 303, 366, 373 nm (ε 11000, 10900, 36400, 16100, 16200 H_2O).

Veterinary Products:

041-061; Mecadox ® Premix 10; *Pfizer*.
092-955; (with Pyrantel Tartrate) Mecadox ® / Banminth ®; *Pfizer*.

237 Methenamine Mandelate

587-23-5 6040 209-597-4

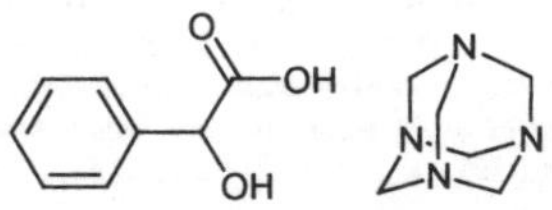

$C_{14}H_{20}N_4O_3$
Hexamethylenetetramine monomandelate.
Mandelamine; hexydraline; hexamethylenetetramine

phenylglycolate; hexamethylenamine mandelate; Mandacon; Purerin; Uro-Cedulamin; Mandamina; Mandurin; Reflux; Renelate; Mandoz; Cedulamin; Mantropine; Uromandelin; Uronamin; component of: Azo-Mandelamine. Used as a urinary antibacterial White crystals; mp = 128-130°; slightly soluble in Me_2CO, Et_2O, soluble in EtOH, $CHCl_3$, very soluble in H_2O; pH of aqueous solution = 4.

Veterinary Products:

030-416; Mesulfin Tablets; *Fort Dodge Animal Health, Divn. AHP.*

238 Mupirocin

12650-69-0 6383

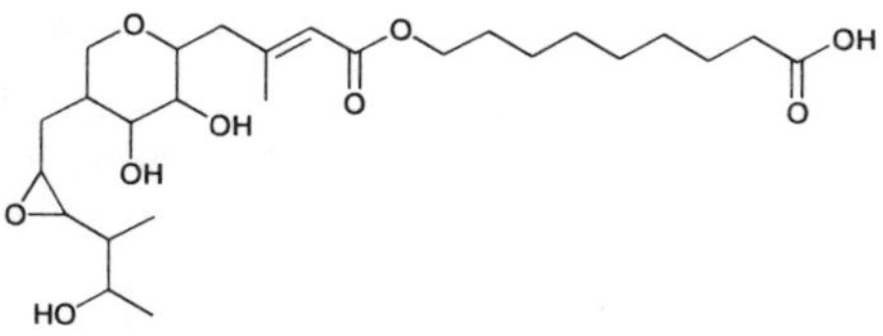

$C_{26}H_{44}O_9$

(E)-(2S,3R,4R,5S)-5-([(2S,3S,4S,5S)-2,3-Epoxy-5-hydroxy-4-methylhexyl]tetrahydro-3,4-dihydroxy-β-methyl-2H-pyran-2-crotonic acid ester with 9-hydroxynonanoic acid. Bactroban; BRL-4910A; Bactoderm; Turixin; pseudomonic acid A. Topical antibacterial. mp = 77-78°; $[\alpha]_D^{20}$ = -19.3° (c = 1 MeOH); λ_m 222 nm (ε 14500 EtOH).

Veterinary Products:

140-839; Bactoderm® Ointment; *Pfizer.*

239 Novobiocin

303-81-1 6818 206-146-3

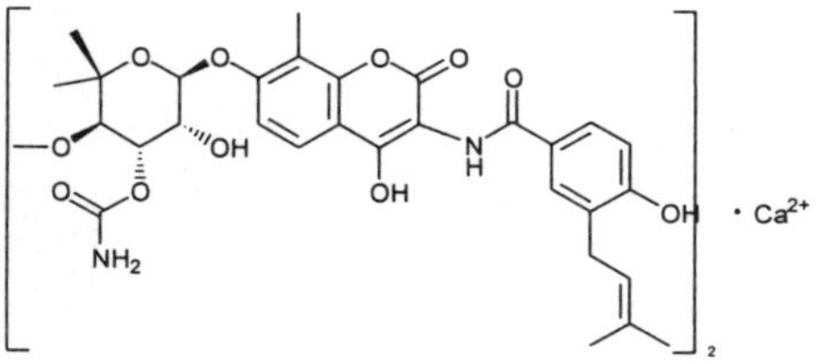

$C_{31}H_{36}N_2O_{11}$

N-[7-[[3-O-(Aminocarbonyl)-6-deoxy-5-C-methyl-4-O-methyl-α-L-lyxo-hexopyranosyl]oxy]-4-hydroxy-8-methyl-2-oxo-2H-1-benzopyran-3-yl]-4-hydroxy-3-(3-methyl-2-butenyl)benzamide.

crystalline acid; streptonivicin; PA-93; U-6591; Albamycin; Biotexin; Cardelmycin; Cathocin; Cathomycin; Inamycin; Spheromycin; Vulcamicina; Vulcamycin; Vulkamycin. Antimicrobial. Used to treat canine respiratory infections and prevent mastitis in dairy cattle. Pale yellow crystals; dec 152-156°; d = 1.3448; pKa_1 = 4.3, pKa_2 = 9.1; $[\alpha]_D^{24}$ = -63.0° (c = 1, EtOH); λ_m = 307 nm ($E^{1\%}_{1\ cm}$ 600, 0.1N NaOH), 324 nm ($E^{1\%}_{1\ cm}$ 390, 0.1N MeOH/HCl), 390 nm ($E^{1\%}_{1\ cm}$ 350, pH 7 phosphate buffer); soluble in H_2O above pH 7.5, soluble in Me_2CO, EtOAc, amyl acetate, lower alcohols, C_5H_5N.

Veterinary Products:

012-375; Albamix® Feed Medication; *Pharmacia & Upjohn.*

240 Novobiocin Sodium

1476-53-5 6818 216-023-6

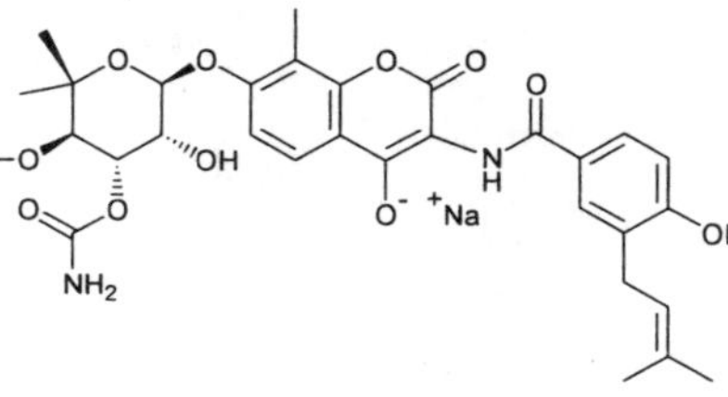

$C_{31}H_{35}NaN_2O_{11}$

N-[7-[[3-O-(Aminocarbonyl)-6-deoxy-5-C-methyl-4-O-methyl-α-L-lyxo-hexopyranosyl]oxy]-4-hydroxy-8-methyl-2-oxo-2H-1-benzopyran-3-yl]-4-hydroxy-3-(3-methyl-2-butenyl)benzamide sodium salt.

Robiocina. Antimicrobial. Used to treat canine respiratory infections and to prevent mastitis in dairy cattle. Minute crystals; dec 220°; $[\alpha]_D^{24}$ = -38° (c = 2.5 in 95% EtOH); soluble in H_2O; pH of aqueous solution = 7.5.

Veterinary Products:

013-467; Albamix ® Susceptibility Disks; *Pharmacia & Upjohn.*
055-072; (with Penicillin G Procaine) Albacillin ® Suspension; Special Formula 17900-Forte ® Suspension; *Pharmacia & Upjohn.*
055-076; (with Tetracycline Hydrochloride) Albaplex® 3x Tablets; Albaplex® Tablets; *Pharmacia & Upjohn.*
055-098; (with Penicillin G Procaine) Albadry Plus ® Suspension; *Pharmacia & Upjohn.*
065-090; (with Prednisolone, Tetracycline Hydrochloride) Delta Albaplex ® 3x Tablets; Delta Albaplex ® Tablets; *Pharmacia & Upjohn.*
065-099; (with Tetracycline Phosphate) Albaplex ® Capsules; *Pharmacia & Upjohn.*
100-808; Albamast ® Suspension; *Pharmacia & Upjohn.*
102-511; Biodry ® Suspension; Drygard ® Suspension; *Pharmacia & Upjohn.*

241 Pirlimycin Hydrochloride

77495-92-2 7655

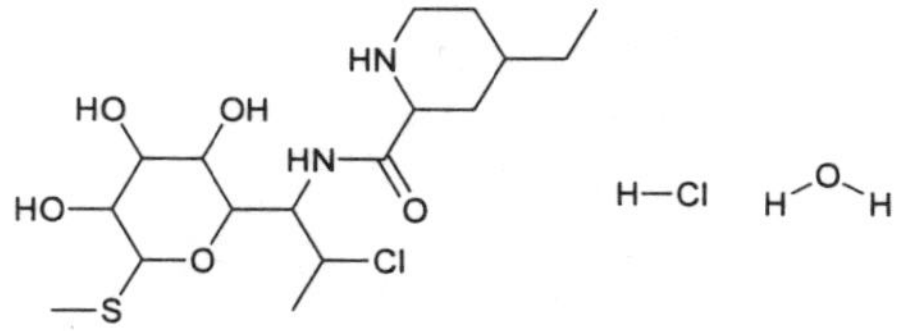

$C_{17}H_{32}Cl_2N_2O_5 \cdot H_2O$

Methyl 7-chloro-6,7,8-trideoxy-6-cis-4-ethyl-L-pipecolamido)-1-thio-L-threo-α-D-galactooctopyranoside monohydrochloride.

U-57930E; Pirsue. Antibiotic. Used to treat mastitis in

lactating dairy cattle. Crystals; mp = 210-212°, 222-224°; $[\alpha]_D^{25}$ = +176°, +181°; LD_{50} (mus ip) = 600 mg/kg.

Veterinary Products:

141-036; Pirsue® Aqueous Gel; *Pharmacia & Upjohn.*

242 Squalene

111-02-4 8924 203-826-1

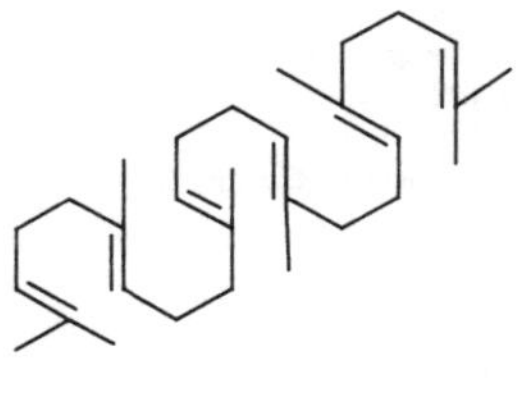

$C_{30}H_{50}$

(all-E)-2,6,10,15,19,23-Hexamethyl-2,6,10,14,18,22-Tetracosahexaene.

trans-squalene; 2,6,10,15,19,23-hexamethyl-2,6,10,14,18,22-tetracosahexaene; spinacene; all-trans-squalene; trans-spinacene; squalen; Supraene; Hexamethyltetracosahexaene. Bactericide. FDA approved for topicals. Used as a chemical intermediate, emollient, protectant (bactericide) and vehicle in topicals. Liquid; mp = -75°; bp_{25} = 285°; insoluble in H_2O, slightly soluble in EtOH, soluble in lipids and nonpolar organic solvents; d = 0.858 - 0.860. Nontoxic.

Veterinary Products:

032-984; (with Piperonyl Butoxide, Pyrethrins) Cerumite; *Evsco.*

055-005; (with Chloramphenicol, Prednisolone, Tetracaine) Liquichlor With Cerumene; *Evsco.*

243 Sulfomyxin

1405-52-3

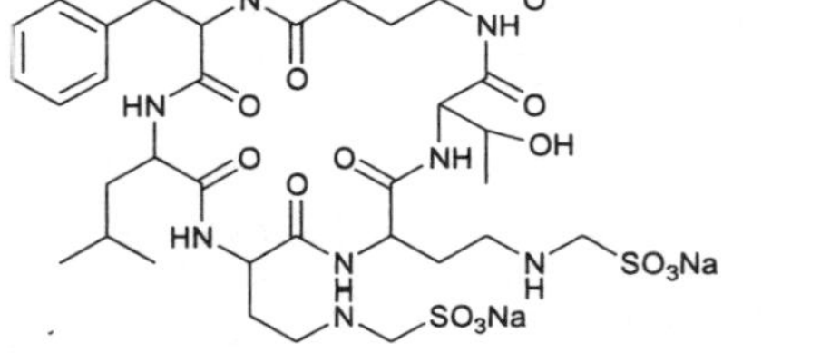

$C_{61}H_{103}N_{16}Na_5O_{28}S_5$

Pentakis(N-sulfomethyl) derivative of N-(3-amino-1-[[1-[[3-amino-1-[[6,9,18-tris-(2-aminoethyl)-15-benzyl-3-(1-hydroxyethyl)-12-isobutyl-2,5,8,11,14,17,20-heptaoxo-14,7,10,13,16,19-heptaazacyclotricos-21-yl]carbamoyl]propyl]carbamoyl]-2-hydroxypropyl]carbamoyl]-propyl]-6-methyloctanamide pentasodium salt.

Sulfomyxin sodium; Dynamyxin.

Veterinary Products:

031-944; Dynamyxin Injectable; *Pfizer.*

244 Tiamulin

55297-95-5 9559 259-580-0

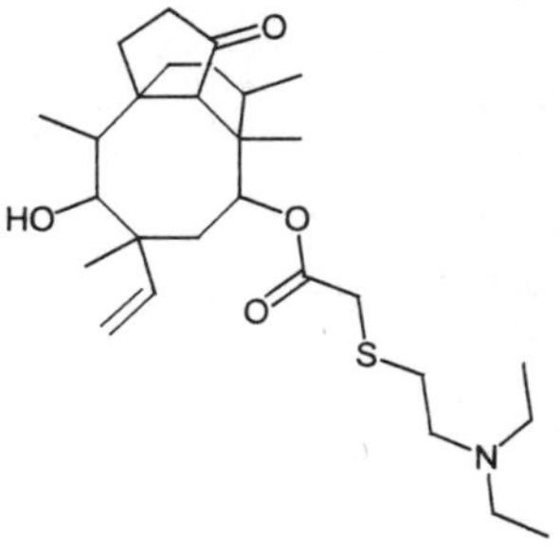

$C_{28}H_{47}NO_4S$

[3αS-(3aα,4β,5α,6α,8β,9α,9aβ,10S*)]-[(2-(Diethylamino)-ethyl)thio]acetic acid 6-ethenyldecahydro-5-hydroxy-4,6,9,10-tetramethyl-1-oxo-3a,9-propano-3aH-cyclopentacycloocten-8-yl ester.

thiamutilin; tiamutin; SQ-14055. Antibacterial. Used to treat pneumonia and dysentery in pigs.

Veterinary Products:

134-644; Denagard ® Soluble Antibiotic; *Boehringer Ingelheim Vetmedica.*

139-472; Denagard ® Premixes (Type A Medicated Articles); *Boehringer Ingelheim Vetmedica.*

245 Tiamulin Hydrogen Fumarate

55297-96-6 9559 259-581-6

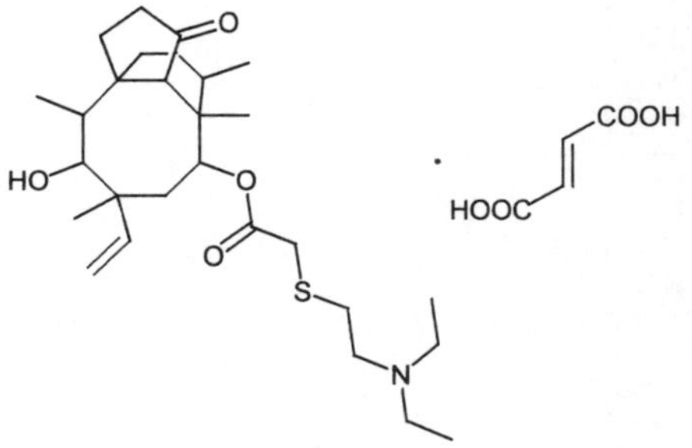

$C_{32}H_{51}NO_8S$

[3αS-(3aα,4β,5α,6α,8β,9α,9aβ,10S*)]-[(2-(Diethylamino)-ethyl)thio]acetic acid 6-ethenyldecahydro-5-hydroxy-4,6,9,10-tetramethyl-1-oxo-3a,9-propano-3aH-cyclopentacycloocten-8-yl ester fumarate ester.

81723 hfu; SQ-22947; Denagard; Dynamutilin; Tiamutin. Antibacterial. Used to treat pneumonia and dysentery in pigs. Crystals; mp = 147-148°.

Veterinary Products:

140-916; Denagard ® Liquid Concentrate; *Boehringer Ingelheim Vetmedica.*
141-011; (with Chlortetracycline Hydrochloride) Denagard ® 10 / Chlortetracycline Premixes; *Boehringer Ingelheim Vetmedica.*

246 Tilmicosin Phosphate

137330-13-3 9580

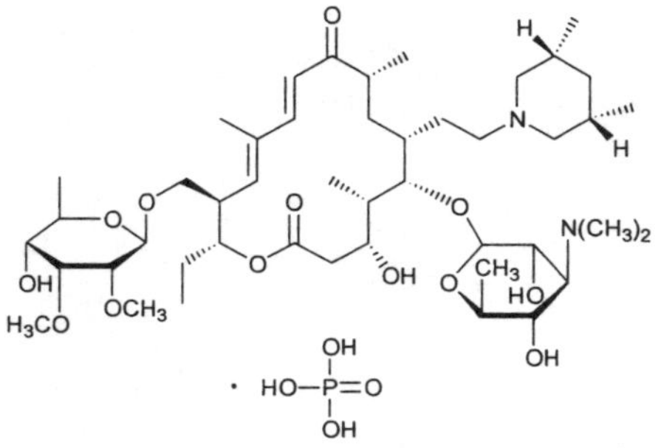

$C_{46}H_{83}N_2O_{17}P$
4^A-O-De(2,6-dideoxy-3-C-methyl-α-L-ribo-hexopyranosyl)-20-deoxo-20-(3,5-dimethyl-1-piperidinyl)tylosin phosphate.
LY177370 phosphate. Antibacterial. Used to treat bovine respiratory diseases.

Veterinary Products:

140-929; Micotil ® 300; *Elanco.*
141-064; Pulmotil ® 90; *Elanco.*

Antibiotics, Polypeptide

247 Amikacin Sulfate

39831-55-5 425 254-648-6

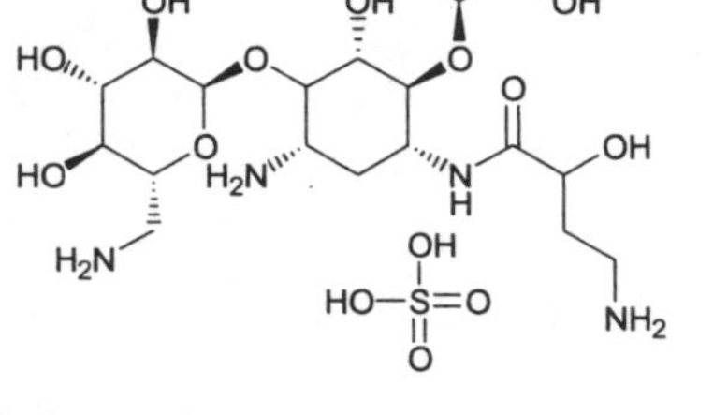

$C_{22}H_{43}N_5O_{13}$
(S)-O-3-Amino-3-deoxy-α-D-glucopyranosyl-(1→6)-O-[(6-amino-6-deoxy-α-D-glucopyranosyl-(1→4)]N^1(4-amino-2-hydroxy-1-oxobutyl)-2-deoxy-D-streptamine.
Lukadin; 1-N-[L-(4-amino-2-hydroxybutyryl]kanamycin A; Amikavet; Amikin; Amiklin; BB-K8; Biklin; Fabianol; Kaminax; Mikavir; Novamin; Pierami; 127-892 Amiglyde-V (Fort Dodge Animal Health), 200-178 Amikacin Sulfate Injection, 200-181 Amikacin Sulfate Solution (Phoenix Scientific). Peptide antibiotic used especially in dogs to deal with serious Gram-negative infections. Amorphous solid; dec 220-230°; $[\alpha]_D^{22}$= +74.75° (H_2O).

Veterinary Products:

127-892; Amiglyde-V; *Fort Dodge Animal Health Divn., Am. Cyanamid.*
200-178; Amikacin Sulfate Injection; *Phoenix Scientific.*
200-181; Amikacin Sulfate Solution; *Phoenix Scientific.*

248 Amphomycin Calcium

1405-31-8 625 215-775-2

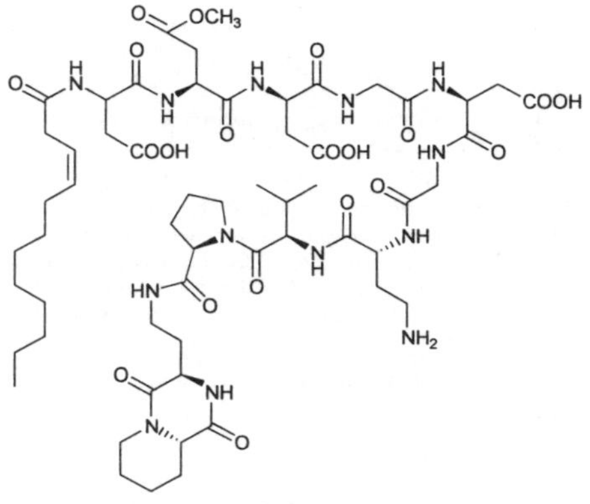

$C_{116}H_{180}CaN_{26}O_{40}$
Amphomycin [1402-82-0]; Amfomycin calcium salt; glumamycin calcium salt. Peptide antibiotic. Crystals, soluble in H_2O, MeOH; LD_{50} (mus iv) = 120.2 mg/kg.

Veterinary Products:

043-784; (with Hydrocortisone Acetate, Kanamycin Sulfate) Kanfosone Ointment; *Fort Dodge Animal Health, Divn. AHP.*
047-997; (with Hydrocortisone Acetate, Kanamycin Sulfate) Amphoderm Ointment; *Fort Dodge Animal Health, Divn. AHP.*

249 Bacitracin Methylene Disalicylic Acid

55852-84-1 965 259-862-3
Bacitracin methylenebis[2-hydroxybenzoate].
bacitracin methylenedisalicylate. Peptide antibiotic.

Veterinary Products:

036-304; (with Amprolium, Ethopabate) Amprol® HI-E Plus; *Merial.*
039-646; (with Carbarsone) Carb-O-Gain; *Alpharma.*
041-541; (with Clopidol, Roxarsone) Coyden® 25/BMD® 15; *Rhône-Poulenc.*
046-592; BMD® 10 Type A Medicated Article; BMD® 25 Type A Medicated Article; BMD® 30 Type A Medicated Article; BMD® 40 Type A Medicated Article; BMD® 45 Type A Medicated Article; BMD® 50 Type A Medicated Article; BMD® 60 Type A Medicated Article; BMD® 75 Type A Medicated Article; *Alpharma.*
049-180; (with Amprolium, Ethopabate, Roxarsone)

Amprol® HI-E/BMD®/Roxarsone; *Merial.*
049-463; (with Monensin Sodium) Monensin & Bacitracin MD; *Elanco.*
049-464; (with Monensin, Roxarsone) Monensin & Bacitracin & Roxarsone; *Elanco.*
065-107; (with Streptomycin Sulfate) Entromycin Powder; *Veterinary Specialties.*
065-280; Soluble Fortracin Concentrate; *Alpharma.*
065-470; BMD® Soluble; BMD® Soluble 50%; Solu-Tracin 200; Solu-tracin 50; *Alpharma.*
097-085; (with Robenidine Hydrochloride) Robenz® Plus Bac MD; *Roche Vitamins.*
098-378; (with Nicarbazin) Nicarbazin/Bacitracin Premix; *Koffolk.*
099-150; (with Clopidol) Coyden® 25 + Fortracin; *Rhône-Poulenc.*
107-996; (with Lasalocid Sodium) Avatec®/Fortracin Premix; *Roche Vitamins.*
116-082; (with Lasalocid Sodium, Roxarsone) BMD®/Avatec®/3-Nitro®; *Alpharma.*
116-088; (with Monensin Sodium, Roxarsone) BMD®/Coban®/3-Nitro®; Fortracin/Coban®/3-Nitro®; *Alpharma.*
131-894; (with Lasalocid Sodium, Roxarsone) Avatec®/Fortracin/3-Nitro® Broiler Premix; *Roche Vitamins.*
135-321; (with Roxarsone, Salinomycin Sodium) Biocox®/3-Nitro®/BMD®; *Roche Vitamins.*
135-746; (with Salinomycin Sodium) Biocox®/BMD®; *Roche Vitamins.*
138-456; (with Monensin Sodium) Coban®/BMD®; *Alpharma.*
140-533; (with Halofunginone Hydrobromide, Roxarsone) Stenorol®/3-Nitro®/BMD®; *Hoechst-Roussel Vet.*
140-584; (with Halofuginone Hydrobromide) Stenorol® + BMD®; *Hoechst-Roussel Vet.*
140-852; (with Narasin, Roxarsone) Monteban®/3-Nitro®/BMD®; *Alpharma.*
140-853; (with Narasin) Monteban®/BMD®; *Alpharma.*
140-919; (with Halofuginone Hydrobromide) Stenorol® + BMD®; *Hoechst-Roussel Vet.*
140-926; (with Narasin, Nicarbazin) BMD®/Maxiban®; *Elanco.*
140-937; (with Monensin Sodium) Coban® and BMD®; *Elanco.*
141-058; (with Roxarsone, Semduramicin Sodium) Aviax™/BMD®/3-Nitro®; *Pfizer.*
141-059; (with Chlortetracycline) BMD®-10,25,30,40,50,60,75 and ChlorMax™-50,65,70/Micro-CTC®100; *Alpharma.*
141-065; (with Semduramicin Sodium) Aviax™/BMD®; *Pfizer.*
141-085; (with Zoalene) Zoamix®/BMD®; *Alpharma.*
141-088; (with Nitarsone) Histostat®/BMD®; *Alpharma.*
141-097; (with Ivermectin) Ivomec® Premix + BMD®; *Merial.*
141-100; (with Decoquinate, Roxarsone) Deccox®/BMD®/3-Nitro®; *Alpharma.*
141-102; (with Decoquinate) Deccox®/BMD®; *Alpharma.*
200-081; (with Roxarsone, Salinomycin Sodium) Sacox® + 3-Nitro® + BMD®; *Hoechst-Roussel Vet.*
200-082; (with Salinomycin Sodium) Sacox® + BMD®; *Hoechst-Roussel Vet.*
200-164; (with Nicarbazin) Nicarmix 25®/BMD ®; *Planalquimica.*
200-207; (with Clopidol, Roxarsone) Coyden 25®/Albac®/3-Nitro®; *Alpharma.*
200-242; (with Chlortetracycline Calcium Complex) Aureomycin®-50,70,80,90,100/BMD®-25,30,40,50,60,70,75; *Roche Vitamins.*

250 Bacitracin Zinc

1405-89-6 10257 215-787-8

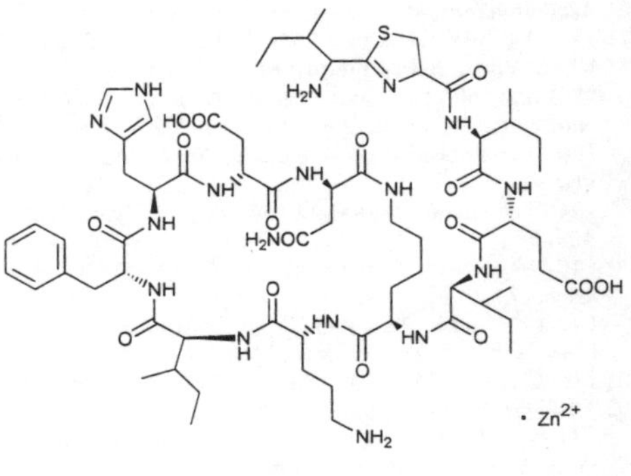

Bacitracin Zinc Complex.

Bacitracin zinc salt; Baciferm. Peptide antibiotic. Soluble in H_2O (0.23 - 0.45 g/100 ml at 28°), MeOH (0.65 g/100 ml), EtOH (0.20 g/100 ml), iPrOH (0.016 g/100 ml), EtOAc (0.13 g/100 ml), $CHCl_3$ (0.001 g/100 ml), petroleum ether (0.0025 g/100 ml).

Veterinary Products:

039-284; (with Amprolium, Ethopabate, Roxarsone) Swisher Super Broiler 300-108; Swisher Super Broiler 400-112; *Swisher.*
044-016; (with Clopidol. Roxarsone) Coyden 25® Medicated Premix; *Rhône-Poulenc.*
045-348; (with Decoquinate) Deccox®/Albac®; Broiler Finisher Medicated; *Alpharma.*
046-920; Baciferm®-10; Baciferm®-25; Baciferm®-40; Baciferm®-50; *Roche Vitamins.*
047-933; (with Monensin Sodium) Monensin & Zinc Bacitracin Premix; *Elanco.*
049-934; (with Clopidol) Coyden 25®; *Rhône-Poulenc.*
065-015; (with Hydrocortisone Acetate, Neomycin Sulfate, Polymyxin B Sulfate) Bacitracin-Neomycin-Polymyxin with Hydrocortison Acetate Ophthalmic Ointment; Vetropolycin HC Ophthalmic Ointment; *Altana.*
065-016; (with Neomycin Sulfate, Polymyxin B Sulfate) Bac-Neo-Poly Ophthalmic Ointment; Vetropolycin Ophthalmic Ointment; *Altana.*
065-114; (with Neomycin Sulfate, Polymyxin B Sulfate) Mycitracin® Sterile Ophthalmic Ointment; *Pharmacia & Upjohn.*
065-313; Baciferm® Soluble 50; *American Cyanamid Divn., AHP Corp.*
065-476; (with Hydrocortisone Acetate, Neomycin Sulfate, Polymyxin B Sulfate) Cortisporin Veterinary Ophthalmic Ointment; *Schering-Plough Animal Health.*
065-485; (with Neomycin Sulfate, Polymyxin B Sulfate) Neosporin Ophthalmic Ointment; *Schering-Plough Animal Health.*
091-326; (with Decoquinate, Roxarsone) Deccox®/3-Nitro®/Albac®; *Alpharma.*
091-646; (with Amprolium, Ethopabate) Rainbow Broiler Base Concentrate; *Stockton Hay & Grain.*
096-933; (with Robenidine Hydrochloride) Robenz® Plus

Zn Bacitracin; *Roche Vitamins.*
098-452; Albac® 50 Type A Medicated Article; *Alpharma.*
105-758; (with Amprolium, Ethopabate, Roxarsone) Zinc Bacitracin & Amprol HI-E®; *Roche Vitamins.*
114-794; (with Amprolium, Ethopabate) Baciferm®/Amprol HI-E® Premix; *Roche Vitamins.*
123-154; (with Monensin Sodium, Roxarsone) Coban®/3-Nitro®10/Baciferm® Premix; *Roche Vitamins.*
126-052; (with Lasalocid Sodium, Roxarsone) Avatec®/3-Nitro®/Baciferm®; *Roche Vitamins.*
128-550; Anchor Zinc Bacitracin; *Boehringer Ingelheim Vetmedica.*
134-830; (with Monensin Sodium) Coban®/Albac®; *Alpharma.*
136-484; (with Carbarsone) Carb-O-Sep®/Baciferm®; *Roche Vitamins.*
137-536; (with Roxarsone, Salinomycin Sodium) Bio-Cox®/3-Nitro® Plus Albac®; *Roche Vitamins.*
138-703; (with Monensin Sodium, Roxarsone) Albac®/Coban®/3-Nitro®; *Alpharma.*
139-190; (with Roxarsone, Salinomycin Sodium) Bio-Cox®/3-Nitro®/Baciferm®; *Roche Vitamins.*
139-235; (with Salinomycin Sodium) Biocox®/Baciferm®; *Roche Vitamins.*
141-109; (with Lasalocid Sodium) Avatec®/Baciferm®; *Roche Vitamins.*
200-086; (with Roxarsone, Salinomycin Sodium) Sacox® + Albac® + 3-Nitro®; *Hoechst-Roussel Vet.*
200-089; (with Salinomycin Sodium) Sacox® + Baciferm®; *Hoechst-Roussel Vet.*
200-143; (with Roxarsone, Salinomycin Sodium) Sacox® + 3-Nitro® + Baciferm®; *Hoechst-Roussel Vet.*
200-203; (with Carbarsone) Carb-O-Sep®/Albac®; *Alpharma.*
200-204; (with Salinomycin Sodium) Bio-Cox®/Albac®; *Alpharma.*
200-205; (with Amprolium, Ethopabate) Amprol HI-E®/Albac®; *Alpharma.*
200-206; (with Decoquinate, Roxarsone) Deccox®/Albac®/3-Nitro®; *Alpharma.*
200-209; (with Roxarsone, Salinomycin Sodium) Sacox®/Albac®/3-Nitro®; *Alpharma.*
200-210; (with Salinomycin Sodium) Sacox®/Albac®; *Alpharma.*
200-211; (with Monensin, Roxarsone) Coban®/Albac®/3-Nitro®; *Alpharma.*
200-212; (with Robenidine Hydrochloride) Robenz®/Albac®; *Alpharma.*
200-213; (with Decoquinate) Deccox®/Albac®; *Alpharma.*
200-214; (with Amprolium, Ethopabate, Roxaresone) Amprol HI-E®/Albac®/3-Nitro®; *Alpharma.*
200-215; (with Roxarsone, Salinomycin Sodium) Bio-Cox®/Albac®/3-Nitro®; *Alpharma.*
200-217; (with Amprolium, Ethopabate, Roxarsone) Amprol HI-E®/Albac®/3-Nitro®; *Alpharma.*
200-218; (with Clopidol) Coyden® 25/Albac®; *Alpharma.*
200-223; Albac® 50 Type A Medicated Article; *Alpharma.*

251 Colistimethate Sodium

8068-28-8 2542 232-516-9

$C_{58}H_{105}N_{16}Na_5O_{28}S_5$

Colistin sodium methanesulfonate.

Alficetin; Methacolimycin. Polypeptide antibiotic. Solid; soluble in H_2O; LD_{50} (mus iv) > 550 mg/kg.

Veterinary Products:

141-069; First Guard® Sterile Powder; *Alpharma.*

252 Polymyxin B Sulfate

1405-20-5 7734 215-774-7

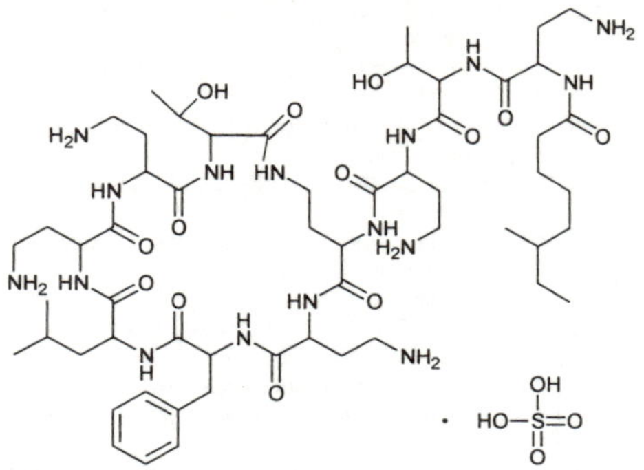

Aerosporin; Mastimyxin. Peptide antibiotic.

Veterinary Products:

008-763; (with Oxytetracycline Hydrochloride) Terramycin ® Ophthalmic Ointment With Polymyxin; *Pfizer.*
038-801; (with Flumethasone, Neomycin Sulfate) Anaprime ® Ophthalmic Solution B Sulfate; *Fort Dodge Animal Health, Divn. AHP.*
040-123; (with Cephalonium, Flumethasone, Iodochlorhydroxyquin, Piperocaine Hydrochloride) Toptic Ointment (15 g); *Elanco.*
049-725; (with Flumethasone, Neomycin Sulfate) Anaprime ® Opthakote Ophthalmic; *Fort Dodge Animal Health, Divn. AHP.*
049-726; (with Neomycin Sulfate) Optiprime ® Opthakote Ophthalmic Solution; *Fort Dodge Animal Health, Divn. AHP.*
065-015; (with Bacitracin Zinc, Hydrocortisone Acetate, Neomycin Sulfate) Bacitracin-Neomycin-Polymyxin With Hydrocortisone Acetate Ophthalmic Ointment; Vetropolycin HC Ophthalmic Ointment; *Altana.*
065-016; (with Bacitracin Zinc, Neomycin Sulfate) Bac-Neo-Poly Ophthalmic Ointment; Vetropolycin ophthalmic Ointment; *Altana.*
065-114; (with Bacitracin Zinc, Neomycin Sulfate) Mycitracin ® Sterile ophthalmic Ointment; *Pharmacia & Upjohn.*
065-119; (with Hydrocortisone Acetate, Neomycin Sulfate, Penicillin G Procaine) Forte Topical ® Ointment; *Pharmacia & Upjohn.*
065-476; (with Bacitracin Zinc, Hydrocortisone Acetate, Neomycin Sulfate) Cortisporin Veterinary Ophthalmic Ointment; *Schering-Plough Animal Health.*
065-485; (with Bacitracin Zinc, Neomycin Sulfate) Neosporin Ophthalmic Ointment; *Schering-Plough Animal Health.*

253 Thiostrepton

1393-48-2 9502 215-734-9

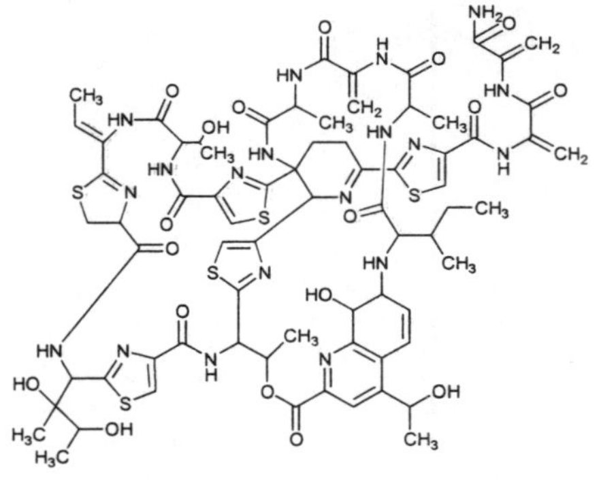

$C_{72}H_{85}N_{19}O_{18}S_5$

Bryamycin; Gargon; Thiactin; Bryamycin. Polypeptide antibiotic produced by *Streptomyces azureus*. Dec 246-256°; $[\alpha]_D^{23}$ = -985.° (AcOH), -61° (dioxane), -20° (C_5H_5N); soluble in $CHCl_3$, dioxane; C_5H_5N, DMF, AcOH; insoluble in H_2O, MeOH, EtOH, nonpolar organic solvents;no uv maxima, shoulders at 225, 250, 280 nm ($E^{1\%}$: 520, 380, 255); [hemisuccinate]: mp = 200-220°.

Veterinary Products:

012-258; (with Neomycin Sulfate, Nystatin, Triamcinolone Acetonide) Panolog ® Ointment; *Fort Dodge Animal Health Divn., Am. Cyanamid.*
096-676; (with Neomycin Sulfate, Nystatin, Triamcinolone Acetonide) Panolog Cream; *Fort Dodge Animal Health Divn., Am. Cyanamid.*
140-810; (with Neomycin Sulfate, Nystatin, Triamcinolone Acetonide) Panavet Ointment; *Med-Pharmex.*
140-847; (with Neomycin Sulfate, Nystatin, Triamcinolone Acetonide) Animax; *Altana.*
140-879; (with Neomycin Sulfate, Nystatin, Triamcinolone Acetonide) Derma 4 Ointment; *Pfizer.*
140-889; (with Neomycin Sulfate, Nystatin, Triamcinolone Acetonide) Derm-Otic Ointment; *Biocraft.*
141-003; (with Neomycin Sulfate, Nystatin, Triamcinolone Acetonide) Derm-Otic Ointment; *Robins.*
200-245; (with Neomycin Sulfate, Nystatin, Triamcinolone Acetonide) Derma-Vet Cream; *Med-Pharmex.*

254 Virginiamycin M_1

21411-53-0 10142 244-376-6

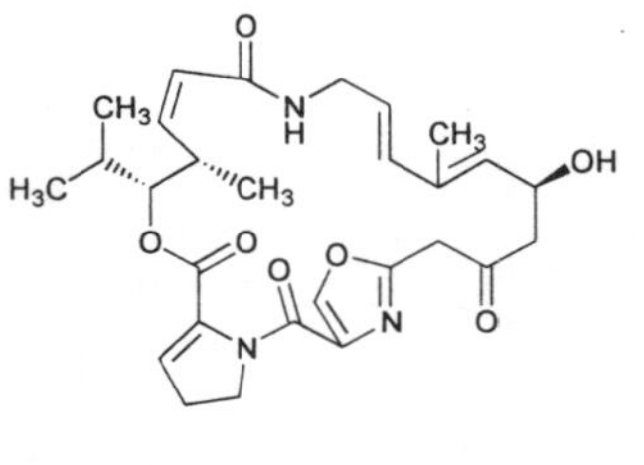

$C_{28}H_{35}N_3O_7$

mikamycin A; ostreogrycin A; pristinamycin II_A; staphylomycin M_1; streptogramin A; vernamycin A. Peptide antibiotic. Used in veterinary medicine as an antibacterial and food additive. mp = 165-167°; $[\alpha]_D$ = -190° ± 2° (c = 0.5 EtOH); λ_m 216 nm ($E^{1\%}_{1\ cm}$ 582 MeOH); soluble in Et_2O (0.1 g/100 ml), C_6H_6 (0.3 g/100 ml), EtOAc (0.5 g/100 ml); Me_2CO (2 g/100 ml), MeOH or EtOH (4 g/100 ml); dioxane and THF (5 g/100 ml), very soluble in $CHCl_3$, DMF, insoluble in H_2O, petroleum ether.

255 Virginiamycin S_1

23152-29-6 10142 245-462-6

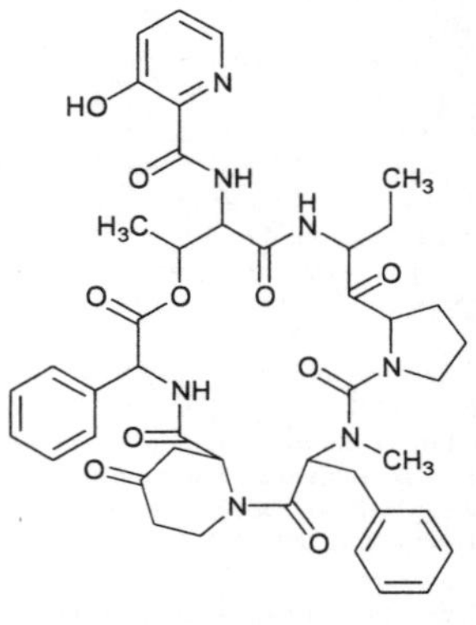

$C_{43}H_{49}N_7O_{10}$

staphylomycin S. Peptide antibiotic. Used in veterinary medicine as an antibacterial and food additive. mp = 240-242°; $[\alpha]_D^{20}$ = -28° (c = 1 EtOH); λ_m 305 nm (log ε 3.85 EtOH); soluble in Et_2O (0.1 g/100 ml), MeOH (0.5 g/100 ml), EtOH (2.5 g/100 ml), C_6H_6 (2.5 g/100 ml), Me_2CO or EtOAc (3 g/100 ml), dioxane (4 g/100 ml), very soluble in $CHCl_3$, DMF, insoluble in H_2O, petroleum ether.

Veterinary Products:

091-467; Stafac ® 500; *Pfizer.*
091-513; Stafac ®; *Pfizer.*
120-724; (with Monensin Sodium, Roxarsone) 3-Nitro ® / Coban ®; *Pfizer.*
122-481; (with Monensin Sodium) Stafac ® / Coban ®; *Pfizer.*
122-608; (with Lasalocid Sodium) Stafac ® / Avatec ®; *Pfizer.*
122-822; (with Amprolium, Ethopabate) Stafac ® / Amprol HI-E ®; *Pfizer.*
133-333; Stafac ® 10; *Akey.*
133-334; Virginiamycin Type A Medicated Article; *Alpharma.*
133-335; Stafac ® Swine Pak 10 g; *Quali-Tech.*
137-483; (with Halofuginone Hydrobromide) Flavomycin ® + Stenorol ®; *Hoechst-Roussel Vet.*
138-828; (with Salinomycin Sodium) Stafac ® 10,20,50, and 500 Type A Medicated Article/ Bio-Cox ® Type A Medicated Article; *Pfizer.*
138-953; (with Roxarsone, Salinomycin Sodium) Bio-Cox ® / Stafac ® / 3-Nitro ® Type A Medicated Articles; *Pfizer.*
139-473; (with Halofuginone Hydrobromide) Stenorol ® + Stafac ®; *Hoechst-Roussel Vet.*
140-998; V-Max ™; *Pfizer.*
141-110; (with Monensin) Coban ® / Stafac ®; *Elanco.*

141-114; (with Semduramicin Sodium) Aviax ™ / Stafac ®; *Pfizer.*
141-150; (with Lasalocid Sodium) Avatec ® / Stafac ®; *Roche Vitamins.*
200-092; (with Salinomycin Sodium) Sacox ® + Stafac ®; *Hoechst-Roussel Vet.*
200-094; (with Roxarsone, Salinomycin Sodium) Sacox ® + Stafac ® + 3-Nitro ®; *Hoechst-Roussel Vet.*

Antibiotics, Quinolone

256 Ciprofloxacin

85721-33-1 2374

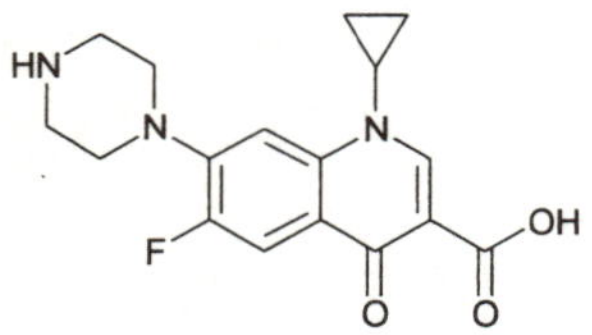

$C_{17}H_{18}FN_3O_3$
1-Cyclopropyl-6-fluoro-1,4-dihydro-4-oxo-7-(1-piperazinyl)-3-quinolinecarboxylic acid.
Cipro IV; Bay q 3939. Quinolone antibiotic. A metabolite of enrofloxacin. Dec 255-257°.

257 Ciprofloxacin Monohydrochloride Monohydrate

86393-32-0 2374

$C_{17}H_{19}ClFN_3O_3.H_2O$
1-Cyclopropyl-6-fluoro-1,4-dihydro-4-oxo-7-(1-piperazinyl)-3-quinolinecarboxylic acid monohydrochloride monohydrate.
Ciloxan; Cipro; Bay o 9867 monohydrate; Baycip; Ciflox; Ciprinol; Ciprobay; Ciproxan; Ciproxin; Flociprin; Septicide; Velmonit. Quinolone antibiotic. A metabolite of enrofloxacin. mp = 318-320°.

258 Difloxacin Hydrochloride

91296-86-5 3187

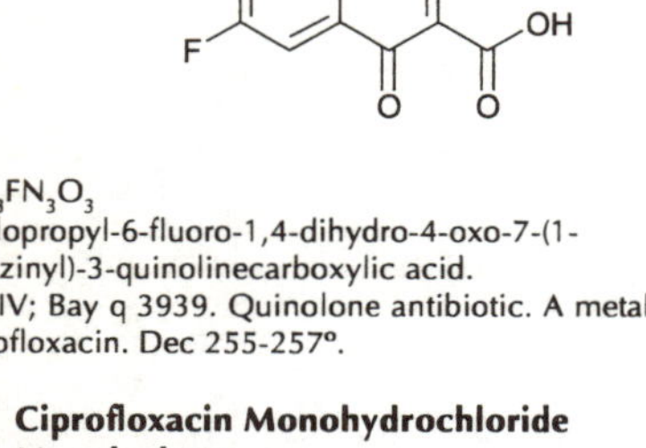

$C_{21}H_{20}ClF_2N_3O_3$
6-Fluoro-1-(p-fluorophenyl)-1,4-dihydro-7-(4-methyl-1-piperazinyl)-4-oxo-3-quinolinecarboxylic acid hydrochloride.
Abbott 56619; A-56619. Quinolone antibiotic. Used to treat bacterial infections in dogs. mp > 275°.

Veterinary Products:

141-096; (with Oxibendazole) Filaribits® Plus Chewable Tablets; *Fort Dodge Animal Health, Divn. AHP.*

259 Enrofloxacin

93106-60-6 3630

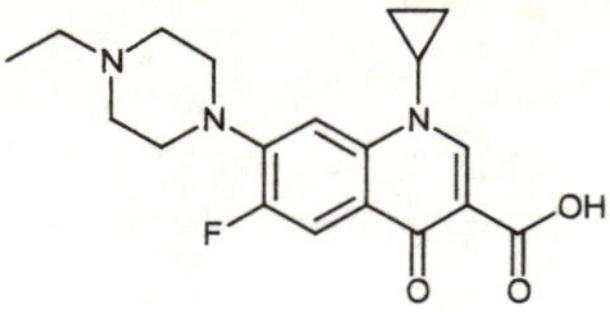

$C_{19}H_{22}FN_3O_3$
1-Cyclopropyl-7-(4-ethyl-1-piperazinyl)-6-fluoro-1,4-dihydro-4-oxo-3-quinolinecarboxylic acid.
CFPQ; Bay Vp; 2674; Baytril. Antibacterial. Used orally as an antibacterial in cats and dogs. Also approved for use in non-dairy cattle, chickens and turkeys. Pale yellow crystals; mp = 219-221°; slightly soluble in H_2O at pH 7; LD_{50} (mmus orl) > 5000 mg/kg, (fmus orl) = 4336 mg/kg, (mmus iv = 200 mg/kg, (fmus iv) = 200 mg/kg, (mrat orl) > 5000 mg/kg, (mrbt orl) = 500-800 mg/kg.

Veterinary Products:

140-441; Baytril ® Antibacterial Tablets; Baytril ® Taste Tabs ™ Antibacterial Tablets; *Bayer.*
140-828; Baytril ® 3.23% Concentrate Antimicrobial Solution; *Bayer.*
140-913; Baytril ® Antibacterial Injectable Solution; *Bayer.*
141-068; Baytril 100 Injectable Solution; *Bayer.*

260 Marbofloxacin

115550-35-1

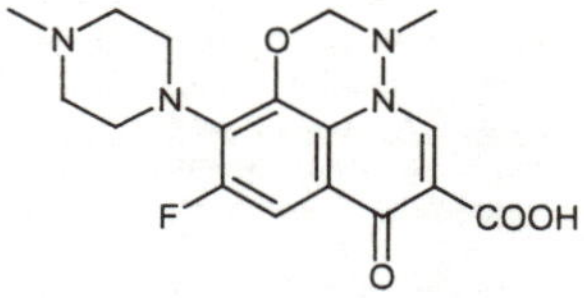

$C_{17}H_{19}FN_4O_4$
9-Fluoro-2,3-dihydro-3-methyl-10-(4-methyl-1-piperazinyl)-7-oxo-7H-pyrido[3,2,1-ij][4,1,2]benzoxadiazine-6-carboxylic acid.
Antibacterial.

Veterinary Products:

141-151; Zeniquin™; *Pfizer.*

261 Orbifloxacin

113617-63-3

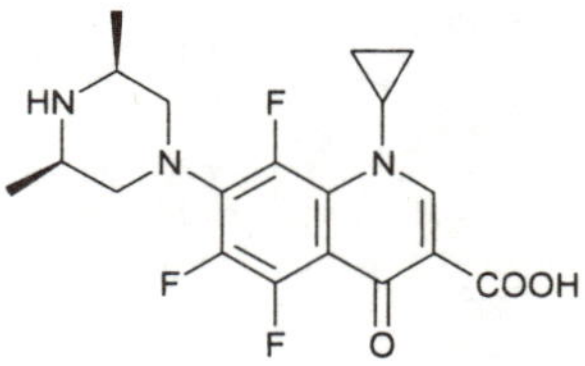

$C_{19}H_{20}F_3N_3O_3$
1-Cyclopropyl-7-(cis-3,5-dimethyl-1-piperazinyl)-5,6,8-trifluoro-1,4-dihydro-4-oxo-4-carboxamide orotate.

AICA. Antibiotic. Used to treat bacterial infections in dogs and cats.

Veterinary products:

141-081; Orbax™ Tablets; *Schering-Plough Animal Health.*

262 Sarafloxacin Hydrochloride
91296-87-6 8517

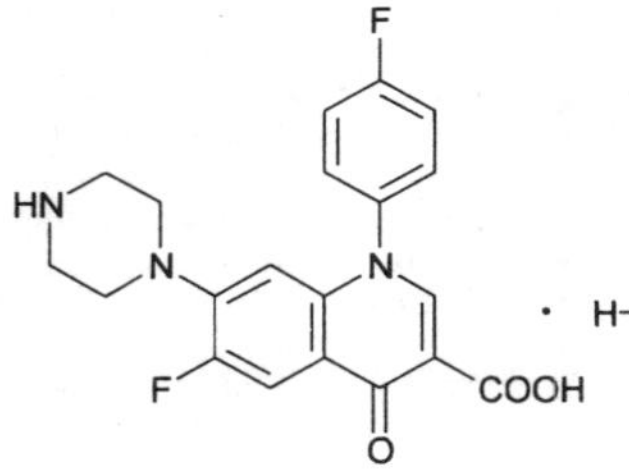

$C_{20}H_{18}ClF_2N_3O_3$
6-Fluoro-1-(4-fluorophenyl)-1,4-dihydro-4-oxo-7-(1-piperazinyl)-3-quinolinecarboxylic acid hydrochloride.
Abbott 56620; A-56620. Antibacterial. Crystals; mp (monohydrate) > 275°.

Veterinary products:

141-017; SaraFlox ® WSP; *Abbott.*
141-018; SaraFlox ® Injection; *Abbott.*

Antibiotics, Sulfonamide

263 Sulfabromomethazine Sodium
9066

$C_{12}H_{12}BrN_4NaO_2S.H_2O$
4-Amino-N-(5-bromo-4,6-dimethyl-2-pyrimidinyl)-benzenesulfonamide sodium salt.
Sulfabrom. Antibiotic. Soluble in H_2O.

Veterinary products:

011-532; Sulfabrom 2.5 g; *Merial.*

264 Sulfachloropyrazine Sodium Monohydrate
23282-55-5 9068 245-553-0

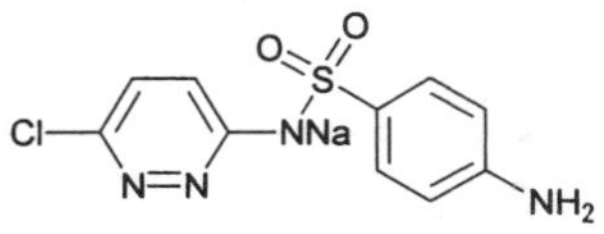

$C_{10}H_8ClN_4NaO_2S$
Sodium sulfachloropyridazine Monohydrate.
Prinzone; Vetisulid. Antibacterial agent used to treat enteric infections.

Veterinary products:

033-127; Prinzone Bolus; Pyradan Bolus; Vetisulid Bolus; *Fort Dodge Animal Health Divn., Am. Cyanamid.*
033-318; Prinzone Injection; Pyradan Injection; Vetisulid Injection; *Fort Dodge Animal Health Divn., Am. Cyanamid.*
033-319; Vetisulid ® Tablets; *Fort Dodge Animal Health Divn., Am. Cyanamid.*
033-373; Prinzone Powder; Pyradan Powder; Vetisulid Powder; *Fort Dodge Animal Health Divn., Am. Cyanamid.*
040-181; Prinzone Oral Suspension; Pyradan Oral Suspension; Vetisulid Oral; *Fort Dodge Animal Health Divn., Am. Cyanamid.*

265 Sulfachlorpyridazine
80-32-0 9068 201-269-9

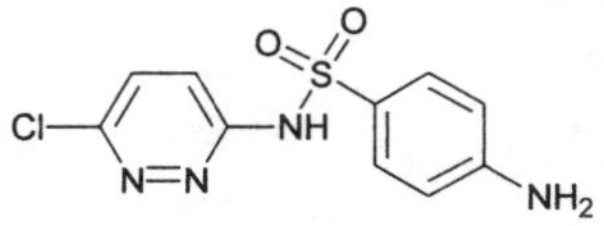

$C_{10}H_9ClN_4O_2S$
N^1-(6-Chloro-3-pyridazinyl)sulfanilamide.
Nefrosul; Sonilyn; Ciba 10370; Ba-10370; Cosulid; Cosumix; Nefrosul; Sonilyn. Sulfonamide antibiotic used to treat diarrhea in calves.

266 Sulfadiazine
68-35-9 9071 200-685-8

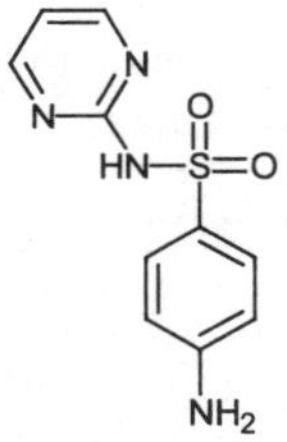

$C_{10}H_{10}N_4O_2S$
N^1-2-Pyrimidinylsulfanilamide.
Coco-Diazine; Eskadiazine; Adiazine; Diazyl; Sulfolex; component of: Sulfonamide Duplex. Sulfonamide antibiotic. Used in conjunction with trimethoprim to treat bacterial infections in dogs and horses. mp = 252-256°; soluble in H_2O (0.013 g/100 ml at pH 5.5, 37°, 0.2 g/100 ml at pH 7.5, 37°), sparingly soluble in EtOH, Me_2CO, freely soluble in dilute acid, alkali.

Veterinary products:

030-330; (with Sulfamerazine, Sulfamethazine) Tylocine Sulfa Tablets 50; *Elanco.*
095-614; (with Trimethoprim) Tribrissen ® 20 mg Tablets; Tribrissen ® 30 mg Tablets; Tribrissen ® 480 mg Tablets; Tribrissen ® 960 mg Tablets; *Schering-Plough Animal Health.*
106-965; (with Trimethoprim) Tribrissen ® 48% Injection; *Schering-Plough Animal Health.*
115-578; (with Trimethoprim) Di-Trim ® Tablets; *Fort Dodge Animal Health, Divn. AHP.*
131-918; (with Trimethoprim) Tribrissen ® 400 Oral Paste; *Schering-Plough Animal Health.*
132-486; (with Trimethoprim) Di-Trim ® 24%; *Fort Dodge Animal Health, Divn. AHP.*
134-778; (with Trimethoprim) Di-Trim ® 48% Injection;

Fort Dodge Animal Health, Divn. AHP.
136-342; (with Trimethoprim) Di-Trim ® 400 Paste; *Fort Dodge Animal Health, Divn. AHP.*
136-741; (with Trimethoprim) Tribrissen ® 60 Oral Suspension; *Schering-Plough Animal Health.*
200-033; (with Trimethoprim) Uniprim ™ Powder; *Macleod.*

267 Sulfadiazine Sodium Salt

547-32-0 9071 208-919-0

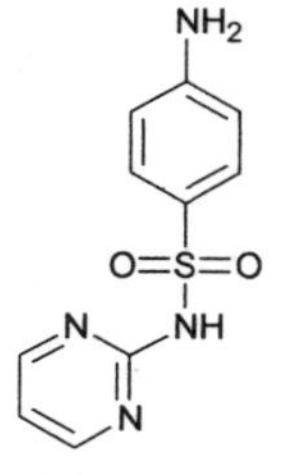

$C_{10}H_9N_4NaO_2S$
N^1-2-Pyrimidinylsulfanilamide monoodium salt.
sulfadiazine sodium; soluble sulfadiazine. Antibiotic. Used with trimethoprim to treat bacterial infections in dogs and horses. Soluble in H_2O (50 g/100 ml).

Veterinary products:

105-093; (with Trimethoprim) Tribrissen® 24% Injection; *Schering-Plough Animal Health.*

268 Sulfadimethoxine

122-11-2 9073 204-523-7

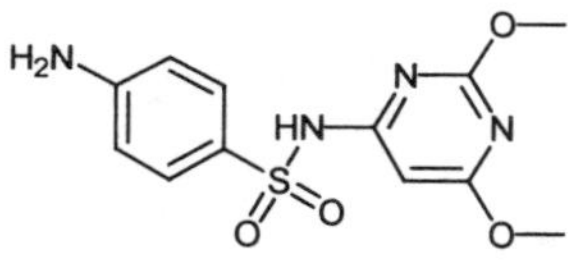

$C_{12}H_{14}N_4O_4S$
N^1-(2,6-Dimethoxy-4-pyrimidinyl)-sulfanilamide.
Agribon; Albon; Madribon; component of: Primor, Rofenaid. Antibiotic. Used to treat respiratory infections in dogs and cats. Used in conjunction with Ormetoprim to treat Staphylococcus aureus infections in dogs. mp = 201-203°; soluble in dilute HCl, $NaHCO_3$, soluble in H_2O (0.0046 g/100 ml at pH 4.10, 0.0295 g/100 ml at pH 6.7, 0.058 g/100 ml at pH 7.06, 5.17 g/100 ml at pH 8.71); LD_{50} (mus orl) > 10000 mg/kg.

Veterinary products:

012-087; Bactrovet Tablets 250 mg; *Schering-Plough Animal Health.*
031-205; Agribon 12.5% Drinking Water Solution; Albon ®; *Pfizer.*
031-715; Agribon Boluses - 2.5, Agribon Bolusus - 5.0; Agribon Boluses - 15.0; Albon ®; *Pfizer.*
037-700; Bactrovet Oral Suspension 12.5%; *Schering-Plough Animal Health.*
040-209; (with Ormetoprim) Rofenaid ® 40; *Roche Vitamins.*
041-245; Agribon Injection 40%; Albon ®; *Pfizer.*
041-984; (with Ormetoprim, Roxarsone) Rofenaid ® Plus Roxarsone.
043-785; Albon Oral Suspension 5%; *Roche Vitamins.*
046-285; Agribon Soluble Powder; Albon ®; *Pfizer.*
093-107; Albon ® S.R. (Sustained Release); *Pfizer.*
098-569; Medacide-SDM Injection 10%; *Boehringer Ingelheim Vetmedica.*
100-929; (with Ormetoprim) Primor ® Tablets; *Pfizer.*
125-933; (with Ormetoprim) Romet ® - 30; *Roche Vitamins.*
200-030; Di-Methox 12.5% Oral Solution; Sulfadimethoxine 12.5% Oral Solution; *Agri Labs.*
200-031; Di-Methox Antibacterial Soluble Powder; Sulfadimethoxine Antibacterial Soluble Powder; *Agri Labs.*
200-038; Di-Methox Injection 40%; Sulfadimethoxine Injection 40%; *Agri Labs.*
200-165; SDM Sulfadimethoxine 12.5% Oral Solution; *Boehringer Ingelheim Vetmedica.*
200-177; Sulfadimethoxine Injection 40%; *Phoenix.*
200-192; Sulfadimethoxine 12.5% Oral Solution; *Phoenix.*
200-238; Sulfasol ® Soluble Powder; *Med-Pharmex.*
200-251; Sulforal ®; *Med-Pharmex.*
200-258; Sulfadimethoxine Soluble Powder; *Phoenix.*

269 Sulfamerazine

127-79-7 9081 204-866-2

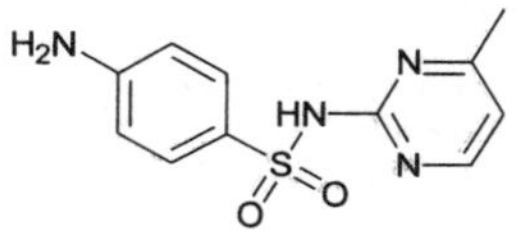

$C_{11}H_{12}N_4O_2S$
N^1-(4-Methyl-2-pyrimidinyl)sulfanilamide.
RP-2632; Mesulfa; Percoccide; component of: Sulfonamide Duplex. Sulfonamide antibiotic. mp = 234-238°; λ_m 243 257 nm ($E^{1\%}_{1\ cm}$ 875 822 H_2O), 243 307 nm ($E^{1\%}_{1\ cm}$ 625 200 0.1M HCl), 271 nm ($E^{1\%}_{1\ cm}$ 835 EtOH); soluble in H_2O (0.035 g/100 ml at pH 5,5, 0.170 g/100 ml at pH 7.5), readily soluble in mineral acid and alkaline solutions; sparingly soluble in Me_2CO, slightly soluble in EtOH, insoluble in Et_2O; $CHCl_3$.

Veterinary products:

030-330; (with Sulfadiazine, Sulfamethazine, Tylosin) Tylocine Sulfa Tablets 50; *Elanco.*
033-950; Sulfamerazine In Fish Grade; *Roche Vitamins.*

270 Sulfamethazine

57-68-1 9083 200-346-4

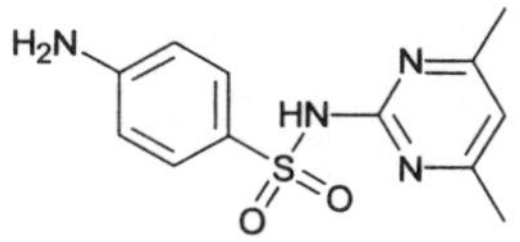

$C_{12}H_{14}N_4O_2S$
N^1-(4,6-Dimethyl-2-pyrimidinyl)sulfanilamide.
Calfspan Tablets; Sulka K Boluses; SulfaSURE SR Bolus;

DiazilSulfadine; S-Dimidine; Dimidin-R; Neazina; Sulmet. Sulfonamide antibiotic. mp = 170-176°, 178-179°, 198-199°, 205-207°; λ_m 241 nm ($E^{1\%}_{1\,cm}$ 670 H_2O, pH 6.6), 243 257 nm ($E^{1\%}_{1\,cm}$ 765 776 0.01N NaOH), 241 297 nm ($E^{1\%}_{1\,cm}$ 561 266 0.01N HCl); soluble in H_2O (0.15 g/100 ml at 29°, 0.192 g/100 ml at 37°, pH 7.00); LD_{50} (mus ip) = 1060 mg/kg.

Veterinary products:

006-084; Sulmet ® Drinking Water Solution; *American Cyanamid Divn., AHP Corp.*
008-774; Sulmet ® Solution Injectable; *American Cyanamid Divn., AHP Corp.*
030-330; (with Sulfadiazine, Sulfamerazine, Tylosin) Tylocine Sulfa Tablets 50; *Elanco.*
035-688; (with Chlortetracycline Calcium Complex, Penicillin G Procaine) Aureo SP-250; Aureomix 500; *Roche Vitamins.*
035-805; (with Chlortetracycline Hydrochloride) Aureomix-S 700 Crumbles; Aureomix-S 700 g; *Roche Vitamins.*
041-275; (with Tylosin Phosphate) Tylan ® 10 Sulfa-G Premix; Tylan ® 40 Sulfa-G Premix; *Elanco.*
041-647; (with Chlortetracycline Complex) Aureomix S 100 A; *Roche Vitamins.*
041-648; (with Chlortetracycline Complex) Aureomix S 700 D; *Roche Vitamins.*
041-649; (with Chlortetracycline Complex) Aureomix S 700 G; *Roche Vitamins.*
041-650; (with Chlortetracycline Complex) Aureomix S 700 E; *Roche Vitamins.*
041-651; (with Chlortetracycline Complex) Aureomix S 100 F; *Roche Vitamins.*
041-652; (with Chlortetracycline Complex) Aureomix S 700 C-2; *Roche Vitamins.*
041-653; (with Chlortetracycline Complex) Aureomix S 700 B; *Roche Vitamins.*
041-654; (with Chlortetracycline Complex) Aureomix S 700 H; *Roche Vitamins.*
042-660; (with Tylosin Phosphate) Purina ® Pork-Plus Medicated; *Purina Mills.*
049-729; Purina ® Sulfa; *Purina Mills.*
049-892; Sulfamethazine Spanbolet II; *Fort Dodge Animal Health Divn., Am. Cyanamid.*
055-012; (with Chlortetracycline Bisulfate) Aureomycin ®-Sulmet Soluble Powder; *American Cyanamid Divn., AHP Corp.*
065-113; (with Chlortetracycline Bisulfate) Aureo Sulfa Soluble Powder; *Purina Mills.*
091-668; (with Chlortetracycline Calcium Complex, Penicillin G Procaine) ChlorMax™-SP 250; ChlorMax™-SP 500; ChlorMax™-SP 1000; Chlorachel 250 Swine; Pficlor 250; *Alpharma.*
091-749; (with Tylosin Phosphate) Tylan ® 40 Plus Sulfa-G; CO-OP Tylosin 40 Plus Sulfamethazine; *Alpharma.*
093-329; HavaSpan Prolonged Release Bolus; SulfaSpan Prolonged Release Bolus; *Bayer.*
097-615; (with Tylosin Phosphate) Swine Med-A-Mix TS 8000 Premix; Tylan ® 5 Sulfa-G; Tylan ® 10 Sulfa-G; Tylan ® 20 Sulfa-G; Tylan ® 40 Sulfa-G; *Golden Sun Feeds.*
097-981; (with Tylosin Phosphate) Quali-Tech Tylan ®-Sulfa Premix 10 -10; Tylan ® 5 Sulfa-G; Tylan ® 10 Sulfa-G; Tylan ® 20 Sulfa-G; Tylan ® 40 Sulfa-G; *Quali-Tech.*
098-639; (with Tylosin Phosphate) Tylan ® 10 Sulfa-G; Tylan ® 20 Sulfa-G; Tylan ® 40 Sulfa-G; Tylan ® 5 Sulfa-G; *Bioproducts.*
099-767; (with Tylosin Phosphate) Purina ® Tylan ®-40 Plus Sulfamethazine; *Purina Mills.*
101-906; (with Tylosin Phosphate) Mill Co Medicator TS-40 Premix; *Waterloo Mills.*
107-002; (with Tylosin Phosphate) Seeco Tylan ®-Sulfa 10 Premix Med.; *Seeco.*
107-957; (with Tylosin Phosphate) Tylan ® 20 Sulfa-G; Tylan ® 40 Sulfa-G; *ADM Animal Health & Nutrition.*
108-484; (with Tylosin Phosphate) HFA Tylosin-10 Plus Sulfa; *ADM Animal Health & Nutrition.*
109-816; (with Tylosin Phosphate) Tylan ® 5 Sulfa-G; Tylan ® 10 Sulfa-G; Tylan ® 20 Sulfa-G; Tylan ® 40 Sulfa-G; *Intl. Nutrition.*
111-069; (with Tylosin Phosphate) Tylan ® 5 Sulfa-G; Tylan ® 10 Sulfa-G; Tylan ® 20 Sulfa-G; Tylan ® 40 Sulfa-G; *Moorman.*
120-614; (with Tylosin Phosphate) Tylan ® 10 Sulfa-G; Tylan ® 20 Sulfa-G; Tylan ® 40 Sulfa-G; Tylan ® 40 Sulfa-G; *Webel Feeds.*
120-615; Sustain III ® Bolus; *Merial.*
122-271; Sulmet ® Oblets; *American Cyanamid Divn., AHP Corp.*
122-272; Sulmet ® Soluble Powder; *American Cyanamid Divn., AHP Corp.*
122-522; (with Tylosin Phosphate) Tylan ® 5 Sulfa-G; Tylan ® 10 Sulfa-G; Tylan ® 20 Sulfa-G; Tylan ® 40 Sulfa-G; *Custom Feed Blenders.*
124-391; (with Tylosin Phosphate) Nutra-Mix Tylan ® - Sulfa Premixes; *Ag-Mark.*
126-232; Calfspan ™; *Fort Dodge Animal Health Divn., Am. Cyanamid.*
127-506; (with Tylosin Phosphate) Heinold Tylan ® 5 Sulfa Premix; Tylan ® 5 Sulfa-G; Tylan ® -10 Sulfa-G; Tylan ® 20 Sulfa-G; Tylan ® 40 Sulfa-G; *Heinold Feeds.*
127-507; (with Tylosin Phosphate) Tylan ® 5 Sulfa-G; Tylan ® 10 Sulfa-G; Tylan ® 20 Sulfa-G; Tylan ® 40 Sulfa-G; *Akey.*
128-411; (with Tylosin Phosphate) Tylan ® 5 Sulfa Premix; *ADM Animal Health & Nutrition.*
129-159; (with Tylosin Phosphate) Tylan ® 5 Sulfa-G; Tylan ® 10 Sulfa-G; Tylan ® 20 Sulfa-G; Tylan ® 40 Sulfa-G; *Custom Feed Services.*
129-161; (with Tylosin Phosphate) Nutra-Blend Tylan ® 5 Sulfa Premix; Tylan ® 5 Sulfa Premix; Tylan ® 10 Sulfa Premix; Tylan ® 20 Sulfa Premix; Tylan ® 40 Sulfa Premix; *Nutra-Blend.*
129-646; (with Tylosin Phosphate) Tylan ® 5Sulfa-G; Tylan ® 10Sulfa-G; Tylan ® 20 Sulfa-G; Tylan ® 40 Sulfa-G; Tylan ® 5 Sulfa Premix; *I.M.S.*
131-956; (with Tylosin) Tylan ® Sulfa-G; *ADM Animal Health & Nutrition.*
131-958; (with Tylosin Phosphate) Tylan ® Sulfa-G; Wayne Feed Divn.*.
138-342; (with Tylosin Phosphate) Tylan ® 5 Sulfa-G Premix; Tylan ® 10 Sulfa-G Premix; Tylan ® 20 Sulfa-G Premix; Tylan ® 40 Sulfa-G Premix; *Feed Service Co.*
140-270; Sulfa Sustained Release Bolus; *Boehringer Ingelheim Vetmedica.*
140-681; (with Tylosin Phosphate) Tylan ® 5 Sulfa-G Premix; Tylan ® 10 Sulfa-G Premix; Tylan ® 20 Sulfa-G Premix; Tylan ® 40 Sulfa-G Premix; *Pennfield Oil.*
140-820; (with Tylosin Phosphate) Tylan ® 5 Sulfa-G Premix; Tylan ® 10 Sulfa-G Premix; Tylan ® 20 Sulfa-G Premix; Tylan ® 40 Sulfa-G Premix; *Furst-McNess.*
140-908; Veta-Meth ™; *Lloyd.*
140-909; Sulka-S ™ Bolus; *Fort Dodge Animal Health Divn., Am. Cyanamid.*

271 Sulfamethizole

144-82-1 9084 205-641-1

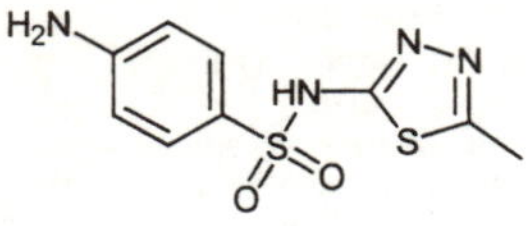

$C_9H_{10}N_4O_2S_2$

N^1-(5-Methyl-1,3,4-thiadiazol-2-yl)sulfanilamide.

Thiosulfil Forte; Famet; Lucosil; Methazol; Renasul; Rufol; Salimol; Sulfapyelon; Thidicur; Thiosulfil; Urolucosil; component of: Thiosulfil-A-Forte. Sulfonamide antibiotic. mp = 208°; soluble in H_2O (0.025 g/100 ml at pH 6.5, 20 g/100 ml at pH 7.5), MeOH (2.5 g/100 ml), EtOH (3,3 g/100 ml), Me_2CO (10 g/100 ml), Et_2) (0.07 g/100 ml), $CHCl_3$ (0.035 g/100 ml), insoluble in C_6H_6.

Veterinary products:

030-416; (with Methenamine Mandelate) Mesulfin Tablets; *Fort Dodge Animal Health, Divn. AHP.*

272 Sulfamethoxazole

723-46-6 9086 211-963-3

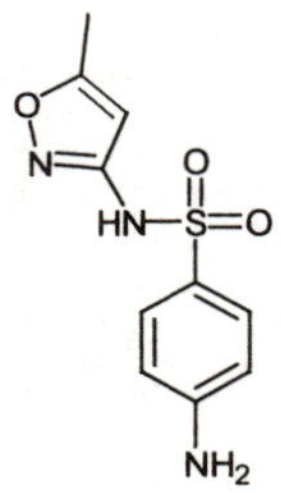

$C_{10}H_{11}N_3O_3S$

N^1-(5-Methyl-3-isoxazolyl)sulfanilamide.

Gantanol; component of: Azo Gantanol, Bactrim, Cotrim, SeptraSMZ/TMP, Sulfatrim. Sulfonamide antibiotic. Used in conjunction with trimethoprim to treat bacterial infections in dogs and horses. Used also to treat pneumocystis. mp = 167°; LD_{50} (mus orl) = 3662 mg/kg.

273 Sulfanitran

122-16-7 9103

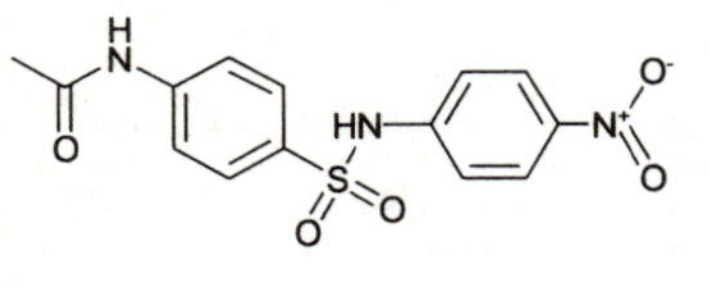

$C_{14}H_{13}N_3O_5S$

4'-[(p-Nitrophenyl)sulfamoyl]acetanilide.

NSC-77120; APNPS; component with Alkomid of 014-250 Novastat Type A Medicated Premix (Am. Cyanamid). Sulfonamide antibiotic. Also used as a coccidiostat in poultry. mp = 239-240°, 264°; freely soluble in Me_2CO, soluble in hot EtOH, MeOH, sparingly soluble in H_2O, Et_2O.

Veterinary products:

011-141; (with Nitromide) Unistat-2 Type A Medicated Premix; *Fort Dodge Animal Health Divn., Am. Cyanamid.*
014-250; (with Alkomide) Novastat Type A Medicated Premix; *Fort Dodge Animal Health Divn., Am. Cyanamid.*
034-537; (with Alkomide, Roxarsone) Novastat-3 Type A Medicated Article; *Fort Dodge Animal Health Divn., Am. Cyanamid.*
035-388; (with Alkomide) Novastat-W; *Fort Dodge Animal Health Divn., Am. Cyanamid.*
039-666; (with Nitromide, Roxarsone) Unistat-3 Type A Medicated Article; *Fort Dodge Animal Health Divn., Am. Cyanamid.*

274 Sulfaquinoxaline

59-40-5 9109 200-423-2

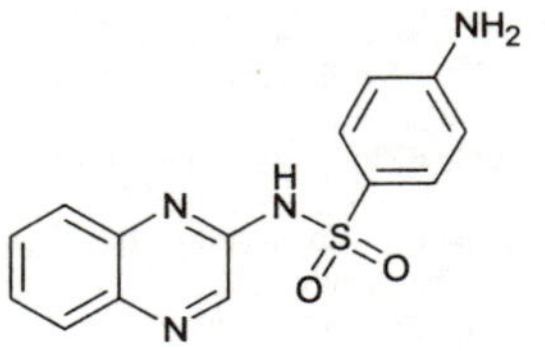

$C_{14}H_{12}N_4O_2S$

N^1-2-Quinoxalinylsulfanilamide.

Compd. 3-120; S.Q.; Sulquin; component of: Sulquin 6-50 Concentrate. Sulfonamide antibiotic. Also used as a coccidiostat in poultry. Crystals; mp = 247-248°; λ_m = 252, 360 nm ($E^{1\%}_{1\ cm}$ 1110, 275, H_2o, pH 6.6); soluble in H_2O at pH 7 (0.00075 g/100 ml), 95% EtOH (0.073 g/100 ml), Me_2CO (0.43 g/100 ml), aqueous $NaHCO_3$, NaOH.

Veterinary products:

006-391; S.Q. 40% Medicated Feed; *Hess & Clark.*
006-677; S.Q. 20% Solution; *Hess & Clark.*
007-087; Sulfaquinoxaline Solubilized; *Hess & Clark.*

275 Sulfaquinoxaline Sodium

9109

$C_{14}H_{11}NaN_4O_2S$

N^1-2-Quinoxalinylsulfanilamide sodium salt.

Aviochina. Sulfonamide antibiotic. Also used as a coccidiostat in poultry. Amorphous, deliquescent solid; very soluble in H_2O, pH of aqueous solution = 10; concerted to the free acid by atmospheric CO_2.

Veterinary products:

006-707; Sulquin ® 6-50; *Fort Dodge Animal Health Divn., Am. Cyanamid.*
006-891; Liquid Sul-Q-Nox; *Russell.*
007-076; Sulfa-Nox Liquid; *PM Resources.*
008-244; Sulfa-Nox Concentrate; *PM Resources.*

276 Sulfathiazole

72-14-0 9115 200-771-5

$C_9H_9N_3O_2S_2$

N[1]-2-Thiazolylsulfanilamide.

RP-2090; M&B-760; Thiazamide; Cibazol; Enterobiocine; Duatok; Sulfamul; Sulfavitina; Sulzol; component of: Sultrin, Trysul. Sulfonamide antibiotic. mp = 202-202.5°; soluble in H_2O (0.06 g/100 ml at pH 6.03), EtOH (0.525 g/100 ml), soluble in Me_2CO, dilute mineral acids, KOH and NaOH solutions, ammonia H_2O.

Veterinary products:

039-077; (with Chlorotetracycline Hydrochloride, Penicilin G Procaine) CSP ™ 250; CSP ™ 500; *Boehringer Ingelheim Vetmedica.*
200-140; (with Chlorotetracycline Hydrochloride, Penicilin G Procaine) Aureozol ®; *Roche Vitamins.*
200-167; (with Chlorotetracycline Calcium Complex, Penicilin G Procaine) Aureozol ® 500 Granular; *Roche Vitamins.*

277 Sulfisoxazole

127-69-5 9125 204-858-9

$C_{11}H_{13}N_3O_3S$

N[1]-(3,4-Dimethyl-5-isoxazolyl)sulfanilamide.

Gantrisin; Sulfalar; sulphafurazole; Sosol; Soxisol; Soxomide; Sulfazin; Sulfoxol; Sulsoxin; component of: Azo Gantrisin. Sulfonamide antibiotic. mp = 194°; soluble in H_2O (0.013 g/100 ml), EtOH; LD_{50} (mus orl) = 6800 mg/kg.

Veterinary products:

007-981; Soxisol Tablets; *Fort Dodge Animal Health, Divn. AHP.*

Antibiotics, Tetracycline

278 Chlortetracycline

57-62-5 2245 200-341-7

$C_{22}H_{23}ClN_2O_8$

7-Chloro-4-(dimethylamino)-1,4,4a,5,5a,6,11,12a-octahydro-3,6,10,12,12a-pentahydroxy-6-methyl-1,11-dioxo-2-naphthacenecarboxamide.

7-chlorotetracycline; Acronize; Aureocina; Aureomycin; Biomitsin; CentrAureo; Chrusomykine; Orospray. Antibacterial, antiamebic and antiprotozoal. Used in veterinary medicine as an antimicrobial agent. mp = 168-169°; $[\alpha]_D^{23}$=-275.0° (MeOH); λ_m= 230, 262.5, 367.5 nm (0.1N HCl), 255, 285, 345 nm (0.1N NaOH); soluble in H_2O (0.5-0.6 mg/ml), cellosolves, dioxane, carbitol, poorly soluble in other organic solvents.

Veterinary products:

048-761; Aureomycin ® Type A Medicated Article; *Roche Vitamins.*
091-647; (with Amprolium, Ethopabate) Rainbow Broiler Base Concentrate; *Stockton Hay & Grain.*
138-935; Chlortetracycline - 50; Chlortetracycline - 60; Chlortetracycline - 70; Chlortetracycline - 80; Chlortetracycline - 100; Chlortetracycline - 100MR; Pennchlor 100MR; Pennchlor 50; Pennchlor 50-G; Pennchlor 70; Pennchlor 80; Pennchlor 90; *Pennfield Oil.*
141-059; (with Bacitracin Methylene Disalicylate) BMD ®-10, 25, 30, 40, 50, 60, 75 and ChlorMax™ -50, 65, 70/ Micro-CTC ®100; *Alpharma.*
200-259; (with Roxarsone, Salinomycin Sodium) ChlorMax ™ / Sacox ® / 3-Nitro ®; ChlorMax ™ / Bio-Cox ® / 3-Nitro ®; *Alpharma.*
200-260; (with Roxarsone, Salinomycin Sodium) ChlorMax ™ / Sacox ® / 3-Nitro ®; ChlorMax ™ / Bio-Cox ® / 3-Nitro ®; *Alpharma.*
200-261; (with Salinomycin Sodium) ChlorMax ™ / Sacox ®; ChlorMax ™ / Bio-Cox ®; *Alpharma.*
200-262; (with Salinomycin Sodium) ChlorMax ™ / Sacox ®; *Hoechst-Roussel Vet.*
200-263; (with Monensin Sodium) ChlorMax™ / Coban ®; *Alpharma.*

279 Chlortetracycline Bisulfate

2245

7-Chloro-4-(dimethylamino)-1,4,4a,5,5a,6,11,12a-octahydro-3,6,10,12,12a-pentahydroxy-6-methyl-1,11-dioxo-2-naphthacenecaroxamide bisulfate.

Antibacterial, antiamebic and antiprotozoal. Used in veterinary medicine as an antimicrobial agent.

Veterinary products:

055-012; (with Sulfamethazine) Aureomycin ®-Sulmet Soluble Powder; *American Cyanamid Divn., AHP Corp.*
055-020; Aureomycin ® Soluble Powder; *American Cyanamid Divn., AHP Corp.*

065-113; (with Sulfamethazine) Aureo Sulfa Soluble Powder; *Purina Mills.*
065-486; Chlortetracycline Bisulfate Soluble Powder; *Boehringer Ingelheim Vetmedica.*

280 Chlortetracycline Calcium Complex

2245

7-Chloro-4-(dimethylamino)-1,4,4a,5,5a,6,11,12a-octahydro-3,6,10,12,12a-pentahydroxy-6-methyl-1,11-dioxo-2-naphthacenecarboxamide, calcium complex.
Antibacterial, antiamebic and antiprotozoal. Used in veterinary medicine as an antimicrobial agent.

Veterinary products:

035-688; (with Penicillin G Procaine, Sulfamethazine) Aureo SP-250; Aureomix 500; *Roche Vitamins.*
036-361; (with Amprolium, Ethopabate, Sodium Sulfate) Amp Ethopabate CTC ® Sodium Sulfate; *Roche Vitamins.*
041-647; (with Sulfamethazine) Aureomix S 700 A; *Roche Vitamins.*
041-648; (with Sulfamethazine) Aureomix S 700 D; *Roche Vitamins.*
041-649; (with Sulfamethazine) Aureomix S 700 G; *Roche Vitamins.*
041-650; (with Sulfamethazine) Aureomix S 700 E; *Roche Vitamins.*
041-651; (with Sulfamethazine) Aureomix S 700 F; *Roche Vitamins.*
041-652; (with Sulfamethazine) Aureomix S 700 C-2; *Roche Vitamins.*
041-653; (with Sulfamethazine) Aureomix S 700 B; *Roche Vitamins.*
041-654; (with Sulfamethazine) Aureomix S 700 H; *Roche Vitamins.*
046-209; (with Clopidol) Coyden 25 ® + CTC ®; *Rhône-Poulenc.*
048-480; Chloratet 50; *ADM Animal Health & Nutrition.*
049-287; Chlorachel-50; *Pfizer.*
091-668; (with Penicillin G Procaine, Sulfamethazine) ChlorMax™-SP 250; ChlorMax™-SP 500; ChlorMax™-SP 1000; Chlorachel 250 Swine; Pficlor 250; *Alpharma.*
092-286; CLTC - 10; CLTC - 20; CLTC - 30; CLTC - 50; CLTC - 70; *Pfizer.*
092-287; CLTC Type A Medicated Article; *Pfizer.*
092-507; (with Robenidine Hydrochloride) Robenz ® with Aureomycin ® 500 Gm; *Roche Vitamins.*
100-901; Pfichlor 100S Milk Replacer Type A Medicated Article; *Roche Vitamins.*
140-859; (with Salinomycin Sodium) Aureomycin ® / Bio-Cox ®; *Roche Vitamins.*
140-867; (with Roxarsone, Salinomycin Sodium) Aureomycin ® / Bio-Cox ® / 3-Nitro ®; *Roche Vitamins.*
200-091; (with Roxarsone, Salinomycin Sodium) Sacox ® + 3-Nitro ® + Aureomycin ®; *Hoechst-Roussel Vet.*
200-167; (with Penicillin G Procaine, Sulfathiazole) Aureozol ® 500 Granular; *Roche Vitamins.*
200-242; (with Bacitracin Methylene Disalicylate) Aureomycin ® - 50, 70, 80, 90, 100 / BMD ® - 25, 30, 40, 50, 60, 75; *Roche Vitamins.*

281 Chlortetracycline Hydrochloride

64-72-2 2245 200-591-7

$C_{22}H_{24}Cl_2N_2O_8$
7-Chloro-4-(dimethylamino)-1,4,4a,5,5a,6,11,12a-octahydro-3,6,10,12,12a-pentahydroxy-6-methyl-1,11-dioxo-2-naphthacenecaroxamide monohydrochloride.
Aureomycin; Fermycin Soluble. Antibacterial, antiamebic and antiprotozoal. Used in veterinary medicine as an antimicrobial agent. Dec > 210°; $[\alpha]_D^{23} = 240°$; soluble in H_2O (0.86 g/100 ml), MeOH (1.74 g/100 ml), EtOH (0.17 g/100 ml); insoluble in Me_2CO, Et_2O, $CHCl_3$, dioxane; LD_{50} (rat orl) = 10300 mg/kg.

Veterinary products:

035-805; (with Sulfamethazine) Aureomix-S 700 Crumbles; Aureomix-S 700 g; *Roche Vitamins.*
039-077; (with Penicillin G Procaine, Sulfathiazole) CSP ™ 250; CSP ™ 500; *Boehringer Ingelheim Vetmedica.*
045-444; (with Decoquinate) Deccox ® / ChlorMax ®; Decoquinate & Chlortetracycline; *Alpharma.*
046-699; ChlorMax™ 10 Type B Medicated Article; ChlorMax™ 100 Type B Medicated Article; ChlorMax™ 50 Type B Medicated Article; ChlorMax™ 60 Type B Medicated Article; ChlorMax™ 75 Type B Medicated Article; Micro CTC 100; *Alpharma.*
055-018; Aureomycin ® Tablets 25 mg; *American Cyanamid Divn., AHP Corp.*
055-039; Aureomycin ® Soluble Oblets; *American Cyanamid Divn., AHP Corp.*
055-040; SF Mix 66; *Roche Vitamins.*
065-071; Aureomycin ® Soluble Powder; *American Cyanamid Divn., AHP Corp.*
065-178; Fermycin Soluble; *Boehringer Ingelheim Vetmedica.*
065-222; Keet Life; *Hartz Mountain Products.*
065-256; Chlortet-Soluble-O; *ADM Animal Health & Nutrition.*
065-440; Aureomycin ® Soluble Powder Concentrate; *American Cyanamid Divn., AHP Corp.*
065-480; Scour-Pneumonia Antibiotic; *Pennfield Oil.*
065-481; Calf Scour Boluses; Chlortetracycline Pneumonia; *Boehringer Ingelheim Vetmedica.*
121-553; (with Monensin Sodium) Coban ® / Aureomycin ®; *Roche Vitamins.*
141-011; (with Tiamulin Hydrogen Fumarate) Denagard ® 10 / Chlortetracycline Premixes; *Boehringer Ingelheim Vetmedica.*
200-140; (with Penicillin G Procaine, Sulfathiazole) Aureozol ®; *Roche Vitamins.*
200-236; Chlortetracycline HCL Soluble Powder; *Phoenix.*

282 Doxycycline Hyclate

24390-14-5 3496

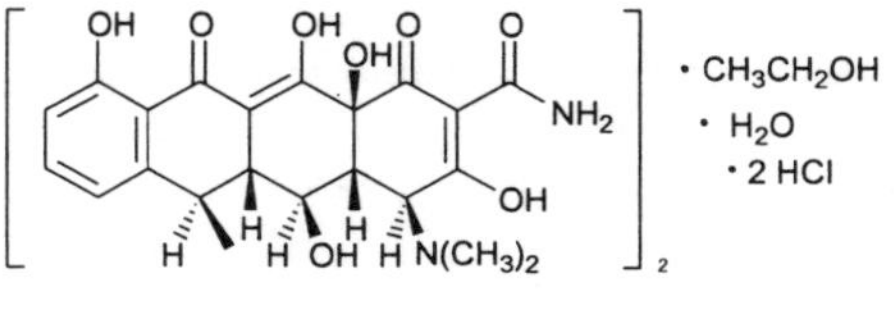

· CH_3CH_2OH · H_2O · 2 HCl

$C_{46}H_{56}Cl_2N_4O_{17}.H_2O$
4-(Dimethylamino)-1,4,4a,5,5a,6,11,12a-octahydro-3,5,10,12,12a-pentahydroxy-6-methyl-1,11-dioxo-2-naphthacenecarboxamide monohydrochloride compound with ethyl alcohol (2:1) monohydrate.
Doryx; Vibra-Tabs; Vivox; Azudoxat; Bassado; Clinofug; Diocimex; Doryx; Doxatet; Doxicrisol; Doxylar; Doxichel hyclate; Doxytem; Duradoxal; Granudoxy; Hydramycin; Mespafin; Nordox; Paldomycin; Retens; Ronaxan; Sigadoxin; Spanor; Tetradox; Unacil; Vibramycin Hyclate; Vibraveineuse; Vibravenös; Zadorin. Tetracycline antibiotic. Used to treat infections in small

animals and psittacosis in birds. Dec 201°; $[\alpha]_D^{25}$= -110° (c = 1 in 0.01N HCl/MeOH); λ_m 267, 351 nm (log ε 4.24 4.12 0.01N HCl/MeOH); soluble in H_2O; LD_{50} (rat ip) = 262 mg/kg.

Veterinary products:

141-082; Heska™ Periceutic™ Gel; Heska™ Periodontal Disease Therapeutic; *Heska.*

283 Oxytetracycline

79-57-2 7111 201-212-8

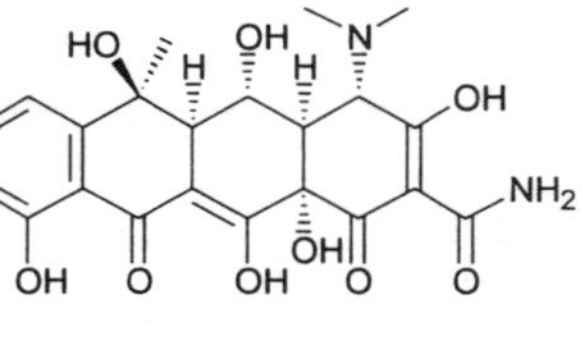

$C_{22}H_{24}N_2O_9$
4-(Dimethylamino)-1,4,4a,5,5a,6,11,12a-octahydro-3,5,6,10,12,12a-hexahydroxy-6-methyl-1,11-dioxo-2-naphthacenecarboxamide.
glomycin; riomitsin; hydroxytetracycline. Tetracycline antibiotic. Used in dogs, cats and large animals.

Veterinary products:

046-718; (with Melengestrol Acetate) MGA ® (liquid) / Terramycin ®; *Pharmacia & Upjohn.*
046-719; (with Melengestrol Acetate) MGA ® (dry) / Terramycin ®; *Pharmacia & Upjohn.*
095-143; OXTC ® - 10; OXTC ® - 30; OXTC ® - 50; OXTC ® - 100; OXTC ® - 200; OXTC ® - 50S; OXTC ® - 100S; OXTC ® - 100MR; *Pfizer.*
113-232; Liquamycin ® LA-200 ®; Terramycin LA-200 ®; *Pfizer.*
138-938; Oxytetracycline - 100; Oxytetracycline - 100MR; Oxytetracycline - 50; Pennox 100; Pennox 100MR; Pennox 200; Pennox 50; *Pennfield Oil.*
200-008; Bio-Mycin ® 200; Oxy-Tet ™ 200; *Boehringer Ingelheim Vetmedica.*
200-095; (with Roxarsone, Salinomycin Sodium) Sacox ® + Aureomycin ®; *Hoechst-Roussel Vet.*
200-096; (with Roxarsone, Salinomycin Sodium) Sacox ® + Terramycin ®; *Hoechst-Roussel Vet.*
200-117; Oxyshot LA; *Cross Vetpharm.*
200-123; Maxim-200; Oxytetracycline Hydrochloride Injection 200mg; *Phoenix.*
200-154; Oxytetracycline 200; *Pennfield Oil.*
200-232; Geomycin 200; *Pliva.*
200-247; Oxytetracycline HCL Soluble Powder-343; *Phoenix.*

284 Oxytetracycline Hydrochloride

2058-46-0 7111 218-161-2
$C_{22}H_{25}ClN_2O_9$
4-(Dimethylamino)-1,4,4a,5,5a,6,11,12a-octahydro-3,5,6,10,12,12a-hexahydroxy-6-methyl-1,11-dioxo-2-naphthacenecarboxamide monohydrochloride.
Medamycin; Oxlopar; Oxyject 100; Oxy WS; Terramycin Hydrochloride; Alamycin; Aquacycline; Bio-Mycin; Duphacycline; Engemycin; Geomycin; Gynamousse; Macocyn; Occrycetin; Oxlopar; Oxybiocycline; Oxybiotic; Oxycyclin; Oxy-Dumocyclin; Oxyject; Oxylag; Oxypan; Oxytetracid; Oxytetrin; Terrafungine; Terraject; Tetramel; Tetran; Tetra-Tablinene; Toxinal; Vendarcin; component of: Terra-Cortril. Tetracycline antibiotic. Used in dogs, cats and large animals. Yellow platelets; very soluble in H_2O; soluble in absolute EtOH (1.2 g/100 ml), 95% EtOH (3.3 g/100 ml).

Veterinary products:

007-879; Terramycin ® Vet Capsules; *Pfizer.*
008-622; Terramycin ® Animal Formula; Terramycin ® Soluble Powder; *Pfizer.*
008-763; (with Polymyxin B Sulfate) Terramycin ® Ophthalmic Ointment With Polymyxin; *Pfizer.*
008-769; Liquamycin ® Injectable; Terramycin ® Injectable; *Pfizer.*
008-804; Terramycin ® 10 Type A Medicated Articles; Terramycin ® 100 Type A Medicated Article; Terramycin ® 100D Type A Medicated Article; Terramycin ® 100SS Type A Medicated Article; Terramycin ® 20 Type A Medicated Articles; Terramycin ® 200 Type A Medicated Article; Terramycin ® 50 Type A Medicated Article; Terramycin ® 50D Type A Medicated Article; Terramycin ® Animal Mix; *Pfizer.*
011-060; Terramycin ® Scour Tablets; *Pfizer.*
013-146; (with Lidocaine) Liquamycin ® Intramuscular; *Pfizer.*
013-293; (with Hydrocortisone) Liqua-Cortril Spray; Terra-Cortril Spray; *Pfizer.*
032-946; (with Carbomycin) Magna Terramycin Soluble Powder; *Pfizer.*
038-200; Medamycin Soluble Antibiotic; OXY WS ™ Soluble Antibiotic; *Boehringer Ingelheim Vetmedica.*
045-143; Oxyject ®; *Boehringer Ingelheim Vetmedica.*
047-278; Bio-Mycin ®; Oxy-Tet ™ 50; *Boehringer Ingelheim Vetmedica.*
048-287; Oxytetracycline-50; *Merial.*
049-948; (with Lidocaine) Aquachel 100 mg; *Pfizer.*
091-127; Rachelle Oxyvet Injection; *Pfizer.*
094-114; Liquamycin ® 100; Terramycin ® 100; *Pfizer.*
095-642; Oxy-Tet ™ Injection; *Boehringer Ingelheim Vetmedica.*
097-452; Oxyject ® 100; *Boehringer Ingelheim Vetmedica.*
099-402; Aquachelle-100; Oxyvet-100; *Pfizer.*
103-758; Terramycin ® Premix; *Pfizer.*
108-963; Medamycin ® Injectable; *Boehringer Ingelheim Vetmedica.*
130-435; Oxy-Tet ™ Soluble; *Russell.*
140-582; Biocyl 100; Biocyl 50; Oxytetracycline Hydrochloride; *Anthony Products.*
141-002; OXY 500 Calf Bolus; OXY 1000 Calf Bolus; *Boehringer Ingelheim Vetmedica.*
200-026; Pennox 343; *Pennfield Oil.*
200-066; Agrimycin ™ - 343 Soluble Powder; *Agri Labs.*
200-068; Oxytetracycline Hydrochloride Injection; *Phoenix.*
200-144; Oxytetracycline HCl Soluble Powder; *Merial.*
200-146; Oxytetracycline HCl Soluble Powder; *Phoenix.*

285 Oxytetracycline Monoalkyl Trimethyl Ammonium Salt

7111
Tetracycline antibiotic. Used in dogs, cats and large animals.

Veterinary products:

038-439; Terramycin ® for Fish; *Pfizer.*
099-006; (with Monensin Sodium) Terramycin ® TM-50;

TM-10 Plus; *Pfizer*.
101-666; (with Robenidine Hydrochloride) Oxytetracycline & Robenz ® Premix; *Pfizer*.
140-448; (with Salinomycin Sodium) Bio-Cox ® / Terramycin ®; *Pfizer*.
140-579; (with Lasalocid Sodium) Bovatec ® / Terramycin ®; *Roche Vitamins*.

286 Tetracycline Phosphate Complex
1336-20-5 9337 215-646-0
(4S,4aS,5aS,12aS)-4-(Dimethylamino)-1,4,4a,5,5a,6,11,12a-octahydro-3,6,10,12,12a-pentahydroxy-6-methyl-1,11-dioxo-2-naphthacenecarboxamide phosphate complex.
Tetrex; Tetrex BidCaps; Panmycin P; Telotrex; Tetradecin; Novum; Upcyclin; component of: Azotrex. Tetracycline antibiotic. Sparingly soluble in H_2O, slightly soluble in EtOH.

Veterinary products:

065-099; (with Novobiocin Sodium) Albaplex® Capsules; *Pharmacia & Upjohn*.

Anticholinergics

287 Aminopentamide Hydrogen Sulfate
60-46-8 479 200-479-8

CONH2 · H2SO4

$C_{19}H_{26}N_2O_5$
(±)-4-Dimethylamino-2,2-diphenylvaleramide hydrogen sulfate.
Amforol®; Centrine; component of 043-078 Centrine Oral Tablets, 043-079 Centrine Injectable, with Attapulgite and Bismuth Subcarbonate of 042-548, Amforol® Suspension, with Ketamine Hydrochloride of 092-116 Ketaset® Plus (Fort Dodge Animal Health). Anticholinergic, antiemetic and anticonvulsant. Used to treat vomiting and diarrhea in dogs and cats. Deliquescent crystals; mp = 185-187°; λ_m = 258.5 nm ($A^{1\%}_{1\ cm}$ 10.3(1% H_2SO_4)freely soluble in H+&2O, EtOH, slightly soluble in $CHCl_3$, insoluble in Et_2O; pH of 2.5% aqueous solution = 1.3-2.2.

Veterinary products:

042-548; (with Attapulgite, Bismuth Subcarbonate, Kanamycin Sulfate, Pectin) Amforol® Suspension; *Fort Dodge Animal Health, Divn. AHP*.
043-078; Centrine Oral Tablets; *Fort Dodge Animal Health, Divn. AHP*.
043-079; Centrine Injectable; *Fort Dodge Animal Health, Divn. AHP*.
092-116; Ketaset® Plus; *Fort Dodge Animal Health, Divn. AHP*.

288 Atropine
51-55-8 907 200-104-8

N O O OH

$C_{17}H_{23}NO_3$
1αH,5αH-Tropan-3α-ol (±)-tropate (ester).
tropine tropate; dl-tropyl tropate. Anticholinergic. Used systemically as a pre-anesthetic, to treat sinus brachycardia, hypersialism or bronchoconstrictive disease, and as an antidote for cholinergic agents or organophosphates. mp = 114-116°; soluble in H_2O (0.22 g/100 ml at 25°, 11.1 g/100 ml at 80°), EtOH (50 g/100 ml at 25°, 83.3 g/100 ml at 60°), glycerol (3.7 g/100 ml), Et_2O (4 g/100 ml), $CHCl_3$ (100 g/100 ml), C_6H_6; LD_{50} (rat orl) = 750 mg/kg.

Veterinary products:

013-248; (with Trichlorfon) Freed No. 25; Freed No. 10; *Fort Dodge Animal Health, Divn. AHP*.

289 Atropine Sulfate [anhydrous]
55-48-1 908 200-235-0
$C_{34}H_{48}N_2O_{10}S$
1αH,5αH-Tropan-3α-ol (±)-tropate (ester) sulfate (2:1) (salt).
Atropisol; Atropisol® Ophthalmic Solution; Atropisol® Ophthalmic Solution. Anticholinergic. Used systemically as a pre-anesthetic, to treat sinus brachycardia, hypersialism or bronchoconstrictive disease, and as an antidote for cholinergic agents or organophosphates. mp = 190-194°; soluble in H_2O (2.5 g/ml), less soluble in organic solvents; LD_{50} (rat orl) = 622 mg/kg.

Veterinary products:

035-650; (with Trichlorfon) Dyrex Powder; *Fort Dodge Animal Health, Divn. AHP*.

290 Atropine Sulfate Monohydrate
5908-99-6 908
$C_{34}H_{48}N_2O_{10}S.H_2O$
1αH,5αH-Tropan-3α-ol (±)-tropate (ester) sulfate (2:1) (salt) monohydrate.
Anticholinergic. Used systemically as a pre-anesthetic, to treat sinus brachycardia, hypersialism or bronchoconstrictive disease, and as an antidote for cholinergic agents or organophosphates.

291 Caramiphen Ethanedisulfonate
125-86-0 1820 204-759-0

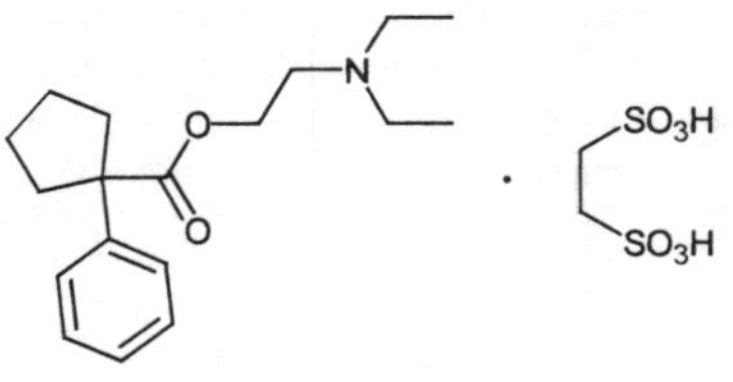

$C_{38}H_{60}N_2O_{10}S_2$
2-Diethylaminoethyl 1-phenylcyclopentane-1-carboxylate ethanedisulfonate.
Alcopon; Taoryl; Toryn. Anticholinergic. mp = 115-116°; soluble in H_2O.

Veterinary products:

009-339; (with Ammonium Chloride) Carafen Cough Syrup; Carafen Cough Tablets; *Fort Dodge Animal Health, Divn. AHP.*

292 Glycopyrrolate
596-51-0 4511 209-887-0

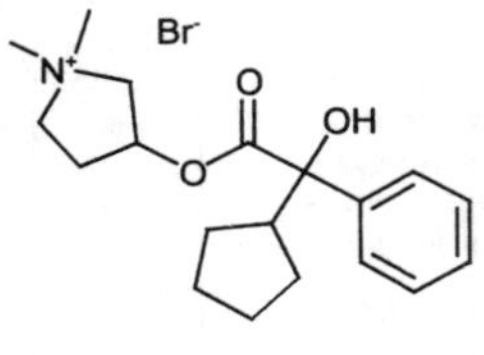

$C_{19}H_{28}BrNO_3$
3-Hydroxy-1,1-dimethylpyrrolidinium bromide α-cyclopentylmandelate.
copyrrolate; Robinul; AHR-504; Nodapton; Robanul; Tarodyl; Tarodyn. Anticholinergic. Used as a preanesthetic anticholinergic agent. mp = 193.2-194.5°; soluble in H_2O; LD_{50} (fmus ip) = 107 mg/kg, (frat ip) = 196 mg/kg, (mrat orl) = 1150 mg/kg.

Veterinary products:

101-777; Robinul®-V Injectable; *Robins.*

293 Isopropamide Iodide
71-81-8 5221 200-766-8

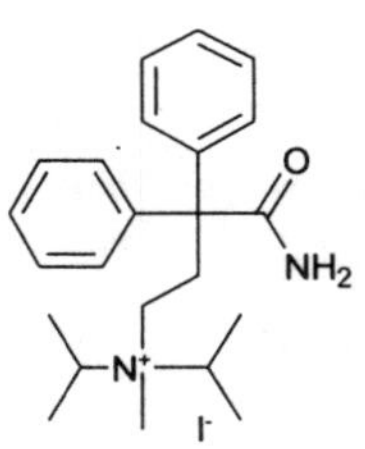

$C_{23}H_{33}IN_2O$
(3-Carbamoyl-3,3-diphenylpropyl)diisopropylmethyl-ammonium iodide.
Darbid; R-79; Priamide; Tyrimide; component of: Combid, Stelabid. Human and veterinary anticholinergic. Potent carbonic anhydrase inhibitor for control of glaucoma. mp = 198-201°, 189.0-191.5°; soluble in hot H_2O, MeOH, EtOH, insoluble in Et_2O.

Veterinary products:

013-201; (with Prochlorperazine Dimaleate) Darbazine Spansule No.1; Darbazine Spansule No.3; *Pfizer.*
015-147; (with Prochlorperazine Edisylate) Darbazine Injectable; *Pfizer.*
031-914; (with Neomycin Sulfate, Prochlorperazine Dimaleate) Neo-Darbazine Spansule Capsule No.1; Neo-Darbazine Spansule Capsule No.3; *Pfizer.*

294 Oxybutynin
5633-20-5 7089

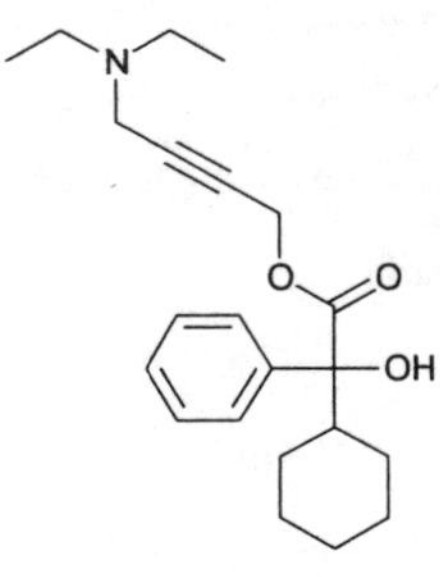

$C_{22}H_{31}NO_3$
4-(Diethylamino)-2-butynyl α-phenylcyclohexane-glycolate.
Anticholinergic. Used as a urinary antispasmodic.

295 Oxybutynin Chloride
1508-65-2 7089 216-139-7

$C_{22}H_{32}ClNO_3$
4-(Diethylamino)-2-butynyl α-phenylcyclohexane-glycolate hydrochloride.
Ditropan; Dridase; MJ-4309-1; 5058; Cystrin; Pollakisu; Tropax. Anticholinergic. Used as a urinary antispasmodic. mp = 129-130°; LD_{50} (rat orl) = 1220 mg/kg.

296 Propantheline Bromide
50-34-0 7989 200-030-6

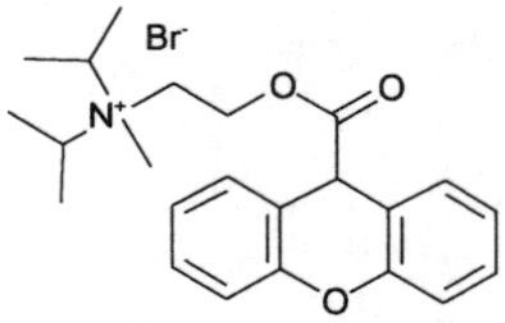

$C_{23}H_{30}BrNO_3$
2-(Hydroxyethyl)diisopropylmethylammonium bromide xanthene-9-carboxylate.
Pro-Banthine; Neo-Metantyl; Pantheline; Corrigast; Ercotina. Anticholinergic. Used as an antispasmodic and antisecretory agent in treatment of diarrhea. mp = 159-161°; very soluble in H_2O, EtOH, $CHCl_3$; insoluble in Et_2O, C_6H_6.

Anticoagulants

297 Heparin

9005-49-6 4685 232-681-7

Heparinic acid.

Arteven; Leparan. Anticoagulant used in veterinary medicine to treat thromboembolism. A glycosaminoglycan with anticoagulant properties. Binds to antithrombin III to accelerate the interaction of antithrombin III with coagulation factors. Catalyzes the inhibition of thrombin by heparin cofactor II. $[\alpha]_D^{20} = 55°$.

298 Polysulfated Glycosaminoglycan

PSGAG

An analog of heparin, used to treat joint disfunction, particularly in horses, and degenerative arthritis in dogs.

Veterinary products:

136-383; Adequan ® Injectable; *Luitpold.*
140-901; Adequan ®; *Luitpold.*
141-038; Adequan ® Canine; *Luitpold.*

299 Warfarin

81-81-2 10174 201-377-6

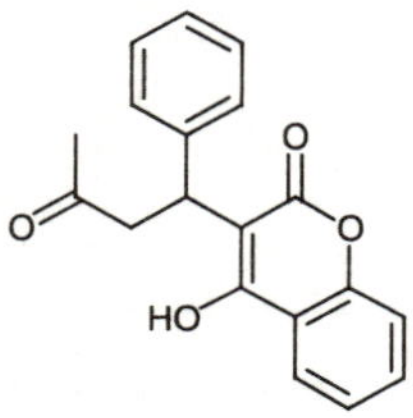

$C_{19}H_{16}O_4$

3-(α-Acetonylbenzyl)-4-hydroxycoumarin.

compound 42; WARF compound 42; Co-Rax; Rodex. Used as a rodenticide and an anticoagulant. Orally effective, fat soluble derivative of 4-hydroxycoumarin that induce hypocoagulability only *in vivo* by inducing the formation of structurally incomplete clotting factors. Used in treatment of thrombotic conditions in dogs, cats and horses. mp = 161°; λ_m = 308 nm (ε 13610 H_2O); soluble in Me_2CO, dioxane; slightly soluble in MeOH, EtOH, iPrOH, some oils; insoluble in H_2O, C_6H_6, C_6H_{12}.

300 Warfarin Sodium

129-06-6 10174 204-929-4

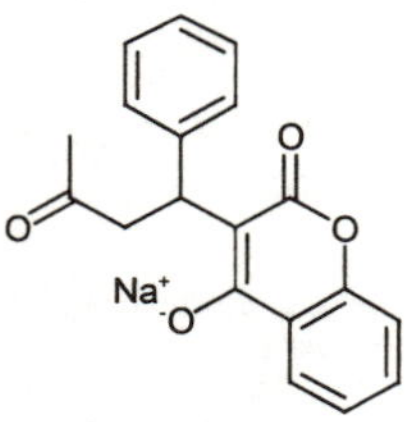

$C_{19}H_{15}NaO_4$

3-(α-Acetonylbenzyl)-4-hydroxycoumarin sodium salt.

Coumadin; Panwarfin; Marevan; Prothromadin; Tintorane; Warfilone; Waran. Used as a rodenticide and an anticoagulant. Orally effective, fat soluble derivative of 4-hydroxycoumarin causing hypocoagulability *in vivo* by inducing formation of structurally incomplete clotting factors. Used in treatment of thrombotic conditions in dogs, cats and horses. Very soluble in H_2O, EtOH; slightly soluble in $CHCl_3$, Et_2O; LD_{50} (mrat orl) = 323 mg/kg, (frat orl) = 58 mg/kg, (mus orl) = 374 mg/kg, (rbt orl) ≅ 800 mg/kg, (mrat orl) = 100.3 mg/kg, (frat orl) = 8.7 mg/kg.

Anticonvulsants

301 Aminopentamide Hydrogen Sulfate

60-46-8 479 200-479-8

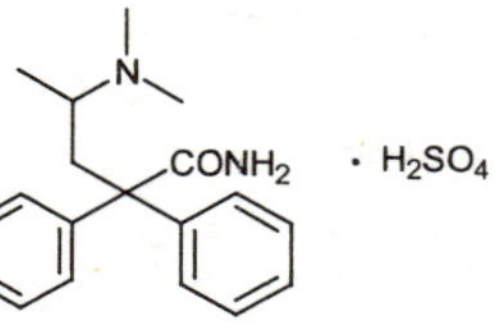

$C_{19}H_{26}N_2O_5$

(±)-4-Dimethylamino-2,2-diphenylvaleramide hydrogen sulfate.

Amforol®; Centrine; component of 043-078 Centrine Oral Tablets, 043-079 Centrine Injectable, with Attapulgite and Bismuth Subcarbonate of 042-548, Amforol® Suspension, with Ketamine Hydrochloride of 092-116 Ketaset® Plus (Fort Dodge Animal Health). Anticholinergic, antiemetic and anticonvulsant. Used to treat vomiting and diarrhea in dogs and cats. Deliquescent crystals; mp = 185-187°; λ_m = 258.5 nm ($A^{1\%}_{1\ cm}$ 10.3 in 1% H_2SO_4) freely soluble in H_2O, EtOH, slightly soluble in $CHCl_3$, insoluble in Et_2O; pH of 2.5% aqueous solution = 1.3-2.2.

Veterinary products:

042-548; (with Attapulgite, Bismuth Subcarbonate, Kanamycin Sulfate, Pectin) Amforol® Suspension; *Fort Dodge Animal Health, Divn. AHP.*
043-078; Centrine Oral Tablets; *Fort Dodge Animal Health, Divn. AHP.*
043-079; Centrine Injectable; *Fort Dodge Animal Health, Divn. AHP.*
092-116; Ketaset® Plus; *Fort Dodge Animal Health, Divn. AHP.*

302 Clonazepam

1622-61-3 2449 216-596-2

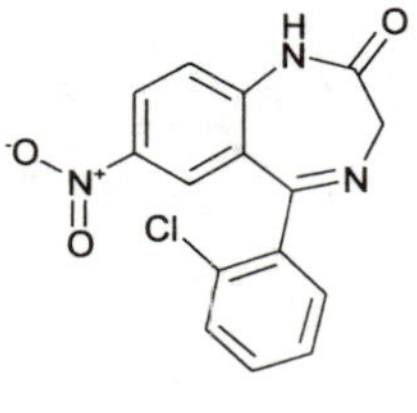

$C_{15}H_{10}ClN_3O_3$

5-(o-Chlorophenyl)-1,3-dihydro-7-nitro-2H-1,4-benzodiazepin-2-one.

Clonopin; Klonopin; Ro-5-4023; Iktorivil; Landsen; Rivotril. Antiepileptic agent with anxiolytic and antimanic properties. Used as an adjunct anti-convulsant in dogs. A benzodiazepine. Used as an an anticonvulsant. mp = 236.5-238.5°; λ_m = 248, 310 nm (ε 14500 11600 MeOH/iPrOH); soluble in Me_2CO (3.1 g/100 ml), $CHCl_3$ (1.5 g/100 ml), MeOH (0.86 g/100 ml), Et_2O (0.07 g/100 ml), C_6H_6 (0.05 g/100 ml), H_2O (< 0.01 g/100 ml); LD_{50} (mus orl) > 4000 mg/kg.

303 Mephenytoin

50-12-4 5898 200-012-8

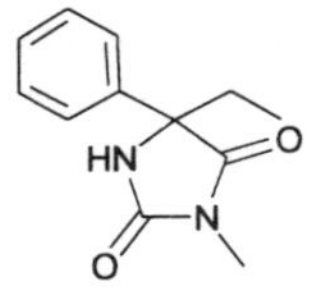

$C_{12}H_{14}N_2O_2$
5-Ethyl-3-methyl-5-phenylhydantoin.
Mesantoin; Mesdontoin; 3-ethylnirvanol; Phenantoin; Sedantional; Gerot-Epilan; Sacerno; component of: Hydantal. An anticonvulsant. Used in dogs as a long-lived second/third line anticonvulsant. mp = 136-137°; insoluble in H_2O; LD_{100} (rat ip) = 270 mg/kg.

304 Phenobarbital

50-06-6 7386 200-007-0

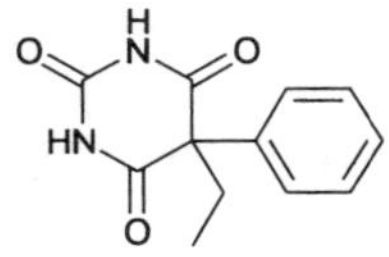

$C_{12}H_{12}N_2O_3$
5-Ethyl-5-phenylbarbituric acid.
Eskabarb; Luminal; Solfoton; Talpheno; Agrypnal; Barbiphenyl; Barbipil; Gardenal; Phenobal; component of: Antrocol, Barbidonna, Bronkotabs, Chardonna-2, Donnatal, Donnazyme, Hydantal, Kinesed, Levsin PB Drops and Tablets, Quadrinal, Tedral. An anticonvulsant, sedative and hypnotic. A long-acting barbiturate with favorable pharmacological profile; the drug of choice to treat epilepsy in dogs and cats. mp = 174-178°; λ_m = 240 nm ($A^{1\%}_{1\ cm}$ 431 pH10), 256 nm ($A^{1\%}_{1\ cm}$ 314 0.1N NaOH); soluble in H_2O (0.1 g/100 ml), EtOH (12.5 g/100 ml), $CHCl_3$ (2.5 g/100 ml), Et_2O (7.7 g/100 ml), C_6H_6 (0.14 g/100 ml); LD_{50} (rat orl) = 162 ± 14 mg/kg.

305 Phenobarbital Sodium

57-30-7 7386 200-322-3

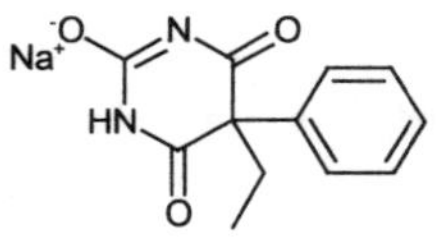

$C_{12}H_{11}N_2NaO_3$
5-Ethyl-5-phenylbarbituric acid sodium salt.
Luminal Sodium; sol phenobarbital; sol phenobarbitone. An anticonvulsant, sedative and hypnotic. A long-acting barbiturate with favorable pharmacological profile, it is the drug of choice to treat epilepsy in dogs and cats. Soluble in H_2O (100 g/100 ml), EtOH (10 g/100 ml); insoluble in Et_2O, $CHCl_3$; LD_{50} (rat orl) = 660 mg/kg.

306 Phenytoin Sodium

630-93-3 7475 211-148-2

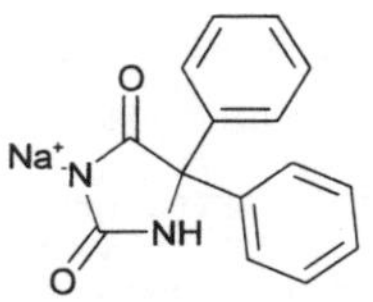

$C_{15}H_{11}N_2NaO_2$
5,5-Diphenylhydantoin sodium salt.
phenytoin soluble; Antisacer; Danten; Diphantoine; Diphenin; Diphenylan sodium; Epanutin; Minetoin; Tacosal; Solantyl; component of: Beuthanasia-D. An anticonvulsant. Has been used in veterinary medicine to manage epilepsy, primarily in dogs. Soluble in EtOH (9.5 g/100 ml), H_2O (1.5 g/100 ml); insoluble in Et_2O, $CHCl_3$; LD_{50} (mus orl) = 490 mg/kg.

Veterinary products:

119-807; (with Pentobarbital Sodium) Beuthanasia-D-Special; *Schering-Plough Animal Health.*
200-071; (with Pentobarbital Sodium) Euthasol ®; *Delmarva.*

307 Primidone

125-33-7 7927 204-737-0

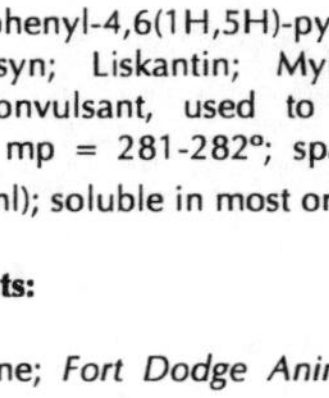

$C_{12}H_{14}N_2O_2$
5-Ethyldihydro-5-phenyl-4,6(1H,5H)-pyrimidinedione.
Mysoline; Neurosyn; Liskantin; Mylepsin; Resimatil; Sertan. An anticonvulsant, used to control epileptic seizures in dogs. mp = 281-282°; sparingly soluble in H_2O (0.06 g/100 ml); soluble in most organic solvents.

Veterinary products:

009-392; Primidone; *Fort Dodge Animal Health, Divn. AHP.*
010-091; Mylepsin Tablets; *Fort Dodge Animal Health, Divn. AHP.*
030-137; Mylepsin; Primidone Medi-Pets; *Fort Dodge Animal Health, Divn. AHP.*
117-689; Neurosyn Tablets; *Boehringer Ingelheim Vetmedica.*

308 Tiletamine Hydrochloride
14176-50-2

$C_{12}H_{18}ClNOS$
2-(Ethylamino)-2-(2-thienyl)cyclohexanone hydrochloride.
CI-634; CL-399; CN-54521-2; component of: Telazol. An anticonvulsant. Used in conjunction with Zolazepam Hydrochloride to provide restraint in cats and analgesia in dogs.

Veterinary products:

106-111; (with Zolazepam Hydrochloride) Telazol®; *Robins.*

309 Valproic Acid
99-66-1 10049 202-777-3

$C_8H_{16}O_2$
Propylvaleric acid.
Depakene; 44089; Mylproin. An anticonvulsant. Third line drug for treatment of seizures in dogs. Clinical efficacy is uncertain. bp_{14} = 120-121°, bp_{20} = 128-130°; d_{25} = 0.904; slightly soluble in H_2O; LD_{50} (rat orl) = 670 mg/kg.

Antidepressants

310 Amitriptyline
50-48-6 511 200-041-6

$C_{20}H_{23}N$
3-(10,11-Dihydro-5H-dibenzo-[a,d]cyclohepten-5-ylidene)-N,N-dimethyl-1-propanamine.
A tricyclic antidepressant. Used to deal with separation anxiety in dogs and excessive grooming, spraying and anxiety in cats.

311 Amitriptyline Hydrochloride
549-18-8 511 208-964-6

$C_{20}H_{24}ClN$
3-(10,11-Dihydro-5H-dibenzo-[a,d]cyclohepten-5-ylidene)-N,N-dimethyl-1-propanamine hydrochloride.
Ro-4-1575; Adepril; Amineurin; Amitid; Amitril; Deprex; Domical; Elavil; Endep; Euplit; Laroxyl; Lentizol; Miketorin; Redomex; Saroten; Sarotex; Sylvemid; Triptizol; Tryptanol; Tryptizol. Antidepressant. Used for separation anxiety in dogs and excessive grooming, spraying and anxiety in cats. mp = 196-197°; freely soluble in H_2O, $CHCl_3$, alcohol; λ_m = 240 nm; pKa = 9.4; LD_{50} (rat orl) = 380 mg/kg.

312 Clomipramine Hydrochloride
17321-77-6 2447 241-344-3

$C_{19}H_{24}Cl_2N_2$
3-Chloro-5-[3-(dimethylamino)propyl]-10,11-dihydro-5H-dibenz-[b,f]azepine monohydrochloride.
Anafranil; G-34586. Antidepressant. Used to manage behavior disorders in dogs. mp = 189-190°.

Veterinary products:

141-120; Clomicalm™; *Novartis Animal Health.*

313 Doxepin
1668-19-5 3492

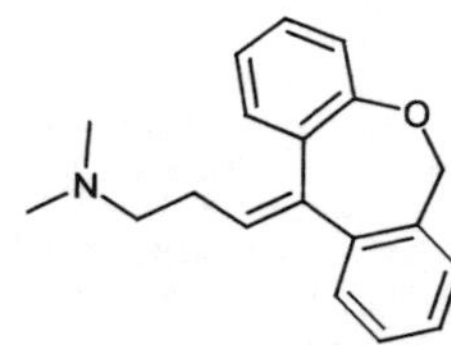

$C_{19}H_{21}NO$
3-Dibenz[b,e]oxepin-11(6H)-ylidene-N,N-dimethyl-1-propanamine.
P-3693A; NSC-108160; Aponal; Curatin; Quitaxon. Tricyclic antidepressant; antipruritic (veterinary). Oily liquid; $bp_{0.03}$ = 154-157°; $bp_{0.2}$ = 260-270°; $:D_{50}$ = (rat iv) = 16 mg/kg, (mus iv) = 26 mg.kg, (rat orl) = 147 mg/kg, (mus orl) = 135 mg/kg.

314 Doxepin Hydrochloride
1229-29-4 3492 214-966-8

$C_{19}H_{22}ClNO$
3-Dibenz[b,e]oxepin-11(6H)-ylidene-N,N-dimethyl-1-propanamine hydrochloride.

Adapin; Aponal; Curatin; Quitaxon; Sinequin. Tricyclic antidepressant; antipruritic (veterinary). A cis-trans (approx. 1:5) mixture. mp = 184-186°.

315 cis-Doxepin Hydrochloride
25127-31-5 3492
$C_{19}H_{22}ClNO$
cis-3-Dibenz[b,e]oxepin-11(6H)-ylidene-N,N-dimethyl-1-propanamine hydrochloride.
cidoxepin hydrochloride; P-4599; doxepin hydrochloride, Z isomer. Tricyclic antidepressant; antipruritic (veterinary). mp = 209-210.5°.

316 trans-Doxepin Hydrochloride
3607-18-9 3492
$C_{19}H_{22}ClNO$
trans-3-Dibenz[b,e]oxepin-11(6H)-ylidene-N,N-dimethyl-1-propanamine hydrochloride.
Tricyclic antidepressant; antipruritic (veterinary). mp = 192-193°.

317 Doxepin Hydrochloride, E Isomer
4698-39-9
$C_{19}H_{22}ClNO$
3-Dibenz[b,e]oxepin-11(6H)-ylidene-N,N-dimethyl-1-propanamine hydrochloride.
Tricyclic antidepressant; antipruritic (veterinary).

318 Imipramine
50-49-7 4955 200-042-1

$C_{19}H_{24}N_2$
10,11-Dihydro-5-(3-(dimethylamino)propyl)-5H-dibenz[b,f]azepine.
G-22355; Imizin. Tricyclic antidepressant. Used to treat urinary incontinence. $bp_{0.1}$ = 160°.

319 Imipramine Hydrochloride
113-52-0 4955 204-030-7
$C_{19}H_{25}ClN_2$
10,11-Dihydro-5-(3-(dimethylamino)propyl)-5H-dibenz[b,f]azepine hydrochloride.
Antideprin; Apo-Imipramine; Berkomine; Censtim; Censtin; Chrytemin; Deprinol; Dimipressin; DIPD; Dyna-Zina; Efuranol; Eupramin; Feinalmin; Imavate; Imidol; Imidobenzyle; Imilanyle; Imiprin; Imizin; Imizinum; Imiprin; Imavate; Impril; Intalpram; Iramil; Irmin; Janimine; Melipramine; Novo-pramine; Praminil; Presamine; Promiben; Pryleugan; SK-Pramine; Surplix; Timolet; Tipramine; Tofranil. Tricyclic antidepressant. Used to treat urinary incontinence. mp = 174-175°; very soluble in H_2O, less soluble in organic solvents; LD_{50} (rat orl) = 490 mg/kg, (rat ip) = 90 mg/kg.

Antidiabetics

320 Chlorpropamide
94-20-2 2239 202-314-5

$C_{10}H_{13}ClN_2O_3S$
4-Chloro-N-[(propylamino)-carbonyl]-benzenesulfonamide.
Diabinese; N-propyl-N'-(p-chlorobenzenesulfonyl)urea; P-607; Adiaben; Asucrol; Catanil; Chloronase; Diabechlor; Diabenal; Diabetoral; Diabinese; Melitase; Millinese; Oradian; Stabinol. A proprietary preparation of chlorpropamide; an oral hypoglycemic agent. A human anti-diabetic agent; also used in treatment of diabetes insipidus in dogs and cats. mp = 127-129°; soluble in H_2O at pH 6; insoluble in H_2O at pH 7.3; soluble in alcohol; moderately soluble in organic solvents; LD_{50} (rat orl) = 2150 mg/kg.

321 Glipizide
29094-61-9 4451 249-427-6

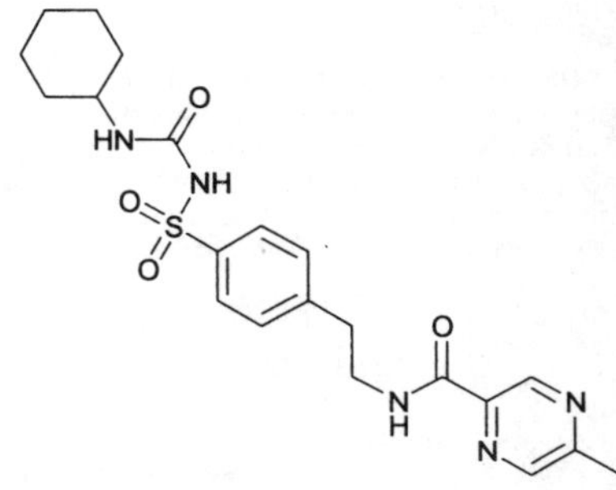

$C_{21}H_{27}N_5O_4S$
1-Cyclohexyl-3-[[p-[2-(5-methylpyrazinecarboxamido)-ethyl]phenyl]sulfonyl]urea.
Glucotrol; Glucotrol XL; CP-28720; K-4024. Antidiabetic agent. Thought to be useful in cats with type II diabetes. mp = 208-209°; LD_{50} (rat ip) = 1.2 g/kg, (mus ip) > 3 g/kg.

322 Insulin
9004-10-8 5011 232-672-8
Antidiabetic agent. Used primarily in small animals. Soluble in dilute acids and alkalies.

Antidiarrheal Agents

323 Bismuth Subsalicylate
14882-18-9 1327 238-953-1

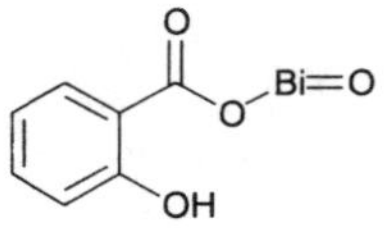

$C_7H_5BiO_4$
2-Hydroxybenzoic acid bismuth (3+) salt.
basic bismuth salicylate; oxo(salicylato)bismuth; Bismogenol Tosse Inj.; Stabisol. Antidiarrheal; antacid; antiulcerative. Also a lupus erythematosus suppressant. Used in animals to manage diarrhea. Insoluble in H_2O, EtOH.

324 Pectin
9000-69-5 7194 232-553-0
Citrus pectin.
Genu® HM USP 100; Genu® HM USP L200; Genu® Pectins; Genu® Pectin (citrus) type USP/100; Genu® Pectin (citrus) type USP/200; Genu® Pectin (citrus) type USP-H; Genu® Pectin (citrus) type USP-L/200; Mexpectin LA 100 Range; Mexpectin LC 700 Range; Mexpectin XSS 100 Range. FDA GRAS, Japan, UK approved, Europe listed, for dentals and topicals, USP/NF compliance. A protective colloid, thickener, emulsifier, adsorbent, in dentals, topicals. In encapsulated drugs, diarrhea treatment; hemostatic formulations, blood plasma substitutes. Used in detoxication, dietary fiber fortification, cholesterol reduction and glucose metabolism. Used in veterinary medicine as an anti-diarrheal. White powder; soluble in H_2O, insoluble in EtOH or other organic solvents. Non-toxic, non-allergenic.

Veterinary products:

042-548; (with aminopentamide hydrogen sulfate, attapulgite; bismuth subcarbonate, kanamycin sulfate) Amforol ® Suspension; *Fort Dodge Animal Health, Divn. AHP.*
042-841; (with Attapulgite, Bismuth Subcarbonate, Kanamycin Sulfate) Amforol ® Veterinary Oral Tablets; *Fort Dodge Animal Health, Divn. AHP.*

Antidiuretics

325 Desmopressin
16679-58-6 2969 240-726-7

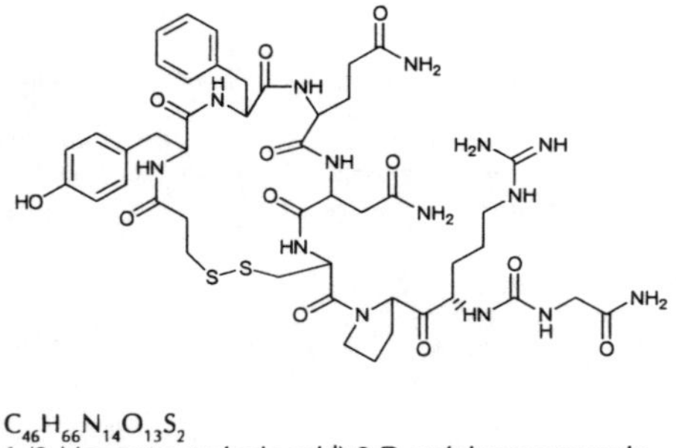

$C_{46}H_{66}N_{14}O_{13}S_2$
1-(3-Mercaptopropionic acid)-8-D-arginine-vasopressin.
Adiuretin SD; DAV Ritter; DDAVP; Desmospray; Minirin. Antidiuretic. Useful in management of diabetes insipidus. $[\alpha]_D^{25} = 85.5° \pm 2°$.

326 Desmopressin Acetate
62357-86-2 2969
$C_{48}H_{68}N_{14}O_{14}S_2.3H_2O$
1-(3-Mercaptopropionic acid)-8-D-arginine-vasopressin monoacetate (salt) trihydrate.
DDAVP; Stimate; Octostim. Antidiuretic. Useful in management of diabetes insipidus.

327 Vasopressin
9034-50-8 10073

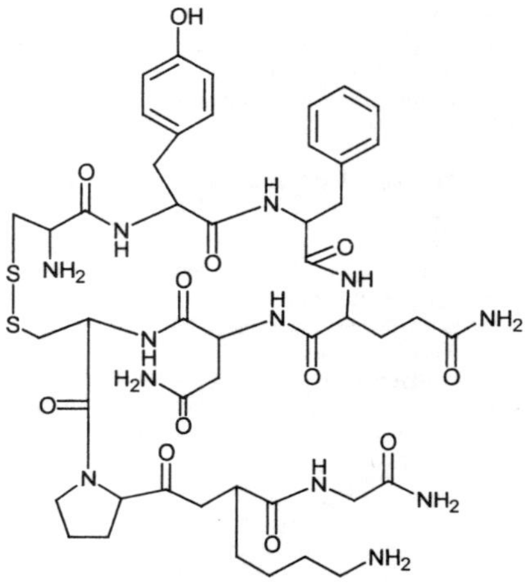

$C_{46}H_{65}N_{15}O_{12}S_2$
β-Hypophamine.
Pitressin; Antidiuretic hormone; beta-hypophamine; Leiormone; Tonephin; Vasophysin. Obtained from posterior lobe of the pituitary of healthy domestic animals or by synthesis. Two vasopressins, which differ in the amino acid at position 8, have been isolated: arginine vasopressin and lysine vasopressin. Antidiuretic and vasopressor hormone. Hemostatic. Used as a diagnostic agent and in treatment of diabetes insipidus in small animals.

328 Vasopressin Injection
11000-17-2 234-236-2
$C_{46}H_{65}N_{13}O_{12}S_2$
Antidiuretic and vasopressor hormone. Hemostatic. Antidiuretic and vasopressor hormone. Hemostatic. Used as a diagnostic agent and in treatment of diabetes insipidus in small animals.

329 Vasopressin, Arginine form
113-79-1 10073 204-035-4
$C_{46}H_{65}N_{15}O_{12}S_2$
8-L-Arginine-vasopressin.
arginine vasopressin; argipressin; rindervasopressin. Antidiuretic and vasopressor hormone. Hemostatic. Antidiuretic and vasopressor hormone. Hemostatic. Used as a diagnostic agent and in treatment of diabetes insipidus in small animals.

330 Vasopressin, Lysine form
50-57-7 10073 200-050-5

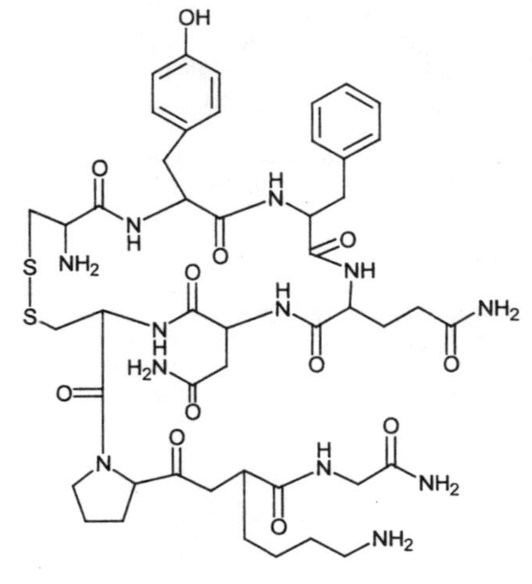

$C_{46}H_{65}N_{13}O_{12}S_2$
8-L-Lysine-vasopressin.
Antidiuretic and vasopressor hormone. Hemostatic. Antidiuretic and vasopressor hormone. Hemostatic. Used as a diagnostic agent and in treatment of diabetes insipidus in small animals.

Antidotes

331 Ammonium Molybdate
12027-67-7 565 234-722-4
$H_{24}Mo_7N_6O_{24}$
Ammonium molybdate(VI).
ammonium paramolybdate; molybdate, hexaammonium; hexaammonium molydbate; molybdic acid ammonium tetrahydrate. Used to treat copper poisoning in sheep. Green-yellow crystals; insoluble in EtOH, soluble in H_2O (43.5 g/100 ml); pH of 5% aqueous solution = 5 - 5.5.

332 Deferoxamine
70-51-9 2914 200-738-5

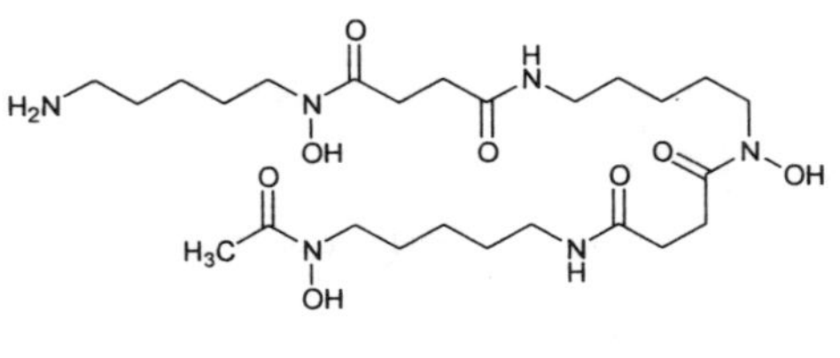

$C_{25}H_{48}N_6O_8$
N-[5-[3-[(5-Aminopentyl)hydroxycarbamoyl]propionamido]pentyl]-3-[[5-(N-hydroxyacetamido)pentyl]-carbamoyl]propionohydroxamic acid.
NSC-527604; Desferioxamine B. Chelating agent for iron. Used to manage iron and aluminum toxicity. mp = 138-140°; soluble in H_2O (1.2 g/100 ml at 20°).

333 Deferoxamine Hydrochloride
1950-39-6 2914 217-767-4
$C_{25}H_{49}ClN_6O_8$
N-[5-[3-[(5-Aminopentyl)hydroxycarbamoyl]propionamido]pentyl]-3-[[5-(N-hydroxyacetamido)pentyl]-carbamoyl]propionohydroxamic acid monohydrochloride.
Ba-29837. Chelating agent for iron. Used to manage iron and aluminum toxicity. mp = 172-175°.

334 Deferoxamine Methanesulfonate
138-14-7 2914 205-314-3
$C_{26}H_{52}N_6O_{11}S$
N-[5-[3-[(5-Aminopentyl)hydroxycarbamoyl]propionamido]pentyl]-3-[[5-(N-hydroxyacetamido)pentyl]-carbamoyl]propionohydroxamic acid monomethanesulfonate.
Desferal mesylate; Ba-33112; DFOM; Desferal. Chelating agent for iron. Used to manage iron and aluminum toxicity. mp = 148-149°; soluble in H_2O (> 20 g/100 ml).

335 Dimercaprol
59-52-9 3255 200-433-7

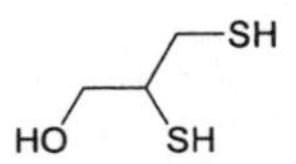

$C_3H_8OS_2$
2,3-Dimercapto-1-propanol.
BAL in Oil; British Anti-Lewisite; Dicaptol; Sulfactin. Antidote to gold, mercury and arsenic poisoning. d_4^{25} = 1.2385; $bp_{0.2}$ = 60°, $bp_{5.6}$ = 100°, bp_{15} = 120°, bp_{25} = 130°, bp_{40} = 140°; soluble in H_2O (8.7 g/100 ml); LD_{50} (rat im) = 86.7 mg/kg.

336 Edetate Calcium Disodium
62-33-9 3555 200-529-9

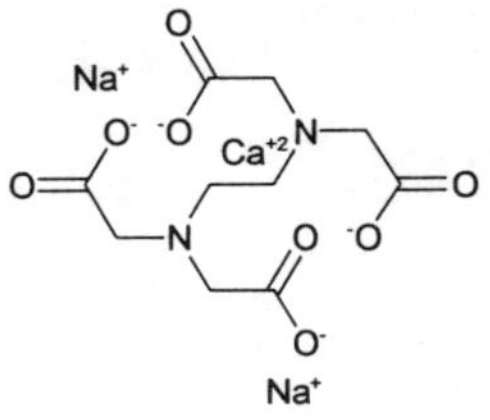

$C_{10}H_{12}CaN_2Na_2O_8 \cdot xH_2O$
Disodium[(ethylenedinitrilo)tetraacetato]calciate(2-) hydrate.
EDTA calcium; edathamiol calcium disodium; sodium calciumedetate; Calcitetracemate Disodium; Calcium Disodium Versenate; Ledclair; Mosatil; Antallin; Sormetal; Versene CA. Complexes lead ions. Used to treat lead poisoning. Soluble in H_2O.

337 Flumazenil
78755-81-4 4169

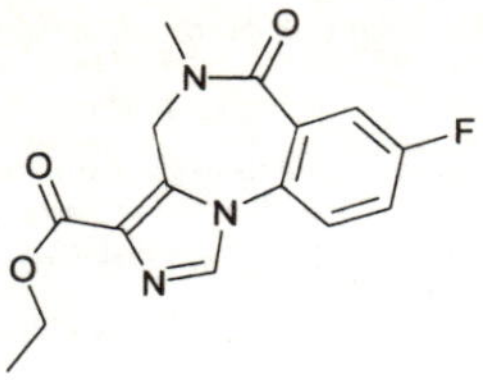

$C_{15}H_{14}FN_3O_3$
Ethyl 8-fluoro-5,6-dihydro-5-methyl-6-oxo-4H-imidazo-[1,5-a][1,4]benzodiazepin-3-carboxylate.
Ro-15-1788/000; flumazepil; Anexate; Lanexat; Mazicon; Romazicon. Benzodiazepine antagonist. Used for reversal of sedation in animals treated (or overdosed) with benzodiazepines. mp = 201-203°; LD_{50} (mus ip) = 4000 mg/kg, (mus orl) = 4300 mg/kg, (rat ip) = 4300 mg/kg, (rat orl) = 6000 mg/kg.

338 Fomepizole
7554-65-6 4257 231-445-0

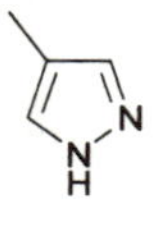

$C_4H_6N_2$
4-Methylpyrazole.
4-MP. Alcohol dehydrogenase inhibitor. Antidote to MeOH and ethylene glycol poisoning. mp = 15.5-18.5°; bp_{18} = 98.5-99.5°; bp_{730} = 204-205°; λ_m = 220 nm (log ε 3.47 EtOH), 226 nm (log ε 3.65 6N HCl); soluble in H_2O, EtOH; LD_{50} (7 day) (mus iv) = 312 mg/kg, (rat iv) = 312 mg/kg, (mus orl) = 640 mg/kg, (rat orl) = 534 mg/kg.

Veterinary products:

141-075; Antizol-Vet®; *Orphan Medical.*

339 Methylene Blue
61-73-4 6137 200-515-2

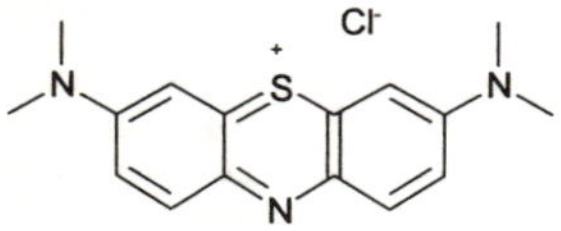

$C_{16}H_{18}ClN_3S.3H_2O$
3,7-Bis(dimethylamino)phenothiazin-5-ium chloride trihydrate.
methylthioninium chloride; C.I. Basic Blue 9 trihydrate; Swiss Blue; Urolene Blue; CI-52015; solvent blue 8. Antimethemoglobinemic in ruminants; antidote to cyanide poisoning. Soluble in H_2O (4 g/100 ml), EtOH (1.54 g/100 ml), $CHCl_3$; insoluble in Et_2O; λ_m = 668, 609 nm.

340 Protamine Sulfate
9009-65-8
Protamine sulphate. Antidote to heparin overdosage which cause hemorrhaging.

341 Sodium Thiosulfate
7772-98-7 8844 231-867-5

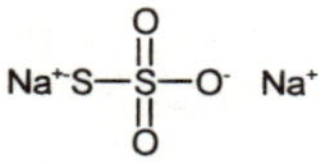

$Na_2S_2O_3$
Disodium thiosulfate.
Sulfactol; sodium hyposulfite; hypo; antichlor; Sodothiol; Sulfothiorine; Ametox; component of: Tinver. Antidote to cyanide poisoning. Also has been used as a topical antifungal agent. d = 1.69; mp = 48°; soluble in H_2O, insoluble in EtOH; LD_{50} (rat iv) > 2500 mg/kg.

Antidyskinetics

342 Selegiline Hydrochloride
14611-52-0 8569

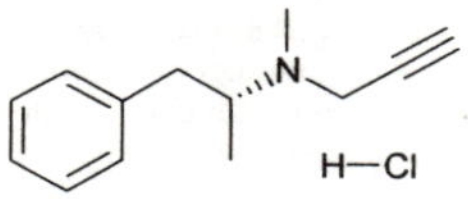

$C_{13}H_{18}ClN$
(-)-(R)-N,α-Dimethyl-N,2-propynylphenethylamine.
Eldepryl; (-)-deprenil; L-deprenyl. Antidyskinetic, antiparkinsonian. Used in dogs for treatment of Cushing's disease. $bp_{0.8}$ = 92-93°; $[\alpha]_D^{2-}$ = -11.2°.

Veterinary products:

141-080; Anipryl®; *Pfizer.*

Antiemetics

343 Aminopentamide Hydrogen Sulfate
60-46-8 479 200-479-8

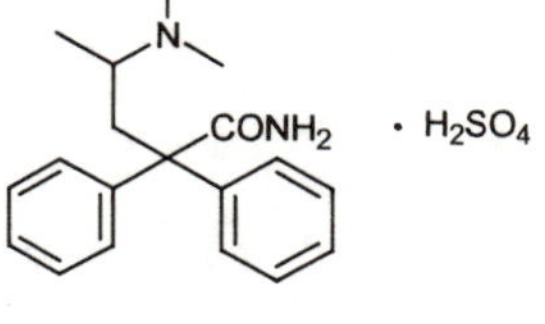

$C_{19}H_{26}N_2O_5$
(±)-4-Dimethylamino-2,2-diphenylvaleramide hydrogen sulfate.
Amforol®; Centrine; component of 043-078 Centrine Oral Tablets, 043-079 Centrine Injectable, with Attapulgite and Bismuth Subcarbonate of 042-548,

Amforol® Suspension, with Ketamine Hydrochloride of 092-116 Ketaset® Plus (Fort Dodge Animal Health). Anticholinergic, antiemetic and anticonvulsant. Used to treat vomiting and diarrhea in dogs and cats. Deliquescent crystals; mp = 185-187°; λ_m = 258.5 nm ($A^{1\%}_{1\ cm}$ 10.3 in 1% H_2SO_4) freely soluble in H_2O, EtOH, slightly soluble in $CHCl_3$, insoluble in Et_2O; pH of 2.5% aqueous solution = 1.3-2.2.

Veterinary products:

042-548; (with Attapulgite, Bismuth Subcarbonate, Kanamycin Sulfate, Pectin) Amforol® Suspension; *Fort Dodge Animal Health, Divn. AHP.*
043-078; Centrine Oral Tablets; *Fort Dodge Animal Health, Divn. AHP.*
043-079; Centrine Injectable; *Fort Dodge Animal Health, Divn. AHP.*
092-116; Ketaset® Plus; *Fort Dodge Animal Health, Divn. AHP.*

344 Attapulgite

1337-76-4 273-218-9

Hydrate magnesium-aluminum silicate.
Palygorskite; Florco®; Florco® -X; Florex®; Florex® Ag-Dri 6/30; LVM 8/16; RVM 8/16; Florigel® H-Y; Flor-Kleen; Min-U-Gel® 100; QA-555; S-60 RVM; Dioctrahedral smectite; Attaclay; Attacote; palygorskite; Attaflow; Attapulgus; Attasorb; Attagel; component with Aminopentamide Hydrogen Sulfate and Bismuth Subcarbonate of 042-548, Amforol® Suspension (Fort Dodge Animal Health); A hydrated aluminum-magnesium silicate, chief ingredient in Fullers earth; drilling fluids, decolorizing oils, filter medium, absorbent. FDA GRAS, USP/NF, BP compliance. TLV:TWA = 10 mg/M^3 (total dust), 5 mg/M^3 (respirable). Used as a suspending agent, viscosity-increasing agent and adsorbent. Used in orals and anti-diarrheal products.

Veterinary products:

042-548; (with Aminopentamide Hydrogen Sulfate, Bismuth Subcarbonate; Kanamycin Sulfate, Pectin) Amforol® Suspension; *Fort Dodge Animal Health, Divn. AHP.*
042-841; (with Bismuth Subcarbonate; Kanamycin Sulfate, Pectin) Amforol® Veterinary Oral Tablets; *Fort Dodge Animal Health, Divn. AHP.*

345 Bismuth Subcarbonate

5892-10-4 1324 227-567-9

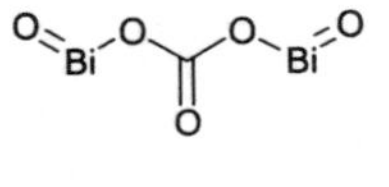

CBi_2O_5
1,3,5-Trioxo-2,4-dioxa-1,5-dibismapentane.
Basic bismuth carbonate; Bismuth oxycarbonate; Bismuth; component with Aminopentamide Hydrogen Sulfate and Attapulgite of 042-548, Amforol® Suspension (Fort Dodge Animal Health). FDA permanently listed. Used as a skin protectant and as a color additive for externally applied pharmaceuticals. Odorless, tasteless powder; insoluble in H_2O, EtOH; d = 6.860.

Veterinary products:

042-548; (with Aminopentamide Hydrogen Sulfate, Kanamycin Sulfate, Pectin) Attapulgite-Amforol ® Suspension; *Fort Dodge Animal Health, Divn. AHP.*
042-841; (with Kanamycin Sulfate, Pectin) Amforol ® Veterinary Oral Tablets; *Fort Dodge Animal Health, Divn. AHP.*

346 Charcoal, Activated

16291-96-6 240-383-3

Animal bone charcoal.
Administered orally, absorbs some drugs and toxins. USP/NF, BP, Ph.Eur. compliance. Used as an adsorbent for pharmaceutical products and in antidiarrheal products. Fine black powder.

347 Dimenhydrinate

523-87-5 3252 208-350-8

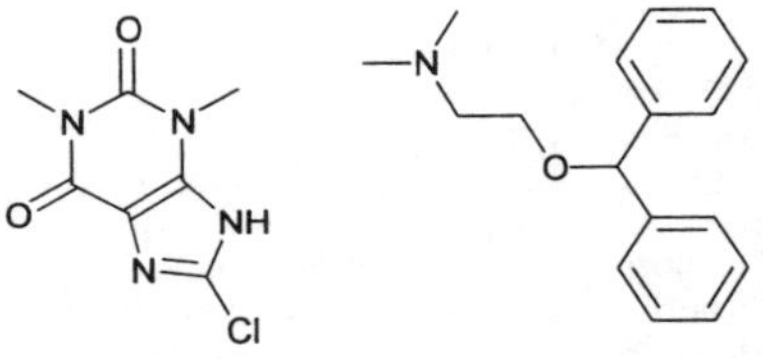

$C_{24}H_{28}ClN_5O_3$
8-Chlorotheophylline compound with 2-(diphenylmethoxy)-N,N-dimethyl-ethylamine (1:1).
Dommanate; chloranautine; Amosyt; Anautine; Andramine; Antemin; Diamarin; Dimate; Dramamine; Dramarin; Dramocen; Dramyl; Emedyl; Emes; Epha; Gravol; Menhydrinate; Reidamine; Removine; Travel-Gum; Travelin; Travelmin; Vomex A; Xamamina; Faston. An anti-emetic used to manage motion sickness in cats and dogs. mp = 102-107°; soluble in H_2O (3 mg/ml), more soluble in organic solvents.

348 Kaolin

1332-58-7 5294 296-473-8

$Al_2O_3.2SiO_2{\cdot}2H_2O$
Native hydrated aluminum silicate.
Aluminum silicate hydroxide; aluminum silicate (hydrated); Anhydrol®; argilla; Bolus alba; China clay; E559; Kaolinite; Kaopectate; Porcelain clay; weisserton; white bole. Specially processed calcined kaolin (aluminum silicate); used as a molecular sieve support and as an adsorbent, suspending agent and tablet and capsule diluent. Used, in combination with Pectin, as an oral anti-diarrheal agent. FDA GRAS; FDA approved for orals. In the FDA list of inactive ingredients, (oral capsules, powders, syrups and tablets). Tablet and capsule diluent; adsorbent in antidiarrheal products, used to treat intestinal disorders, used in orals. Greyish-white, unctuous powder; insoluble in H_2O, dilute acids, dilute alkali; d = 1.8-2.6; viscosity of a 70% w/v aqueous suspension = 300 mPa. Nuisance dust, may cause granulomas.

349 Meclizine
569-65-3 5817 209-323-3

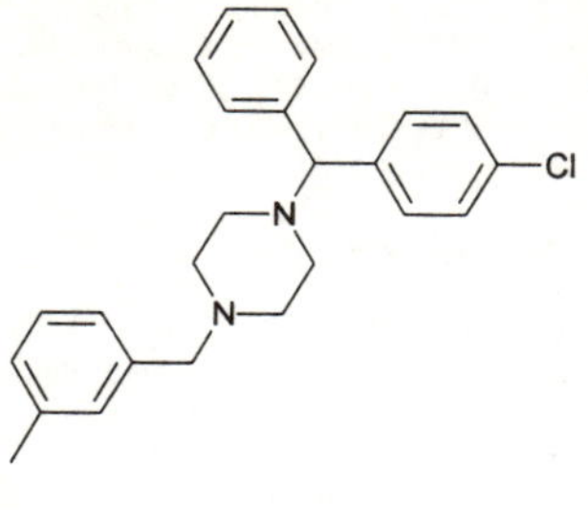

$C_{25}H_{27}ClN_2$
1-(p-Chloro-α-phenylbenzyl)-4-(m-methylbenzyl)-piperazine.
meclozine; parachloramine. Anti-emetic. Used in treatment of motion sickness. $bp_2 = 230°$.

350 Meclizine Hydrochloride
31884-77-2 5817 214-164-8

$C_{25}H_{29}Cl_3N_2.2H_2O$
1-(p-Chloro-α-phenylbenzyl)-4-(m-methylbenzyl)-piperazine dihydrochloride monohydrate.
UCB-5062; Ancolan; Antivert; Bonamine; Bonine; Calmonal; Diadril; Histametizine; Navicalm; Neo-Istafene; Peremesin; Postafene; Sabari; Sea-Legs; Veritab. Anti-emetic. Used in treatment of motion sickness. Insoluble in H_2O (0.1 g/100 ml); freely soluble in $CHCl_3$, C_5H_5N.

351 Metoclopramide
364-62-5 6226 206-662-9

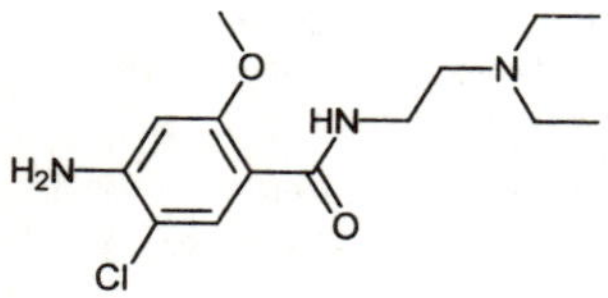

$C_{14}H_{22}ClN_3O_2$
4-Amino-5-chloro-N-[2-(diethylamino)ethyl]-2-methoxybenzamide.
Metoclopramide; 4-Amino-5-chloro-N-[2-(diethylamino)-ethyl]-o-anisamide.HCl.H2O; AHR-3070-C; Cerucal; Clopromate; Draclamid; Emperal; Eucil; Gastrese; Gastrobid; Gastromax; Gastrosil; Gastro-tablinen; Gastrotem; Gastro-Timelets; Maxeran; Maxolon; MCP-ratiopharm; Meclopran; Metamide; Metocobil; Metramid; Parmid; Paspertin; Plasil; Regla. A dopamine D_2 receptor antagonist. Promotes gastric motility and emptying. Used as a GI stimulator and an anti-emetic. mp = 146-148°; soluble in H_2O (0.2 g/100 ml), more soluble in organic solvents.

352 Metoclopramide Hydrochloride
54143-57-6 6226 206-662-9

$C_{14}H_{23}Cl_2N_3O_2.H_2O$
4-Amino-5-chloro-N-[2-(diethylamino)ethyl]-2-methoxybenzamide monohydrochloride monohydrate.
Maxolon; Reglan; AHR-3070-C; Emetid; Gastronerton; Primperan; AHR-3070-C; Cerucal; Clopromate; Draclamid; Emperal; Eucil; Gastrese; Gastrobid; Gastromax; Gastrosil; Gastro-tablinen; Gastrotem; Gastro-Timelets; Maxeran; Maxolon; MCP-ratiopharm; Meclopran; Metamide; Metoclol; Metocobil; Metramid; Moriperan; Mygdalon; Parmid; Paspertin; Peraprin; Plasil; Pramiel; Reglan. A proprietary preparation of metoclopramide. A dopamine D_2 receptor antagonist. Promotes gastric motility and emptying. Used as a GI stimulator and an anti-emetic. mp = 182.5-184°; slightly soluble in H_2O, more soluble in organic solvents.

353 Prochlorperazine Edisylate
1257-78-9 7942 215-019-1

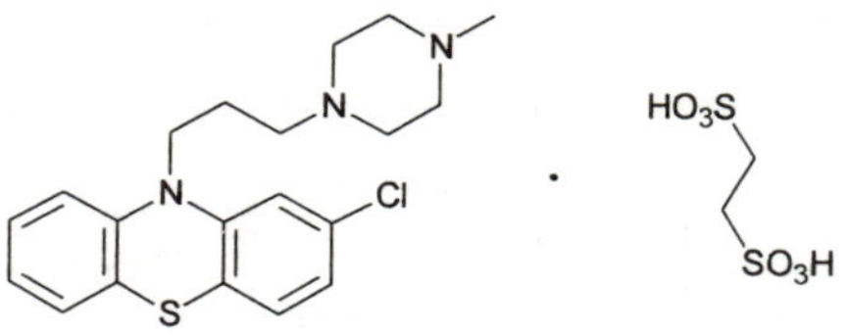

$C_{22}H_{30}ClN_3O_6S_3$
2-Chloro-10[3-(4-methyl-1-piperazinyl)propyl]-phenothiazine 1,2-ethanedisulfonate.
Compazine; Novamin; Tementil. Anti-emetic and antipsychotic used in treatment of vertigo and as an anti-emetic in dogs.

Veterinary products:

015-147; (with Isopropamide Iodide) Darbazine Injectable; *Pfizer*.

354 Prochlorperazine Maleate
84-02-6 7942 201-511-3

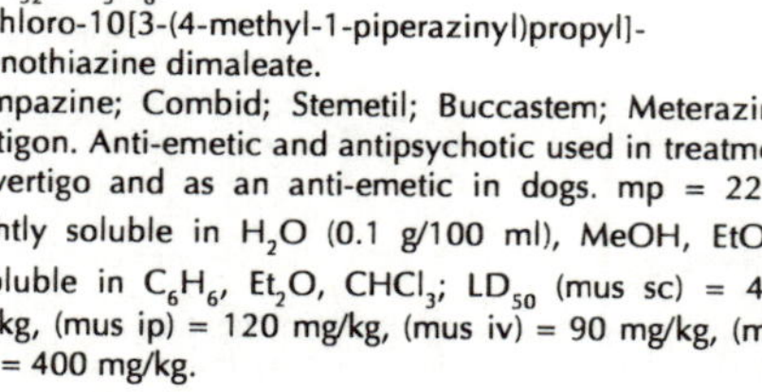

$C_{28}H_{32}ClN_3O_8S$
2-Chloro-10[3-(4-methyl-1-piperazinyl)propyl]-phenothiazine dimaleate.
Compazine; Combid; Stemetil; Buccastem; Meterazine; Vertigon. Anti-emetic and antipsychotic used in treatment of vertigo and as an anti-emetic in dogs. mp = 228°; slightly soluble in H_2O (0.1 g/100 ml), MeOH, EtOH; insoluble in C_6H_6, Et_2O, $CHCl_3$; LD_{50} (mus sc) = 400 mg/kg, (mus ip) = 120 mg/kg, (mus iv) = 90 mg/kg, (mus orl) = 400 mg/kg.

Veterinary products:

013-201; (with Isopropamide Iodide) Darbazine Spansule Capsule No. 1; Neo-Darbazine Spansule Capsule No. 3; *Pfizer*.

Antifungals

355 Amphotericin B

1397-89-3 627 215-742-2

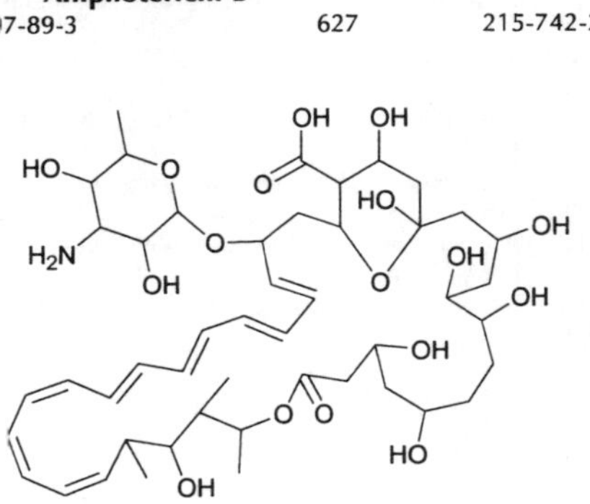

$C_{47}H_{73}NO_{17}$

[1R-1R*,3S*,5R*,6R*,9R*,11R*,15S*,16R*,17R*,18S*, 19E,21E,23E,25E,27E,29E,31E,33R*,35S*,36R*,37S*)]33-[(3-Amino-3,6-dideoxy-β-D-mannopyranosyl)oxy]-1,3,5,6,9,11,17,37-octahydroxy-15,16,18-trimethyl-13-oxo-14,39-dioxabicyclo[33.3.1]nonatriaconta-19,21,23,25,27,29,31-heptaene-36-carboxylic acid.

Amphocin; Fungizone; Ambisome; Amphozone; Fungilin; Ampho-Moronal; component of: Mysteclin-F. Antifungal agent. Used only for potentially fatal fungal infections in dogs because it is itself toxic. mp > 170° (dec); λ_m = 406, 382, 363, 345 nm (MeOH); $[\alpha]_D^{24}$= 333° (acidic DMF), -33.6° (0.1N MeOH/HCl); insoluble in H_2O at pH 6-7, at pH 2 or pH 11, 0.001 g/100 ml); soluble in DMF (0.2 - 0.4 g/100 ml), DMSO (3-4 g/100 ml); LD_{50} (mus ip) = 88 mg/kg, (mus iv) = 4 mg/kg.

356 Benzoic Acid

65-85-0 1122 200-618-2

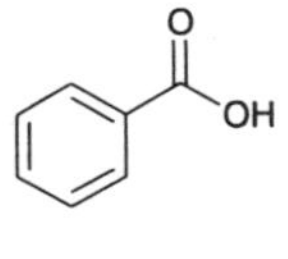

$C_7H_6O_2$

Benzenecarboxylic acid.

benzene formic acid; benzenecarboxylic acid; benzenemethonic acid; Carboxybenzene; Diacylic acid; Dracylic acid; E210; Oracylic acid; phenyl carboxylic acid; phenylformic acid; Retarded BA; Retardex; Salvo; Tennplas; Used in the manufacture of sodium and butyl benzoates, plasticizers, benzoyl chlorides, food preservatives, flavors, perfumes, antifungal agent. FDA GRAS, included in the FDA Inactive Ingredients Guide (im and iv injections, irrigation and oral solutions, suspensions, syrups and tablets, rectal, topical and vaginal preparations). Used as an antimicrobial preservative in cosmetics, foods and pharmaceuticals and as a topical antifungal. White crystals; mp = 122.4°; bp = 249°; d^{24} = 1.311; soluble in H_2O (0.29 g/100 ml), EtOH (37.0 g/100 ml at 15°, 45.4 g/100 ml at 37°), Me_2CO (43.5 g/100 ml), C_6H_6 (10.6 g/100 ml), $CHCl_3$ (22.2 g/100 ml), Et_2O (33.3 g/100 ml) and other organic solvents; LD_{50} (rat orl) = 1700 mg/kg, 2530 mg/kg, (mus orl) = 1940 mg/kg, (mus ip) = 1460 mg/kg, (dog orl) = 2000 mg/kg, (cat orl) = 2000 mg/kg.

Veterinary products:

013-663; (with Amprolium) Purina® Liquid Amprol; *Purina Mills.*

357 Clotrimazole

23593-75-1 2478 245-764-8

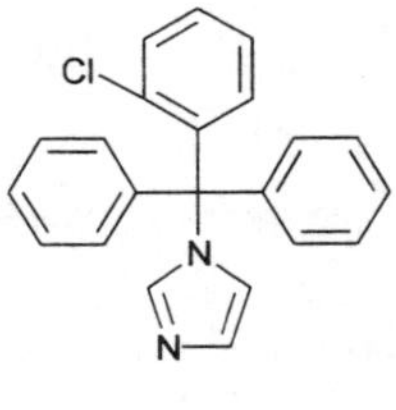

$C_{22}H_{17}ClN_2$

1-(o-Chloro-α,α-di[henylbenzyl)imidazole.

Femcare; Gyne-Lotrimin; Lotrimin; Lotrimin AF Cream; Lotrimin AF Solution; Lotrimin Jock-Itch Cream; Lotrimin Jock-Itch Lotion; Mycelex; Mycelex 7; Mycelex G; Mycelex OTC; Mycelex Troche; Veltrim; BAY 5907; Canesten; Canifug; Empecid; Monobaycuten; Mycofug; Mycosporin; Pedisafe; Rimazole; Tibatin; Trimysten; component of: Lotrimax, Lotrisone, Otomax. Antifungal. mp = 147-149°; slighlty soluble in H_2O, C_6H_6, C_7H_8; soluble in Me_2CO, $CHCl_3$, EtOAc, DMF; LD_{50} (mmus orl) = 923 mg/kg, (rat orl) = 708 mg/kg; [hydrochloride]: mp = 159°.

Veterinary products:

116-089; Veltrim 1% Dermatologic Cream; *Bayer.*
140-896; (with Betamethasone Valerate, Gentamicin Sulfate) Otomax ®; *Schering-Plough Animal Health.*
200-229; (with Betamethasone Valerate, Gentamicin Sulfate) Tri-Otic Ointment; *Med-Pharmex.*

358 Cuprimyxin

28069-65-0

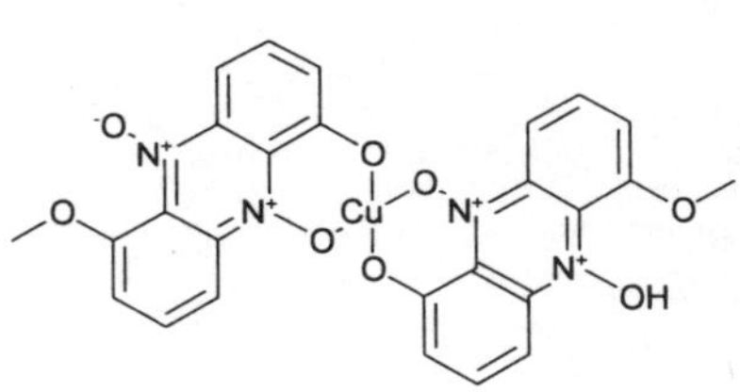

$C_{26}H_{18}CuN_4O_8$

Bis(6-methoxy-1-phenazinol 5,10-dioxidato)copper.

Unitop; Ro-7-4488/1. Antifungal.

Veterinary products:

093-029; Unitop Topical 0.5% Cream; *Roche Vitamins.*

359 Fluconazole
86386-73-4 4158

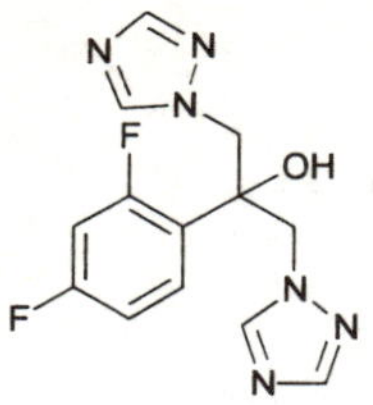

$C_{13}H_{12}F_2N_6O$
2,4-Difluoro-α,α-bis(1H-1,2,4-triazol-1-ylmethyl)benzyl alcohol.
Diflucan; UK-49858; Biozolene; Elazor; Triflucan. Antifungal. Used to treat systemic mycoses, particularly CNS-related conditions in dogs. mp = 138-140°.

360 Flucytosine
2022-85-7 4161 217-968-7

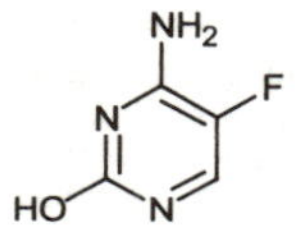

$C_4H_4FN_3O$
4-Amino-5-fluoro-2(1H)-pyrimidinone.
5-fluorocytosine; Ancobon; Ro-2-9915; Ancotil; Alcobon. Antifungal active against *Cryptococcus and Candida* strains. mp = 295-297°; λ_m = 285 nm (ε 8900 0.1N HCl); soluble in H_2O (1.5 g/100 mlat 25°); LD_{50} (mus orl, sc) > 2000 mg/kg, (mus ip) = 1190 mg/kg, (mus iv) = 500 mg/kg.

361 Griseofulvin
126-07-8 4571 204-767-4

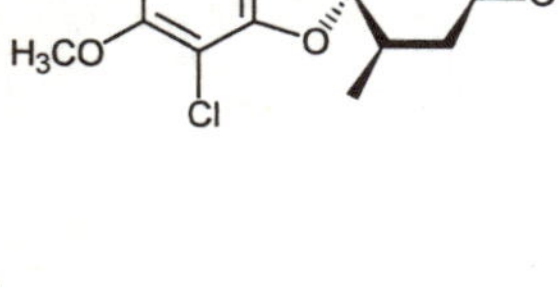

$C_{17}H_{17}ClO_6$
7-Chloro-2',4,6-trimethoxy-6'β-methylspiro[benzofuran-2(3H),1'-[2]cyclohexene]-3,4'-dione.
Fulvicin Bolus; Fulvicin-P/G; Fulvicin-U/F; Grifulvin V; Grisactin; Gris-PEG; Fulvidex; Grysio; amudane; Curling factor; Fulcin; Fulvicin; Grifulvin; Griséfulin; Grisovin; Lamoryl; Likuden; Neo-Fulcin; Polygris; Poncyl-FP; Spirofulvin; Sporostatin. Antifungal. Used to dermatological fungal infections in dogs and cats and ringworm in horses. mp = 220°; $[\alpha]_D^{17}$ = 370° ($CHCl_3$, saturated); λ_m = 286, 325 nm; soluble in DMF (12-14 g/100 ml); slightly solube in MeOH, EtOH, Me_2CO, C_6H_6, $CHCl_3$, EtOAc, AcOH; insoluble in H_2O, petroleum ether.

Veterinary products:

012-227; Fulvicin-U/F ® Tablets; *Schering-Plough Animal Health.*
013-755; Grysio (Microsize); *Fort Dodge Animal Health, Divn. AHP.*
031-447; Fulvicin U/F ® Bolus Veterinary; *Schering-Plough Animal Health.*
039-792; Fulvicin U/F ® Powder; *Schering-Plough Animal Health.*

362 Itraconazole
84625-61-6 5262

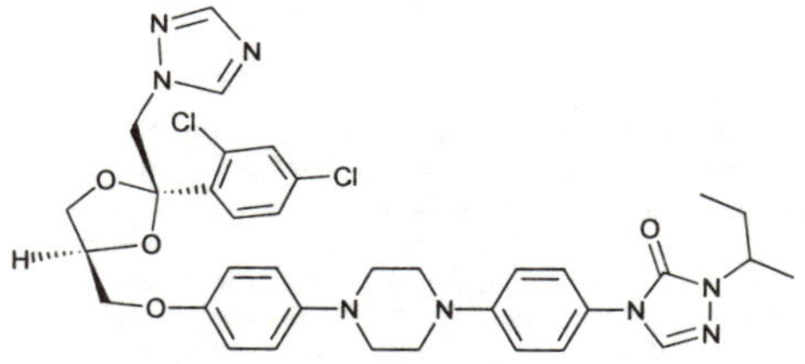

$C_{35}H_{38}Cl_2N_8O_4$
(±)-1-sec-Butyl-4-p-[4-[p-[[(2R*,4S*)-2-(2,4-dichlorophenyl)-2-(1H-1,2,4-triazol-1-ylmethyl)-1,3-dioxolan-4-yl]methoxy]phenyl]-1-piperazinyl]phenyl]-Δ^2-1,2,4-triazolin-5-one.
Sporanox; oriconazole; R-51211; Itrizole; Triasporin. Orally active antifungal. Structurally related to ketoconazole. Used to treat systemic mycoses. mp = 166.2°; insoluble in H_2O; LD_{50} (14 day) (mus orl) > 320 mg/kg, (rat orl) > 320 mg/kg, (dog orl) > 200 mg/kg.

363 Ketoconazole
65277-42-1 5313 265-667-4

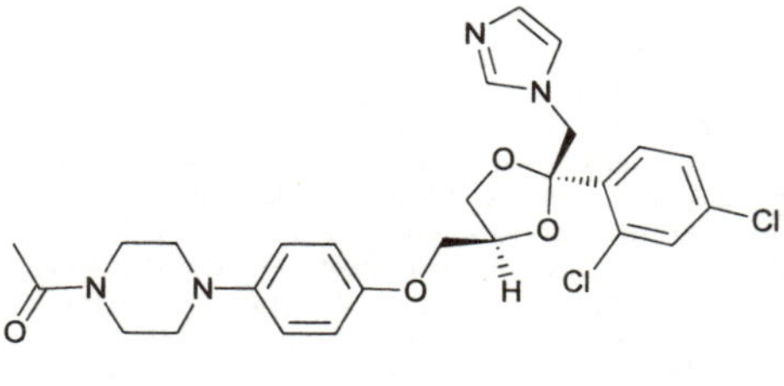

$C_{26}H_{28}Cl_2N_4O_4$
(±)-cis-1-Acetyl-4-[p-[[2-(2,4-dichlorophenyl)-2-(imidazol-1-ylmethyl)-1,3-dioxolan-4-yl]methoxy]-phenyl]piperazine.
Nizoral, R-41400; Ketoisdin; Fungarest; Fungoral; Ketoderm; Orifungal M; Panfungol. Orally active, broad-spectrum antimycotic. mp = 146°; LD_{50} (mus iv) = 44

mg/kg, (mus orl) = 702 mg/kg, (rat iv) = 86 mg/kg, (rat orl) = 227 mg/kg, (gpg iv) = 28 mg/kg, (gpg orl) = 202 mg/kg, (dog iv) = 49 mg/kg, (dog orl) = 780 mg/kg.

364 2-Mercaptobenzothiazole

149-30-4 5916 205-736-8

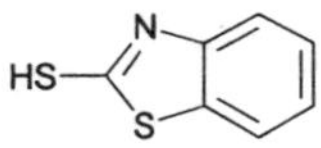

$C_7H_5NS_2$
2(3H)-Benzothiazolethione.
2-benzothiazolethiol; MBT; Captax; Dermacid; Mertax; Thiotax. Vulcanization accelerator. Zinc and sodium salts used as a fungicide. mp = 179°; d = 1.4200; insoluble in H_2O, soluble in organic solvents.

Veterinary products:

005-236; Sulfodene Medication for Dogs; *Combe, Inc.*

365 Miconazole Nitrate

22832-87-7 6266 245-256-6

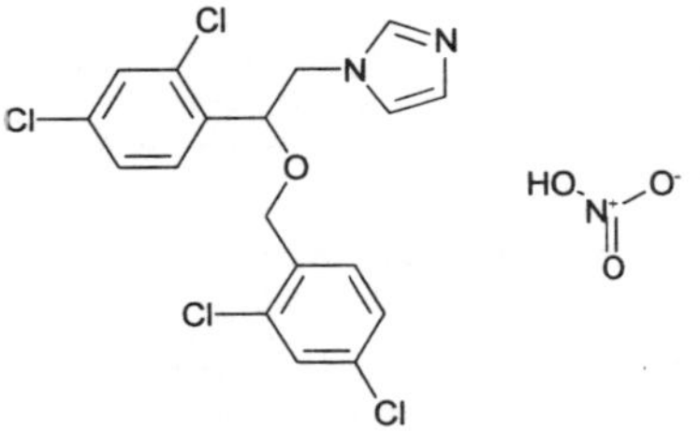

$C_{18}H_{15}Cl_4N_3O_4$
1-[2,4-Dichloro-β-[(2,4-dichlorobenzyl]oxy]phenethyl]-imidazole nitrate.
R-14889; Antifungal Cream; Lotrimin AF Powder; Lotrimin AF Powder Aerosol; Lotrimin AF Spray Liquid; Loptrimin AF Jock-Itch Powder Aerosol; Micatin; Monistat Cream and Suppositories; Monistat-Derm; Zeasorb-AF; Aflorix; Albistat; Andergin; Brentan; Conoderm; Conofite; Daktar; Daktarin; Deralbine; Dermonistat; Epi-Monistat; Florid; Fungiderm; Fungisdin; Gyno-Daktarin; Gyno-Monistat; Miconal; Ecobi; Micotef; Monistat; Prilagin; Vodol. Antifungal. Used with animals as a topical antifungal. mp = 170.5, 184-185°; [(+)-form nitate]: mp = 135.3°; $[\alpha]_D^{20}$ = 59° (MeOH); [(-)-form nitate]: mp = 135°; $[\alpha]_D^{20}$ = - 58° (MeOH).

Veterinary products:

095-183; Conofite ® Cream 2%; *Schering-Plough Animal Health.*
095-184; Conofite ® Lotion 1%; *Schering-Plough Animal Health.*
200-196; Miconosol; *Med-Pharmex.*

366 Monensin

17090-79-8 6329 241-154-0

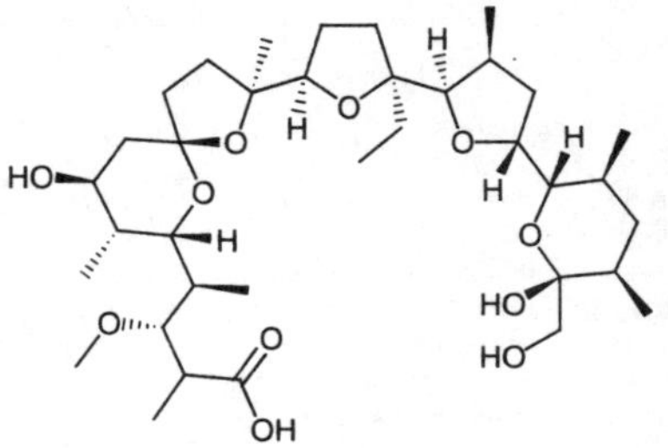

$C_{36}H_{62}O_{11}$
2-[2-Ethyloctahydro-3'-methyl-5'-[tetrahydro-6-hydroxy-6-(hydroxymethyl)-3,5-dimethyl-2H-pyran-2-yl][2,2'-bifuran-5-yl]]-9-hydroxy-β-methoxy-α,σ,2,8-tetramethyl-1,6-dioxapsiro[4.5]decan-7-butanoic acid.
A-3823A; 63714; monensic acid. Antibiotic produced by *Streptomyces cinnamonensis*. Antifungal, antibiotic and antiprotozoal. Used as a coccidiostat in chickens. mp = 103-105°; $[\alpha]_D$ = 47.7°; slightly soluble in H_2O, more soluble in hydrocarbons, very soluble in organic solvents; LD_{50} (mus orl) = 43.8 ± 5.2 mg/kg, (chicks orl) = 284 ± 47 mg/kg.

Veterinary products:

049-464; (with Bacitracin Methylene Disalicylate, Roxarsone) Monensin & Bacitracin & Roxarsone; *Elanco.*
092-522; (with Lincomycin Hydrochloride) Lincomix ® + Coban ® with or without Roxarsone; Lincomix ® / Coban ® / Roxarsone; *Pharmacia & Upjohn.*
098-341; (with Bambermycins, Roxarsone) Flavomycin ® + 3-Nitro ® + Coban ®; *Hoechst-Roussel Vet.*
141-110; (with Virginiamycin) Coban ® / Stafac ®; *Elanco.*
200-211; (with Bacitracin Zinc, Roxarsone); *Alpharma.*

367 Monensin Sodium

22373-78-0 6329 244-941-7

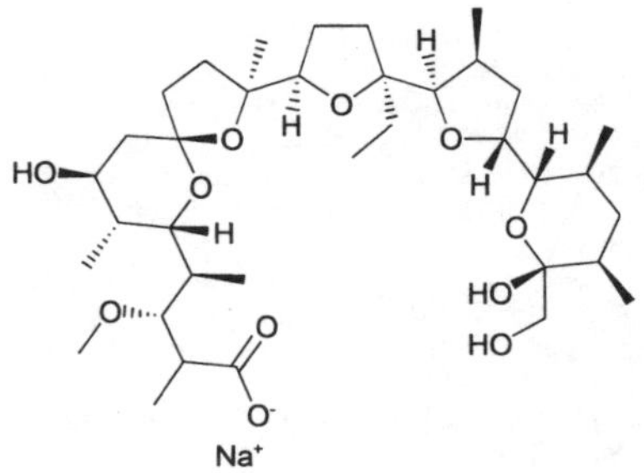

$C_{36}H_{61}NaO_{11}$
Sodium 2-[2-ethyloctahydro-3'-methyl-5'-[tetrahydro-6-hydroxy-6-(hydroxymethyl)-3,5-dimethyl-2H-pyran-2-yl][2,2'-bifuran-5-yl]]-9-hydroxy-β-methoxy-α,σ,2,8-tetramethyl-1,6-dioxapsiro[4.5]decan-7-butanoate.

Rumensin; Romensin; Coban. Antibiotic produced by *Streptomyces cinnamonensis*. Antifungal, antibiotic and antiprotozoal. Used as a coccidiostat in chickens. mp = 267-269°; $[\alpha]_D$ = 57.3°; slightly soluble in H_2O, more soluble in hydrocarbons, very soluble in organic solvents.

Veterinary products:

038-878; Coban ® - 45; Coban ® - 60; Coban ® - 110; Elancoban-100; *Elanco.*
041-500; (with Roxarsone) Coban ® - 1; *Elanco.*
047-933; (with Bacitracin Zinc) Monensin & Zinc Bacitracin Premix; *Elanco.*
049-463; (with Bacitracin Methylene Disalicylate) Monensin & Bacitracin MD; *Elanco.*
092-482; (with Lincomycin Hydrochloride) Coban ® / Lincomix ®; Lincomix ® / Coban ®; *Pharmacia & Upjohn.*
095-735; Rumensin ®; Rumensin ® 80; *Elanco.*
098-340; (with Bambermycins) Flavomycin ® + Monensin; *Hoechst-Roussel Vet.*
099-006; (with Oxytetracycline Monoalkyl Trimethylammonium Salt) Terramycin ® TM-50; TM-10 Plus; *Pfizer.*
104-646; (with Tylosin Phosphate) Tylan ® / Rumensin ®; *Elanco.*
109-471; Staley Sweetlix With Rumensin ®; *PM Ag Products.*
115-581; Moorman's ® Mintrate Blonde Block; *Moorman.*
116-088; (with Bacitracin Methylene Disalicylate) BMD ® / Coban ® / 3-Nitro ®; Fortracin / Coban ® / 3-Nitro ®; *Alpharma.*
118-509; Pasture Gainer Block-37 R350; *Farmland.*
119-253; Cattle Block M; *Co-op. Res. Farms.*
120-724; (with Roxarsone, Virginiamycin) Stafac ® / 3-Nitro ® / Coban ®; *Pfizer.*
121-553; (with Chlortetracycline Hydrochloride) Coban ® / Aureomycin ®; *Roche Vitamins.*
122-481; (with Virginiamycin) Stafac ® / Coban ®; *Pfizer.*
123-154; (with Bacitracin Zinc, Roxarsone) Coban ® / 3-Nitro ® 10 / Baciferm ® Premix; *Roche Vitamins.*
124-309; (with Melengestrol Acetate) MGA ® 100-200 / Rumensin; *Pharmacia & Upjohn.*
125-476; (with Melengestrol Acetate) MGA ® 500 / Rumensin; *Pharmacia & Upjohn.*
130-736; Coban ®; *Elanco.*
134-830; (with Bacitracin Zinc) Coban ® / Albac ®; *Alpharma.*
138-456; (with Bacitracin Methylene Disalicylate) Coban ® / BMD ®; *Alpharma.*
138-703; (with Bacitracin Zinc, Roxarsone) Albac® / Coban® / 3-Nitro ®; *Alpharma.*
138-792; (with Melengestrol Acetate, Tylosin Phosphate) MGA ® 100/200 / Rumensin / Tylan; *Pharmacia & Upjohn.*
138-870; (with Melengestrol Acetate, Tylosin Phosphate) MGA ® (liquid) / Rumensin / Tylan; MGA ® 100-200 Premix / MGA 500 Liquid Premix / Rumensin ® / Tylan ®; *Pharmacia & Upjohn.*
140-937; (with Bacitracin Methylene Disalicylate) Coban ® and BMD ®; *Elanco.*
140-939; (with Tylosin Phosphate) Liquid Type B Medicated Cattle Feed R/T 150; Liquid Type B Medicated Cattle Feed R/T 400; *Elanco.*
200-263; (with Chlortetracycline) ChlorMax™ / Coban ®; *Alpharma.*

368 Nystatin

1400-61-9 6834 215-749-0

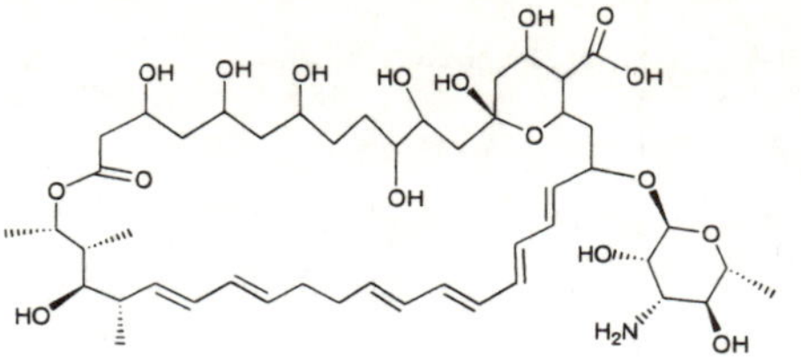

Mycostatin; Mycostatin Pastilles; Nystex; O-V Statin; Fungicidin; Biofanal; Diastatin; Candex; Candio-Hermal; Moronal; Nystavescent; component of: Mycolog II, Myco-Triacet II, Mytrex, Nystaform, Nystaform HC, Panolog Cream, Terrastatin. Antifungal. Mixture of Nystatin A_1, A_3 and A_3, biologically active polyene antibiotics. Dec 250°; $[\alpha]_D^{25}$ = -10° (AcOH), 21° (C_5H_5N), 12° (DMF), -7° (0.1N HCl/EtOH); λ_m = 290, 307, 322 nm; soluble in H_2O (0.4 g/100 ml), MeOH (1.12 g/100 ml), EtOH (0.12 g/100 ml), CCl_4 (0.123 g/100 ml), $CHCl_3$ (0.048 g/100 ml), C_6H_6 (0.028 g/100 ml), ethylene glycol (0.875 g/100 ml); LD_{50} (mus ip) ≅ 200 mg/kg.

Veterinary products:

012-258; (with Neomycin Sulfate, Thiostrepton, Triamcinolone Acetonide) Panolog ® Ointment; *Fort Dodge Animal Health Divn., Am. Cyanamid.*
012-680; Pharmastatin ® 20 Type A Medicated Article; *Alpharma.*
096-676; (with Neomycin Sulfate, Thiostrepton, Triamcinolone Acetonide) Panolog Cream; *Fort Dodge Animal Health Divn., Am. Cyanamid.*
140-810; (with Neomycin Sulfate, Thiostrepton, Triamcinolone Acetonide) Panavet Ointment; *Med-Pharmex.*
140-847; (with Neomycin Sulfate, Thiostrepton, Triamcinolone Acetonide) Animax; *Altana.*
140-879; (with Neomycin Sulfate, Thiostrepton, Triamcinolone Acetonide) Derma 4 Ointment; *Pfizer.*
140-889; (with Neomycin Sulfate, Thiostrepton, Triamcinolone Acetonide) Derm-Otic Ointment; *Biocraft.*
141-003; (with Neomycin Sulfate, Thiostrepton, Triamcinolone Acetonide) Derm-Otic Ointment; *Robins.*
200-245; (with Neomycin Sulfate, Thiostrepton, Triamcinolone Acetonide) Derma-Vet Cream; *Med-Pharmex.*

369 Salicylic Acid

69-72-7 8484 200-712-3

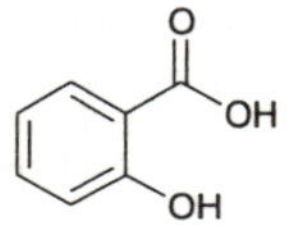

$C_7H_6O_3$
2-Hydroxybenzoic acid.
Keralyt; Occlusal; Verrugon. Keratolytic. Used externally as an antiseptic and in treatment of skin disorders. Crystals; mp = 157-159°; bp_{20} = 211°; d = 1.44; soluble in H_2O (0.22 g/100 ml), boiling H_2O (6.7 g/100 ml), EtOH (37 g/100 ml), Me_2CO (33.3 g/100 ml), $CHCl_3$ (2.4

g/100 ml), Et_2O (33.3 g/100 ml), C_6H_6 (0.74 g/100 ml), oil turpentine (1.9 g/100 ml), glycerol (1.67 g/100 ml); LD_{50} (mus iv) = 500 mg/kg.

Veterinary products:

010-481; Shurjets; *Jorgensen.*

370 Selenium Sulfide

7488-56-4 8580 231-303-8

SeS_2

Selenium disulfide.

Exsel; Seleen; Selsun; Selsun Blue. Antifungal; antiseborrheic. Insoluble in H_2O, 0.01N HCl; LD_{50} (rat orl) = 138 mg/kg.

Veterinary products:

008-422; Seleen ® Suspension; *Merial.*
111-349; SE S2 Shampoo; *Schering-Plough Animal Health.*
115-579; Dandrex; *Farnam.*
121-556; Selenium Sulfide Suspension; *Happy Jack.*

371 Tolnaftate

2398-96-1 9656 219-266-6

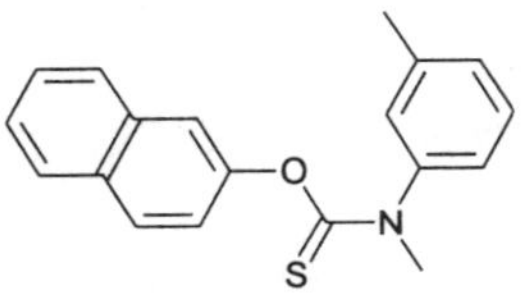

$C_{19}H_{17}NOS$

O-2-Naphthyl m,N-dimethylthiocarbanilate.

Aftate; Dr. Scholl's Athlete's Foot Spray; Tinactin; Tritin; Sch-10144; component of: Tinavet; naphthiomate T; Sch-10144; Chinofungin; Dungistop; Hi-Alarzin; Sporiline; Timoped; Tonoftal; Tniaderm. Antifungal used in veterinary medicine. mp = 110.5-111.5°; insoluble in H_2O; sparingly soluble in MeOH, EtOH; soluble in $CHCl_3$ (66 g/100 ml), Me_2CO (12.5 g/100 ml), CCl_4 (10 g/100 ml); LD_{50} (mus orl) > 10000 mg/kg, (mus sc) > 6000 mg/kg, (rat oprl) > 6000 mg/kg, (rat sc) > 4000 mg/kg.

Veterinary products:

037-502; Tinavet® Cream 1%; *Schering-Plough Animal Health.*

Antigout Agents

372 Allopurinol

315-30-0 287 206-250-9

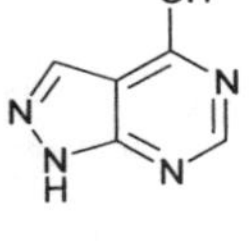

$C_5H_4N_4O$

1H-Pyrazolo[3,4-d]pyrimidin-4-ol.

HPP; BW-15658; NSC-1390; Adenock; Aloral; Alositol; Sllo-puren; Allozym; Allurtal; Anoprolin; Anzief; Apulonga; Apurol; Apurin; Bleminol; Bloxanth; Caplenal; Cellidrin; Cosuric; Dabroson; Embarin; Epidropal; Foligan; Geapur; Gichtex; Hamarin; Hexanurat; Ketanrift; Ketobun-A; Ledopur; Lopurin; Lysuron; Miniplanor; Monarch; Nektrohan; Remid; Riball; Sigapurol; Suspendol; Takanarumin; Urbol; Uricemil; Uripurinol; Urobenyl; Urosin; Urtias; Xanturat; Zyloprim; Zyloric. Xanthine oxidase inhibitor. Used to treat gout and, in veterinary medicine, to treat recurrent uric acid and hyperuricosuric calcium oxalate uroliths in small animals. mp > 350°; λ_m = 257 nm (ε 7200 0.1N NaOH), 250 nm (ε 7600 0.1N HCl), 252 nm (ε 7600 MeOH); soluble in H_2O (0.048 g/100 ml), $CHCl_3$ (0.060 g/100 ml), EtOH (0.030 g/100 ml), DMSO (0.46 g/100 ml), n-octanol (< 0.001 g/100 ml).

373 Carprofen

53716-49-7 1912 258-712-4

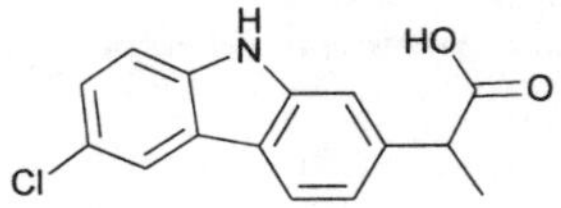

$C_{15}H_{12}ClNO_2$

(±)-6-Chloro-α-methylcarbazole-2-acetic acid.

Ro-20-5720/000; C-5720; Imadyl; Rimadyl. Anti-inflammatory; also used to treat gout. mp = 197-198°; LD_{50} (mus orl) = 400 mg/kg.

Veterinary products:

141-053; Rimadyl ®; *Pfizer.*
141-111; Rimadyl ® Chewable Tablets; *Pfizer.*

374 Colchicine

64-86-8 2536 200-598-5

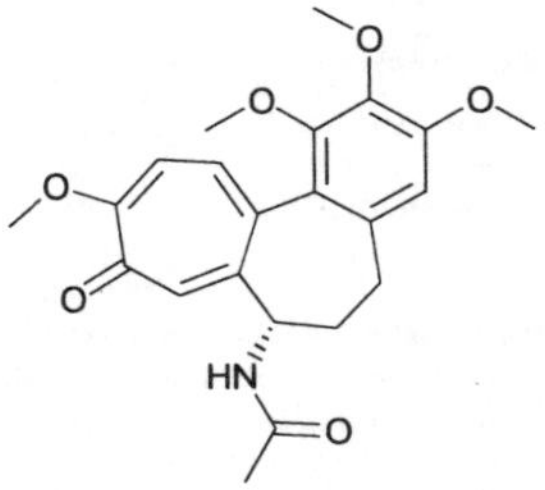

$C_{22}H_{25}NO_6$

[N-(5,6,7,9-Tetrahydro-1,2,3,10-tetramethoxy-9-oxobenzo[a]heptalen-7-yl]acetamide.

component of: Colbenemid. Used to treat gout. May be useful in treatment of amyloidosis in animals. mp = 142-150°, 157°; $[\alpha]_D^{17}$ = -429° (c = 1.72), $[\alpha]_D^{17}$ = -121° (c = 0.9 $CHCl_3$); λ_m = 350.5, 243 nm (log ε 4.22 4.47 EtOH); soluble in H_2O (4.5 g/100 ml), Et_2O (0.45 g/100 ml), C_6H_6 (1 g/100 ml); freely soluble in EtOH, $CHCl_3$; insoluble in petrolerum ether; LD_{50} (rat iv) = 1.6 mg/kg, (mus iv) = 4.13 mg/kg.

Antihistaminics

375 Chlorpheniramine
132-22-9 2232 205-054-0

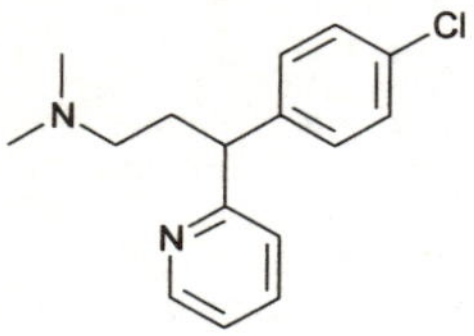

$C_{16}H_{19}ClN_2$
2-[p-Chloro-α-[2-(dimethylamino)ethyl]benzyl]pyridine.
chlorprophenpyridamine; chlorphenamine; Haynon. Antihistaminic. Used to treat histamine-related conditions. $bp_{1.0}$ = 142°.

376 Chlorpheniramine, d-Form
25523-97-1 2232 247-073-7

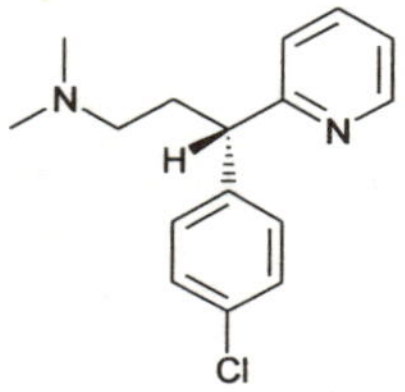

$C_{16}H_{19}ClN_2$
γ-(4-Chlorophenyl)-N,N-dimethyl-2-pyridinepropanamine.
dexchlorpheniramine; d-chlorpheniramine. Antihistaminic. Used to treat histamine-related conditions. Oily liquid; $bp_{1.0}$ = 142°.

377 Chlorpheniramine Maleate
113-92-8 2232 204-037-5

$C_{20}H_{23}ClN_2O_4$
2-[p-Chloro-α-[2-(dimethylamino)ethyl]benzyl]pyridine maleate (1:1).
chlorphenamine hydrogen maleate; Allergisan; Antagonate; Chlor-Trimeton; Chlor-Tripolon; Cloropiril; C-Meton; Histadur; Histaspan; Lorphen; Piriton; Pyridamal-100; Teldrin; component of: Allarest, A.R.M; Azimycin, Cerose-DM, Children's Tylenol Cold, Comhist, Contac, Coricidin, Corilin, Deconamine, Demazin, Diathal, Dristan Cold, Drize, Fedahist, Histabid Duracap, Histalet Forte, Histaspan Plus, Hycomine, Intensin, Isoclor, Metrevet, Naldecon, Novahistine, Ornade, PediaCare Cold-Cough, PV Tussin Syrup, Rhinex D-Lay, Ru-Tuss, Robitussin, Sinarest, Sine-Off, Sudafed Plus, Theraflu, Triaminic, Trind, Trind-DM, Tussar-2, Tussar DM, Tussar SF, TussarViro-Med, Tylenol Cold and Flu, Tylenol Cold Medication. Antihistaminic. Used to treat histamine-related conditions. mp = 130-135°; λ_m = 261 nm (ε 5760 H_2O); soluble in EtOH (330 g/l), $CHCl_3$ (240 g/l), H_2O (160 g/l), MeOH (130 g/l); less soluble in non-polar solvents; LD_{50} (mus orl) = 162 mg/kg.

378 Chlorpheniramine Maleate, d-Form
2438-32-6 2232 219-450-6

$C_{20}H_{23}ClN_2O_4$
γ-(4-Chlorophenyl)-N,N-dimethyl-2-pyridinepropanamine maleate.
Fortamine; Isomerine; Phenamin; Phendextro; Polamin; Polaramine; Polaronil; Sensidyn. Antihistaminic. Used to treat histamine-related conditions. mp = 113-115°; $[\alpha]_D^{25}$ = 44.3° (c = 1 DMF).

379 Clemastine
15686-51-8 2405

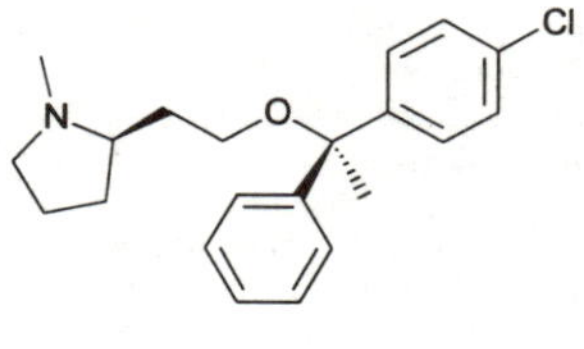

$C_{21}H_{26}ClNO$
(+)-(2R)-2-[2-[[(R)-p-Chloro-α-methyl-α-phenylbenzyl]oxy]ethyl]-1-methylpyrrolidine.
meclastine. Antihistaminic. Used for treatment of histamine-related allergic conditions. $bp_{0.02}$ = 154°; $[\alpha]_D^{20}$ = 33.6° (EtOH).

380 Clemastine Fumarate
14976-57-9 2405 239-055-2

$C_{25}H_{30}ClNO_5$
(+)-(2R)-2-[2-[[(R)-p-Chloro-α-methyl-α-phenylbenzyl]-oxy]ethyl]-1-methylpyrrolidine fumarate.
Tavist; HS-592; Aloginan; Alphamin; Anhistan; Fuluminol; Inbestan; Kinotomin; Lacretin: Lecasol; Maikohis; Mallermin-F; Marsthine; Masletine; Piloral; Reconin; Tavegil; Tavegyl; Telgin-G; Trabest; Xolamin. Antihistaminic. Used for treatment of histamine-related allergic conditions. mp = 177-178°; $[\alpha]_D^{21}$ = 16.9° (MeOH); LD_{50} (mus orl) = 730 mg/kg, (mus iv) = 43 mg/kg, (rat orl) = 3550 mg/kg, (rat iv) = 82 mg/kg.

381 Diphenhydramine
58-73-1 3367 200-396-7

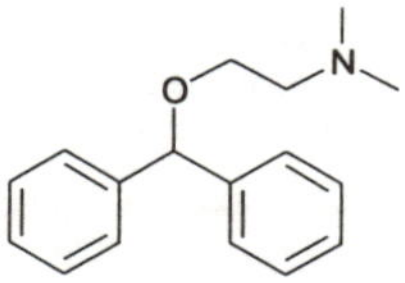

$C_{17}H_{21}NO$
2-(Diphenylmethoxy)-N,N-dimethylethylamine.
benzhydramine. Antihistaminic with antiemetic and sedative effects. Used to treat motion sickness. $bp_{2.0}$ = 150-165°.

382 Diphenhydramine Citrate
88637-37-0 3367

$C_{23}H_{29}NO_8$
2-(Diphenylmethoxy)-N,N-dimethylethylamine citrate.
Excedrin. Antihistaminic with antiemetic and sedative effects. Used to treat motion sickness.

383 Diphenhydramine Hydrochloride

147-24-0 3367 205-687-2

$C_{17}H_{22}ClNO$

2-(Diphenylmethoxy)-N,N-dimethylethylamine hydrochloride.

Actifed; Alledryl; Allergina; Amidryl; Bagodryl; Bax; Benadryl; Benocten; Benodine; Benzantin; Dibondrin; Dihydral; Diphantine; Dolestan; Fenylhist; Halbmond; Histacyl; Noctomin; S 8; Sedopretten; Sekundal-D; Syntedril; Wehydryl; component of: Benacine, Benylin, Caladryl, Contac, Coricidin, Midol, Sleep-eze, Tylenol PM, Ziradryl. Antihistaminic with antiemetic and sedative effects. Used to treat motion sickness. mp = 166-170°; soluble in H_2O (1 g/ml), EtOH (0.5 g/ml), $CHCl_3$ (0.5 g/ml), Me_2CO (0.02 g/ml); insoluble in non-polar solvents; LD_{50} (rat orl) = 500 mg/kg.

384 Doxylamine Succinate

562-10-7 3497 209-228-7

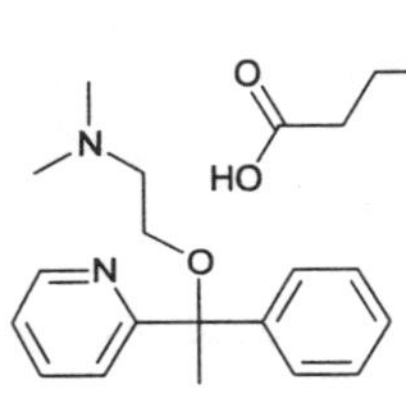

$C_{21}H_{28}N_2O_5$

2-[α[2-(Dimethylamino)ethoxy]-α-methylbenzyl]pyridine succinate.

Mereprine; Alsodorm; Decapryn succinate; Gittalun; Hoggar N; Sedaplus; Unisom. Antihistaminic and hypnotic. mp = 100-104°; soluble in H_2O (1 g/ml), EtOH (0.5 g/ml), $CHCl_3$ (0.5 g/ml); less soluble in C_6H_6, Et_2O; LD_{50} (mus orl) = 470 mg/kg, (mus iv) = 62 mg/kg, (rbt orl) = 250 mg/kg, (rbt iv) = 49 mg/kg, (mus sc) = 460 mg/kg, (mrat sc) = 440 mg/kg, (frat sc) = 445 mg/kg.

Veterinary products:

006-602; A-H-Tablets 100 mg; A-H-Tablets 25 mg; *Schering-Plough Animal Health.*
006-983; (with Chlorobutanol) A-H Injection; *Schering-Plough Animal Health.*

385 Ethylisobutrazine Hydrochloride

3737-33-5 3938 223-111-8

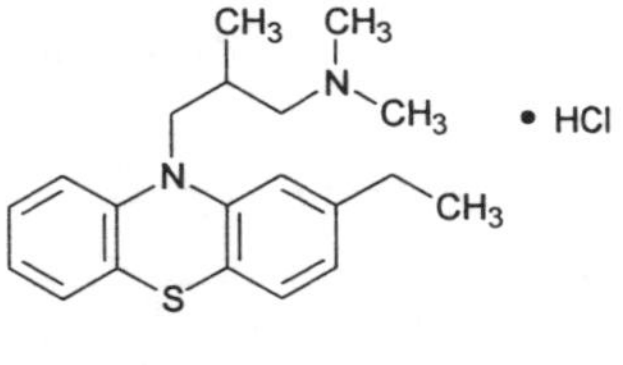

$C_{20}H_{27}ClN_2S$

2-ethyl-N,N,β-trimethyl-10H-phenothiazine-10-propanamine hydrochloride.

Nuital; Sergetyl. Antihistaminic used in veterinary medicine as a tranquillizer Crystals; mp = 160-163°; soluble in H_2O, EtOH, MeOH, Me_2CO, $CHCl_3$, insoluble in Et_2O.

Veterinary products:

011-222; Diquel Tablet; *Schering-Plough Animal Health.*
035-265; Diquel Solution; *Schering-Plough Animal Health.*

386 Pyrilamine

91-84-9 8168 202-102-2

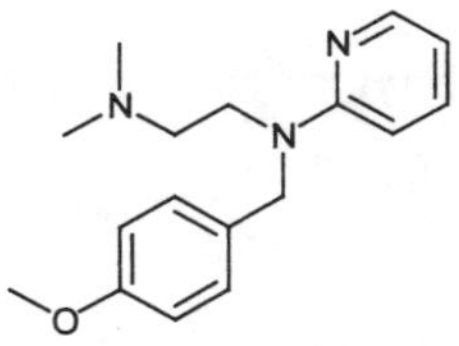

$C_{17}H_{23}N_3O$

2-[[2-(Dimethylamino)ethyl](p-methoxybenzyl)amino]-pyridine.

mepyramine; pyranisamine; RP-2786. Antihistaminic. Used to control histamine mediated adverse effects. bp_5 = 201°; LD_{50} (mus orl) = 312 mg/kg.

387 Pyrilamine Maleate

59-33-6 8168 200-422-7

$C_{21}H_{27}N_3O_5$

2-[[2-(Dimethylamino)ethyl](p-methoxybenzyl)amino]-pyridine maleate.

Histavet-P; Pymafed; Histalet Forte; Midol; PV Tussin Syrup; Robitussin; Antamine; Antisan; Dorantamin; Enrumay; Histalon; Histan; Histapyran; Histatex; Neo-Antergan; Paraminyl; Parmal; Pyramal; Stamine; Stangen; Thylogen. Antihistaminic. Used to control histamine mediated adverse effects. mp = 100-101°; λ_m = 244 nm ($E^{1\%}_{1\ cm}$ 420); soluble in H_2O (2.5 g/ml), less soluble in organic solvents; LD_{50} (mus orl) = 338 mg/kg.

Veterinary products:

046-288; Histavet-P ®; *Schering-Plough Animal Health.*
138-405; Pyrilamine Maleate Injection; *Anthony Products.*

388 Tripelennamine Hydrochloride

154-69-8 9868 205-833-5

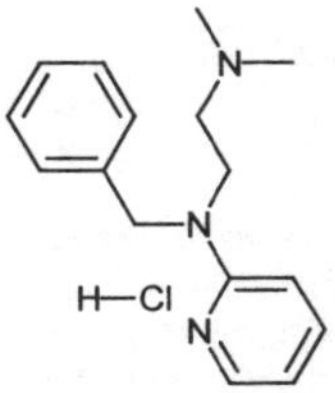

$C_{16}H_{22}ClN_3$

2-[Benzyl[2-(dimethylamino]ethyl]amino]pyridine hydrochloride.

Re Covr; Dehistin; Azaron; PBZ; Vetobenzamina; Pyribenzamine Hydrochloride; Vetibenzamine. Antihistaminic. Has also been used in cattle as a CNS stimulant. mp = 192-193°; λ_m = 244, 305 nm (ε 14470, 4780 H_2O); soluble in H_2O (1.3 g/ml), less soluble in organic solvents; LD_{50} (mus ip) = 47 mg/kg.

Veterinary products:

006-417; Recovr Injectable; *Fort Dodge Animal Health Divn., Am. Cyanamid.*
200-162; Tripelennamine Hydrochloride Injection; *Phoenix.*

Antihyperlipoproteinemics

389 Carnitine
461-06-3 1898

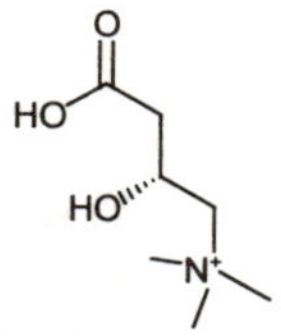

$C_7H_{15}NO_3$
L-(3-Carboxy-2-hydroxypropyl))trimethylammonium hydroxide inner salt.
3-Hydroxy-4-trimethylammoniobutanoate; γ-Trimethyl-β-hydroxybutyrobetaine; levocarnitine; vitamin B_7; Cardiogen; Carnitene; Carnicor; Carnum; Carrier; Miocor; Miotonal; Vitacarn. Hypolipidemic agent. Used in dogs as an adjunct in treatment of dilated cardiomyopathy. mp = 197-198° (dec); $[\alpha]_D^{30}$ = -23.9° (H_2O c = 0.86); soluble in H_2O, hot EtOH; insoluble in Me_2CO, Et_2O, C_6H_6.

390 Levocarnitine
541-15-1 1898 208-768-0

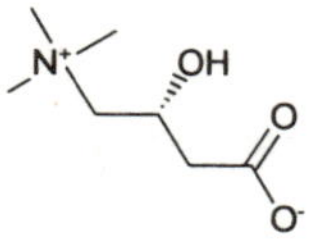

$C_7H_{15}NO_3$
L-(3-Carboxy-2-hydroxypropyl))trimethylammonium hydroxide inner salt.
3-Hydroxy-4-trimethylammoniobutanoate; γ-Trimethyl-β-hydroxybutyrobetaine; levocarnitine; vitamin B_7; Cardiogen; Carnitene; Carnicor; Carnum; Carrier; Miocor; Miotonal; Vitacarn. Hypolipidemic agent. Used in dogs as an adjunct in treatment of dilated cardiomyopathy. mp = 197-198° (dec); $[\alpha]_D^{30}$ = -23.9° (H_2O c = 0.86); soluble in H_2O, hot EtOH; insoluble in Me_2CO, Et_2O, C_6H_6.

Antihypertensives

391 Amlodipine
88150-42-9 516

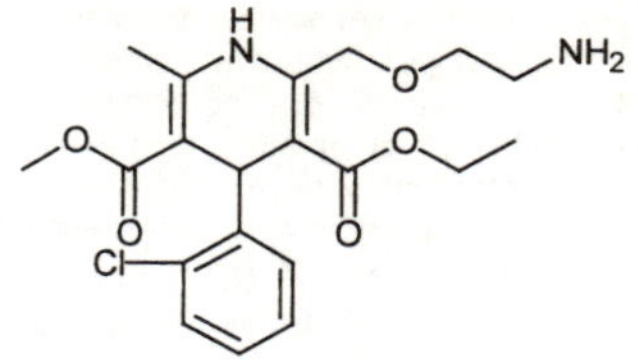

$C_{20}H_{25}ClN_2O_5$
3-Ethyl-5-methyl (±)-2-[(2-aminoethoxy)methyl]-4-(o-chlorophenyl)-1,4-dihydro-6-methyl-3,5-pyridinedicarboxylate.
Dihydropyridine calcium channel blocker. Has anti-hypertensive properties and is used as an anti-anginal agent. Used, particularly as the besylate to treat hypertension in cats.

392 Amlodipine Besylate
111470-99-6 516

$C_{26}H_{31}ClN_2O_8S$
3-Ethyl-5-methyl (±)-2-[(2-aminoethoxy)methyl]-4-(o-chlorophenyl)-1,4-dihydro-6-methyl-3,5-pyridinedicarboxylate monobenzenesulfonate.
Norvasc; UK-48340-26; Antacal; Istin; Monopina; component of Lotrel. Dihydropyridine calcium channel blocker. Has anti-hypertensive properties and is used as an anti-anginal agent. Used to treat hypertension in cats. mp = 178-179°.

393 Benazepril
86541-75-5 1058

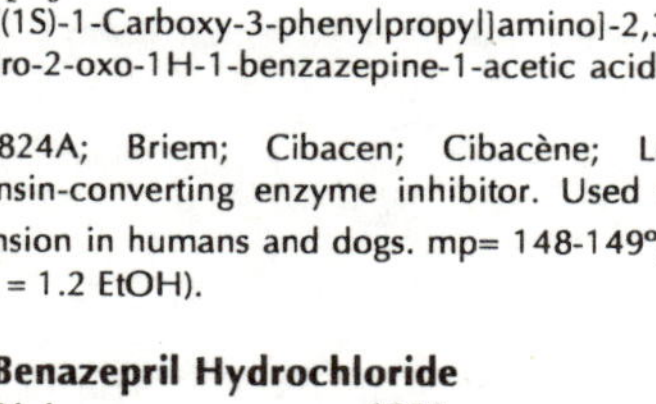

$C_{24}H_{28}N_2O_5$
(3S)-3-[[(1S)-1-Carboxy-3-phenylpropyl]amino]-2,3,4,5-tetrahydro-2-oxo-1H-1-benzazepine-1-acetic acid 3-ethyl ester.
CGS-14824A; Briem; Cibacen; Cibacène; Lotensin. Angiotensin-converting enzyme inhibitor. Used to treat hypertension in humans and dogs. mp= 148-149°; $[\alpha]_D$ = -159° (c = 1.2 EtOH).

394 Benazepril Hydrochloride
86541-74-4 1058

$C_{24}H_{29}ClN_2O_5$
(3S)-3-[[(1S)-1-Carboxy-3-phenylpropyl]amino]-2,3,4,5-tetrahydro-2-oxo-1H-1-benzazepine-1-acetic acid 3-ethyl ester hydrochloride.
Lotensin; CGS-14824A HCl; component of Lotensin-HCT, Lotrel, Lotrel capsules. Angiotensin-converting enzyme

inhibitor. Used to treat hypertension in humans and dogs. mp = 188-190°; $[\alpha]_D$ = -141° (c = 0.9 EtOH).

395 Captopril

62571-86-2 1817 263-607-1

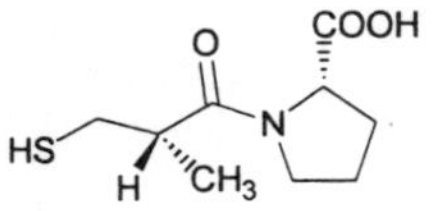

$C_9H_{15}NO_3S$
1-[(2S)-3-Mercapto-2-methylpropionyl]-L-proline.
Capoten; SQ-14225; Acediur; Acepril; Aceplus; Alopresin; Acepress; Capoten; Captolane; Captoril; Cesplon; Dilabar; Garranil; Hipertil; Lopirin; Lopril; Tensobon; Tensoprel; component of: Capozide, Acezide, Captea, Ecazide. Angiotensin-converting enzyme inhibitor. Orally active peptidyldipeptide hydrolase inhibitor. Used to treat hypertension and congestive heart failure. mp = 103-104°, 86°, 87-88°, 104-105°; $[\alpha]_D^{22}$= -131.0° (c = 1.7 EtOH); freely soluble in H_2O, EtOH, $CHCl_3$, CH_2Cl_2; LD_{50} (mus iv) = 1040 mg/kg, (mus orl) = 6000 mg/kg.

396 Diazoxide

364-98-7 3051 206-668-1

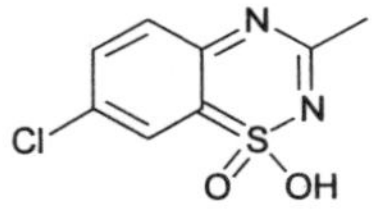

$C_8H_7ClN_2O_2S$
7-Chloro-3-methyl-2H-1,2,4-benzothiadiazine 1,1-dioxide.
SRG-95213; Sch-6783; NSC-64198; Eudemine injection; Proglicem; Hyperstat; Hypertonalum; Mutabase; Proglycem. An ATP-dependent potassium-channel opener. A benzothiadiazine derivative similar in structure to the thiazides. An antihypertensive and potent vasodilator. Used orally for the treatment of hypoglycemia in dogs. mp = 330-331°; λ_m = 268 nm (ε = 11300 MeOH); soluble in EtOH, insoluble in H_2O.

397 Enalapril Maleate

76095-16-4 3605 278-375-7

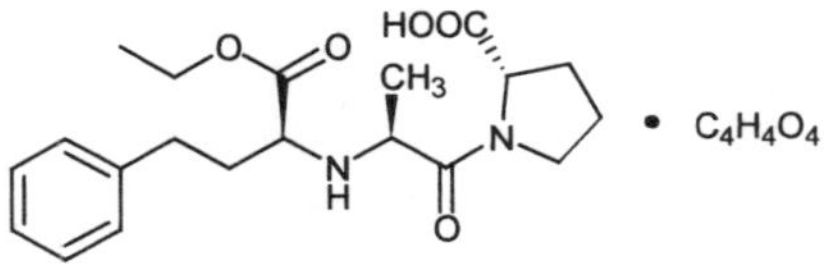

$C_{24}H_{32}N_2O_9$
1-[N-[(S)-1-Carboxy-3-phenylpropyl]-L-alanyl]-L-proline 1'-ethyl ester maleate (1:1).
Enacard; Renitec; Vasotec; MK-421; Amprace; Bitensil; Cardiovet; Enaloc; Enapren; Glioten; Hipoartel; Innovace; Lotrial; Olivin; Pres; Reniten; Renivace; Xanef; component of: Vaseretic, Acesistem, Co-Renitec, Innozide, Renacor, Xynertec. Angiotensin-converting enzyme inhibitor used in veterinary medicine as a vasodilator to treat heart failure and hypertension. mp = 143-144.5°; $[\alpha]_D^{25}$= -42.2° (c = 1 MeOH); soluble in H_2O (2.5 g/100 ml), EtOH (8 g/100 ml), MeOH (20 g/100 ml).

Veterinary products:

141-015; Enacard ®; Enacard ® Tablets For Dogs; *Merial.*

398 Hydralazine

86-54-4 4800 201-680-3

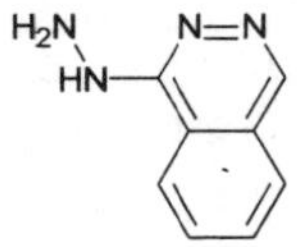

$C_8H_8N_4$
1-Hydrazinophthalazine.
Apresoline; Hypophthalin; Hipoftalin; C-5968; Präparat 5968; Ciba-5968; 1-hydrazinophthalazine. Antihypertensive agent used in treatment of congestive heart failure. mp= 172-173°; soluble in 2N AcOH (33.3 g/100 ml), warm MeOH (8.4 g/100 ml); LD_{50} (mus orl) = 122 mg/kg, (mus ip) = 101 mg/kg, (rat orl) = 90 mg/kg, (rat ip) = 40 mg/kg.

399 Hydralazine Hydrochloride

304-20-1 4800 206-151-0

$C_8H_9ClN_4$
1-Hydrazinophthalazine hydrochloride.
Apresoline hydrochloride; Lopres; component of: Apresazide, BiDil, H.H. 25/25, H.H. 50/50, Ser-Ap-Es, Unipres. Antihypertensive agent used in treatment of congestive heart failure. mp = 273° (dec); soluble in H_2O (3.01 g/100 ml at 15°, 4.42 g/100 ml at 25°), EtOH (0.2 g/100 ml); slightly soluble in Et_2O; λ_m = 211, 240, 260, 304, 315 nm.

400 Hydrochlorothiazide

58-93-5 4822 200-403-3

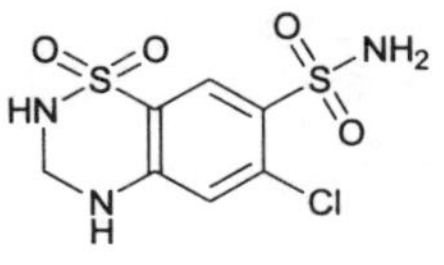

$C_7H_8ClN_3O_4S_2$
6-Chloro-3,4-dihydro-2H-1,2,4-benzothiadiazine-7-sulfonamide 1,1-dioxide.
Esidrex; Dichlotride; HydroDIURIL; Hydrozide; Oretic; Thiuretic; Acuretic; Aldactazide; Aldoril; Apresazide; Caplaril; Capozide; Dyazide; Esimil; H.H. 25/25; H.H. 50/50; Hydropres; Hyzaar; Inderide; Lopressor HCT; Lotensin HCT; Maxzide; Moduretic; Prinzide; Ser-Ap-Es; Timolide; Unipres; Vaseretic; Ziac. Diuretic. Used to treat edema in dogs and cattle. mp= 273-275°; λ_m = 317, 271, 226 nm ($A^{1\%}_{1\,cm}$ 130, 654, 1280 MeOH/HCl); soluble in MeOH, EtOH, Me_2CO; insoluble in H_2O; LD_{50} (mus iv) = 590 mg/kg, (mus orl) > 8000 mg/kg.

Veterinary products:

013-674; Hydrozide Injection; *Merial.*

401 Metoprolol
37350-58-6 6235 253-483-7

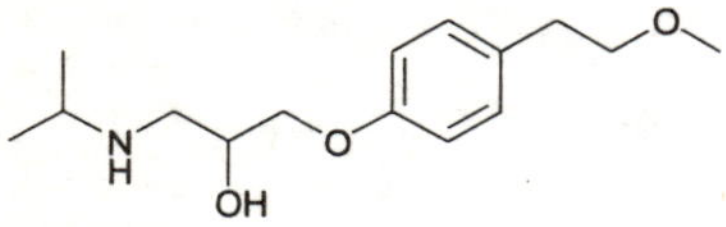

$C_{15}H_{25}NO_3$
1-(Isopropylamino)-3-[p-(2-methoxyethyl)phenoxy]-2-propanol.
H 23/96. Class II antiarrhythmic agent with antihypertensive and antianginal activity. A β-adrenergic blocker which lacks intrinsic sympathomimetic activity. Favored over propranolol in treatment of bronchospastic disease.

402 Metoprolol Fumarate
119637-66-0 6235
$C_{34}H_{54}N_2O_{10}$
1-(Isopropylamino)-3-[p-(2-methoxyethyl)phenoxy]-2-propanol fumarate (2:1) (salt).
Lopressor OROS; CGP 2175C. Class II antiarrhythmic with antihypertensive, antianginal activity. A β-adrenergic blocker lacking intrinsic sympathomimetic activity. Favored over propranolol in treatment of bronchospastic disease. Insoluble in EtOAC, Me_2CO, diethylether, heptane.

403 Metoprolol Succinate
98418-47-4 6235
$C_{34}H_{56}N_2O_{10}$
1-(Isopropylamino)-3-[p-(2-methoxyethyl)phenoxy]-2-propanol succinate (2:1) salt.
Toprol XL; H 93/26 succinate. Class II antiarrhythmic; β_1-selective (cardioselective) adrenoceptor blocker, for oral administration, available as extended release tablets. Also has antihypertensive and antianginal activity. Favored over propranolol in treatment of bronchospastic disease. Freely soluble in H_2O; soluble in MeOH; sparingly soluble in EtOH; slightly soluble in CH_2Cl_2, iPrOH; practically insoluble in EtOAc, Me_2CO, Et_2O, and C_7H_{16}.

404 Metoprolol Tartrate
56392-17-7 6235 260-148-9
$C_{34}H_{56}N_2O_{12}$
1-(Isopropylamino)-3-[p-(2-methoxyethyl)phenoxy]-2-propanol (2:1) dextro-tartrate salt.
HCTCGP 2175E; Beloc; Betaloc; Lopresor; Prelis; Seloken; Selopral; Selo-Zok; component of: Lopressor. Class II antiarrhythmic. A β_1-selective (cardioselective) adrenoceptor blocking agent, for oral administration, available as extended release tablets. Also has antihypertensive and antianginal activity. Favored over propranolol in treatment of bronchospastic disease. Soluble in H_2O (> 100 g/100 ml), MeOH (>50 g /100 ml), $CHCl_3$ (49.6 g/100 ml), Me_2CO (0.11 g/100 ml), CH_3CN (0.089 g/100 ml); insoluble in C_6H_{14}; LD_{50} ((fmus iv) = 118 mg/kg, (fmus orl) = 2090 mg/kg, (mrat iv) ≅ 90 mg/kg, (mrat orl) = 3090 mg/kg.

Antihyperthyroids

405 Methimazole
60-56-0 6049 200-482-4

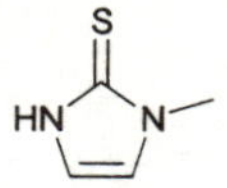

$C_4H_6N_2S$
1-Methylimidazole-2-thiol.
Tapazole; mercazolyl; thiamazole; Basolan; Danantizol; Favistan; Frentirox; Mercazole; Metazolo; Thacapzol; Thycapsol; Strumazol. Antihyperthyroid. Agent of choice for feline hyperthyroidism. mp = 146-148°; bp = 280° (dec); λ_m = 211, 251.5 nm ($E^{1\%}_{1cm}$ 593, 1528, 0.1N H_2SO_4); freely soluble in H_2O; soluble in EtOH, $CHCl_3$; sparingly soluble in Et_2O, C_6H_6, petroleum ether.

Antihypothyroids

406 Liothyronine Sodium
55-06-1 5535 200-223-5

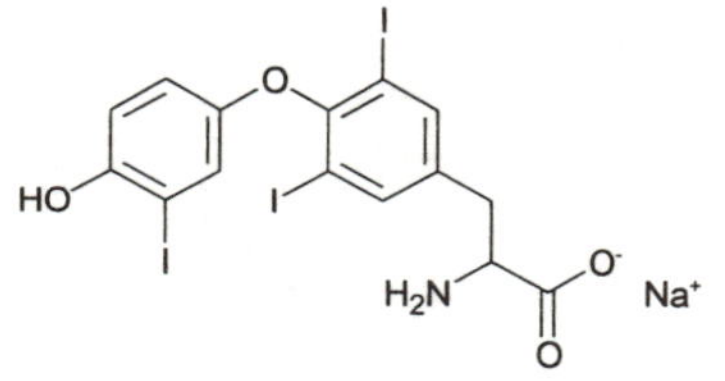

$C_{15}H_{11}I_3NNaO_4$
L-3-[4-(4-Hydroxy-3-iodophenoxy)-3,5-diiodophenyl]alanine monosodium salt.
Cytomel; Triostat; Cyomel; Cytobin; Cytomine; Cynomel; Tertroxin; Trithyrone. Thyroid hormone.

Veterinary products:

014-366; Cytobin Tablets; *Pfizer.*

Anti-inflammatories, Nonsteroidal

407 Etodolac
41340-25-4 3920

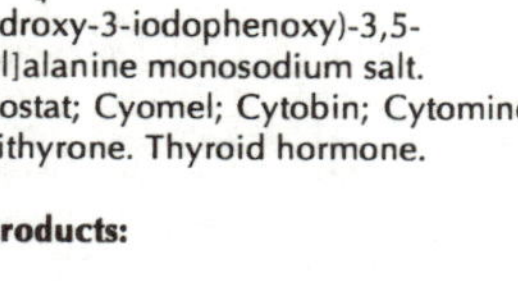

$C_{17}H_{21}NO_3$
1,8-Diethyl-1,3,4,9-tetrahydropyranol[3,4-b]indole-1-acetic acid.
etodolic acid; AY-24236; Edolan; Lodine; Ramodar; Tedolan; Ultradol; Zedolac. Arylacetic acid derivative. Anti-inflammatory; analgesic. Used to manage pain resulting from osteoarthritis in dogs. mp = 145-148°.

Veterinary products:

141-108; EtoGesic™; *Fort Dodge Animal Health, Divn. AHP.*

408 Ketoprofen

22071-15-4 5316 244-759-8

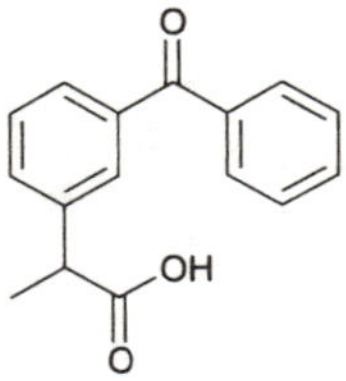

$C_{16}H_{14}O_3$
m-Benzoylhydratropic acid.
RP-19583; Alrheumat; Alrheumun; Capisten; Dexal; Epatec; Fastum; Iso-K; Kefenid; Ketopron; Lertus; Menamin; Meprofen; Orudis; Orugesic; Oruvail; Oscorel; Profenid; Toprec; Toprek. Anti-inflammatory; analgesic. Used as an anti-inflammatory in horses. mp = 94°; λ_m = 255 nm (log ε 4.33 in MeOH); soluble in Et_2O, alc, Me_2CO, $CHCl_3$, DMF, EtOAc; slightly soluble in H_2O; LD_{50} (rat orl) = 101 mg/kg.

Veterinary products:

140-269; Ketofeb™; *Fort Dodge Animal Health, Divn. AHP.*

409 Meclofenamic Acid

644-62-2 5819 211-419-5

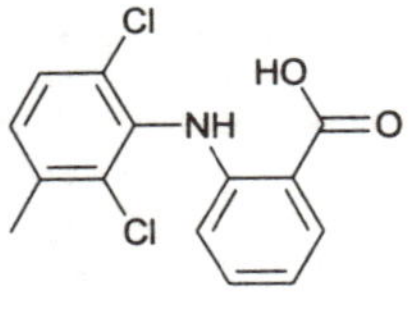

$C_{14}H_{11}Cl_2NO_2$
2-[(2,6-Dichloro-3-methylphenyl)amino]benzoic acid.
meclophenamic acid; CI-583; INF-4668; Arquel. Aminoarylcarboxylic acid derivative. Anti-inflammatory; antipyretic. Used to relieve inflammation in dogs. mp = 257-259°, 248-250°; slightly soluble in H_2O (0.03 mg/ml); more soluble in 0.1N NaOH (28 mg/ml); pH (saturated aqueous solution) 6.9.

Veterinary products:

095-641; Arquel Granules; *Fort Dodge Animal Health, Divn. AHP.*
110-201; Arquel Tablets; *Fort Dodge Animal Health, Divn. AHP.*

410 Naproxen

22204-53-1 6504 244-838-7

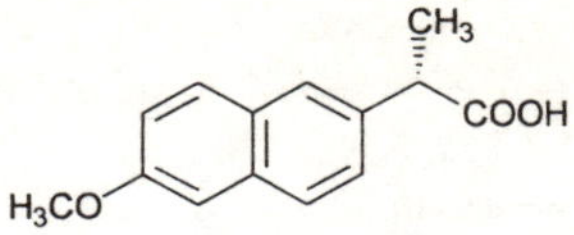

$C_{14}H_{14}O_3$
(+)-6-Methoxy-α-methyl-2-napthaleneacetic acid.
MNPA; RS-3540; Bonyl; Diocodal; Dysmenalgit; Equiproxen; Floginax; Laraflex; Laser; Naixan; Napren; Naprium; Naprius; Naprosyn; Naprosyne; Naprux; Naxen; Nycopren; Panoxen; Prexan; Proxen; Proxine; Reuxen; Veradol; Xenar. Arylpropionic acid derivative. Anti-inflammatory; analgesic; antipyretic. Used in horses to minimize pain associated with soft tissue diseases. mp = 152-154°; $[\alpha]_D$ = +66° (c = 1 in $CHCl_3$); nearly insoluble in H_2O; soluble in organic solvents; LD_{50} (mus orl) = 1234 mg/kg, (rat orl) = 534 mg/kg, (rat ip) = 575 mg/kg.

Veterinary products:

096-674; Equiproxen Granules; *Fort Dodge Animal Health, Divn. AHP.*
096-675; Equiproxen 10% Solution; *Fort Dodge Animal Health, Divn. AHP.*

411 Orgotein

9016-01-7 9177 232-771-6
Artolasi; Ormetein; Ontosein; Oximorm; Palosein; Peroxinorm. Member of the superoxide dismutase family, a group of naturally occuring enzymes that act as oxygen free radical scavengers to protect against the effects of biologically generated superoxide oxygen radicals through dismutation to hydrogen peroxide. Water soluble protein congeners derived from red blood cells, liver, and other tissues. Molecular weight 33000. Copper-zinc chelate with superoxide dismutase activity. Used as an anti-inflammatory and antirheumatic. LD_{50} (mus sc) > 5800 mg/kg, (rat sc) > 400 mg/kg, (mus ip) > 60 mg/kg, (rat ip) > 284 mg/kg, (mus iv) > 4000 mg/kg.

Veterinary products:

045-863; Palosein; *Oxis.*

412 Phenylbutazone

50-33-9 7431 200-029-0

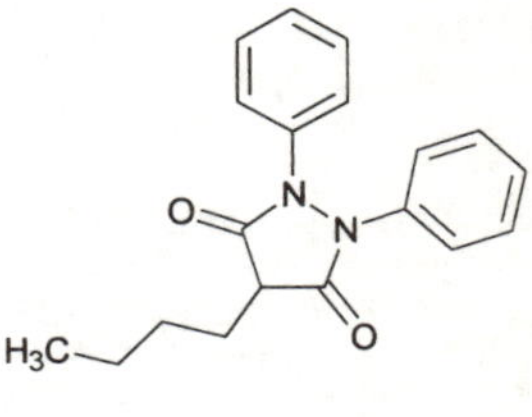

$C_{19}H_{20}N_2O_2$
4-Butyl-1,2-diphenyl-3,5-pyrazolidinedione.
flexazone; diphebuzol; fenibutazona; G-13871; R-3-ZON; Ambene; Artrizin; Azolid; Bizolin; Butacote; Butadion; Buitapirazol; Butadiona; Butatron; Butoz; Butazolidin; Buzon; Ecobutazone; Equipalazone;

Exrheudon N; Fenibutol; Intrabutazone; Intrazone; Mepha-Butazon; Phenyzene; Robizone-V; Tevcodyne; Uzone. Analgesic and anti-inflammatory. Used to treat lameness in horses. Crystals; mp = 105°; soluble in H_2O (0.07 - 0.22 g/100 ml at 22.5°; λ_m = 239.5 nm (log ε 4.19, acidic MeOH).

Veterinary products:

010-987; Butazolidin Bolus; Butazolidin Tablets; *Schering-Plough Animal Health.*
011-575; Butazolidin Injectable 20%; *Schering-Plough Animal Health.*
038-800; Butazolidin Granules; *Schering-Plough Animal Health.*
044-756; Tevcodyne; *Merial.*
045-416; Tevcodyne Injectable; *Merial.*
045-514; EquiBute Injection; *Fort Dodge Animal Health, Divn. AHP.*
045-515; EquiBute Tablets 100 mg; *Fort Dodge Animal Health, Divn. AHP.*
045-848; Phenylbutazone Injection; *Steris.*
046-780; Phen-Buta Vet Injection; *Anthony Products.*
047-712; Bizolin 200; *Boehringer Ingelheim Vetmedica.*
048-646; Therazone Injection; *Wendt.*
048-647; Therazone Tablets; *Wendt.*
049-187; Phen-Buta Vet Tablets; *Anthony Products.*
091-065; Robizone-V; *Robins.*
091-818; Phenylbutazone Tablets, USP 1 gram; *Phoenix.*
093-105; Robizone-V; *Robins.*
093-516; Bizolin Injection 20%; *Boehringer Ingelheim Vetmedica.*
094-170; Phenylbutazone Tablets, USP 100 mg; *Phoenix.*
096-671; Phen-Buta-Vet Injection; *Anthony Products.*
096-672; Phen-Buta-Vet Tablets; *Anthony Products.*
098-640; Robizone Injectable 20%; *Robins.*
099-618; Bizolin; *Boehringer Ingelheim Vetmedica.*
102-824; Phenylbutazone Tablets; *Pegasus.*
113-510; Equipalazone; *Merial.*
116-087; Butazolidin; Butazolidin Paste; Bute; Phenylzone Paste; *Schering-Plough Animal Health.*
118-979; Butatron Gel; *Merial.*
140-958; Equiphen ® Paste; *Luitpold.*
200-126; Phenylbutazone 20% Injection; *Phoenix.*

413 Piroxicam
36322-90-4 7661 252-974-3

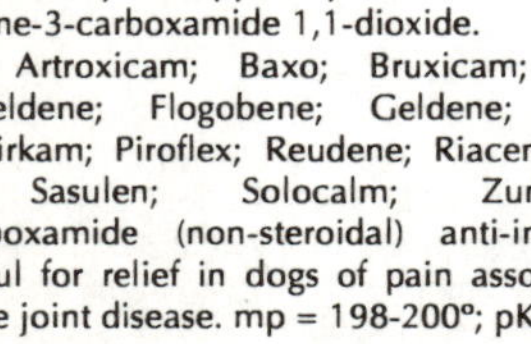

$C_{15}H_{13}N_3O_4S$
4-Hydroxy-2-methyl-N-2-pyridinyl-2H-1,2-benzothiazine-3-carboxamide 1,1-dioxide.
CP-16171; Artroxicam; Baxo; Bruxicam; Caliment; Erazon; Feldene; Flogobene; Geldene; Improntal; Larapam; Pirkam; Piroflex; Reudene; Riacen; Roxicam; Roxiden; Sasulen; Solocalm; Zunden. A thiazinecarboxamide (non-steroidal) anti-inflammatory agent. Useful for relief in dogs of pain associated with degenerative joint disease. mp = 198-200°; pKa = 6.3 (2:1 dioxane:H_2O); LD_{50} (mus orl) = 360 mg/kg.

414 Piroxicam Cinnamate
87234-24-0 7661

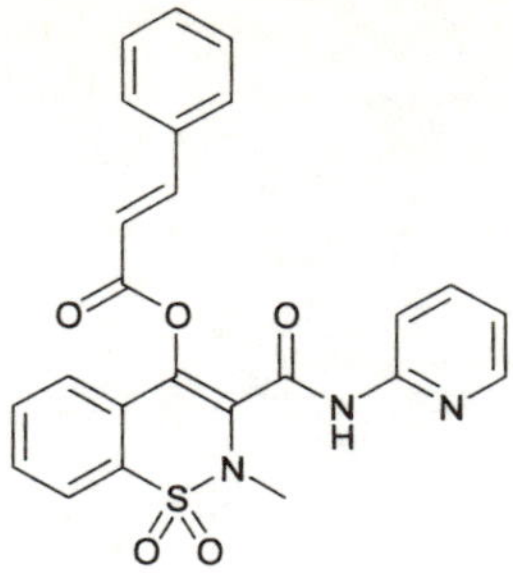

$C_{24}H_{19}N_3O_5S$
4-Hydroxy-2-methyl-N-2-pyridinyl-2H-1,2-benzothiazine-3-carboxamide 1,1-dioxide cinnamic acid ester.
cinnoxicam; SPA-S-510; Sinartol; Zelis; Zen. A thiazinecarboxamide (non-steroidal) anti-inflammatory agent. Useful for relief in dogs of pain associated with degenerative joint disease.

415 Piroxicam Compound with β-Cyclodextrin
121696-62-6 7661
$C_{57}H_{83}N_3O_{39}S$
4-Hydroxy-2-methyl-N-2-pyridinyl-2H-1,2-benzothiazine-3-carboxamide 1,1-dioxide compound with β-cyclodextrin.
Brexin; Cicladol; Cycladol. A thiazinecarboxamide (non-steroidal) anti-inflammatory agent. Useful for relief in dogs of pain associated with degenerative joint disease.

416 Piroxicam Olamine
85056-47-9 7661
$C_{17}H_{20}N_4O_5S$
4-Hydroxy-2-methyl-N-2-pyridinyl-2H-1,2-benzothiazine-3-carboxamide 1,1-dioxide compound with 2-aminoethanol.
CP-16171-85. A thiazinecarboxamide (non-steroidal) anti-inflammatory agent. Useful for relief in dogs of pain associated with degenerative joint disease.

Anti-inflammatories, Steroidal

417 Dexamethasone Isonicotinate
2265-64-7 2986 218-866-5

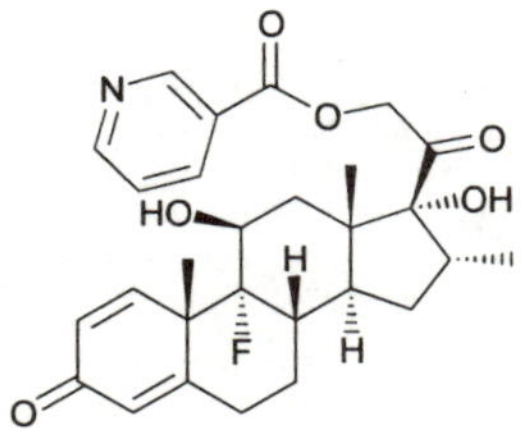

$C_{28}H_{32}FNO_6$
9-Fluoro-11β,17,21-trihydroxy-16α-methylpregna-1,4-diene-3,20-dione 21-(4-pyridinecarboxylate).

dexamethasone 21-isonicotinate; Auxiloson; Ausixone; Voren. An antiasthmatic glucocorticoid used in diagnosis of Cushing's syndrome and depression and in many animal species. mp = 250-252°; $[\alpha]_D^{27}$ = +183.5° (in dioxane).

Veterinary products:

093-600; Voren® Suspension; *Boehringer Ingelheim Vetmedica.*

418 Flumethasone

2135-17-3 4173 218-370-9

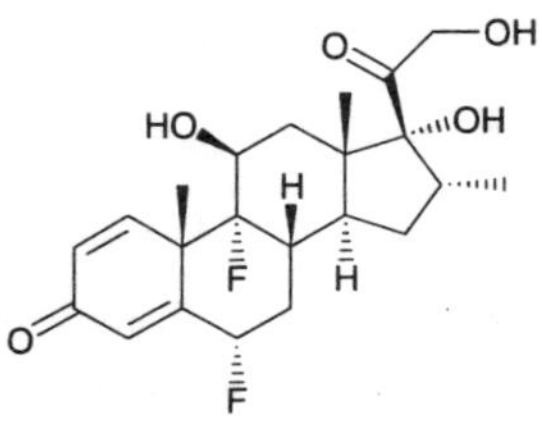

$C_{22}H_{28}F_2O_5$
(6α,11β,16α)-6,9-Difluoro-11,17,21-trihydroxy-16-methylpregna-1,4-diene-3,20-dione.
flumetasone; 6α-fluorodexamethazone; U-10974; NSC-5402; Aniprome; Cortexilar; Flucort; Methagon. Glucocorticoid; anti-inflammatory. Used to manage musculoskeletal inflammation and some dermatological conditions in dogs, cats and horses.

Veterinary products:

030-414; Flucort ® Solution; *Fort Dodge Animal Health, Divn. AHP.*
030-415; Flucort ® Tablets; *Fort Dodge Animal Health, Divn. AHP.*
036-211; Aniprime ® Suspension; *Fort Dodge Animal Health, Divn. AHP.*
038-801; (with Neomycin Sulfate, Polymyxin B Sulfate) Anaprime ® Ophthalmic Solution; *Fort Dodge Animal Health, Divn. AHP.*
040-123; (with Cephalonium, Iodochlorhydroxyquin, Piperocaine Hydrochloride, Polymyxin B Sulfate) Toptic Ointment (15 g); *Elanco.*
049-725; (with Neomycin Sulfate, Polymyxin B Sulfate) Anaprime ® Opthakote Ophthalmic; *Fort Dodge Animal Health, Divn. AHP.*

419 Flumethasone Acetate

4173

$C_{24}H_{30}F_2O_6$
(6α,11β,16α)-6,9-Difluoro-11,17,21-trihydroxy-16-methylpregna-1,4-diene-3,20-dione 21-acetate.
flumetasone 21-acetate. Glucocorticoid; anti-inflammatory. Used to manage musculoskeletal inflammation and some dermatological conditions in dogs, cats and horses. mp = 260-264°; $[\alpha]_D$ = 91° (in EtOH); λ_m = 237 nm (log ε 4.16).

Veterinary products:

036-212; (with Neomycin Sulfate, Polymyxin B Sulfate) Anaprime® Opthakote Ophthalmic; *Fort Dodge Animal Health, Divn. AHP.*

420 Isoflupredone Acetate

338-98-7 5190 206-423-9

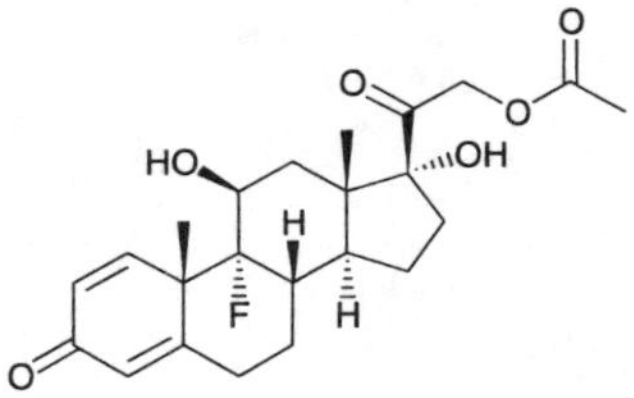

$C_{23}H_{29}FO_6$
(11β)-9-Fluoro-11,17,21-trihydroxypregna-1,4-diene-3,20-dione 21-acetate.
U-6013; Predef. Anti-inflammatory (veterinary). mp = 244-246° (dec); $[\alpha]_D^{23}$ = +108° (c = 0.735 dioxane); λ_m = 240 nm (ε 16250 EtOH).

Veterinary products:

011-789; Predef ® 2x Sterile Aqueous Suspension; *Pharmacia & Upjohn.*
015-433; (with Myristyl σ-Picolinium Chloride, Neomycin Sulfate, Tetracaine Hydrochloride) Neo-Predef ® with Tetracaine Ointment; *Pharmacia & Upjohn.*
030-025; (with Neomycin Sulfate, Tetracaine Hydrochloride) Neo-Predef ® with Tetracaine Top. Ointment; Tritop ® Topical Ointment; *Pharmacia & Upjohn.*
034-872; (with Neomycin Sulfate) Neo Predef ® Sterile Ointment; *Pharmacia & Upjohn.*

Antimalarials

421 Pyrimethamine

58-14-0 8169 200-364-2

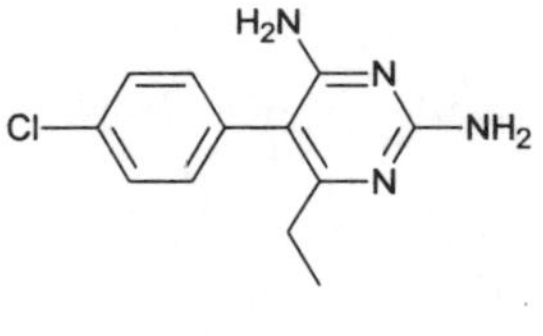

$C_{12}H_{13}ClN_4$
2,4-Diamino-5-(p-chlorophenyl)-6-ethylpyrimidine.
RP-4753; Chloridin; Malocide; Tindurin; Daraprim; Fansidar. Antimalarial and antiprotozoal. Used (often with sulfonamides) to combat toxoplasmosis in small animals. Used in horses to treat equine toxoplasmosis. mp = 233-234°; soluble in EtOH 0.9 g/100 ml at 25°, 2.5 g/100 ml at 76°), dil. HCl (0.5 g/100 ml).

422 Quinacrine
83-89-6 8225 201-508-7

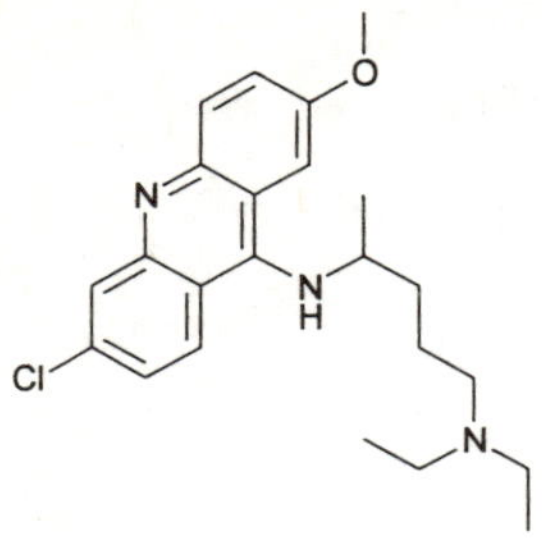

$C_{23}H_{30}ClN_3O$
6-Chloro-9-[[4-(diethylamino)-1-methylbutyl]amino]-2-methoxyacridine.
Mepacrine; Atabrine. Anthelmintic and antimalarial. Used to treat *Giardia and Trichomonas* infections.

423 Quinacrine Dihydrochloride Dihydrate
69-05-6 8225 200-700-8
$C_{23}H_{31}Cl_2N_3O{\cdot}2H_2O$
6-Chloro-9-[[4-(diethylamino)-1-methylbutyl]amino]-2-methoxyacridine dihydrochloride dihydrate.
Atabrine hydrochloride; RP-866; SN-390. Anthelmintic and antimalarial. Used to treat *Giardia and Trichomonas* infections. Dec 248-250°; soluble in H_2O (2.8 g/100 ml); slightly soluble in EtOH, MeOH; insoluble in C_6H_6, Et_2O, Me_2CO.

424 Quinacrine Methanesulfonate Hydrate
316-05-2 8225 206-256-1
$C_{25}H_{38}ClN_3O_7S_2.H_2O$
6-Chloro-9-[[4-(diethylamino)-1-methylbutyl]amino]-2-methoxyacridine methanesulfonate monohydrate.
Anthelmintic and antimalarial. Used to treat *Giardia and Trichomonas* infections. Soluble in H_2O (33 g/100 at 15°), EtOH (2.7 g/100 ml at 15°).

425 Quinidine
56-54-2 8244 200-279-0

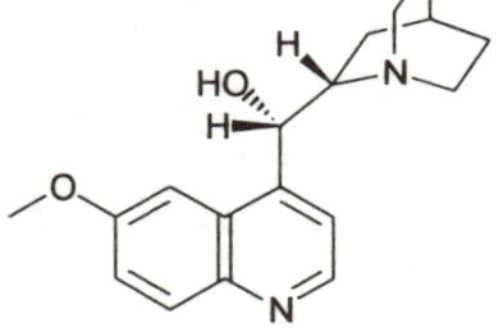

$C_{20}H_{24}N_2O$
(8R,9S)-6'-Methoxycinchonan-9-ol.
conquinine; pitayine; β-quinine. Antimalarial and Class IA anti-arrhythmic agent. A dextrorotatory stereoisomer of quinine. Found in cinchona bark. Used to treat ventricular arrhythmias and and ventricular tachycardia. mp = 174-175°; $[\alpha]_D^{15}$ = 230° (c = 1.8 $CHCl_3$), $[\alpha]_D^{17}$ = 258° (EtOH), $[\alpha]_D^{17}$ = 322° (c = 1.6 2M HCl); soluble in H_2O (0.05 g/100 ml at 20°, 0.12 g/100 ml at 100°), EtOH (2.8 g/100 ml), Et_2O (1.8 g/100 ml), $CHCl_3$ (62.5 g/100 ml); very soluble in MeOH; insoluble in petroleum ether; LD_{50} (rat iv) = 30 mg/kg, (rat orl) = 263 mg/kg.

426 Quinidine Gluconate
7054-25-3 8244 230-333-9

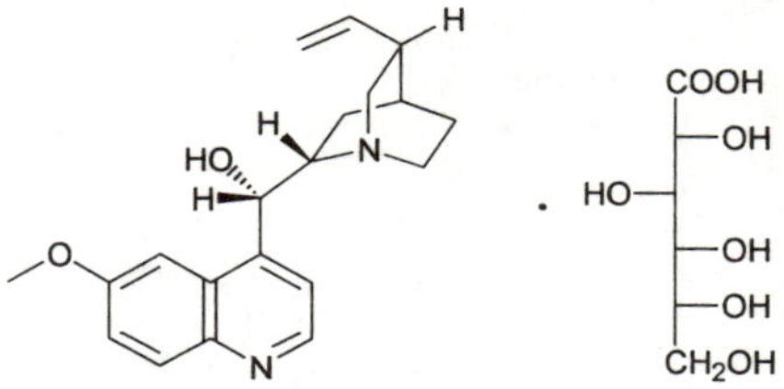

$C_{26}H_{36}N_2O_9$
(8R,9S)-6'-Methoxycinchonan-9-ol gluconate.
gluconic acid quinidine salt; Duraquin; Quinaglute. Antimalarial and Class IA anti-arrhythmic agent. Used to treat ventricular arrhythmias and and ventricular tachycardia. mp = 175-176.5°; soluble in 9 parts H_2O, 6 parts alcohol.

427 Quinidine Hydrogen Sulfate Tetrahydrate
747-45-5 8244 212-020-9
$C_{20}H_{24}N_2O{\cdot}H_2SO_4.4H_2O$
(8R,9S)-6'-Methoxycinchonan-9-ol hydrogen sulfate tetrahydrate.
quinidine bisulfate; Chinidin-Duriles; Kiditard; Kinichron; Kinidin Durules; Quiniduran. Antimalarial and Class IA anti-arrhythmic agent. Used to treat ventricular arrhythmias and and ventricular tachycardia. Soluble in H_2O.

428 Quinidine Polygalacturonate
7681-28-9 8244
$(C_{20}H_{24}N_2O{\cdot}C_6H_{10}O_7{\cdot}H_2O)_x$
(8R,9S)-6'-Methoxycinchonan-9-ol polygalacturonate.
Cardioquin; Galactoquin; Naticardina. Antimalarial and Class IA anti-arrhythmic agent. Used to treat ventricular arrhythmias and and ventricular tachycardia. mp = 180° (dec); soluble in hot alcohols, H_2O; LD_{50} (rat orl) = 3200 ±350, (mus orl) = 2680 ±210.

429 Quinidine Sulfate Dihydrate
6591-63-5 8244
$(C_{20}H_{24}N_2O)_2{\cdot}H_2SO_4{\cdot}2H_2O$
(8R,9S)-6'-Methoxycinchonan-9-ol sulfate dihydrate.
Cin-Quin; Quinidex Extentabs; Quinicardine; Quinora. Antimalarial and Class IA anti-arrhythmic agent. Used to treat ventricular arrhythmias and and ventricular tachycardia. $[\alpha]_D^{25}$ +212° (95% alcohol), +260° (dilute HCl); pKa 4.2, 8.8; pH (1% aqueous solution) = 6.0-6.8; slightly soluble in H_2O (0.01 mg/ml), boiling H_2O (0.07 mg/ml), alcohol (0.1 mg/ml), MeOH (0.33 mg/ml), $CHCl_3$ (0.08 mg/ml); insoluble in EtOH, C_6H_6l LD_{50} (mus orl) = 700 mg/kg, (rat orl) = 455.8, (mus iv) = 83 mg/kg, (rat iv) 56 mg/kg.

Antineoplastics

430 Asparaginase

9015-68-3 871 232-765-3

L-Asparaginase aminohydrolase.

colaspase; L-asnase; E.C. 3.5.1.1.; MK-965; Crasnitin; Elspar; Kidrolase; Leunase; NSC-109229. Antineoplastic agent used to treat lymphosarcoma in dogs. $[\alpha]_D^{20} = -31°$; λ_m = 278 nm ($A_{1\ cm}^{12\%}$ 7.1 0.03M sodium phosphate at pH 7.3); soluble in H_2O, insoluble in organic solvents.

431 Bleomycin Sulfate

9041-93-4 1351 232-925-2

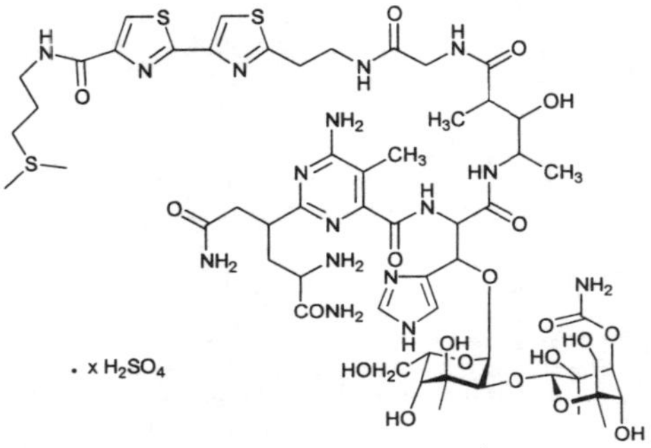

$C_{55}H_{84}N_{17}O_{21}S_3$

N[1]-[3-(Dimethylsulfonio)-propyl]bleomycinamide.

Blenoxane. Antineoplastic agent. A mixture of glycopeptide antibiotics isolated from a strain of *Streptomyces verticillus* and converted into sulfates. Useful against a variety of small tumors, such as thyroid tumors, in dogs and cats.

432 Busulfan

55-98-1 1529 200-250-2

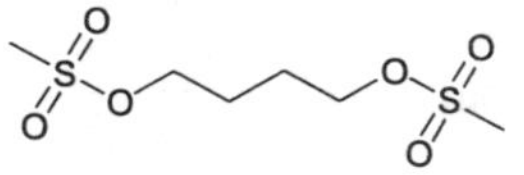

Antineoplastic agent used in adjunct therapy of acute granulocytic leukemias in small animals.

433 Carboplatin

41575-94-4 1870 255-446-0

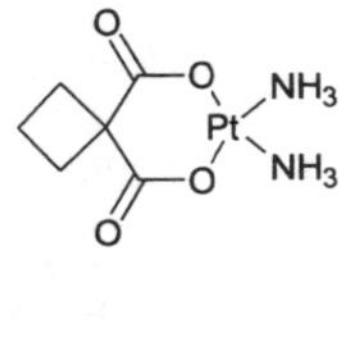

$C_6H_{12}N_2O_4Pt$

cis-Diammine(1,1-cyclobutanedicarboxylato)platinum.

Paraplatin; JM-8; NSC-241240. Antineoplastic agent. Used in post-operative treatment of osteogenic sarcoma. Soluble in H_2O; LD_{50} (mus ip) = 150 mg/kg, (mus iv) = 140 mg/kg; (rat iv) = 85 mg/kg.

434 Chlorambucil

305-03-3 2116 206-162-0

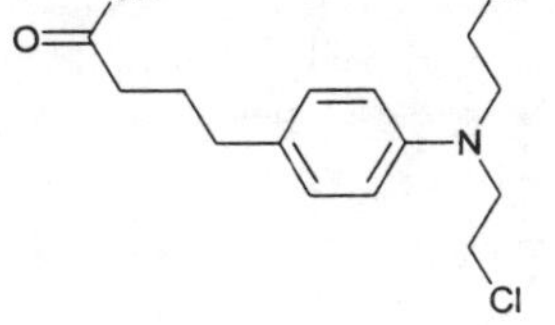

$C_{14}H_{19}Cl_2NO_2$

4-[Bis(2-chloroethyl)amino]benzenebutanoic acid.

Leukeran Tablets; 4-[p-[bis(2-chloroethyl)amino]-phenyl]butyric acid; Ambochlorin; Leukeran; chloraminophene; CB-1348; NSC-3088. A proprietary formulation of chlorambucil; an antineoplastic agent used for treatment of chronic lymphocytic leukemia, Hodgkins disease, certain forms of non-Hodgkins lymphoma, Walderstroms macroglobuliremia and advanced ovarian adenocarcinoma. Used to treat various neoplastic diseases in animals. mp = 64-66°; soluble in Et_2O, alcohol, $CHCl_3$, Me_2OH; insoluble in H_2O; LD_{50} (rat ip) = 17.7 mg/kg.

435 Cisplatin

15663-27-1 2378 239-733-8

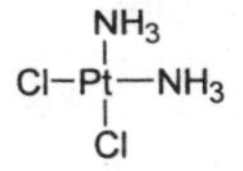

$Cl_2H_6N_2Pt$

cis-Diamminedichloroplatinum.

cis-diamminedichloroplatinum; cis-platinum II; cis-DDP; CACP; CPDC; DDP; Briplatin; Cismaplat; Cisplatyl; Citoplatino; Lederplatin; Neoplatin; Platamine; Platinex; Platiblastin; Platinol; Platinoxan; Platistin; Platosin; Rand; NSC-119875. Antineoplastic agent. Used to treat neoplastic diseases in dogs. mp = 270° (dec); soluble in H_2O (253 mg/100g), insoluble in organic solvents; LD_{50} (gpg ip) = 9.7 mg/kg.

436 Cyclophosphamide

50-18-0 2816 200-015-4

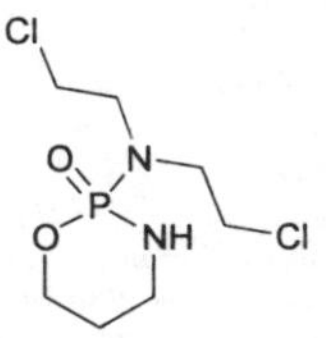

$C_7H_{15}Cl_2N_2O_2P$

(Bis(chloro-2-ethyl)amino)-2-tetrahydro-3,4,5,6-oxazaphosphorine-1,3,2-oxide-2 hydrate.

ASTA-B-518; Clafen; Claphene; Cyclophosphamid;

Cyclophosphamide; Cyclophosphamidum; Cyclophosphan; Cyclophosphane; Cyclostin; Cytophosphan; Cytoxan; NSC-26271; CB-4564; CP; CPA; CTX; CY; Endoxan; Endoxan R; Endoxan-Asta; Endoxana; Endoxanal; Endoxane; Enduxan; Genoxal; Hexadrin; Mitoxan; Neosar; NCI-C04900; Procytox; Semdoxan; Sendoxan; Senduxan; SK 20501; Zyklophosphamid. Used in veterinary medicine as both an antineoplastic agent and immunosuppressant. Also used as a chemical shearing agent in sheep. mp = 41-45°; soluble in H_2O (40 g/l), less soluble in oganic solvents; LD_{50} (rat orl) = 94 mg/kg.

437 Cyclophosphamide, Hydrated

6055-19-2 2816

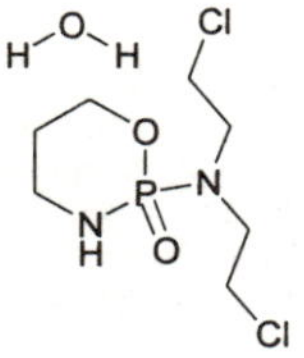

$C_7H_{15}Cl_2N_2O_2P{\cdot}H_2O$

N,N-Bis(2-chloroethyl)tetrahydro-2H-1,3,2-oxazaphosphorin-2-amine 2-oxide.

2-[bis(2-chloroethyl)amino]tetrahydro-2H-1,3,2-oxazophosphorine 2-oxide; 1-bis-(2-chloroethyl)amino-1-oxa-2-aza-5-oxaphosphoridin; B 518; Cycloblastin; Cyclostin; Endoxan; Procytox; Sendoxan; Cytoxan; NSC-26271. Used in veterinary medicine as both an antineoplastic agent and immunosuppressant. Also used as a chemical shearing agent in sheep. mp = 41-45°; soluble in H_2O (40 g/l), less soluble in oganic solvents; LD_{50} (rat orl) = 94 mg/kg.

438 Dacarbazine

4342-03-4 2866 224-396-1

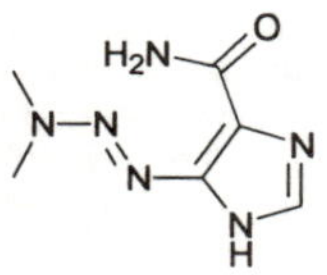

$C_6H_{10}N_6O$

5-(3,3-Dimethyl-1-triazenyl)-1H-imidazole-4-carboxamide.

(Dimethyltriazeno)imidazolecarboxamide; Dacarbazine; Deticene; NSC-45388; Dimethyltriazenoimidazolecarboxamide; DIC; DTIC; DTIC-Dome; DTIE; Imidazole carboxamide; ICDMT; ICDT; NCI-C04717. Antineoplastic agent. Has been used to manage lymphoreticular neoplasms in dogs. Dec (explosive) 250-255°; λ_m = 237 nm (ε 11200 pH 7); insoluble in H_2O (10 mg/100 ml), organic solvents; LD_{50} (rat orl) = 2147 mg/kg.

439 Dactinomycin

50-76-0 2867 200-063-6

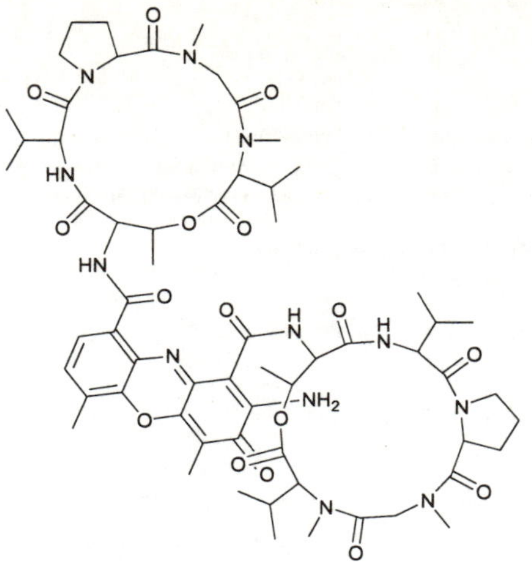

$C_{62}H_{86}N_{12}O_{16}$

Di-.xi.-lactone N,N'-[(2-amino-4,6-dimethyl-3-oxo-3H-phenoxazine-1,9-diyl)bis[carbonylimino[2-(1-hydroxyethyl)-1-oxo-2,1-ethanediyl]imino[2-(1-methylethyl)-1-oxo-2,1-ethanediyl]-1,2-pyrrolidinediylcarbonyl(methylimino)(1-oxo-2,1-ethanediyl]bis[N-methyl]-L-valine.

Actinomycin 7; Actinomycindioic D acid, dilactone; NSC-3053; Actactinomycin A IV; Actinomycin AIV; Actinomycin D; Actinomycin IV; Actinomycin X 1; actinomycin[thr-val-pro-sar-meval]; Cosmegen; C1; Dactinomycin; Dactinomycin D; Dilactone actinomycin D acid; Dilactone actinomycindioic D acid; HBF-386; Lyovac cosmegen; meractinomycin; NCI-C04682; Oncostatin K. Antibiotic from *Streptomyces parvullus*. Also an antineoplastic agent used in treament of bone and soft tissue sarcomas in small animals. mp = 241-243° (dec); $[\alpha]_D^{28}$ = -315° (MeOH, c = 0.25); λ_m = 244, 441 nm ($A_{1\ cm}^{1\%}$ 281, 206); soluble in organic solvents; light sensitive; LD_{50} (mus orl) = 13.0 mg/kg, (rat orl) = 7.2 mg/kg.

440 Doxorubicin

23214-92-8 3495 245-495-6

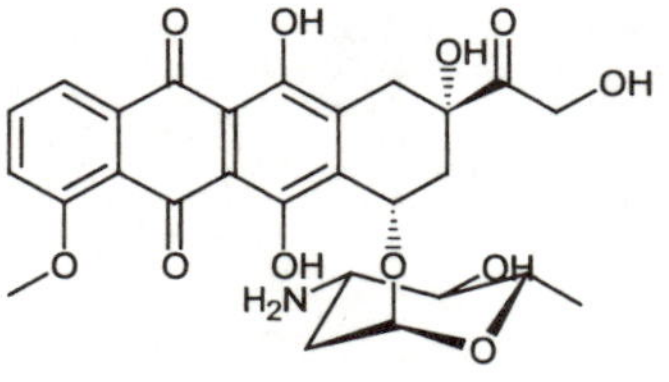

$C_{27}H_{29}NO_{11}$

α-3b-Glycoloyl-1,2,3,4,6,11-hexahydro-3,5,12-trihydroxy-10-methoxy-6,11-dioxo-1a-naphthacenyl 3-amino-2,3,6-trideoxy-L-lyxo-hexopyranoside.

adriblastina; FI-106. Antineoplastic agent. Widely used in small animal medicine. mp = 229-231°.

441 Doxorubicin Hydrochloride
25316-40-9 3495 246-818-3

$C_{27}H_{30}ClNO_{11}$
α-3b-Glycoloyl-1,2,3,4,6,11-hexahydro-3,5,12-trihydroxy-10-methoxy-6,11-dioxo-1a-naphthacenyl 3-amino-2,3,6-trideoxy-L-lyxo-hexopyranoside hydrochloride.
Adriacin; Adriamycin hydrochloride; Adriamycin, hydrochloride (8CI); Adriblastin; NSC-123127; Adriblastina; ADM hydrochloride; ADR; Doxorubicin; Doxorubicin hydrochloride; DOX HCl; FI 106; FI 6804; Hydroxydaunorubicin hydrochloride. Antineoplastic agent. Widely used in small animal medicine. mp = 204-205°; $[\alpha]_D^{20}$ = 248° (MeOH c= 0.1); λ_m = 233, 252, 288, 479, 496, 529 nm; soluble in H_2O, polar organic solvents; LD_{50} (mus iv) = 21.1 mg/kg.

442 Hydroxyurea
127-07-1 4896 204-821-7

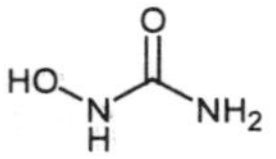

$CH_4N_2O_2$
N-(Aminocarbonyl)hydroxylamine.
Biosupressin; Hidrix; Hydrea; Hydreia; N-carbamoylhydroxylamine; Hydroxylurea; NSC-32065; Hydroxyurea; Hydroxyurea; Hydroxyurea; Hydura; Hydurea; HU; Litaler; Litalir; N-Hydroxyurea; NCI-C04831; Oncocarbide; Oxyurea; SK 22591; SQ-1089; Urea, hydroxy-. Used to treat leukemias in dogs and cats. mp = 133-136°; soluble in H_2O, EtOH.

443 Interferon alfa-2a
76543-88-9 5016
Interferon alfa-a.
alferon; alfa interferon; IFN-α; LeIF; leukocyte interferon; lymphoblastoid interferon; Ro-22-8181; roferon A. Antineoplastic agent and immunomodulator. Alpha interferon, natural (injectable form); used for the treatment of non-neoplastic feline leukemia virus.

444 Isotretinoin
4759-48-2 8333 225-296-0

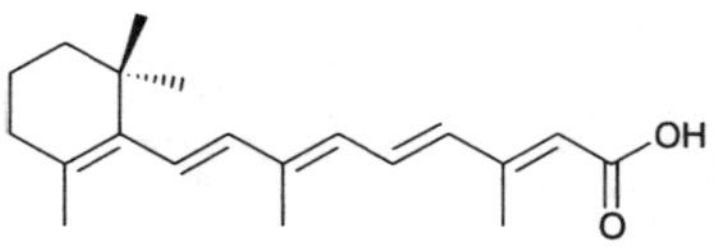

$C_{20}H_{28}O_2$
3,7-Dimethyl-9-(2,6,6-trimethyl-1-cyclohexen-1-yl)-2,(E),4,6,8(Z,Z,Z)-nonatetraenoic acid.
Accutane; 13-cis-Vitamin A acid; 13-cis-Retinoic acid; cis-retinoic acid; neovitamin a acid; 13-RA; Ro-4-3780; retinoic acid, 9Z form; Isotretinoin; Accure; IsotrexGel; Roaccutane. Synthetic retinoid, used to treat a variety of dermatological conditions, particularely in dogs. mp = 174-175°; λ_m = 354 nm (ε 39800); LD_{50} (mus ip 20 day) = 904 mg/kg, (rat ip 20 day) = 901 mg/kg, (mus orl 20 day) = 3389 mg/kg, (rat orl 20 day) > 4000 mg/kg.

445 Lomustine
13010-47-4 5594 235-859-2

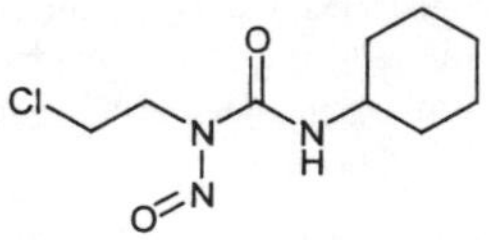

$C_9H_{16}ClN_3O_2$
N-(2-Chloroethyl)-N'-cyclohexyl-N-nitroso-urea.
Belustine; Cecenu; CeeNU; Chloroethylcyclohexylnitrosourea; CiNu; CCNU; ICIG-1109; NSC-79037; NCI-C04740; SRI-2200. Antineoplastic agent. Used to treat CNS neoplasms in dogs. mp = 90°; soluble in H_2O, organic solvents; LD_{50} (mus orl) = 51 mg/kg, (mus ip) = 56 mg/kg, (mus sc) = 61 mg/kg.

446 Mechlorethamine
51-75-2 5815 200-120-5

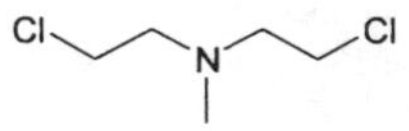

$C_5H_{11}Cl_2N$
N,N-Bis(2-chloroethyl)methylamine.
MBA; Nitrogen Mustard; Mechloroethamine; HN-2; Mustine Note; Dichloren; Caryolysine. Antineoplastic. Uncertain value in small animals. mp = -60°; bp_{18} = 87°; d_4^{25}= 1.118; slightly soluble in H_2O, soluble in organic solvents; LD_{50} (rat iv) = 1.1 mg/kg (hydrochloride).

447 Mechlorethamine Hydrochloride
55-86-7 5815 200-246-0
$C_5H_{12}Cl_3N$
N,N-Bis(2-chloroethyl)methylamine monohydrochloride.
Azotoyperite; C-6866; Caryolysine hydrochloride; Chloramin hydrochloride; Chlorethamine; Chlorethazine; Chlormethine hydrochloride; Chlormethinum; Dema; Dichloren hydrochloride; Dichloromethyldiethylamine hydrochloride; Dimitan; Embechine; Embichin hydrochloride; NSC-762; Embikhine; Erasol hydrochloride; Erasol-Ido; HN2 hydrochloride; Mechlorethamine hydrochloride; Mitoxine; Mustargen hydrochloride; Mustine hydrochloride; MBA hydrochloride; N-Lost; Nitol; Nitol takeda; Nitrogen mustard hydrochloride; Nitrogranulogen hydrochloride; NCI-C56382; NM; SK-101. Antineoplastic agent. Of uncertain value in small animals.

448 Melphalan
148-82-3 5871 205-726-3

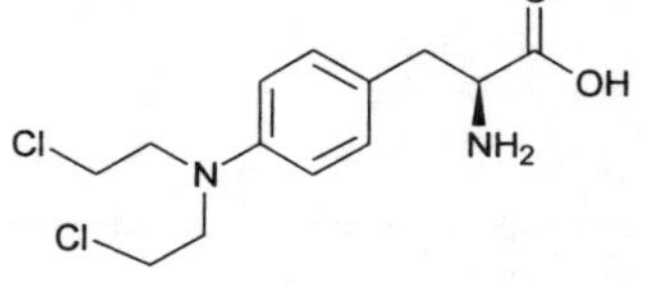

$C_{13}H_{18}Cl_2N_2O_2$
4-[Bis(2-chloroethyl)amino]-L-phenylalanine.
CB-3025; alanine nitrogen mustard; L-phenylalanine

mustard hydrochloride; L-PAM; melfalan; L-sarcolysine; Alkeran; Sarcoclorin; NSC-8806 [as hydrochloride]. Antineoplastic agent. Used to treat a variety of neoplasms such as osteosarcome and ovarian carcinoma. Carcinogen. mp = 182-183°; $[\alpha]_D^{25}$= 7.5° (1.0N HCl); $[\alpha]_D^{22}$= -31.5° (MeOH, c = 0.67); insoluble in H_2O, soluble in EtOH; LD_{50} (rat ip) = 4.5 mg/kg.

449 Melphalan, DL-form

5871

$C_{13}H_{18}Cl_2N_2O_2$
4-[Bis(2-chloroethyl)amino]-phenylalanine.
merphalan; sarcolysine. Antineoplastic agent. Used to treat a variety of neoplasms such as osteosarcome and ovarian carcinoma. Carcinogen. mp = 180-181°.

450 Melphalan, D-form

5871

$C_{13}H_{18}Cl_2N_2O_2$
4-[Bis(2-chloroethyl)amino]-D-phenylalanine.
D-sarcolysine; medphalan; CB-3026. Antineoplastic agent. Used to treat a variety of neoplasms such as osteosarcome and ovarian carcinoma. Carcinogen. mp = 181.5-182° (dec); $[\alpha]_D^{21}$= -7.5° (c = 1.26 in 1.0 N HCl).

451 6-Mercaptopurine

50-44-2 5919 200-037-4

$C_5H_4N_4S$
1,7-Dihydro-6H-purine-6-thione.
Mercaptopurine; Purine-6-thiol; Hypoxanthine, thio-; Ismipur; Leukeran; Leukerin; Leupurin; Mercaleukim; Mercaleukin; Mercapurin; Mern; Puri-Nethol; Purimethol; Purinethiol; U-4748; 3H-Purine-6-thiol; 6 MP; 6-Purinethiol; 6-Thioxopurine; 7-Mercapto-1,3,4,6-tetrazaindene; 6-Purinethiol hydrate; NSC-755. Antineoplastic, immunosuppressant. Used to treat acute leukemias and rheumatoid arthritis. Dec 313-314°; λ_m = 230, 312 nm (ε 14000, 19600 0.1N NaOH); insoluble in H_2O, organic solvents; slightly soluble in EtOH; LD_{50} (mus ip) = 157 mg/kg.

452 Mitotane

53-19-0 6302 200-166-6

$C_{14}H_{10}Cl_4$
1,1-Dichloro-2-(o-chlorophenyl)-2-(p-chlorophenyl)-ethane.
Lysodren; o,p'-DDE; NSC-38721. Antineoplastic agent. Used to treat hyperadrenocorticism in dogs. mp = 76-78°; soluble in EtOH, isooctane, CCl_4.

453 Mitoxantrone

65271-80-9 6303

$C_{22}H_{28}N_4O_6$
1,4-Dihydroxy-5,8-bis[[2-[(2-hydroxyethyl)amino]ethyl]-amino]-9,10-anthracenedione.
Dihydroxyanthraquinone; DHAQ; Mitoxantrone [as free base]; NSC-279836. Antineoplastic agent. Used in treatment of several neoplastic diseases in dogs. mp = 160-162°; λ_m = 244, 279, 525, 620, 660 nm (log ε 4.64, 4.31, 3.70, 4.37, 4.38 EtOH); sparingly soluble in H_2O, EtOH; insoluble in organic solvents.

454 Mitoxantrone Dihydrochloride

70476-82-3 6303 274-619-1

$C_{22}H_{30}Cl_2N_4O_6$
1,4-Dihydroxy-5,8-bis[[2-[(2-hydroxyethyl)amino]ethyl]-amino]-9,10-anthracenedione dihydrochloride.
DHAQ; CL232315; NSC-279836; Novantrone; Immunex; CL-232315. Antineoplastic agent. Used in treatment of several neoplastic diseases in dogs. mp = 203-205°; λ_m = 241, 273, 608, 658 nm (ε 41000, 12000, 19200, 20900 H_2O); sparingly soluble in H_2O, MeOH; insoluble in organic solvents.

455 Thioguanine Hemihydrate

5580-03-0

$C_5H_5N_5S \cdot 0.5H_2O$
2-Amino-1,7-dihydro-6H-purine-6-thione.
Lanvis; BW-5071; Tabloid; Thioguanine; Tioguanin; Tioguanine; TG; Wellcome U3B; X 27; NSC-752. Used as adjunct therapy for acute lymphocytic or granulocytic leukemia in dogs and cats. mp > 360°.

456 Thiotepa

52-24-4 9805 200-135-7

$C_6H_{12}N_3PS$
1,1',1''-Phosphinothioylidynetris-aziridine.
CBC-806495; Girostan; NCI-C01649; Oncotepa;

Oncothio-tepa; SK-6882; STEPA; Tespa; NSC-6396; Tespamin; Tespamine; Thio-tepa; Thio-tepa S; Thio-Tep; Thio-Tepa; Thiofozil; Thiotef; Thiotepa; Thioplex; Tifosyl; Tio-tef; Tiofosfamid; Tiofosyl; Tiofozil; TESPA; TIO TEF; TSPA. Antineoplastic agent. Used as an adjunct in treatment of carcinomas. mp = 51°; soluble in H_2O (190 g/l), organic solvents; LD_{50} (rat iv) = 15 mg/kg.

457 Vinblastine

865-21-4 10119 212-734-0

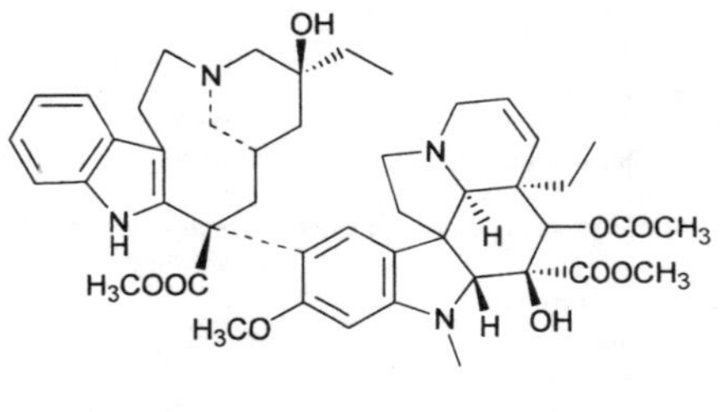

$C_{46}H_{58}N_4O_9$

Vincaleukoblastine.

vinblastine; NSC-49842. Antineoplastic agent. Used in treatment of various tumors in small animals. mp = 211-216°; $[\alpha]_D^{26}$ = 42° ($CHCl_3$); insoluble in H_2O, soluble in organic solvents.

458 Vinblastine Sulfate

143-67-9 10119 205-606-0

$C_{46}H_{50}N_4O_{13}S$

Vincaleukoblastine sulfate.

vinblastine sulfate; 29060-LE; Velsar; Belvan, VLB; Exal; Vincaleukoblastine sulfate (1:1) (salt);VLB monosulfate; Velbe; NSC-49842. Antineoplastic agent. Used in copmbination therapy of various tumors in small animals. mp = 284-285°; $[\alpha]_D^{26}$ = -28° (c = 1.01 in MeOH).

459 Vincristine Sulfate

2068-78-2 10124 218-190-0

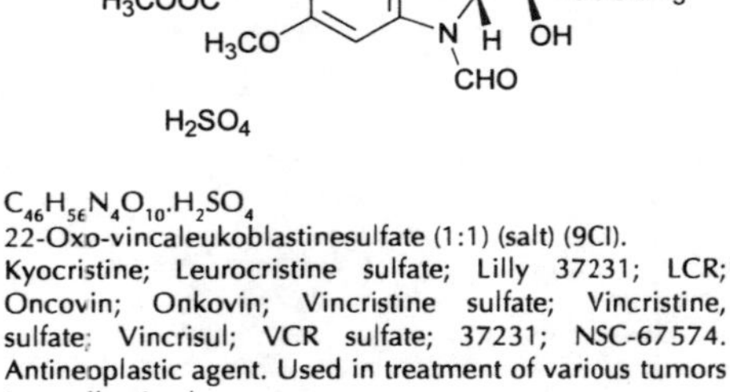

$C_{46}H_{56}N_4O_{10}.H_2SO_4$

22-Oxo-vincaleukoblastinesulfate (1:1) (salt) (9CI).

Kyocristine; Leurocristine sulfate; Lilly 37231; LCR; Oncovin; Onkovin; Vincristine sulfate; Vincristine, sulfate; Vincrisul; VCR sulfate; 37231; NSC-67574. Antineoplastic agent. Used in treatment of various tumors in small animals.

Antiosteoporotics

460 Etidronic Acid Disodium Salt

7414-83-7 3908 231-025-7

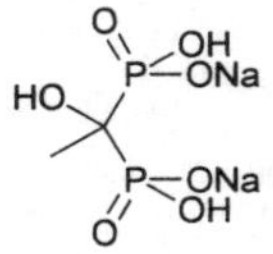

$C_2H_6Na_2O_7P_2$

Disodium dihydrogen (1-hydroxyethylidene)-diphosphonate.

Didronel; Calcimux; Diphos; Etidron. Calcium regulator. Bone resorption inhibitor used to treat acute hypercalcemia. Soluble in H_2O.

Antiparkinsonians

461 Bromocriptine

25614-03-3 1437 247-128-5

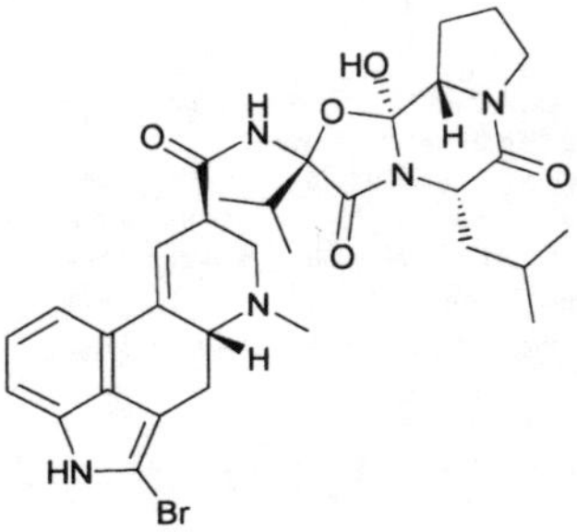

$C_{32}H_{40}BrN_5O_5$

2-Bromo-12'-hydroxy-2'-(1-methylethyl)-5'-(2-methyl-propyl)-5'α-ergotaman-3',6',18-trione.

CB-154. Antiparkinsonian. Dopamine receptor agonist and prolactin inhibitor. Used to treat acromegaly and pituitary adenomas and presudopregnancy in dogs and horses. mp = 215-218° (dec); $[\alpha]_D^{20}$ = -195° (c = 1 CH_2Cl_2); LD_{50} (rbt orl) > 1000 mg/kg, (rbt iv) = 12 mg/kg.

462 Bromocriptine Methanesulfonate

22260-51-1 1437 244-881-1

$C_{33}H_{44}BrN_5O_8S$

2-Bromo-12'-hydroxy-2'-(1-methylethyl)-5'-(2-methyl-propyl)-5'α-ergotaman-3',6',18-trione methanesulfonate.

CB-154 mesylate; Parlodel; Bagren; Pravidel. Antiparkinsonian. Dopamine receptor agonist and prolactin inhibitor. Used to treat acromegaly and pituitary adenomas and presudopregnancy in dogs and horses. mp = 192-196° (dec); $[\alpha]_D^{20}$ = 95° (c = 1 MeOH/CH_2Cl_2); soluble in MeOH (91 g/100 ml); EtOH (2.3 g/100 ml), H_2O (0.08 g/100 ml), $CHCl_3$ (0.045 g/100ml)C_6H_6 (< 0.01 g/100 ml).

Antipheochromocytoma Agents

463 Phenoxybenzamine
59-96-1 7409 200-446-8

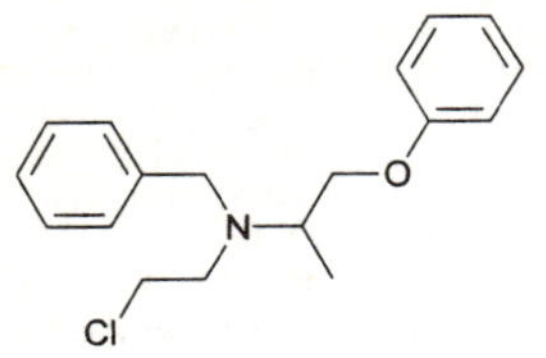

$C_{18}H_{22}ClNO$
N-(2-Chloroethyl)-N-(1-methyl-2-phenoxyethyl)-benzylamine.
668-A; N-phenoxyisopropyo-N-benzyl-β-chloroethylamine; bensylyt. Antihypertensive. Antipheochromocytoma. α-Adrenergic blocker. CAUTION: The hydrochloride may be a carcinogen. mp = 38-40°; soluble in C_6H_6.

464 Phenoxybenzamine Hydrochloride
63-92-3 7409 200-569-7
$C_{18}H_{23}Cl_2NO$
N-(2-Chloroethyl)-N-(1-methyl-2-phenoxyethyl)-benzylamine hydrochloride.
Dibenzyline; Dibenzylin; Dibenyline; Dibenzyran. Antihypertensive. Antipheochromocytoma. α-Adrenergic blocker. CAUTION: The hydrochloride may be a carcinogen. mp = 137.5-140°; soluble in EtOH, propylene glycol; sparingly soluble in H_2O.

Antiprotozoals

465 Bicyclohexylammonium Fumagillin
41567-78-6 4308

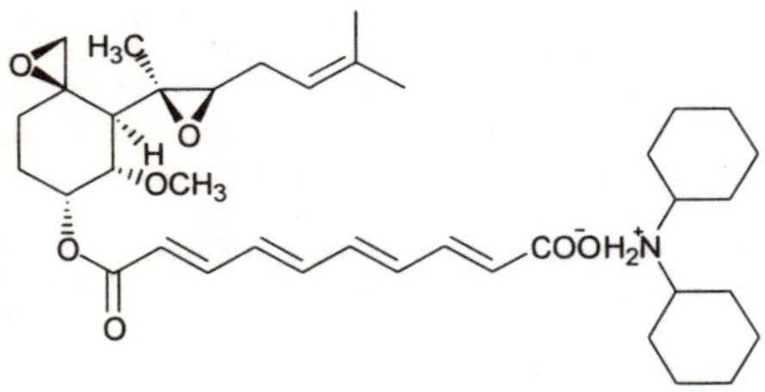

Derivative of an antibiotic substance produced by *Aspergillus fumigatus*. Antiprotozoal used for control of *Nosema apis* in honey bees.

Veterinary products:

009-252; Fumidil B; *Mid-Continent Agrimarketing.*

466 Carnidazole
42116-76-7 1897 255-663-0

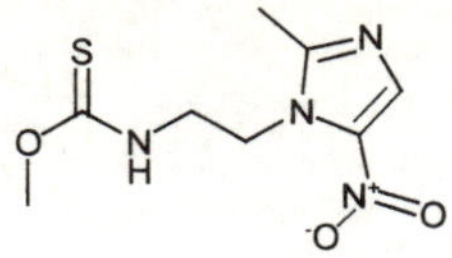

$C_8H_{12}N_4O_3S$
O-Methyl [2-(2-methyl-5-nitroimidazol-1-yl)-ethyl]thiocarbamate.
R-25831; Spartrix. Antiprotozoal against Trichomonas. mp = 142.4°.

Veterinary products:

139-879; Carnidazole; Spartrix®; *Wildlife.*

467 Castor Oil
1323-38-2 1904(11) 232-293-8
triglyceride of ricinoleic, oleic, linoleic, palmitic, stearic acids; *Ricinus communis*; Ricinus oil; Aceite de ricino; Huile de ricini; Oleum ricini; Ricini oleum; Oil of Palma Christi; tangantangan oil. FDA, FEMA GRAS, FDA approved for injectables, orals, topicals, USP/NF, BP, JP, Ph.Eur. compliance. Used as an oleaginous vehicle, emollient, lubricant, plasticizer, laxative, solvent in intramuscular injectables, solid oral dosage forms, topical pharmaceuticals, capsules and emulsions. Soothing to the eyes, cathartic and purgative. Pale yellow, viscous liquid; mp = -12°; bp = 313°; d = 0.961; soluble in EtOH, AcOH, $CHCl_3$, Et_2O, insoluble in H_2O. Moderately toxic by ingestion.

Veterinary products:

031-555; (with Balsam Peru Oil, Trypsin) Trypzyme Aerosol; *Farnam.*
039-583; (with Balsam Peru Oil, Trypsin) Granulex Aerosol Spray; *Hickam.*

468 Chlortetracycline Bisulfate
2245
7-Chloro-4-(dimethylamino)-1,4,4a,5,5a,6,11,12a-octahydro-3,6,10,12,12a-pentahydroxy-6-methyl-1,11-dioxo-2-naphthacenecaroxamide bisulfate.
Antibacterial, antiamebic and antiprotozoal. Used in veterinary medicine as an antimicrobial agent.

Veterinary products:

055-012; (with Sulfamethazine) Aureomycin ®-Sulmet Soluble Powder; *American Cyanamid Divn., AHP Corp.*
055-020; Aureomycin ® Soluble Powder; *American Cyanamid Divn., AHP Corp.*
065-113; (with Sulfamethazine) Aureo Sulfa Soluble Powder; *Purina Mills.*
065-486; Chlortetracycline Bisulfate Soluble Powder; *Boehringer Ingelheim Vetmedica.*

469 Chlortetracycline Calcium Complex
2245
7-Chloro-4-(dimethylamino)-1,4,4a,5,5a,6,11,12a-octahydro-3,6,10,12,12a-pentahydroxy-6-methyl-1,11-dioxo-2-naphthacenecarboxamide, calcium complex.
Antibacterial, antiamebic and antiprotozoal. Used in veterinary medicine as an antimicrobial agent.

Veterinary products:

035-688; (with Penicillin G Procaine, Sulfamethazine) Aureo SP-250; Aureomix 500; *Roche Vitamins.*
036-361; (with Amprolium, Ethopabate, Sodium Sulfate) Amp Ethopabate CTC ® Sodium Sulfate; *Roche Vitamins.*
041-647; (with Sulfamethazine) Aureomix S 700 A; *Roche Vitamins.*
041-648; (with Sulfamethazine) Aureomix S 700 D; *Roche Vitamins.*
041-649; (with Sulfamethazine) Aureomix S 700 G; *Roche Vitamins.*
041-650; (with Sulfamethazine) Aureomix S 700 E; *Roche Vitamins.*
041-651; (with Sulfamethazine) Aureomix S 700 F; *Roche Vitamins.*
041-652; (with Sulfamethazine) Aureomix S 700 C-2; *Roche Vitamins.*
041-653; (with Sulfamethazine) Aureomix S 700 B; *Roche Vitamins.*
041-654; (with Sulfamethazine) Aureomix S 700 H; *Roche Vitamins.*
046-209; (with Clopidol) Coyden 25 ® + CTC ®; *Rhône-Poulenc.*
048-480; Chloratet 50; *ADM Animal Health & Nutrition.*
049-287; Chlorachel-50; *Pfizer.*
091-668; (with Penicillin G Procaine, Sulfamethazine) ChlorMax™-SP 250; ChlorMax™-SP 500; ChlorMax™-SP 1000; Chlorachel 250 Swine; Pficlor 250; *Alpharma.*
092-286; CLTC - 10; CLTC - 20; CLTC - 30; CLTC - 50; CLTC - 70; *Pfizer.*
092-287; CLTC Type A Medicated Article; *Pfizer.*
092-507; (with Robenidine Hydrochloride) Robenz ® with Aureomycin ® 500 Gm; *Roche Vitamins.*
100-901; Pfichlor 100S Milk Replacer Type A Medicated Article; *Roche Vitamins.*
140-859; (with Salinomycin Sodium) Aureomycin ® / Bio-Cox ®; *Roche Vitamins.*
140-867; (with Roxarsone, Salinomycin Sodium) Aureomycin ® / Bio-Cox ® / 3-Nitro ®; *Roche Vitamins.*
200-091; (with Roxarsone, Salinomycin Sodium) Sacox ® + 3-Nitro ® + Aureomycin ®; *Hoechst-Roussel Vet.*
200-167; (with Penicillin G Procaine, Sulfathiazole) Aureozol ® 500 Granular; *Roche Vitamins.*
200-242; (with Bacitracin Methylene Disalicylate) Aureomycin ® - 50, 70, 80, 90, 100 / BMD ® - 25, 30, 40, 50, 60, 75; *Roche Vitamins.*

470 Chlortetracycline Hydrochloride

64-72-2 2245 200-591-7

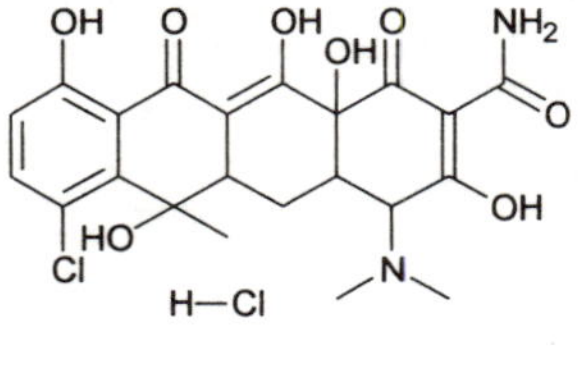

$C_{22}H_{24}Cl_2N_2O_8$
7-Chloro-4-(dimethylamino)-1,4,4a,5,5a,6,11,12a-octahydro-3,6,10,12,12a-pentahydroxy-6-methyl-1,11-dioxo-2-naphthacenecaroxamide monohydrochloride.
Aureomycin; Fermycin Soluble. Antibacterial, antiamebic, antiprotozoal. Antimicrobial (veterinary). Dec > 210°; $[\alpha]_D^{23}$ = 240°; soluble in H_2O (0.86 g/100 ml), MeOH (1.74 g/100 ml), EtOH (0.17 g/100 ml); insoluble in Me_2CO, Et_2O, $CHCl_3$, dioxane; LD_{50} (rat orl) = 10300 mg/kg.

Veterinary products:

035-805; (with Sulfamethazine) Aureomix-S 700 Crumbles; Aureomix-S 700 g; *Roche Vitamins.*
039-077; (with Penicillin G Procaine, Sulfathiazole) CSP™ 250; CSP ™ 500; *Boehringer Ingelheim Vetmedica.*
045-444; (with Decoquinate) Deccox ® / ChlorMax ®; Decoquinate & Chlortetracycline; *Alpharma.*
046-699; ChlorMax™ 10 Type B Medicated Article; ChlorMax™ 100 Type B Medicated Article; ChlorMax™ 50 Type B Medicated Article; ChlorMax™ 60 Type B Medicated Article; ChlorMax™ 75 Type B Medicated Article; Micro CTC 100; *Alpharma.*
055-018; Aureomycin ® Tablets 25 mg; *American Cyanamid Divn., AHP Corp.*
055-039; Aureomycin ® Soluble Oblets; *American Cyanamid Divn., AHP Corp.*
055-040; SF Mix 66; *Roche Vitamins.*
065-071; Aureomycin ® Soluble Powder; *American Cyanamid Divn., AHP Corp.*
065-178; Fermycin Soluble; *Boehringer Ingelheim Vetmedica.*
065-222; Keet Life; *Hartz Mountain Products.*
065-256; Chlortet-Soluble-O; *ADM Animal Health & Nutrition.*
065-440; Aureomycin ® Soluble Powder Concentrate; *American Cyanamid Divn., AHP Corp.*
065-480; Scour-Pneumonia Antibiotic; *Pennfield Oil.*
065-481; Calf Scour Boluses; Chlortetracycline Pneumonia; *Boehringer Ingelheim Vetmedica.*
121-553; (with Monensin Sodium) Coban ® / Aureomycin ®; *Roche Vitamins.*
141-011; (with Tiamulin Hydrogen Fumarate) Denagard ® 10 / Chlortetracycline Premixes; *Boehringer Ingelheim Vetmedica.*
200-140; (with Penicillin G Procaine, Sulfathiazole) Aureozol ®; *Roche Vitamins.*
200-236; Chlortetracycline HCL Soluble Powder; *Phoenix.*

471 Furazolidone

67-45-8 4320 200-653-3

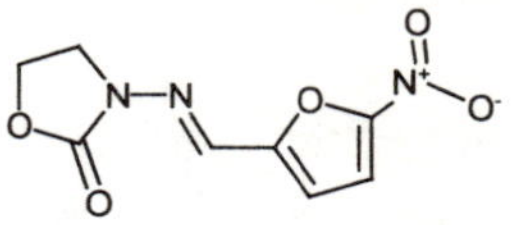

$C_8H_7N_3O_5$
3-[(5-Nitrofurfurylidene)amino]-2-oxazolidinone.
Fiurox Aerosol Powder; Furoxone. Antiprotozoal against Trichomonas. Used to treat enteric infections in small animals. mp = 256-257°; soluble in H_2O (0.004 g/100 ml).

Veterinary products:

032-319; Furox ® Aerosol Powder; Topazone Aerosol Powder; *Fort Dodge Animal Health Divn., Am. Cyanamid.*
111-104; Furall; *Farnam.*

472 Gelatin

9000-70-8 4388 232-554-6
Gelatin, granular.
Alkaline Processed Gelatin; Bone Gelatin Type B 200 Bloom; Croda 60 Bloom; Croda 50 Bloom Gelatin; Croda 160 Bloom Limed Gelatin; Croda 190 Bloom Acid Ossein

Gelatin; Croda 250 Bloom Acid Ossein Gelatin; Crodyne BY-19; Edible Beef Gelatin; Gelatin XF; Gelatin USP/NF Type A; Gelatins; HSA Minispheres; Liquid Fish Gelatin Conc.; P-4 Pharmaceutical Gelatin; P-5 Pharmaceutical Gelatin; P-6 Pharmaceutical Gelatin; P-7 Pharmaceutical Gelatin; P-8 Pharmaceutical Gelatin; component of: Canthaxanthin Beadlets 10%, Dry β-Carotene Beadlets 10%CWS No. 65633; Dry β-Carotene Beadlets 10%CWS No. 65661; Dry Vitamin D_3 Beadlets Type 850 No. 652550401, 652550601, Dry Vitamin D_3 Type 100CWS No. 65242, Dry Vitamin E Acetate 50% type CWS/F No. 652530001, Palma-Sperse® Type 250-S No. 65322, Palma-Sperse® Type 250A/50 D-S No. 65221. Spray Dried Fish Gelatin/Maltodextrin; Coating agent, film former, gelling agent, suspending agent, tablet binder and viscosity-increasing agent. FDA GRAS, Japan approved, FDA approved for dentals, inhalants, intramuscular injectables, intravenous, IV infusions, orals, topicals, USP/NF, BP, Ph.Eur. compliance. In the FDA list of inactive ingredients, (dental preparations, inhalations, injections, oral capsules, pastilles, solutions, syrups and tablets, topicals and vaginals). Used as an emulsifier, vehicle, binder, suspending agent, table binder/coating, (hard/soft capsules and microencapsulated products), wound/burn healing, surgical sponges, ointments, suppositories, dentals, inhalants, IM injectables, intravenous, orals, topicals, vaginals. Used in veterinary medicine as a plasma expander and in hemostasis. Yellowish, brittle solid; d = 1.325 (type A), 1.283 (type B); soluble in warm H_2O, glycerol, AcOH, insoluble in organic solvents; viscosity of 6.67% w/v aqueous solution at 60° = 4.3 - 4.7 mPa, for a 12.5% w/v aqueous solution at 60° = 18.5 - 20.5 mPa. LD_{50} (rat orl) = 5000 mg/kg, may cause anaphyllaxis.

Veterinary products:

006-281; (with Sodium Chloride) Intragel; *Fort Dodge Animal Health, Divn. AHP.*

473 Halofuginone Hydrobromide

64924-67-0 4627

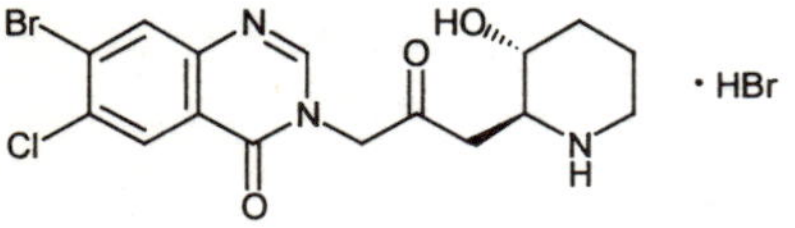

$C_{16}H_{18}Br_2ClN_3O_3$
(±)-trans-7-Bromo-6-chloro-3-[3-(3-hydroxy-2-piperidyl)-acetonyl]-4(3H)-quinazolinone monohydrobromide.
Stenorol; RU-19110. Antiprotozoal (coccidiostat). mp = 247° (dec).

Veterinary products:

130-951; Stenorol ®; *Hoechst-Roussel Vet.*
137-483; (with Virginiamycin) Flavomycin ® + Stenorol ®; *Hoechst-Roussel Vet.*
139-473; (with Virginiamycin) Stenorol ® + Stafac ®; *Hoechst-Roussel Vet.*
140-340; (with Lincomycin Hydrochloride) Stenorol ® + Lincomix ®; *Hoechst-Roussel Vet.*
140-533; (with Bacitracin Methylene Disalicylate, Roxarsone) Stenorol ® + 3-Nitro ® + BMD ®; *Hoechst-Roussel Vet.*
140-584; (with Bacitracin Methylene Disalicylate) Stenorol ® + BMD; *Hoechst-Roussel Vet.*
140-824; Stenorol ® Type A Medicated Article; *Hoechst-Roussel Vet.*
140-918; (with Bambermycins) Stenorol ® + Flavomycin ®; *Hoechst-Roussel Vet.*
140-919; (with Bacitracin Methylene Disalicylate) Stenorol ® + BMD ®; *Hoechst-Roussel Vet.*

474 Metronidazole

443-48-1 6242 207-136-1

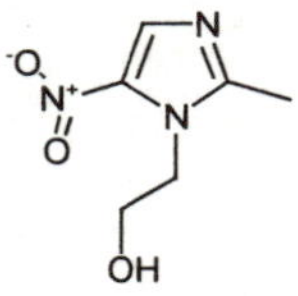

$C_6H_9N_3O_3$
2-Methyl-5-nitroimidazole-1-ethanol.
Metro Cream & Gel; Protostat; Satric; Bayer 5360; RP-8823; NSC-50364; Arilin; Clont; Deflamon; Elyzol; Flagyl; Fossyol; Gineflavir; Klion; MetroGel; Metrolag; Metrolyl; Metrotop; Orvagil; Rathimed; Sanatrichom; Trichazol; Tricocet; Trichocide; Tricho Cordes; Tricho-Gynaedron; Trivazol; Vagilen; Vagimid; Zadstat; component of: Flagyl I.V. RTU, Metro I.V. Antiprotozoal against Trichomonas. Anti-amebic and antibacterial. Use for control of *Giardia* in dogs and cats. mp =158-160°; soluble in H_2O (1.0 g/100 ml), EtOH (0.5 g/100 ml), Et_2O (< 0.05 g/100 ml), $CHCl_3$ (< 0.05 g/100 ml); sparingly soluble in DMF.

475 Metronidazole Hydrochloride

69198-10-3 6242

$C_6H_{10}ClN_3O_3$
2-Methyl-5-nitroimidazole-1-ethanol.
Flagyl I.V.; SC-326421. Antiprotozoal against Trichomonas. Anti-amebic and antibacterial. Use for control of *Giardia* in dogs and cats.

476 Metronidazole Phosphate

73334-05-1

$C_6H_{10}N_3O_6P$
2-Methyl-5-nitroimidazole-1-ethanol dihydrogen phosphate (ester).
U-54555. Antiprotozoal against Trichomonas. Anti-amebic and antibacterial. Use for control of *Giardia* in dogs and cats.

477 Nitarsone

98-72-6 6659 202-695-8

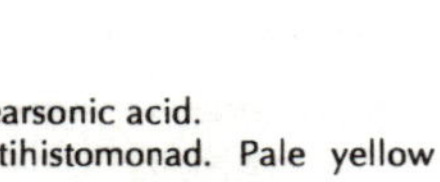

$C_6H_6AsNO_5$
p-Nitrobenzenearsonic acid.
NSC-5085. Antihistomonad. Pale yellow crystals; dec 298-300°; slightly soluble in H_2O, EtOH at 25°; more soluble in warm H_2O, EtOH.

Veterinary products:

007-616; Histostat®-50 Type A Medicated Article; *Alpharma*.
141-088; (with Bacitracin Methylene Disalicylate) Histostat ® / BMD ®; *Alpharma*.

Antipruritics

478 Cyproheptadine
129-03-3 2842 204-928-9

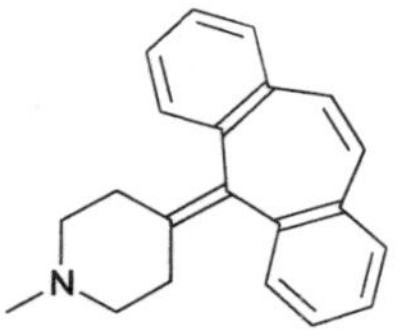

$C_{21}H_{21}N$
4-(5H-Dibenzo[a,d]cyclohepten-5-ylidene-1-methylpiperidine.
Antipruritic. Has possible value as an antihistamine and an appetite stimulant (cats). mp = 112.3-113.3°.

479 Cyproheptadine Hydrochloride Monohydrate
6032-06-0 2842
$C_{21}H_{22}ClN{\cdot}H_2O$
4-(5H-Dibenzo[a,d]cyclohepten-5-ylidene-1-methylpiperidine hydrochloride monohydrate.
Antipruritic. Has possible value as an antihistamine and an appetite stimulant (cats). mp = 214-216°; soluble in H_2O (0.5 g/100 ml).

480 Cyproheptadine Hydrochloride Sesquihydrate
41354-29-4 2842
$C_{21}H_{22}ClN \cdot 1.5H_2O$
4-(5H-Dibenzo[a,d]cyclohepten-5-ylidene-1-methylpiperidine hydrochloride sesquihydrate.
Anarexol; Antegan; Ifrasarl; Nuran; Periactin; Vimicon; Cipractin; Peritol; Periactin. Antipruritic. Has possible value as an antihistamine and an appetite stimulant (cats). Dec 252.6-253.6°; λ_m = 224, 285 nm ($E^{1\%}_{1\ cm}$ 1656 355 0.1N H_2SO_4); soluble in MeOH (66.6 g/100 ml), $CHCl_3$ (6.25 g/100 ml), EtOH (2.88 g/100 ml), H_2O (0.36 g/100 ml); insoluble in Et_2O; LD_{50} (mus orl) = 74.2 mg/kg.

481 Glycine
56-40-6 4500 200-272-2

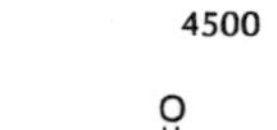

$C_2H_5NO_2$
Aminoacetic acid.
Gly; G; Glycolixir; glycocoll; Gyn-Hydralin; Glycosthène; component of: Corilin. Antipruritic. mp = 290°; d = 1.1607; soluble in H_2O (25.0 g/100 ml at 25°, 39.1 g/100 ml at 50°, 54.4 g/100 ml at 75°, 67.2 g/100 ml at 100°), EtOH (0.06 g/100 ml, C_5H_5N (0.60 g/100 ml); insoluble in Et_2O.

Veterinary products:

125-961; (with Citric Acid, Dextrose, Potassium Citrate, Potassium Phosphate, Sodium Chloride) Re-Sorb; Vy'trate; *Pfizer*.

482 Trimeprazine Tartrate
4330-99-8 9834 224-368-9

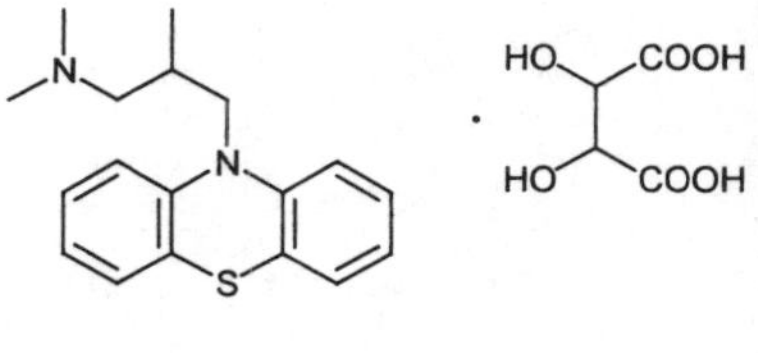

$C_{40}H_{50}N_4O_6S_2$
10-[3-(Dimethylamino)-2-methylpropyl]-phenothiazinetartrate.
Temaril16; Panectyl; Repeltin; Temaril; Theralene; Vallergan. Antipruritic. Used as an antipruritic in allergic conditions. Soluble in H_2O, slightly soluble in EtOH.

Veterinary products:

012-437; (with Prednisolone) Temaril-P-Tablets; *Pfizer*.
035-161; (with Prednisolone) Temaril-P Spansule Capsule No.1; Temaril-P Spansule; *Pfizer*.

Antipsoriatics

483 Acitretin
55079-83-9 114 259-474-4

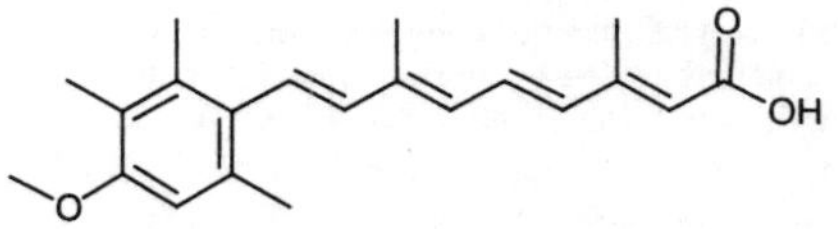

$C_{21}H_{26}O_3$
(all E)-9-(4-Methoxy-2,3,6-trimethylphenyl)-3,7-dimethyl-2,4,6,8-nonatrienoic acid.
Soriatane; etretin; Neotigason; Ro-10-1670/000. Antipsoriatic. Used to treat seborrhea and keratosis, particularly in dogs. mp = 228-230°; LD_{50} (mus ip) > 400 mg/kg (1 day), = 700 (10days), = 700 (20 days).

484 Etretinate
54350-48-0 3935 259-119-3

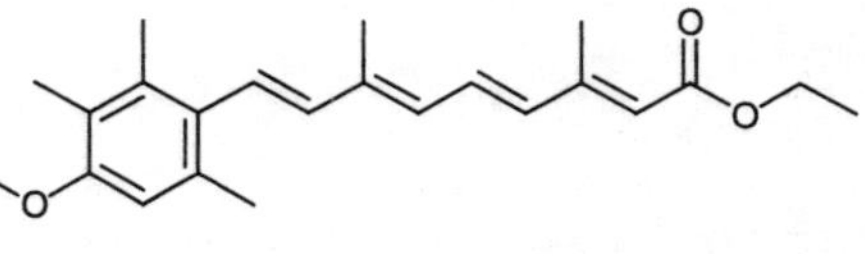

$C_{23}H_{30}O_3$
Ethyl (all E)-9-(4-methoxy-2,3,6-trimethylphenyl)-3,7-dimethyl-2,4,6,8-nonateraenoate.
Ro-10-9359; Tegison; Tigason. Antipsoriatic. No longer

marketed; replaced by Acitretin. mp = 104-105°; LD_{50} (mus ip 1 day) > 4000 mg/kg, (mus ip 20 day) = 1176 mg/kg, (rat ip 20 day) > 2000 mg/kg, (mus orl 20 day) > 2000 mg/kg, (rat orl 20 day) > 4000 mg/kg.

Antipsychotics

485 Chlorpromazine

50-53-3 2238 200-045-8

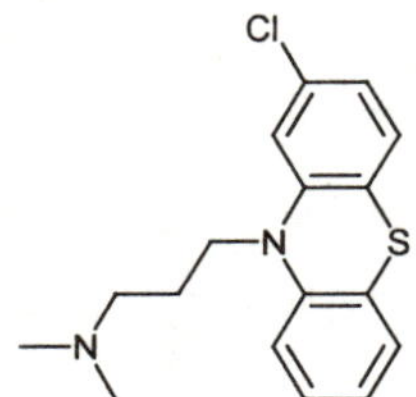

$C_{17}H_{19}ClN_2S$

2-Chloro-10-[3-(dimethyamino)propyl]phenothiazine.

2601-A; HL-5746; RP-4560; SKF-2601-A; Aminazine; Ampliactil; Amplictil; Chlorderazin; Chlorpromados; Elmarin; Esmind; Fenactil; Novomazina; Promactil; Promazil; Proma; Prozil; Plegomazin; Sanopron; Wintermin. A phenothiazine with antiemetic and antipsychotic properties. Used as a tranquillizer and anti-emetic in veterinary medicine. Oily liquid; $bp_{0.8}$ = 200-205°.

486 Chlorpromazine Hydrochloride

69-09-0 2238 200-701-3

$C_{17}H_{20}Cl_2N_2S$

2-Chloro-10-[3-(dimethyamino)propyl]phenothiazine monohydrochloride.

Chloractil; Chlorazin; Hebanil; Hibanil; Hibernal; Klorpromex; Largactil; Largaktyl; Marazine; Megaphen; Promacid; Propaphenin; Sonazine; Taroctyl; Thorazine; Torazina. A phenothiazine with antiemetic and antipsychotic properties. Used as a tranquillizer and anti-emetic in veterinary medicine. Dec 179-180°; pH (5% aq) = 4.0-5.5; soluble in H_2O, EtOH; practically insoluble in organic solvents; LD_{50} (rat orl) = 2225 mg/kg.

487 Droperidol

548-73-2 3505 208-957-8

$C_{22}H_{22}FN_3O_2$

1-[1-[3-(p-Fluorobenzoyl)propyl]-1,2,3,6-tetrahydro-4-pyridyl]-2-benzimidazolinone.

McN-JR-4749; R-4749; dehydrobenzperidol; Dridol; Droleptan; Inapsine; component of: Innovar, Thalamonal. A butyrophenone with antipsychotic properties; Used as a veterinary tranquillizer mp = 145-146.5°; λ_m = 245, 280 nm (ε 15600, 7500); soluble in $CHCl_3$ and DMF; barely soluble in C_6H_6, EtOH, H_2O; pKa = 7.64; heat and light sensitive; LD_{50} (mus sc) 125 mg/kg, (mus ip) = 43 mg/kg.

Veterinary products:

030-438; (with Fentanyl Citrate) Innovar®-Vet Injection; *Schering-Plough Animal Health.*

488 Promazine Hydrochloride

53-60-1 7966 200-179-7

$C_{17}H_{21}ClN_2S$

10-[3-(Dimethylamino)propyl]phenothiazine monohydrochloride.

Liranol; Promwill; Prazine; Prolactyl; Sparine; Talofen; component of: Ketaset Plus (veterinary). A phenothiazine used as an antipsychotic and veterinary tranquilizer. Dec 181°; hygroscopic; soluble in H_2O, EtOH, $CHCl_3$, practically insoluble in Et_2O, C_6H_6; incompatible with alkalies, oxidizing agents, heavy metals.

Veterinary products:

010-782; Sparine Injection; *Wyeth.*

011-241; Promazine HCl Injectable; *Fort Dodge Animal Health, Divn. AHP.*

012-656; Promazine Granules; *Fort Dodge Animal Health, Divn. AHP.*

092-116; (with Aminopentamide Hydrogen Sulfate, Ketamine Hydrochloride) Ketaset ® Plus; *Fort Dodge Animal Health, Divn. AHP.*

119-141; Tranquazine Injection; *Anthony Products.*

489 Triflupromazine Hydrochloride

440-17-5 9814 214-149-6

$C_{18}H_{20}ClF_3N_2S$

10-[3-(Dimethylamino)propyl]-2-(trifluoromethyl)-phenothiazine monohydrochloride.

Adazine; Fluorofen; Psyquil; Syquil; Vespral; Vesprin; Vetame. A phenothiazine used as an antipsychotic. Dec

173-174°; λ_m = 255, 305 nm (E 700, 90); soluble in H_2O, EtOH, Me_2CO.

Veterinary products:

011-482; Vetame Injectable; Vetame Tablets; *Fort Dodge Animal Health, Divn. Am. Cyanamid.*

Antipyretics

490 Acetaminophen

103-90-2 45 203-157-5

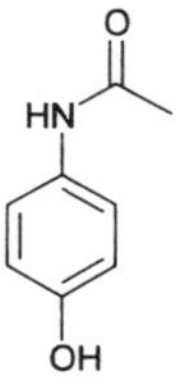

$C_8H_9NO_2$
4'-Hydroxyacetanilide.
p-hydroxyacetanilide; p-acetamidophenol; p-acetaminophenol; N-acetyl-p-aminophenol; paracetamol; Abensanil; Acamol; Acetalgin; Alpiny; Amadil; Anaflon; Anhiba; Apamide; APAP; Banesin; Ben-u-ron; Bickie-mol; Calpol; Captin; Claratal; Cetadol; Dafalgan; Datril; Dirox; Disprol; Doliprane; Dolprone; Dymadon; Enelfa; Eneril; Eu-Med; Exdol; Febrilex; Finimal; Gelocatil; Hedex; Homoolan; Korum; Momentum; Naprinol; Nebs; Nobedon; Ortensan; Pacemol; Paldesic; Panadol; Panaleve; Panasorb; Panets; Panodil; Parelan; Paraspen; Parmol; Pasolind N; Phenaphen; Salzone; Tabalgin; Tapar; Tempra; Tralgon; Tylenol; Valadol; component of: Actifed Plus, Allerest Sinus Pain Formula, Anexsia, Aspirin-Free Anacin, Children's Tylenol Cold Tablets, Contac Cough & Sore Throat Formula, Contac Jr Non-drowsy Formula, Contac Nighttime Cold Medicine, Contac Severe Cold Formula, Coricidin, Darvoset-N, Dristan Cold (Multisymptom), Dristan Cold (No Drowsiness), Empracet, Endecon, Gemnisyn, Headache Strength Allerest, Hycomine Compound, Hy-Phen, Intensin, Liquiprin, Maximum Strength Sine-Aid, Midol, Midol Maximum Strength, Midol PMS, Naldegesic, Naldetuss, Ornex, Percocet, Percogesic with Codeine, Propacet, Quiet World, Rhinex D-Lay Tablets, Sinarest, Sine-Off Maximum Strength Allergy/Sinus, Sine-Off Maximum Strength No Drowsiness Formula Caplets, Sinubid, St Joseph's Cold Tablets for Children, Sudafed Sinus, Supac, Teen Midol, TheraFlu, Tylenol Allergy Sinus, Tylenol Cold and Flu Multi-Symptom, Tylenol Cold Medication Caplets, Liquid and Tablets, Tylenol Cold Night TIme Liquid, Tylenol Cold No Drowsiness, Tylenol PM Tablets and Caplets, Tylenol with Codeine, Tylox, Vanquish, Vicodin, Wygesic, Zydone. Analgesic, antipyretic, anti-inflammatory. Used as an oral analgesic, sometimes with oral codeine phosphate, in dogs. mp = 169-170.5°; d_4^{21}= 1.293; λ_m = 250 nm (ε 13800 in EtOH); slightly soluble in cold H_2O, Et_2O; more soluble in hot H_2O; soluble in alcohol, dimethylformamide, ethylene dichloride, Me_2CO, EtOAc; nearly insoluble.

491 Flunixin Meglumine

42461-84-7 4182 255-836-0

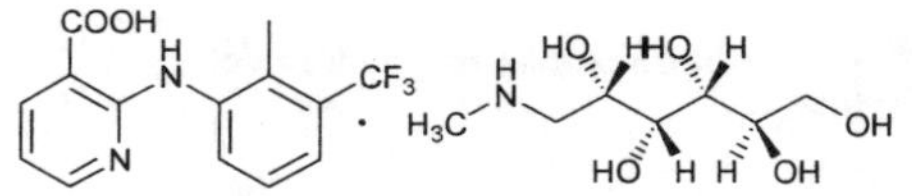

$C_{21}H_{28}F_3N_3O_7$
2-(2-Methyl-3-trifluoromethylanilino)nicotinic acid compound with 1-deoxy-1-(methylamino)-D-glucitol (1:1).
Banamine; Finadyne. Cyclooxygenase inhibitor. Used as an analgesic, anti-inflammatory and antipyretic. Used to alleviate inflammatory pain in horses and cattle and, except in the US, in dogs. mp = 135-139°; soluble in H_2O.

Veterinary products:

101-479; Banamine ® Injectable Solution; *Schering-Plough Animal Health.*
106-616; Banamine ® Granules; *Schering-Plough Animal Health.*
137-409; Banamine ® Paste; *Schering-Plough Animal Health.*
200-061; Flunixin Meglumine Injection; *Agri Labs.*
200-124; Flunixin Meglumine; Flunixin Meglumine Injection; *Phoenix.*
200-142; Flunixin Meglumine Solution; *Fort Dodge Animal Health, Divn. AHP.*

Antiseptics

492 Chlorhexidine Acetate

56-95-1 2140 200-302-4

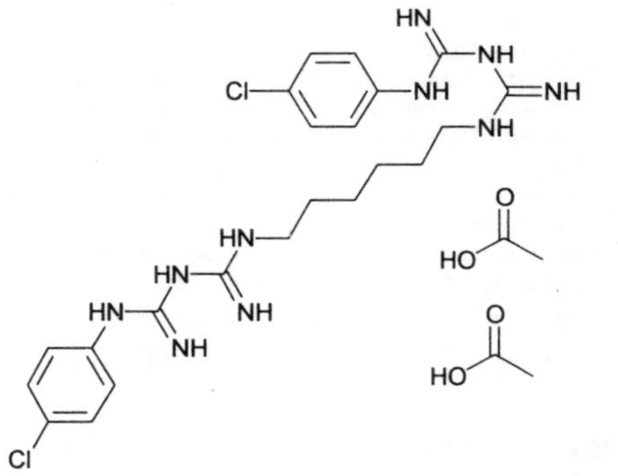

$C_{26}H_{38}Cl_2N_{10}O_4$
1,1-Hexamethylenebis[5-(p-chloropnehyl)biguanide] diacetate.
Chlorasept 2000; Nolvasan. Antiseptic; disinfectant. mp = 154-155°; neutral reaction; soluble in H_2O at 20° (1.9 g/100 ml); aqueous solutions decompose above 70°; soluble in alcohol, glycerol; propylene glycol; polyethylene glycols; LD_{50} (mus orl) = 2 g/kg.

Veterinary products:

009-782; Nolvasan® Cap-Tabs®; *Fort Dodge Animal Health, Divn. AHP.*

493 Chlorhexidine Hydrochloride

3697-42-5 2140 223-026-6

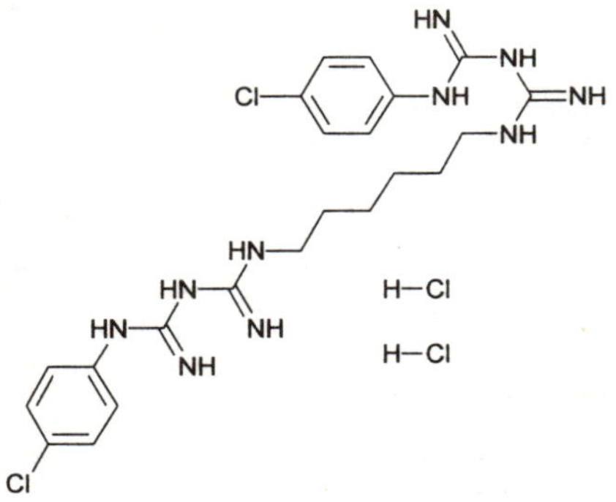

$C_{22}H_{32}Cl_4N_{10}$
1,1-Hexamethylenebis[5-(p-chloropnehyl)biguanide] dihydrochloride.
AY-5312; Lisium. Topical anti-infective. Dec 260-262°; soluble in H_2O at 20° (0.06 g/ 100 ml).

Veterinary products:

009-809; Nolvasan ® Cap-Tabs ®; *Fort Dodge Animal Health, Divn. AHP.*
010-434; Nolvasan ® Suspension; *Fort Dodge Animal Health, Divn. AHP.*

494 Ethanol

64-17-5 3806 200-578-6

C_2H_6O
Methylcarbinol.
alcohol; Algrain; Anhydrol; Cologne; Cologne spirits (alcohol); dehydrated alcohol; Denatured alcohol; ethanol 200 proof; Ethanol absolute; Ethanol; ethyl hydroxide; ethyl hydrate; fermentation alcohol; grain alcohol; Jaysol S; Jaysol; molasses alcohol; potato alcohol; spirit; spirits of wine; Synasol; Tecsol. Used in veterinary medicine to treat methanol or ethylene glycol poisoning. Clear, mobile liquid; mp = -114.1°; bp = 78.5°; d_4^{20} = 0.789; soluble in H_2O and most organic solvents; LD_{50} (young rat orl) = 10600 mg/kg, (old rat orl) = 7060 mg/kg.

495 Formalin

50-00-0 4262 200-001-8

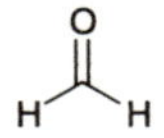

CH_2O
Formaldehyde.
formalin; formic aldehyde; oxymethylene; methanal; methyl aldehyde; aldehyde C_1; oxomethane; methylene oxide; Hercules® 37M6-8; Used in urea and melamine resins, polyacetal resins, phenolic resins, fertilizers, preservatives, reducing agent, corrosive inhibitor. FDA approved for topicals, BP compliance (solutions). Used as an antimicrobial in biologics, topicals, hepatitis B vaccine, sterilizer for kidney dialysis membranes. mp = -92°; soluble in water, alcohol, ether. LD_{50} (rat orl) = 800 mg/kg; TLV = 1 ppm in air.

Veterinary products:

137-687; Formalin-F; *Natchez Animal Supply.*
140-831; Paracide-F; *Argent.*
140-989; Parasite-S ®; *Western Chemical.*

Antispasmodics

496 Aminopropazine Fumarate

3688-62-8 487 222-987-9

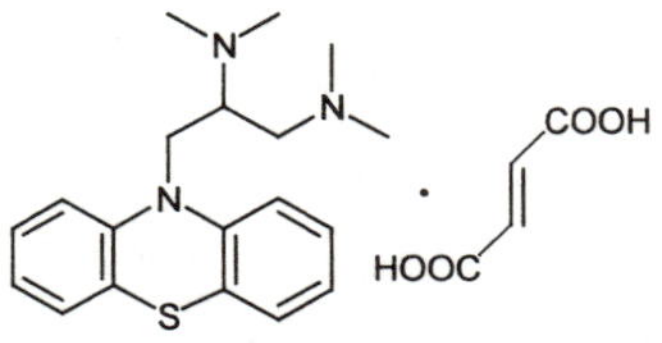

$C_{42}H_{54}N_6O_4S_2$
10-[2,3-Bis(dimethylamino)propyl]phenothiazine fumarate.
aminopromazine fumarate; proquamezine fumarate; Lispamol; Lorusil; Spamol. Antispasmodic agent. Used to control excessive smooth muscle contractions in cats, dogs and horses. Crystals, light sensitive; dec 166-170°; insoluble in C_6H_6, Et_2O, soluble in H_2O (9 g/100 ml), MeOH (5 g/100 ml), EtOH (0.5 g/100 ml); pH of 2% aqueous solution = 5.0 - 7.0.

Veterinary products:

011-877; Jenotone Tablets; *Schering-Plough Animal Health.*
013-181; (with Neomycin Sulfate) Jenomycin Tablets; *Schering-Plough Animal Health.*
034-477; Jenotone Solution; *Schering-Plough Animal Health.*

Antiulceratives

497 Cimetidine

51481-61-9 2337 257-232-2

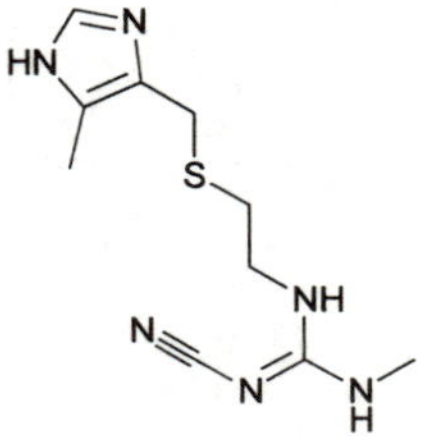

$C_{10}H_{16}N_6S$
2-Cyano-1-methyl3-[2-[[(5-methylimidazol-4-yl)-methyl]thio]ethyl]guanidine.

SKF-92334; Acibilin; Acinil; Cimal; Cimetag; Cimetum; Edalene; Dyspamet; Eureceptor; Gastromet; Peptol; Tagamet; Tametin; Tratul; Ulcedin; Ulcedine; Ulcerfen; Ulcimet; Ulcofalk; Ulcomedina; Ulcomet; Ulhys. Histamine H_2-receptor antagonist used as an anti-ulcerative. Used to treat ulcers and gastritis in dogs. mp = 141-143°; soluble in H_2O (1.14%); LD_{50} (rat orl) = 5000 mg/kg, (rat iv) = 106 mg/kg, (rat ip) = 650 mg/kg, (mus orl) = 2600 mg/kg, (mus iv) = 150 mg/kg, (mus ip) = 650 mg/kg.

498 Cimetidine Hydrochloride

70059-30-2 2337 274-297-2

$C_{10}H_{17}ClN_6S$

2-Cyano-1-methyl3-[2-[[(5-methylimidazol-4-yl)-methyl]thio]ethyl]guanidine monohydrochloride.

Tagamet Injection; Tagamet Liquid; Aciloc; Biomag; Brumetidina, Notul. Histamine H_2-receptor antagonist used as an anti-ulcerative. Used to treat ulcers and gastritis in dogs.

499 Cisapride

81098-60-4 2377 279-689-7

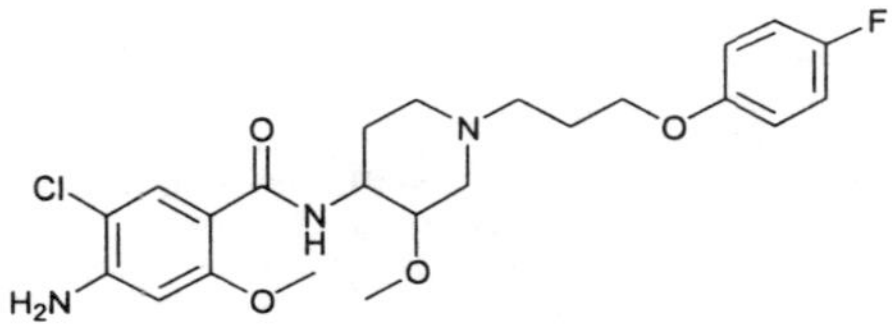

$C_{23}H_{29}ClFN_3O_4$

cis-4-Amino-5-chloro-N-[1-[3-(p-fluorophenoxy)-propyl]-3-methoxy-4-piperidyl]-o-anisamide.

R-51619; Propulsid; Acenalin; Alimix; Cipril; Prepulsid; Propulsin; Risamal. Peristaltic stimulant. Used as a gastrointestinal prokinetic agent which ameliorates gastric reflux in small animals. mp = 109.8°.

500 Famotidine

76824-35-6 3972

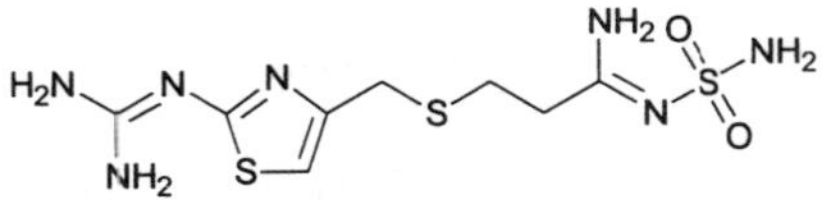

$C_8H_{15}N_7O_2S_3$

[1-Amino-3-[[[2-[(diaminomethylene)amino]-4-thiazolyl]-metyhl]thio]propylidene]sulfamide.

Pepcid®; Pepcid AC; MK-208; YM-11170; Amfamox; Dispromil; Famodil; Famodine; Famosan; Famoxal; Fanosin; Fibonel; Ganor; Gaster; Gastridin; Gastropen; Ifada; Lecedil; Motiax; Muclox; Nulcerin; Pepcidina; Pepcidine; Pepdine; Pepdul; Peptan; Ulcetrax; Ulfamid; Ulfinol. Histamine H_2 receptor antagonist used as an antiulcerative and to treat gastritis. mp = 163-164°; soluble in DMF (80 g/100 ml), AcOH (50 g/100 ml), MeOH (0.3 g/100 ml), H_2O (0.1 g/100 ml); insoluble in EtOH, EtOAc, $CHCl_3$; LD_{50} (mus iv) = 244.4 mg/kg.

501 Misoprostol

59122-46-2 6297

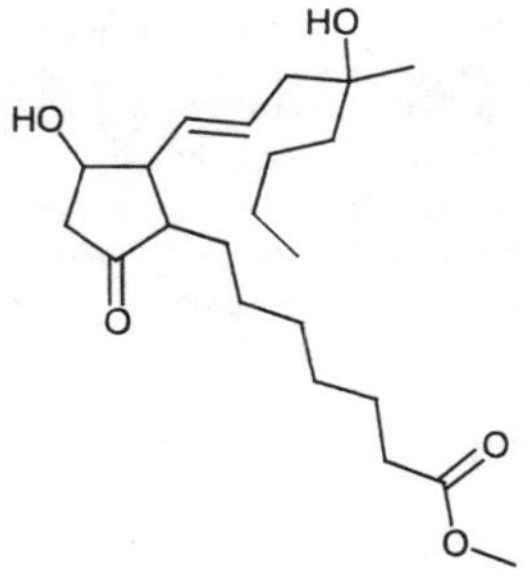

$C_{22}H_{38}O_5$

(±)-Methyl(1R,2R,3R)-3-hydroxy-2-[(E)-(4RS)-4-hydroxy-4-methyl-1-octenyl]-5-oxocyclopentaneheptanoate.

Cytotec; SC-29333. Antiulcerative. Cytoprotective prostaglandin PGE_1 analog. Soluble in H_2O; LD_{50} (rat ip)= 40-62 mg/kg, (rat orl) = 81-100 mg/kg, (mus ip) = 70-160 mg/kg, (mus orl) = 27-138 mg/kg.

502 Omeprazole

73590-58-6 6977

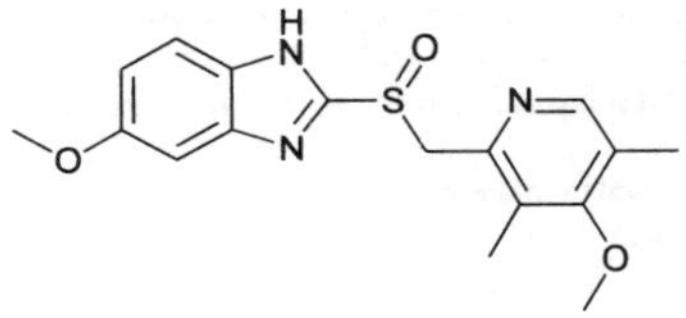

$C_{17}H_{19}N_3O_3S$

5-Methoxy-2-[[(4-methoxy-3,5-dimethyl-2-pyridyl)-methyl]sulfonyl]benzimidazole.

Prilosec. Gastric antisecretory and anti-ulcerative. Used in treatment of Zollinger-Ellison syndrome. Used to treat gastric ulcers in domestic animals. mp = 156°; LD_{50} (mus iv) = 0.08 g/kg, (mus orl) > 4 g/kg, rat iv) > 0.05 g/kg, (rat orl) > 4 g/kg.

Veterinary products:

141-123; Gastrogard™; *Merial.*

503 Ranitidine

66357-35-5 8286 266-332-5

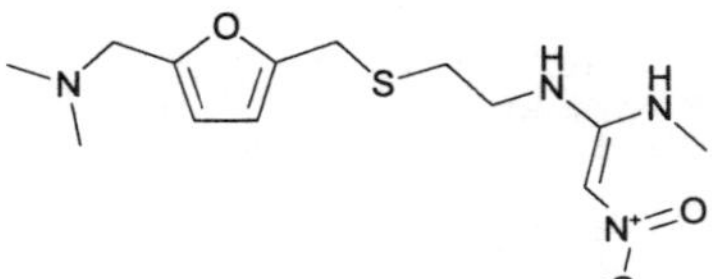

$C_{13}H_{22}N_4O_3S$

N-[2-[[5-[(Dimethylamino)methyl]furfuryl]thio]ethyl]-N'-methyl-2-nitro-1,1-ethenediamine.

Antiulcerative. Histamine H_2 receptor antagonist; inhibits gastric secretion. mp = 69-70°.

504 Ranitidine Hydrochloride
66357-59-3 8286 266-333-0

$C_{13}H_{23}ClN_4O_3S$

N-[2-[[5-[(Dimethylamino)methyl]furfuryl]thio]ethyl]-N'-methyl-2-nitro-1,1-ethenediamine hydrochloride.

AH-19065; Azantac; Melfax; Noctone; Raniben; Ranidil; Raniplex; Sostril; Taural; Terposen; Trigger; Ulcex; Ultidine; Zantac; Zantic. Antiulcerative. Histamine H_2 receptor antagonist; inhibits gastric secretion. Soluble in AcOH, H_2O, MeOH; less soluble in EtOH; insoluble in $CHCl_3$.

505 Sucralfate
54182-58-0 9049 259-018-4

$R = SO_3[Al_2(OH)_5]$

$C_{12}H_mAl_{16}O_nS_8$

Sucrose octakis(hydrogen sulfate) aluminum complex.

Carafate; Antepsin; Citogel; Hexagastron; Keal; Succosa; Sucralfin; Sucrate; Sugast; Sulcrate; Ulcar; Ulcerlmin; Ulcogant. Gastrointestinal antiulcerative used in treatment of ulcers and other gastric erosions. Insoluble in H_2O, EtOH, soluble in dilute HCl and NaOH solutions.

506 Sulfasalazine
599-79-1 9112 209-974-3

$C_{18}H_{14}N_4O_5S$

5-[[p-(2-Pyridylsulfamoyl)phenyl]azo]salicylic acid.

Azulfidine. Antiulcerative and antibacterial. Has been used in the treatment of Crohn's disease. Used in dogs and cats to treat inflammatory bowel disease. mp = 240-245°; λ_m = 237 nm ($E^{1\%}_{1\,cm}$ 658) and 359 nm; insoluble in H_2O, C_6H_6, $CHCl_3$, Et_2O; soluble in EtOH.

Antivirals

507 Acyclovir
59277-89-3 148 261-685-1

$C_8H_{11}N_5O_3$

2-Amino-1,9-dihydro-9-[(2-hydroxyethoxy)methyl]-6H-purin-6-one.

Azone; Acycloguanosine; BW-248U; Wellcome 248U; Aciclofal; Cargosil; Laurocapram; Poviral; Virorax; Zovirax; Vipral; Aciclovir; Acyclo-V; Zyclir. Antiviral agent. Antiviral used in treatment of herpes virus. A nucleoside analog that is preferentially taken up by infected cells and then inhibits viral DNA synthesis by interfering with transcription. Used to treat herpes infections in birds and cats. mp = 256.5-257°; LD_{50} (mus orl) >10,000 mg/kg.

508 Acyclovir Sodium
69657-51-8 148

$C_8H_{10}N_5NaO_3$

2-Amino-1,9-dihydro-9-[(2-hydroxyethoxy)methyl]-6H-purin-6-one monosodium salt.

Antiviral agent. Antiviral used in treatment of herpes virus. A nucleoside analog that is preferentially taken up by infected cells and then inhibits viral DNA synthesis by interfering with transcription. Used to treat herpes infections in birds and cats.

509 Cytarabine
147-94-4 2853 205-705-9

$C_9H_{13}N_3O_5$

4-Amino-1-β-D-arabinofuranosyl-2(1H)-pyrimidinone.

1-β-D-arabinofuranosylcytosine; β-cytosine arabino-side; CHX-3311; U-19920; Alexan; Arabitin; Aracytidine; Aracytine; Ara-C; Cytosar; Cytosar U; Erpalfa; Iretin; Udicil. A cytotoxic drug. Also has antiviral activity. Used in small animals as an antineoplastic agent. mp = 212-213°; $[\alpha]_D^{23}$ = 158° (c = 0.5, H_2O); λ_m = 281, 212.5 nm (ε 13171, 10230, pH 2).

510 Cytarabine Hydrochloride
69-74-9 2853 200-713-9

$C_9H_{13}N_3O_5$

4-Amino-1-β-D-arabinofuranosyl-2(1H)-pyrimidinone.

NSC-63878. A cytotoxic drug. Also has antiviral activity. Used in small animals as an antineoplastic agent.

Anxiolytics

511 Acepromazine Maleate
3598-37-6 32 222-748-9

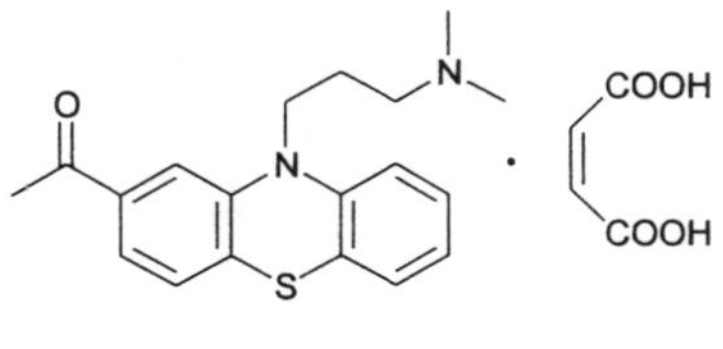

$C_{23}H_{26}N_2O_5S$
1-[10-[3-(Dimethylamino)propyl]-10H-phenothiazin-2-yl]ethanone maleate.
Atravet; Calmivet; Notensil; Plegicil; Sedalin; Soprontin; component of: 015-030, PromAce Injectable, 032-702, PromAce® Tablets (Fort Dodge Animal Health), 117-531, Acepromazine Maleate Inj., 117-532, Acepromazine Maleate Tablets (Boehringer Ingelhein Vetmedica Inc.). Animal tranquillizer. Yellow crystals; mp = 135-136°; soluble in H_2O; pH of 1% aqueous solution = 5.2; LD_{50} (rat orl) = 94 mg/kg, (rat iv) = 50 mg/kg.

Veterinary products:

015-030; PromAce Injectable; *Fort Dodge Animal Health, Divn. AHP.*
032-702; PromAce® Tablets; *Fort Dodge Animal Health, Divn. AHP.*
117-531; AcePromazine Maleate Inj.; *Boehringer Ingelheim Vetmedica.*
117-532; AcePromazine Tablets; *Boehringer Ingelheim Vetmedica.*

512 Buspirone
36505-84-7 1528 253-072-2

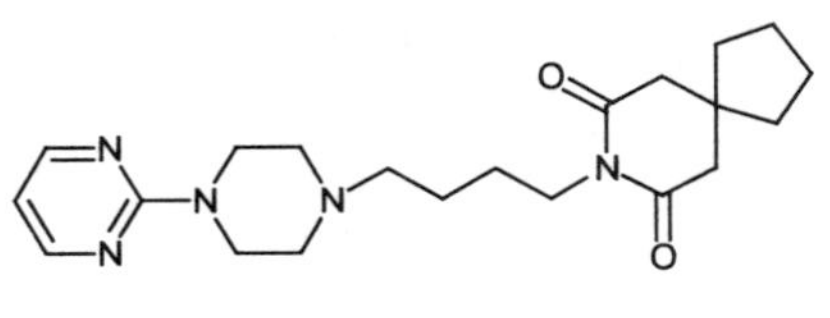

$C_{21}H_{31}N_5O_2$
N-[4-[4-(2-Pyrimidinyl)-1-piperazinyl]butyl]-1,1-cyclopentanediacetamide hydrochloride.
Non-benzodiazepine anxiolytic, used as a minor tranquillizer. Used to treat fear and phobia-related disorders in dogs and cats.

513 Buspirone Hydrochloride
33386-08-2 1528 251-489-4
$C_{21}H_{32}ClN_5O_2$
N-[4-[4-(2-Pyrimidinyl)-1-piperazinyl]butyl]-1,1-cyclopentanediacetamide hydrochloride.
Ansial; Ansiced; Axoren; Bespar; Buspar; Buspimem; Buspinol; Buspisal; Censpar; Lucelan; Narol; Travin. Non benzodiazepine anxiolytic, used as a minor tranquillizer. Used to treat fear and phobia-related disorders in dogs and cats. mp = 201.5-202.5°; LD_{50} (rat ipr) = 136 mg/kg.

514 Clorazepate Dipotassium
57109-90-7 2465 260-565-6

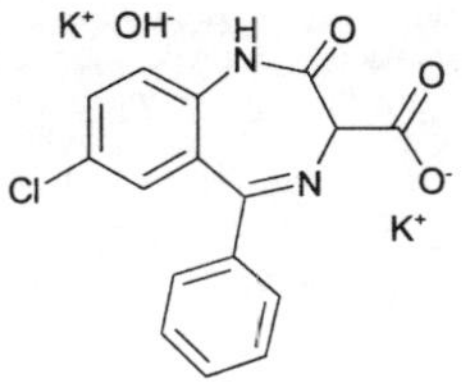

$C_{16}H_{11}ClK_2N_2O_4$
Potassium 7-chloro2,3-dihydro-2-oxo-5-phenyl-1H-1,4-benzodiazepine-3-carboxylate compound with KOH (1:1).
Tranxene; 4306 CB; Abbott 35616; Belseren; Mendon; Tranxilene; Tranxilium; Transene. Anxiolytic and anticonvulsant. Used as a minor tranquillizer for treatment of behavior disorders. Poorly soluble in EtOH; insoluble in Et_2O, $CHCl_3$; λ_m = 231, 311 nm (ε 33500, 2450 H_2O); LD_{50} (mus orl) =700 mg/kg, (mus ip) = 290 mg/kg, (rat orl) > 1000 mg/kg.

515 Clorazepate Monopotassium
5991-71-9 2465 227-817-7

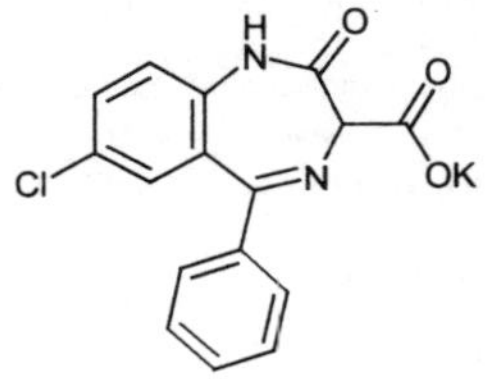

$C_{16}H_{10}ClKN_2O_3$
Potassium 7-chloro2,3-dihydro-2-oxo-5-phenyl-1H-1,4-benzodiazepine-3-carboxylate.
4311 CB; Abbott 39083; Azene. Anxiolytic and anticonvulsant. Used as a minor tranquillizer for treatment of behavior disorders.

516 Diazepam
439-14-5 3042 207-122-5

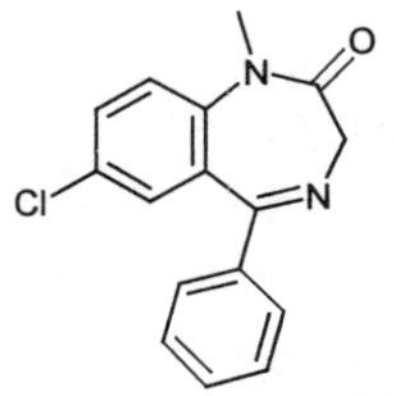

$C_{16}H_{13}ClN_2O$
7-Chloro-1,3-dihydro-1-methyl-5-phenyl-2H-1,4-benzodiazepin-2-one.
Alupram; Valium; Valrelease; LA 111; Ro- 5-2807; Wy-3467; NSC-77518; Apaurin; Atensine; Atilen; Bialzepam; Calmpose; Ceregulart; Dialar; Diazemuls; Dipam; Eridan; Eurosan; Evacalm; Faustan; Gewacalm; Horizon; Lamra; Lembrol; Levium; Mandrozep; Neurolytril; Noan; Novazam; Paceum; Pacitran; Paxate; Paxel; Pro-Pam; Q-

Pam; Relanium; Sedapam; Seduxen; Servizepam; Setonil; Solis; Stesolid; Tranquase; Tranquo-Puren; Tranquo-Tablinen; Unisedil; Valaxona; Valiquid; Valium; Valrelease; Vival; Vivol. Anxiolytic and skeletal muscle relaxant. Pharmaceutical preparation for the treatment of depression. Used in veterinary medicine for its anxiolytic, muscle relaxant, hypnotic, appetite stimulant and anticonvulsant activities. mp = 125-126°; soluble in DMF, $CHCl_3$, C_6H_6; Me_2CO, EtOH; slightly soluble in H_2O; LD_{50} (rat orl) = 710 mg/kg.

Veterinary products:

140-848; Veteeze® Injection; Roche Vitamins.

517 Ethylisobutrazine Hydrochloride

3737-33-5 3938 223-111-8

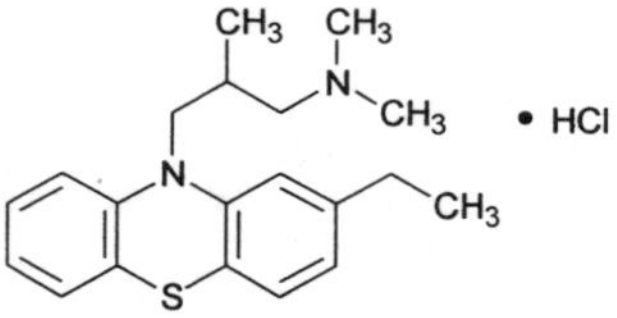

$C_{20}H_{27}ClN_2S$

2-Ethyl-N,N,β-trimethyl-10H-phenothiazine-10-propanamine hydrochloride.

Nuital; Sergetyl. Antihistaminic used in veterinary medicine as a tranquillizer Crystals; mp = 160-163°; soluble in H_2O, EtOH, MeOH, Me_2CO, $CHCl_3$; insoluble in Et_2O.

Veterinary products:

011-222; Diquel Tablet; *Schering-Plough Animal Health.*
035-265; Diquel Solution; *Schering-Plough Animal Health.*

518 Fentanyl Citrate

990-73-8 4043 213-588-0

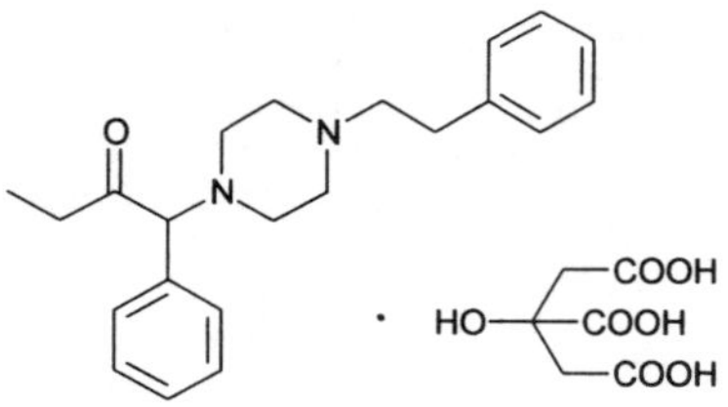

$C_{28}H_{36}N_2O_8$

N-(1-Phenylethyl-4-piperidinyl)propionanilide citrate (1:1).

Fentanest; Leptanol; Pentanyl; Sublimaze. Analgesic, narcotic; tranquilizer (veterinary). Federally controlled substance (opiate). mp = 149-151°; soluble in H_2O (1 g/40 ml), less soluble in organic solvents; LD_{50} (mus iv)= 11.2 mg/kg, (msu sc) = 62 mg/kg.

Veterinary products:

030-438; (with Droperidol) Innovar®-Vet Injection; *Schering-Plough Animal Health.*

519 Oxazepam

604-75-1 7059 210-076-9

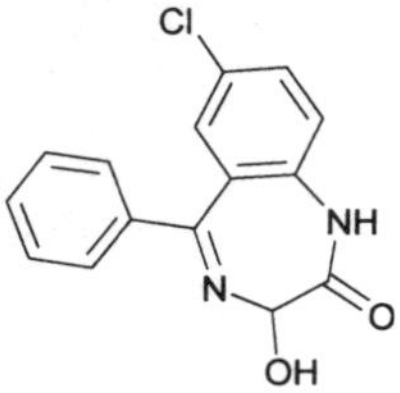

$C_{15}H_{11}ClN_2O_2$

7-Chloro-1,3-dihydro-3-hydroxy-5-phenyl-2H-1,4-benzodiazepin-2-one.

Wy-3498; Adumbran; Aplakil; Azutranquil; Bonare; Durazepam; Enidrel; Hilong; Isodin; Lederpam; Limbial; Serax; Nesontil; Noctazepam; Oxanid; Oxa-puren; Praxiten; Propax; Quilibrex; Rondar; Serax; Serenal; Serenid; Serepax; Seresta; Sigacalm; Sobril; Tazepam; Uskan; Zaxopam. A benzodiazepine anxiolytic. Used as an appetite stimulant in cats and dogs. mp = 205-206°; insoluble in H_2O; soluble in EtOH, $CHCl_3$, dioxane; LD_{50} (mus, rat orl) > 5010 mg/kg.

520 Propiopromazine Hydrochloride

8014

$C_{20}H_{25}ClN_2OS$

1-[10-[3-(Dimethylamino)propyl]-10H-phenothiazin-2-yl]-1-propanone hydrochloride.

Tranvet. Veterinary tranquillizer.

Veterinary products:

041-665; Tranvet Chewable Tablets (20 mg); *Fort Dodge Animal Health, Divn. AHP.*
045-716; Tranvet Injectable Solution; *Fort Dodge Animal Health, Divn. AHP.*

Bronchodilators

521 Albuterol

18559-94-9 217 242-424-0

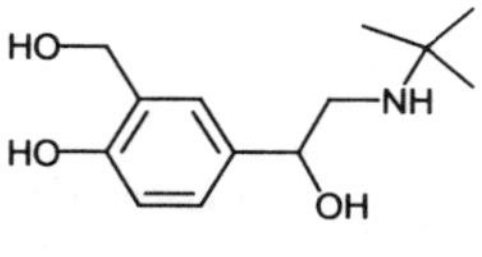

$C_{13}H_{21}NO_3$

2-(tert-Butylamino)-1-(4-hydroxy-3-hydroxymethylphenyl)ethanol.

salbutamol; Proventil Inhaler; Ventalin Inhaler. Bronchodilator. Ephedrine derivative used as a bronchodilator in dogs and cats. mp = 151° (also as 157-158°); soluble in most organic solvents.

522 Albuterol Sulfate

51022-70-9 217 256-916-8

$C_{26}H_{44}N_2O_{10}S$

2-(tert-Butylamino)-1-(4-hydroxy-3-hydroxymethylphenyl)ethanol sulfate (2:1).

Sch-13949W Sulfate; Aerolin; Asmaven; Broncovaleas;

Cetsim; Cobutolin; Ecovent; Loftan; Proventil; Salbumol; Salbutard; Salbutine; Salbuvent; Sultanol; Ventelin; Ventodiscks; Ventolin; Volma. Bronchodilator. Ephedrine derivative used as a bronchodilator in dogs and cats.

523 Aminophylline

317-34-0 485 206-264-5

$C_{16}H_{24}N_{10}O_4$
3,7-Dihydro-1,3-dimethyl-1H-purine-2,6-dione compound with ethylenediamine (2:1).
Aminophyllin; Phyllocontin; Rectalad Aminophylline; Somophyllin; component of: Mudrane Tablets, Mudrane 2 Tablets, Mudrane GG Tablets, Mudrane GG-2 Tablets; theophyllamine; Carena; Inophylline; Metaphyllin; Theophyldine; Aminocardol; Aminodur; Ammophyllin; Cardiofilina; Cardophylin; Phylcardin; Tefamin; Cardiomin; Grifomin; Minaphil; Pecram; Peterphyllin; Phyllocontin; Somophyllin; Stenovasan; Theodrox; Cardophyllin; Diophyllin; Euphyllin CR; Genophyllin; Phyllindon; Theolamine; Theomin. Smooth muscle relaxant, used in animals as a bronchodilator. Soluble in H_2O (20 g/100 ml); insoluble in EtOH, Et_2O; LD_{50} (mus orl) = 540 mg/kg.

524 Epinephrine Acetate

3656

$C_{11}H_{15}NO_4$
(R)-4-[1-Hydroxy-2-(methylamino)ethyl]-1,2-benzenediol actetate.
Used in veterinary medicine as a vasoconstrictor and cardiostimulant.

Veterinary products:

045-578; (with Lidocaine Hydrochloride) Lidocaine Hydrochloride with Epinephrine; *Steris.*

525 Theophylline

58-55-9 9421 200-385-7

$C_7H_8N_4O_2$
3,7-Dihydro-1,3-dimethyl-1H-purine-2,6-dione.
1,3-Dimethylxanthine; Accurbron; Aerobin; Aerolate; Afonilm; Armophylline; Austyn; Bilordyl; Brochoretard; Bronkodyl; Cétraphylline; Constant-T; Duraphyl; Duraphyllin; Diffumal; Elixophyllin; Etheophyl; Euphyllin; Euphylong; LaBID; Lasma; Physpan; Pro-Vent; PulmiDur; Pulmo-Timelets; Respbid; Slo-Bid; Slo-Phyllin; Solosin; Somophyllin-CRT; Somophyllin-T; Sustaire; Talotren; Tesona; Theobid; Theo-24; Theobid Duracap; Theoclear; Theochron; Theo-Dur; Theograd; Theolair; Theon; Theophyl; Theoplus; Theo-Sav; Theostat; Theovent; Unifyl; Uniphyl; Uniphyllin; Xanthium; component of: Bronkotabs, Dicurin Procaine, Mudrane GG Elixer, Primatene Tablets, Quibron, Quibron Plus, Quibron-T/SR, Slo-Phyllin GG, Tedral, Theolair Plus, Theo-Organdin. Used in animals as a bronchodilator. Xanthine derivative with diuretic, cardiac stimulant and smooth muscle relaxant activities; isomeric with theobromine; small amounts found in tea.

Calcium Regulators

526 Calcitonin, Salmon Synthetic

47931-85-1 1680 256-342-8

$C_{145}H_{240}N_{44}O_{48}S_2$
H-Cys-Gly-Asn-Leu-Ser-Thr-Cys-Met-Leu-Gly-Thr-Tyr-Thr-Gln-Asp-Phe-Asn-Lys-Phe-His-Thr-Phe-Pro-Gln-Thr-Ala-Ile-Gly-Val-Gly-Ala-Pro-NH_2 (1→7) cyclic disulfide.
Salcatonin; Calciben; Calcimar; Calsyn; Calsynar; Catonin; Karil; Miacalcic; Miacalcin; Miadenil; Osteocalcin; Prontocalcin; Rulicalcin; Salmotonin; Stalcin; Tonocalcin. A polypeptide hormone that lowers the calcium concentration in the plasma of mammals. Calcium regulator; used to control hypercalcemia in small animals.

527 Dihydrotachysterol

67-96-9 3223 200-672-7

$C_{28}H_{46}O$
9,10-Secoergosta-5,7,22-trien-3β-ol.
Hytakerol; AT-10; Antitanil; Calcamine; Dygratyl; Dihydral; Parterol; Tachyrol. Calcium regulator used in veterinary medicine to treat hypercalcemia. mp = 125-127°; $[\alpha]_D^{22}$= 07.5° ($CHCl_3$); λ_m = 242, 251, 261 nm ($E_{1\ cm}^{1\%}$ 870, 1010, 650); insoluble in H_2O, readily soluble in organic solvents.

Carbonic Anhydrase Inhibitors

528 Dichlorphenamide

120-97-8 3127 204-440-6

$C_6H_6Cl_2N_2O_4S_2$
4,5-Dichloro-m-benzenedisulfonamide.

Daranide; Antidrasi; Oratrol. Carbonic anhydrase inihibitor used to treat glaucoma in humans and possibly in animals. mp = 239-241°, 228.5-229°; insoluble in H_2O, soluble in alklaine solutions.

Cardiotonics

529 Amrinone

60719-84-8 634 262-390-0

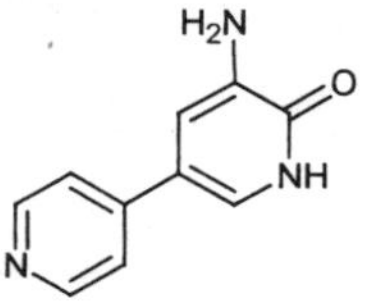

$C_{10}H_9N_3O$

5-Amino[3,4'-bipyridin]-6(1H)-one.

Inocor; Win-40680; Vesistol; Cartonic; Wincoram. Phosphodiesterase inhibitor used as a cardiotonic. Used as a second line treatment for short term management of congestive heart failure in dogs and cats. mp = 294-297° (dec).

530 Digitoxin

71-63-6 3206 200-760-5

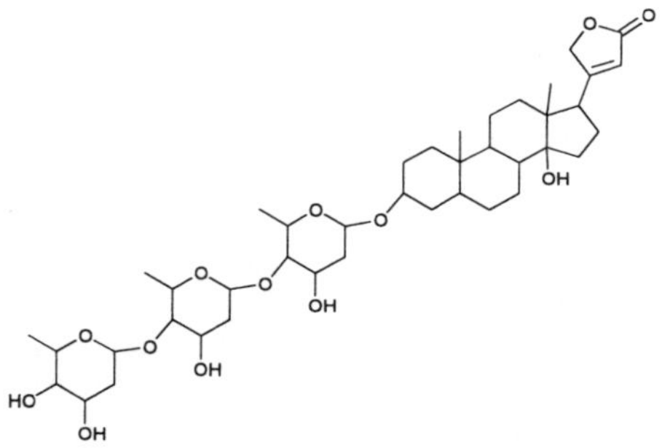

$C_{41}H_{64}O_{13}$

3-[(O-2,6-Dideoxy-β-D-ribohexopyranosyl-(1→4)-O-2,6-dideoxy-β-D-ribohexopyranosyl-(1→4)-2,6-dideoxy-β-D-ribohexopyranosyl)oxy]-14-hydroxycard-20(22)-enolide.

Cardidigin; Crystodigin; Digisidin; Unidigin; digitophyllin; Cardigin; Carditoxin; Coramedan; Cristapurat; Digicor; Digilong; Digimerck; Digimed; Digipural; Ditaven; Digisidin; Digitaline Nativelle; Lanatoxin; Myodigin; Purodigin; Purpurid; Tradigal. Secondary glycoside from the dried leaves of *Digitalis purpurea L. Scrophulariaceae*. Used as a cardiotonic. mp = 256-257°; $[\alpha]_D^{20}$ = 4.8° (c = 1.2 dioxane); soluble in $CHCl_3$ (2.5 g/100 ml), EtOH (1.67 g/100 ml), EtOAc (0.25 g/100 ml), Me_2CO, amyl alcohol, C_5H_5N; sparingly soluble in Et_2O, petroleum ether, H_2O 0.001 g/100 ml at 20°; LD_{50} (gpg orl) = 60 mg/kg, (cat orl) = 0.18 mg/kg.

531 Digoxin

20830-75-5 3210 244-068-1

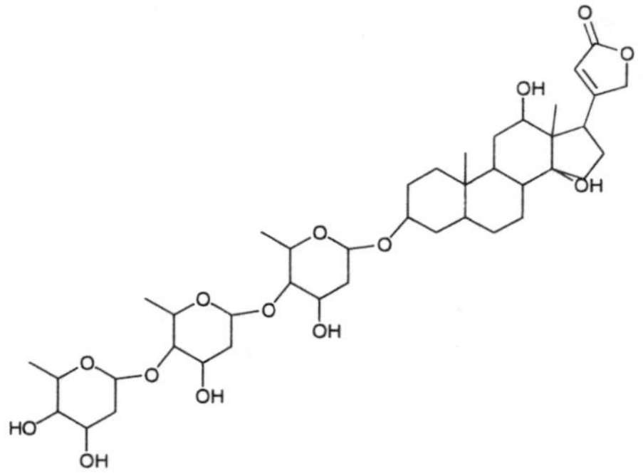

$C_{41}H_{64}O_{14}$

3-[(O-2,6-Dideoxy-β-D-ribohexopyranosyl-(1→4)-O-2,6-dideoxy-β-D-ribohexopyranosyl-(1→4)-2,6-dideoxy-β-D-ribohexopyranosyl)oxy]-14-hydroxycard-20(22)-enolide.

Lanoxicaps; Lanoxin; Cordioxil; Davoxin; Digacin; Dilanacin; Dixina; Dokim; Dynamos; Eudigox; Lanacordin; Lanicor; Lenoxicaps; Lenoxin; Longdigox; NeoDioxanin; Rougoxin; Stillacor; Vanoxin. Secondary glycoside from the dried leaves of *Digitalis purpurea L. Scrophulariaceae*. Used as a cardiotonic. mp = 230-265°; $[\alpha]_{Hg}^{25}$ = 13.4 - 13.8° (c = 10 C_5H_5N); λ_m = 220 nm (ε 12800 EtOH); soluble in EtOH, C_5H_5N; insoluble in $CHCl_3$, Me_2CO, EtOAc, H_2O, Et_2O.

Cholelitholytics

532 Ursodiol

128-13-2 10026 204-879-3

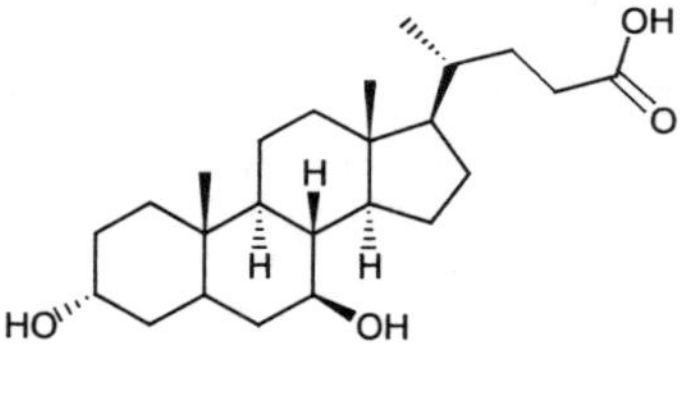

$C_{24}H_{40}O_4$

3α,7β-Dihydroxy-5β-cholan-24-oic acid.

ursodeoxycholic acid; Actigall; Arsacol; Cholit-Ursan; Delursan; Desol; Destolit; Deursil; Litursol; Lyeton; Paptarom; Solutrat; Urdes; Ursacol; Urso; Ursobilin; Ursochol; Ursodamor; Ursofalk; Ursolvan. Anticholelithogenic. Helps in removal of cholesterol-containing gallstones. mp = 203°; $[\alpha]_D^{20}$ = 57° (c = 2 EtOH); freely soluble in EtOH, AcOH; soluble in $CHCl_3$, Et_2O; insoluble in H_2O; LD_{50} (mus iv) = 100 mg/kg, 260 mg/kg, (mus sc = 6000 mg/kg, (mus ip) = 1200 mg/kg, (rat sc) = 2000 mg/kg, (rat ip) = 1000 mg/kg, (rat iv) = 310 mg/kg; [Diformate ($C_{26}H_{40}O_6$)]: mp = 170°; [diacetate ($C_{28}H_{44}O_6$)]: mp = 98-102°.

Cholinergics

533 Dexpanthenol
81-13-0 2988 201-327-3

$C_9H_{19}NO_4$
D-(+)-2,4-Dihydroxy-N-(3-hydroxypropyl)-3,3-dimethylbutyramide.
D-Panthenol 50; Ilopan; Motilyn; pantothenol; pantothenyl alcohol; Alcopan-250; Intrapan; Pantenyl; Panthoderm; Bepanthen; Cozyme; Urupan; component of: Ilopan-Choline. Cholinergic agent. The dl form is used as a vitamin. Has been proposed for modulation of smooth muscle disorders. $bp_{0.02}$ = 118-120°; d_{20}^{20} = 1.2; $[\alpha]_D^{20}$= 29.5° (c = 5); freely soluble in H_2O, EtOH, MeOH; slightly soluble in Et_2O.

534 Edrophonium Chloride
116-38-1 3562 204-138-4

$C_{10}H_{16}ClNO$
Ethyl(m-hydroxyphenyl)dimethylammonium chloride.
Antirex; Enlon; Reversol; Tensilon; edrophone bromide [as bromide]; component of: Enlon Plus. Cholinergic agent. Curare antidote used in diagnosis of myasthenia gravis. mp = 162-163°; soluble in H_2O, EtOH; insoluble in $CHCl_3$, Et_2O.

535 Neostigmine
59-99-4 6553

$[C_{12}H_{19}N_2O_2]^+$
3-[[(Dimethylamino)carbonyl]oxy]-N,N,N-trimethylbenzenaminium.
synstigmin; proserine. Cholinergic agent. Also a miotic and used as an antidote for curare poisoning. Used in cattle, horses, sheep and pigs to initiate peristalsis and bladder evacuation.

536 Neostigmine Bromide
114-80-7 6553 204-054-8

$C_{12}H_{19}BrN_2O_2$
3-[[(Dimethylamino)carbonyl]oxy]-N,N,N-trimethylbenzenaminium bromide.
Juvastigmin (tabl.); Neoesserin; Neostigmin (tabl.); Normastigmin (tabl.); Prostigmin. Cholinergic agent. Also a miotic and used as an antidote for curare poisoning. Used in cattle, horses, sheep and pigs to initiate peristalsis and bladder evacuation. mp = 167° (dec); soluble in H_2O (100 g/100 ml), soluble in EtOH.

537 Neostigmine Methylsulfate
51-60-5 6553 200-109-5
$C_{13}H_{22}N_2O_6S$
3-[[(Dimethylamino)carbonyl]oxy]-N,N,N-trimethylbenzenaminium methyl sulfate.
Intrastigmina; Juvastigmin (amp.); Metastigmin; Neostigmin (inj.); Normastigmin (amp.); Prostigmin (amp.); Stiglyn. Cholinergic agent. Also a miotic and used as an antidote for curare poisoning. Used in cattle, horses, sheep and pigs to initiate peristalsis and bladder evacuation. mp = 142-145°; soluble in H_2O (10 g/100 ml), less soluble in EtOH; LD_{50} (mus iv) = 0.16 mg/kg, (mus sc) = 0.42 mg/kg, (mus orl) = 7.5 mg/kg.

Veterinary products:

008-097; Stiglyn 1:500; *Schering-Plough Animal Health.*

Cholinesterase Inhibitors

538 Pyridostigmine
155-97-5 8161

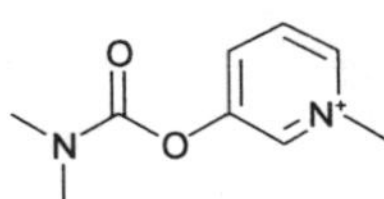

$C_9H_{13}N_2O_2^+$
3-Hydroxy-1-methylpyridinium dimethylcarbamate.
Cholinesterase Inhibitor. Used in treatment of myasthenia gravis in dogs.

539 Pyridostigmine Bromide
101-26-8 8161 202-929-9
$C_9H_{13}BrN_2O_2$
3-Hydroxy-1-methylpyridinium bromide dimethylcarbamate.
Mestinon; Regonol; Kalymin; Ro-1-5130. Cholinesterase Inhibitor. Used in treatment of myasthenia gravis in dogs. mp = 152-154°; soluble in H_2O, EtOH; insoluble in Et_2O, Me_2CO, C_6H_6.

Cholinesterase Reactivators

540 Pralidoxime Chloride

51-15-0 7884 200-080-9

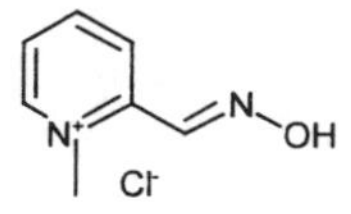

$C_7H_9ClN_2O$

2-[(Hydroxyimino)methyl]-1-methylpyridinium chloride.

2-pyridine aldoxime methyl chloride; 2-PAM chloride; Protopam Chloride. Cholinesterase reactivator, used as an antidote in cases of poisoning by nerve gas and cholinesterase inhibitory insecticides. Crystals; mp = 235-238° (dec); soluble in H_2O (65.5 g/100 ml), MeOH (8.5 g/100 ml), EtOH (0.89 g/100 ml), iPrOH (0.09 g/100 ml), insoluble in Me_2CO; LD_{50} (rat iv) = 96 mg/kg, (rbt iv) = 95 mg/kg, (mus iv) = 115 mg/kg, (musiv) = 115 mg/kg, (mus ip) = 205 mg/kg, (mus orl) = 4100 mg/kg.

Veterinary products:

039-204; Protopam; *Fort Dodge Animal Health, Divn. AHP.*

541 Pralidoxime Iodide

94-63-3 7884 202-349-6

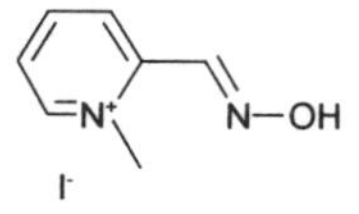

$C_7H_9IN_2O$

2-[(Hydroxyimino)methyl]-1-methylpyridinium iodide.

2-PAM. Cholinesterase reactivator, used as an antidote in cases of poisoning by nerve gas and cholinesterase inhibitory insecticides. mp = 225-226°; very soluble in H_2O; soluble in EtOH (4.8 g/100 ml at 25°); insoluble in Et_2O, Me_2CO; LD_{50} (mus iv) = 140-178 mg/kg, (mus ip) = 136-260 mg/kg, (mus sc) = 290-340 mg/kg, (mus orl) = 1500-4000 mg/kg.

Coccidiostats

542 Aklomide

3011-89-0 198 221-143-7

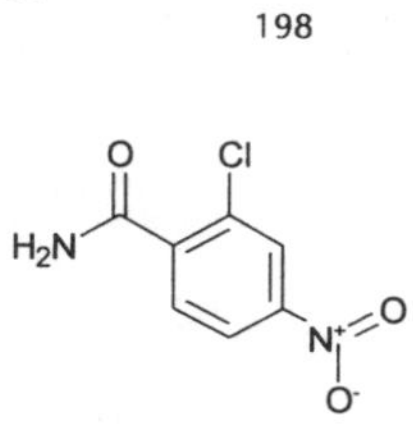

$C_7H_5ClN_2O_3$

2-Chloro-4-nitrobenzamide.

Aklomix; component with Sulfanitran of: 014-250 Novastat Type A Medicated Premix, with Roxarsone of 034-536 Aklomix Type A Medicated Article; Aklomix-3, with Roxarsone and sulfanitran of 034-537 Novastat-3 Type A Medicated Article, with Sulfanitran of 035-388 Novastat-W (Am. Cyanamid). Coccidiostat. Grey crystals; mp = 172°.

Veterinary products:

014-250; (with Sulfanitran) Novastat Type A Medicated Premix; *Fort Dodge Animal Health Divn., Am. Cyanamid.*
034-536; (with Roxarsone) Aklomix Type A Medicated Article: Aklomix-3; *Fort Dodge Animal Health Divn., Am. Cyanamid.*
034-537; (with Roxarsone, Sulfanitran) Novastat-3 Type A Medicated Premix; *Fort Dodge Animal Health Divn., Am. Cyanamid.*
035-388; (with Sulfanitran) Novastat-W; *Fort Dodge Animal Health Divn., Am. Cyanamid.*

543 Amprolium

121-25-5 631 204-458-4

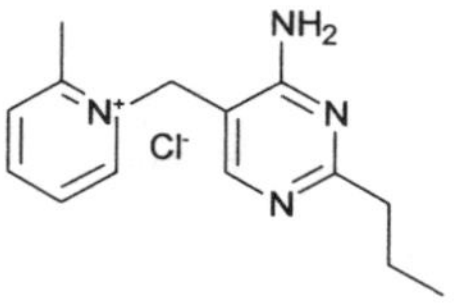

$C_{14}H_{19}ClN_4$

1-[(4-Amino-2-propyl-5-pyrimidinyl)methyl]-2-methylpyridinium chloride.

1-[(4-amino-2-propyl-5-pyrimidinyl)methyl]-2-picolinium chloride; Corid. Coccidiostat. Also used as an anti-parasitic in cattle. Crystals; dec 248-249°; freely soluble in H_2O, MeOH, 95% EtOH, DMF, sparingly soluble in EtOH, insoluble in iPrOH, BuOH, dioxane, Me_2CO, EtOAc, CH_3CN, isooctane; pH (10% aqueous solution) = 2.5 - 3.0.

Veterinary products:

012-350; Amprovine 25%; *Merial.*
013-149; Amprovine 9.6% Solutions; *Merial.*
013-461; (with Ethopabate, Roxarsone) Broiler PMX No. 1620; *Merial.*
013-663; (with Benzoic Acid) Purina+sl Liquid Amprol; *Purina Mills.*
033-165; Amprovine 20% Soluble Powder; Corid 20% Soluble Powder; *Merial.*
036-304; (with Bacitracin Methylene Disalicylate, Ethopabate) Amprol HI-E® Plus; *Merial.*
036-361; (with Chlortetracycline Calcium Complex, Ethopabate, Sodium Sulfate) Amp Ethopabate CTC® Sodium Sulfate; *Roche Vitamins.*
038-242; (with Arsanilic Acid, Erythromycin Thiocyanate, Ethopabate) Erythro® (Low Lev)/Amp + Etho; *Merial.*
039-284; (with Bacitracin Zinc, Ethopabate, Roxarsone) Swisher Super Broiler.
300-108; Swisher Super Broiler 400-112; *Swisher.*
040-920; Chick Grower-Developer Fortified; *Honeggers.*
041-178; (with Ethopabate, Lincomycin Hydrochloride Monohydrate, Roxarsone) Lincomix®/Amprol Plus/Roxarsone; *Pharmacia & Upjohn.*
044-820; (with Ethopabate, Lincomycin Hydrochloride

Monohydrate) Lincomix® & Amprol Plus; *Pharmacia & Upjohn*.
049-179; (with Ethopabate, Roxarsone) Amprol HI-E®/Roxarsone; *Merial*.
049-180; (with Bacitracin Methylene Disalicylate, Ethopabate, Roxarsone) Amprol HI-E®/BMD/Roxarsone; *Merial*.
049-462; (with Arsanilic Acid, Ethopabate, Penicillin G Procaine, Streptomycin) Rainbrook Broiler Premix No. 1; *Stockton Hay & Grain*.
091-646; (with Bacitracin Zinc, Ethopabate) Rainbow Broiler Base Concentrate; *Stockton Hay & Grain*.
091-647; (with Chlortetracycline, Ethopabate) Rainbow Broiler Base Concentrate; *Stockton Hay & Grain*.
095-543; (with Bambermycins, Ethopabate) Amprol HI-E® + Flavomycin®; *Hoechst-Roussel Vet.*
095-547; (with Bambermycins, Ethopabate, Roxarsone) Amprol HI-E® + Flavomycin® + 3-nitro®; *Hoechst-Roussel Vet.*
095-548; (with Bambermycins, Roxarsone) Amprol HI-E® + Flavomycin®; *Hoechst-Roussel Vet.*
095-549; (with Bambermycins, Ethopabate, Roxarsone) Amprol + 3-nitro® + Flavomycin®; *Hoechst-Roussel Vet.*
105-758; (with Bacitracin Zinc, Ethopabate, Roxarsone) Zinc Bacitracin & Amprol HI-E®; *Roche Vitamins*.
114-794; (with Bacitracin Zinc, Ethopabate) Baciferm®/ & Amprol HI-E® Premix; *Roche Vitamins*.
118-507; (with Carbarsone) Amprol/Carb-O-Sep®; *Merial*.
122-822; (with Ethopabate, Virginiamycin) Stafac®/Amprol HI-E®; *Pfizer*.
130-185; (with Bambermycins) Flavomycin® + Amprolium; *Hoechst-Roussel Vet.*
200-205; (with Bacitracin Zinc, Ethopabate) Amprol HI-E®/Albac®; *Alpharma*.
200-214; (with Bacitracin Zinc, Ethopabate, Roxarsone) Amprol HI-E®/Albac®/3-nitro®; *Alpharma*.
200-217; (with Bacitracin Zinc, Ethopabate, Roxarsone) Amprol HI-E®/Albac®/3-nitro®; *Alpharma*.

544 Buquinolate

5486-03-3 1524 226-814-8

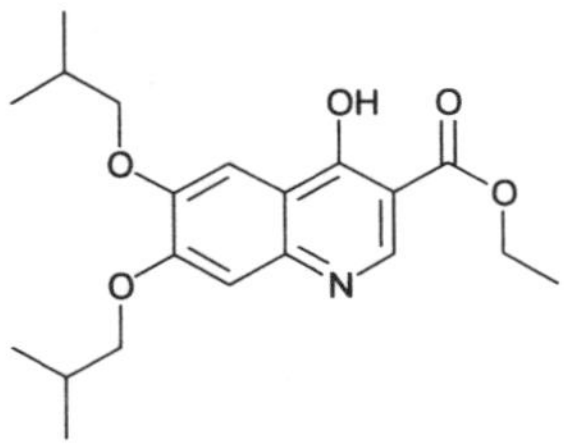

$C_{20}H_{27}NO_5$
4-Hydroxy-6,7-bis(2-methylpropoxy)-3-quinoline carboxylic acid ethyl ester.
Bonaid. Coccidiostat. Crystals; mp = 288-291°.

Veterinary products:

045-738; (with Lincomycin Hydrochloride Monohydrate) Lincomix ® / Bonaid; Lincomycin & Buquinolate; *Pharmacia & Upjohn*.

545 Clopidol

2971-90-6 2458 221-008-2

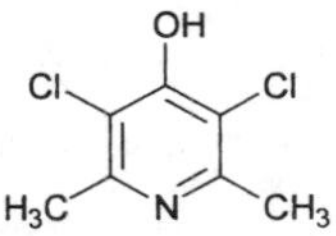

$C_7H_7Cl_2NO$
3,5-Dichloro-2,6-dimethyl-4-pyridinol.
meticlorpindol; clopindol; Coyden. Coccidiostat. Solid; mp > 320°; insoluble in H_2O; LD_{50} (rat orl) = 18000 mg/kg.

Veterinary products:

034-393; Coyden 25 ®; Lerbek 25; *Rhône-Poulenc*.
040-264; (with Roxarsone) Coyden 25 ® with 3-Nitro ®; *Rhône-Poulenc*.
041-541; (with Bacitracin Methylene Disalicylate, Roxarsone) Coyden 25 ® / BMD ® 15; *Rhône-Poulenc*.
044-016; (with Bacitracin Zinc, Roxarsone) Coyden 25 ® Medicated Premix; *Rhône-Poulenc*.
044-972; (with Lincomycin Hydrochloride Monohydrate) Lincomix ® / Coyden; *Pharmacia & Upjohn*.
046-209; (with Chlortetracycline Calcium Complex) Coyden 25 ® + CTC ®; *Rhône-Poulenc*.
049-934; (with Bacitracin Zinc) Coyden 25 ®; *Rhône-Poulenc*.
099-150; (with Bacitracin Methylene Disalicylate) Coyden 25 ® + Fortracin; *Rhône-Poulenc*.
200-207; (with Bacitracin Methylene Disalicylate, Roxarsone) Coyden 25 ® / Albac ® / 3-Nitro ®; *Alpharma*.
200-218; (with Bacitracin Zinc) Coyden 25 ® / Albac ®; *Alpharma*.

546 Decoquinate

18507-89-6 2910 242-389-1

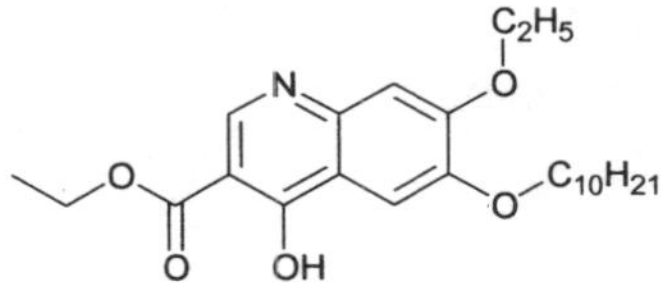

$C_{24}H_{35}NO_5$
6-Decyloxy-7-ethoxy-4-hydroxy-3-quinolinecarboxylic acid ethyl ester.
Deccox; M&B 15497. Coccidiostat. Used to prevent coccidiosis in cattle and poultry.

Veterinary products:

039-417; Deccox ® Type A Medicated Article; *Alpharma*.
040-435; (with Roxarsone) Deccox ® / 3-Nitro ®; Decoquinate & 3 Nitro ®; *Alpharma*.
045-348; (with Bacitracin Zinc) Deccox ® / Albac ®; Broiler Finisher Medicated; *Alpharma*.
045-444; (with Chlortetracycline Hydrochloride) Deccox ® / ChlorMax ®; Decoquinate & Chlortetracycline; *Alpharma*.
047-261; (with Lincomycin Hydrochloride Monohydrate) Lincomix ® / Deccox ®; *Pharmacia & Upjohn*.

047-262; (with Lincomycin) Deccox ® / Lincomycin; Decoquinate / Lincomycin; *Alpharma*.
091-326; (with Bacitracin Zinc, Roxarsone) Deccox ® / 3-Nitro ® / Albac ®; *Alpharma*.
141-060; Deccox ® - M Medicated Powder for Whole Milk; *Alpharma*.
141-100; (with Bacitracin Methylene Disalicylate, Roxarsone) Deccox ® / BMD ® / 3-Nitro ®; *Alpharma*.
141-102; (with Bacitracin Methylene Disalicylate) Deccox ® / BMD ®; *Alpharma*.
200-206; (with Bacitracin Zinc, Roxarsone) Deccox ® / Albac ® / 3-Nitro ®; *Alpharma*.
200-213; (with Bacitracin Zinc) Deccox ® / Albac ®; *Alpharma*.

547 Diclazuril

101831-37-2 3130

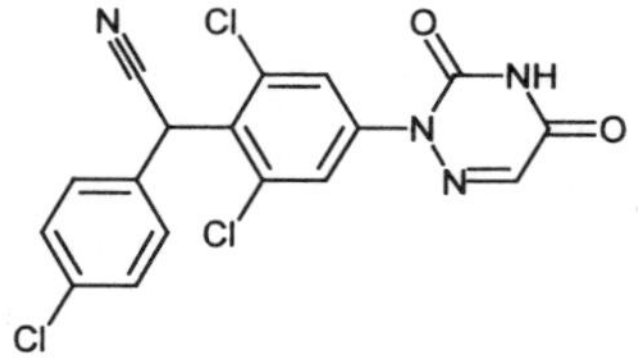

$C_{17}H_9Cl_3N_4O_2$
2,6-Dichloro-α-(4-chlorophenyl)-4-(4,5-dihydro-3,5-dioxo-1,2,4-triazin-2(3H)-yl)benzeneacetonitrile.
Clinacox; DCL; DCZL; α-(p-Chlorophenyl)(2,6-dichloro-4-(4-5-dihydro-3,5-dioxo-as-triazine-2-(3H)-yl)phenyl) acetonitrile; (p-Chlorophenyl) (2,6-dichloro-4-(4,5-dihydro-3,5-dioxo-as- triazin-2-(3H)-yl)phenyl)acetonitrile. Coccidiostat. Crystals; mp = 290.5°.

Veterinary products:

140-951; Clinacox ™; *Schering-Plough Animal Health*.

548 Ethopabate

59-06-3 3791 200-414-3

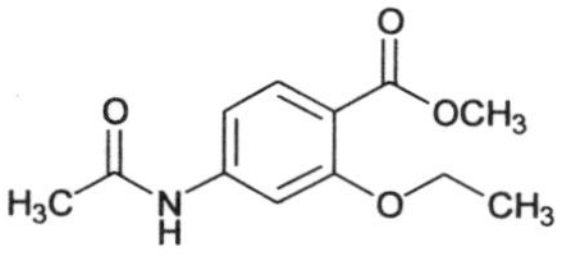

$C_{12}H_{15}NO_4$
Methyl 4-acetamido-2-ethoxybenzoate.
ethyl pabate. Coccidiostat, generally as a mixture with Amprolium. White-pink crystals; mp = 148-149°; λ_m = 267, 298 nm ($A^{1\%}_{1\ cm}$ 365, 805, MeOH); soluble in MeOH, EtOH, Me_2CO, CH_3CN, sparingly soluble in iPrOH, dioxane, EtOAc, CH_2Cl_2, insoluble in H_2O, isooctane.

Veterinary products:

013-461; (with Amprolium, Roxarsone) Broiler PMX No.1620; *Merial*.
036-304; (with Amprolium, Bacitracin Methylene Disalicylate) Amprol HI-E ® Plus; *Merial*.
036-361; (with Amprolium, Chloretracycline Calcium Complex, Sodium Sulfate) Amp Ethopabate CTC ® Sodium Sulfate; *Roche Vitamins*.
038-242; (with Amprolium, Arsanilic Acid, Erythromycin Thiocyanate) Erythro ® (Low Lev) / Amp + Etho; *Merial*.
039-284; (with Amprolium, Bacitracin Zinc, Roxarsone) Swisher Super Broiler 300-108; Swisher Super Broiler 400-112; *Swisher*.
041-178; (with Amprolium, Lincomycin Hydrochloride Monohydrate, Roxarsone) Lincomix ® / Amprol Plus / Roxarsone; *Pharmacia & Upjohn*.
044-820; (with Amprolium, Lincomycin Hydrochloride Monohydrate) Lincomix ® & Amprol Plus; *Pharmacia & Upjohn*.
049-179; (with Amprolium, Roxarsone) Amprol HI-E ® + Roxarsone; *Merial*.
049-180; (with Amprolium, Bacitracin Methylene Disalicylate, Roxarsone) Amprol HI-E ® / BMD ® / Roxarsone; *Merial*.
049-462; (with Amprolium, Arsanilic Acid, Penicillin G Procaine, Streptomycin) Rainbrook Broiler Premix No.1; *Stockton Hay & Grain*.
091-646; (with Amprolium, Bacitracin Zinc) Rainbow Broiler Base Concentrate; *Stockton Hay & Grain*.
091-647; (with Amprolium, Chloretracycline) Rainbow Broiler Base Concentrate; *Stockton Hay & Grain*.
095-543; (with Amprolium, Bambermycins) Amprol HI-E ® + Flavomycin ®; *Hoechst-Roussel Vet*.
095-547; (with Amprolium, Bambermycins, Roxarsone) Amprol HI-E ® + Flavomycin ® + 3-Nitro ®; *Hoechst-Roussel Vet*.
095-549; (with Amprolium, Bambermycins, Roxarsone) Amprol + 3-Nitro + Flavomycin ®; *Hoechst-Roussel Vet*.
105-758; (with Amprolium, Bacitracin Zinc, Roxarsone) Zinc Bacitracin & Amprol HI-E ®; *Roche Vitamins*.
114-794; (with Amprolium, Bacitracin Zinc) Baciferm ® / Amprol HI-E ® Premix; *Roche Vitamins*.
122-822; (with Amprolium, Virginiamycin) Stafac ® / Amprol HI-E ®; *Pfizer*.
200-205; (with Amprolium, Bacitracin Zinc) Amprol Hi-E ® / Albac ®; *Alpharma*.
200-214; (with Amprolium, Bacitracin Zinc, Roxarsone) Amprol Hi-E ® / Albac ® / 3-Nitro ®; *Alpharma*.
200-217; (with Amprolium, Bacitracin Zinc, Roxarsone) Amprol Hi-E ® / Albac ® / 3-Nitro ®; *Alpharma*.

549 Lasalocid Sodium

25999-20-6 5384 247-400-3

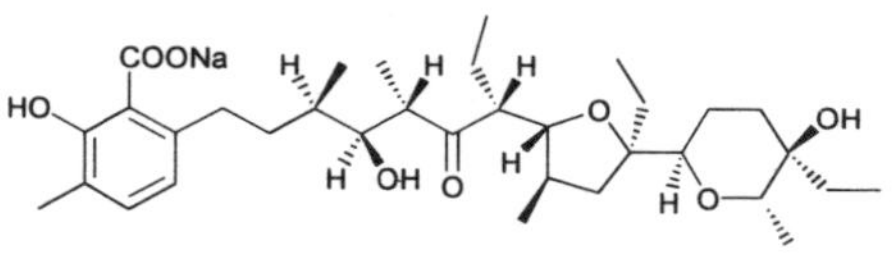

$C_{34}H_{53}NaO_8$
Sodium [2R-[2α-[2S*(3R*,4S*,5S*,7R*)-3S*,5S*],5α,6β]]-6-[7-[5-ethyl-5-(5-ethyltetrahydro-5-hydroxy-6-methyl-2H-pyran-2-yl)tetrahydro-3-methyl-2-furanyl]-4-hydroxy-3,5-dimethyl-6-oxononyl]-2-hydroxy-3-methylbenzoate.
antibiotic X-537A; ionophore X-4537A; Ro-2-2985; X-537A; Bovatec; Bovatec. Used in poultry as a coccidiostat. Crystals; mp = 168-171° (dec), 191-192° (dec); $[\alpha]_D^{25}$ = -30° (c = 1, MeOH); λ_m = 308 nm (ε 4100, 50% aqueous iPrOH).

Veterinary products:

096-298; Avatec ®; Bovatec ® Premix; Bovatec ® Type A Medicated Article; *Roche Vitamins.*
101-689; (with Lincomycin Hydrochloride) Lincomix ® / Avatec ®; *Pharmacia & Upjohn.*
102-485; (with Roxarsone) Avatec ®/ 3-Nitro ®; *Roche Vitamins.*
107-996; (with Bacitracin Methylene Disalicylate) Avatec ®/ Fortracin Premix; *Roche Vitamins.*
112-661; (with Lincomycin Hydrochloride, Roxarsone) Avatec ® / Lincomycin / 3-Nitro ®; *Roche Vitamins.*
112-687; (with Bambermycins, Roxarsone) Avatec ® / 3-Nitro ® / Flavomycin ®; *Roche Vitamins.*
116-082; (with Bacitracin Methylene Disalicylate, Roxarsone) BMD ® / Avatec ® / 3-Nitro ®; *Alpharma.*
122-608; (with Virginiamycin) Stafac ® / Avatec ®; *Pfizer.*
126-052; (with Bacitracin Zinc, Roxarsone) Avatec ® / 3-Nitro ® / Baciferm ®; *Roche Vitamins.*
131-894; (with Bacitracin Methylene Disalicylate, Roxarsone) Avatec ® / Fortracin / 3-Nitro ® Broiler Premix; *Roche Vitamins.*
138-904; (with Melengestrol Acetate, Tylosin Phosphate) MGA ® 100-200 Premix / MGA ® 500 Liquid Premix / Bovatec ® / Tylan ®; MGA ® 100 / Bovatec / Tylan; *Pharmacia & Upjohn.*
138-992; (with Melengestrol Acetate, Tylosin Phosphate) MGA ® (liquid) / Bovatec / Tylan; *Pharmacia & Upjohn.*
138-993; Moorman's ® Cattle Minerals BT; *Moorman.*
139-876; (with Melengestrol Acetate) MGA ® 100 Bovatec; MGA ® 200 Bovatec; *Pharmacia & Upjohn.*
140-288; (with Melengestrol Acetate) MGA ® 500 / Bovatec; *Pharmacia & Upjohn.*
140-579; (with Oxytetracycline Monoalkyl Trimethylammonium Salt) Bovatec ® / Terramycin ®; *Roche Vitamins.*
141-109; (with Bacitracin Zinc) Avatec ® / Baciferm ®; *Roche Vitamins.*
141-129; (with Bambermycins) Avatec ® / Flavomycin ®; *Hoechst-Roussel Vet.*
141-150; (with Virginiamycin) Avatec ® / Stafac ®; *Roche Vitamins.*

550 Maduramicin Ammonium

84878-61-5 5682

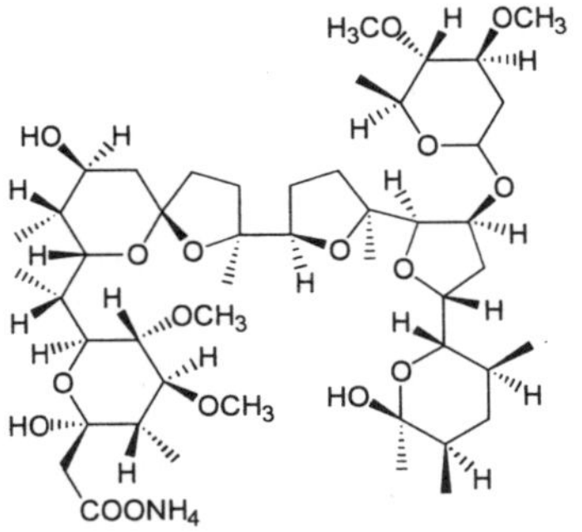

$C_{47}H_{83}NO_{17}$
(3R,4S,5S,6R,7S,22S)-23,27-Didemethoxy-2,6,22-tridemethyl-11-O-demethyl-22-[(2,6-dideoxy-3,4-di-O-methyl-β-L-arabinohexopyranosyl)oxy]-6-methoxylono,ycin A monoammonium salt.
antibiotic X-14868A ammonium salt; CL-273703; Cygro. Coccidiostat.

Veterinary products:

139-075; Cygro ®; *Roche Vitamins.*

551 Monensin

17090-79-8 6329 241-154-0

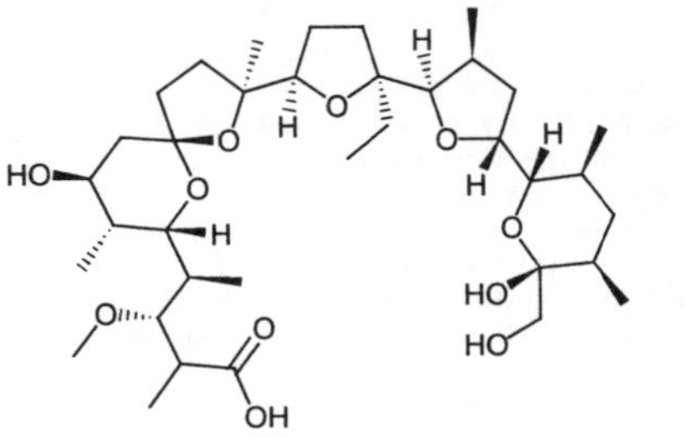

$C_{36}H_{62}O_{11}$
2-[2-Ethyloctahydro-3'-methyl-5'-[tetrahydro-6-hydroxy-6-(hydroxymethyl)-3,5-dimethyl-2H-pyran-2-yl][2,2'-bifuran-5-yl]]-9-hydroxy-β-methoxy-α,σ,2,8-tetramethyl-1,6-dioxapsiro[4.5]decan-7-butanoic acid.
A-3823A; 63714; monensic acid. Antibiotic produced by *Streptomyces cinnamonensis*. Antifungal, antibiotic and antiprotozoal. Used as a coccidiostat in chickens. mp = 103-105°; $[\alpha]_D$ = 47.7°; slightly soluble in H_2O, more soluble in hydrocarbons, very soluble in organic solvents; LD_{50} (mus orl) = 43.8 ± 5.2 mg/kg, (chicks orl) = 284 ± 47 mg/kg.

Veterinary products:

049-464; (with Bacitracin Methylene Disalicylate, Roxarsone) Monensin & Bacitracin & Roxarsone; *Elanco.*
092-522; (with Lincomycin Hydrochloride) Lincomix ® + Coban ® with or without Roxarsone; Lincomix ® / Coban ® / Roxarsone; *Pharmacia & Upjohn.*
098-341; (with Bambermycins, Roxarsone) Flavomycin ® + 3-Nitro ® + Coban ®; *Hoechst-Roussel Vet.*
141-110; (with Virginiamycin) Coban ® / Stafac ®; *Elanco.*
200-211; (with Bacitracin Zinc, Roxarsone); *Alpharma.*

552 Monensin Sodium

22373-78-0 6329 244-941-7

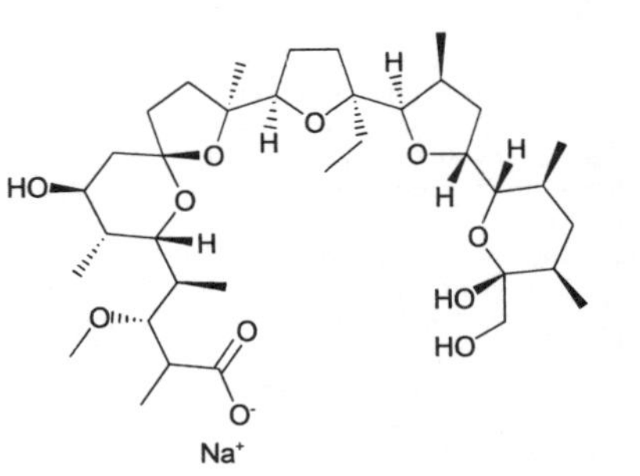

$C_{36}H_{61}NaO_{11}$
Sodium 2-[2-ethyloctahydro-3'-methyl-5'-[tetrahydro-6-hydroxy-6-(hydroxymethyl)-3,5-dimethyl-2H-pyran-2-yl]-[2,2'-bifuran-5-yl]]-9-hydroxy-β-methoxy-α,σ,2,8-tetramethyl-1,6-dioxapsiro[4.5]decan-7-butanoate.

Rumensin; Romensin; Coban. Antibiotic produced by *Streptomyces cinnamonensis*. Antifungal, antibiotic and antiprotozoal. Used as a coccidiostat in chickens. mp = 267-269°; $[\alpha]_D$ = 57.3°; slightly soluble in H_2O, more soluble in hydrocarbons, very soluble in organic solvents.

Veterinary products:

038-878; Coban ® - 45; Coban ® - 60; Coban ® - 110; Elancoban-100; *Elanco.*
041-500; (with Roxarsone) Coban ® - 1; *Elanco.*
047-933; (with Bacitracin Zinc) Monensin & Zinc Bacitracin Premix; *Elanco.*
049-463; (with Bacitracin Methylene Disalicylate) Monensin & Bacitracin MD; *Elanco.*
092-482; (with Lincomycin Hydrochloride) Coban ® / Lincomix ®; Lincomix ® / Coban ®; *Pharmacia & Upjohn.*
095-735; Rumensin ®; Rumensin ® 80; *Elanco.*
098-340; (with Bambermycins) Flavomycin ® + Monensin; *Hoechst-Roussel Vet.*
099-006; (with Oxytetracycline Monoalkyl Trimethylammonium Salt) Terramycin ® TM-50; TM-10 Plus; *Pfizer.*
104-646; (with Tylosin Phosphate) Tylan ® / Rumensin ®; *Elanco.*
109-471; Staley Sweetlix With Rumensin ®; *PM Ag Products.*
115-581; Moorman's ® Mintrate Blonde Block; *Moorman.*
116-088; (with Bacitracin Methylene Disalicylate) BMD ® / Coban ® / 3-Nitro ®; Fortracin / Coban ® / 3-Nitro ®; *Alpharma.*
118-509; Pasture Gainer Block-37 R350; *Farmland.*
119-253; Cattle Block M; *Co-op. Res. Farms.*
120-724; (with Roxarsone, Virginiamycin) Stafac ® / 3-Nitro ® / Coban ®; *Pfizer.*
121-553; (with Chlortetracycline Hydrochloride) Coban ® / Aureomycin ®; *Roche Vitamins.*
122-481; (with Virginiamycin) Stafac ® / Coban ®; *Pfizer.*
123-154; (with Bacitracin Zinc, Roxarsone) Coban ® / 3-Nitro ® 10 / Baciferm ® Premix; *Roche Vitamins.*
124-309; (with Melengestrol Acetate) MGA ® 100-200 / Rumensin; *Pharmacia & Upjohn.*
125-476; (with Melengestrol Acetate) MGA ® 500 / Rumensin; *Pharmacia & Upjohn.*
130-736; Coban ®; *Elanco.*
134-830; (with Bacitracin Zinc) Coban ® / Albac ®; *Alpharma.*
138-456; (with Bacitracin Methylene Disalicylate) Coban ® / BMD ®; *Alpharma.*
138-703; (with Bacitracin Zinc, Roxarsone) Albac ® / Coban ® / 3-Nitro ®; *Alpharma.*
138-792; (with Melengestrol Acetate, Tylosin Phosphate) MGA ® 100/200 / Rumensin / Tylan; *Pharmacia & Upjohn.*
138-870; (with Melengestrol Acetate, Tylosin Phosphate) MGA ® (liquid) / Rumensin / Tylan; MGA ® 100-200 Premix / MGA 500 Liquid Premix / Rumensin ® / Tylan ®; *Pharmacia & Upjohn.*
140-937; (with Bacitracin Methylene Disalicylate) Coban ® and BMD ®; *Elanco.*
140-939; (with Tylosin Phosphate) Liquid Type B Medicated Cattle Feed R/T 150; Liquid Type B Medicated Cattle Feed R/T 400; *Elanco.*
200-263; (with Chlortetracycline) ChlorMax™ / Coban ®; *Alpharma.*

553 Narasin

55134-13-9 6506

$C_{43}H_{72}O_{11}$
α-Ethyl-6-[5-[2-(5-ethyltetrahydro-5-hydroxy-6-methyl-2H-pyran-2-yl)-15-hydroxy-2,10,12-trimethyl-1,6,8-trioxadispiro[4.1.5.3]pentadec-13-en-9-yl]-2-hydroxy-1,3-dimethyl-4-oxoheptyl]tetrahydro-3,5-dimethyl-2H-pyran-2-acetic acid.
Natacyn; CL-12625; Antibiotic A-5283; Monteban; C-7819B. Veterinary growth stimulant and coccidostat. mp = 98-100°, 198-200°; λ_m 285 nm (ε 58 EtOH); $[\alpha]_D^{25}$ = -54° (c = 0.2 MeOH); insoluble in H_2O, soluble in organic solvents; LD_{50} (mus ip) = 7.15 mg/kg.

Veterinary products:

118-980; Monteban ®; *Elanco.*
138-952; (with Nicarbazin) Maxiban ® 72 / Narasin / Nicarbazin; *Elanco.*
140-445; (with Roxarsone) Monteban ® and Roxarsone; *Elanco.*
140-843; (with Bambermycins, Roxarsone) Monteban ® + Flavomycin ® + 3-Nitro ®; *Hoechst-Roussel Vet.*
140-845; (with Bambermycins) Flavomycin ® + Monteban ®; *Hoechst-Roussel Vet.*
140-852; (with Bacitracin Methylene Disalicylate, Roxarsone) Monteban ® / 3-Nitro ® / BMD ®; *Alpharma.*
140-853; (with Bacitracin Methylene Disalicylate) Monteban ® / BMD ®; *Alpharma.*
140-926; (with Bacitracin Methylene Disalicylate, Nicarbazin) BMD ® / Maxiban ®; *Elanco.*
140-947; (with Lincomycin, Nicarbazin) Maxiban ® / Lincomix ®; *Elanco.*
141-112; (with Nicarbazin, Roxarsone) 3-Nitro ® / BMD ® / Maxiban ®; *Alpharma.*
141-113; (with Nicarbazin, Roxarsone) 3-Nitro ® / Maxiban ®; *Elanco.*

554 Nequinate

13997-19-8 6558 237-796-6

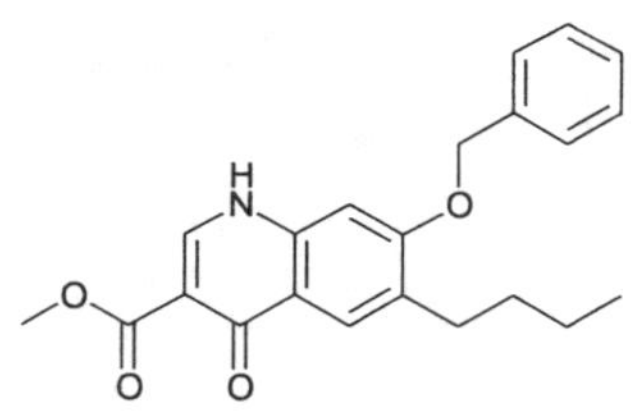

$C_{22}H_{23}NO_4$
6-Butyl-1,4-dihydro-4-oxo-7-(phenylmethoxy)-3-quinolinecarboxylic acid methyl ester.

methyl benzoquate; ICI-55052; Statyl. Coccidiostat. Used in poultry feedstuffs. Crystals; mp = 287-288°.

Veterinary products:

046-700; Statyl; *Purina Mills.*

555 Nicarbazin

330-95-0 6578 206-359-1

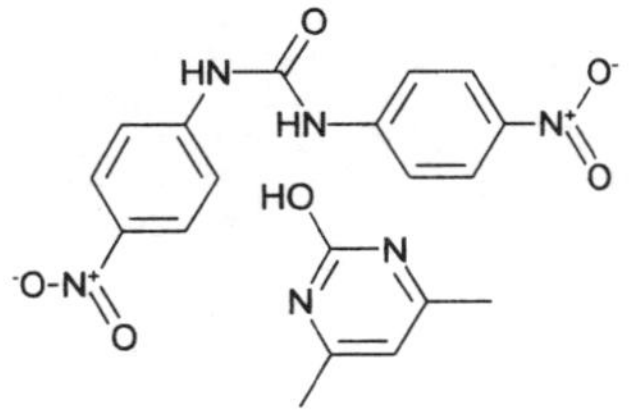

$C_{19}H_{18}N_6O_6$
N,N'-Bis(4-nitrophenyl)urea.
Nicarb; Nicoxin; Nicrazin. Coccidiostat. Used with poultry. Crystals; dec 265-275°; λ_m = 298 nm ($A^{1\%}_{1\ cm}$ = 670 conc. H_2SO_4); insoluble in H_2O; some crystals are electrostatic, creating difficulties with dry mixing.

Veterinary products:

009-476; Nicarbazin; *Koffolk.*
098-378; (with Bacitracin Methylene Disalicylate) Nicarbazin / Bacitracin Premix; *Koffolk.*
107-997; (with Lincomycin Hydrochloride) Medicated Broiler Feed; *Koffolk.*
108-115; (with Roxarsone) Nicarbazin Roxarsone Premix; *Koffolk.*
108-116; (with Lincomycin Hydrochloride) Nicarbazin Lincomycin Premix; *Koffolk.*
135-468; Carbigran 25; Nicarbazin; *Elanco.*
138-952; (with Narasin) Maxiban ® 72 / Narasin / Nicarbazin; *Elanco.*
140-339; (with Bambermycins) Flavomycin ® / Nicarb ®; *Hoechst-Roussel Vet.*
140-926; (with Bacitracin Methylene Disalicylate, Narasin) BMD ® / Maxiban ®; *Elanco.*
140-947; (with Lincomycin, Narasin) Maxiban ® / Lincomix ®; *Elanco.*
141-112; (with Bacitracin Methylene Disalicylate, Narasin, Roxarsone) 3-Nitro ® / BMD ® / Maxiban ®; *Alpharma.*
141-113; (with Narasin, Roxarsone) 3-Nitro ® / Maxiban ®; *Elanco.*
200-027; Nicarmix 25 ®; *Planalquimica.*
200-164; (with Bacitracin Methylene Disalicylate) Nicarmix 25 ® plus BMD ®; *Planalquimica.*
200-170; (with Lincomycin, Roxarsone) Nicarmix 25 ® / 3-Nitro ® / Lincomix ®; *Planalquimica.*
200-171; (with Lincomycin) Nicarmix 25 ® / Lincomix ®; *Planalquimica.*
200-172; (with Roxarsone) Nicarmix 25 ® / 3-Nitro ®; *Planalquimica.*

556 Nitromide

121-81-3 6709 204-499-8

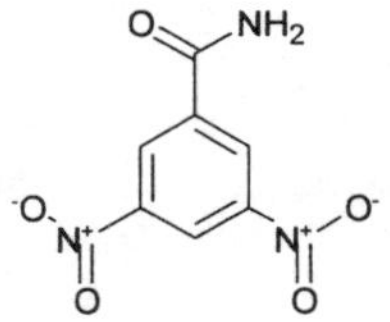

$C_7H_5N_3O_5$
3,5-Dinitrobenzamide.
Antibacterial and poultry coccidiostat. Leaflets; mp = 183°; soluble in hot H_2O, slightly soluble in cold H_2O.

Veterinary products:

011-141; (with Sulfanitran) Unistat-2 Type A Medicated Premix; *Fort Dodge Animal Health, Divn. Am. Cyanamid.*
039-666; (with Sulfanitran, Roxarsone) Unistat-3 Type A Medicated Article; *Fort Dodge Animal Health, Divn. Am. Cyanamid.*

557 Robenidine Hydrochloride

25875-50-7 8403 247-307-8

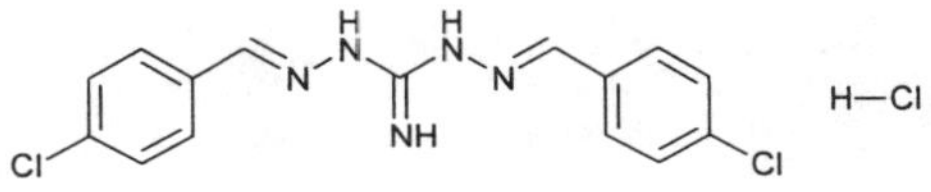

$C_{15}H_{14}Cl_3N_5$
Bis[(4-chlorophenyl)methylene]carbonimidic dihydrazide hydrochloride.
1,3-bis[(p-chlorobenzylidene)amino]guanidine hydrochloride; robenzidine; Cyclostat; Robenz. Coccidiostat. Crystals; mp = 289-290°.

Veterinary products:

048-486; Robenz ®; *Roche Vitamins.*
092-507; (with Chlortetracycline Calcium Complex) Robenz ® with Aureomycin ® 500 Gm; *Roche Vitamins.*
093-106; (with Lincomycin Hydrochloride) Lincomix ® + Robenz ®; *Pharmacia & Upjohn.*
095-546; (with Roxarsone) Robenz ® Plus Roxarsone; *Roche Vitamins.*
096-933; (with Bacitracin Zinc) Robenz ® Plus Zn Bacitracin; *Roche Vitamins.*
097-085; (with Bacitracin Methylene Disalicylate) Robenz ® Plus Bac MD; *Roche Vitamins.*
101-666; (with Oxytetracycline Monoalkyl Trimethylammonium Salt) Oxytetracycline & Robenz ® Premix; *Pfizer.*
200-212; (with Bacitracin Zinc) Robenz ® / Albac ®; *Alpharma.*

558 Roxarsone
121-19-7 8430 204-453-7

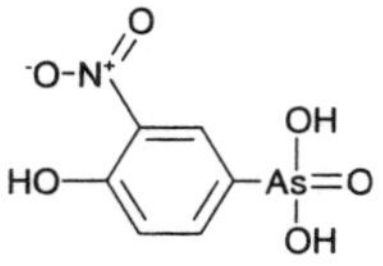

$C_6H_6AsNO_6$
4-Hydroxy-3-nitrobenzenearsonic acid.
NSC-2101; Ren-o-sal. Coccidiostat. Used to control enteric infections and to promote growth and feed efficiency. Pale yellow needles; slightly soluble in cold H_2O, more soluble in hot H_2O (3.3 g/100 ml), freely soluble in MeOH, EtOH, AcOH, Me_2CO, alkalies, insoluble in Et_2O, EtOAc; LD_{50} (rat orl) = 155 mg/kg, (rat ip) = 66 mg/kg, (ckn orl) = 110-123 mg/kg, (ckn ip) = 34 mg/kg.

Veterinary products:

005-414; Ren-O-Sal ® Tablets; *Alpharma.*
006-019; Zoco Poultry Tablets; Zuco Tablets for Poultry; *Russell.*
006-081; Korum Improved Formula; *Russell.*
007-891; 3-Nitro ® 10 Type A Medicated Article; 3-Nitro ® 20 Type A Medicated Article; 3-Nitro ® 50 Type A Medicated Article; 3-Nitro ® 80 Type A Medicated Article; *Alpharma.*
008-274; Pig Scour Tablets; *Fort Dodge Animal Health Divn., Am. Cyanamid.*
013-461; (with Amprolium, Ethopabate) Broiler PMX No.1620; *Merial.*
034-536; (with Aklomide) Aklomix Type A Medicated Article; Aklomix-3; *Fort Dodge Animal Health Divn., Am. Cyanamid.*
034-537; (with Aklomide, Sulfanitran) Novastat-3 Type A Medicated Article; *Fort Dodge Animal Health Divn., Am. Cyanamid.*
039-284; (with Amprolium, Bacitracin Zinc, Ethopabate); *Swisher.*
039-666; (with Nitromide, Sulfanitran) Unistat-3 Type A Medicated Article; *Fort Dodge Animal Health Divn., Am. Cyanamid.*
040-264; (with Clopidol) Coyden 25 ® with 3-Nitro ®; *Rhône-Poulenc.*
040-435; (with Decoquinate) Deccox ® / 3-Nitro ®; Decoquinate & 3 Nitro ®; *Alpharma.*
041-178; (with Amprolium, Ethopabate, Lincomycin Hydrochloride Monohydrate) Lincomix ® / Amprol Plus / Roxarsone; *Pharmacia & Upjohn.*
041-500; (with Monensin Sodium) Coban ® - 1; *Elanco.*
041-541; (with Bacitracin Methylene Disalicylate, Clopidol) Coyden 25 ® / BMD ® 15; *Rhône-Poulenc.*
041-984; (with Ormetoprim, Sulfadimethoxine) Rofenaid ® Plus Roxarsone; *Roche Vitamins.*
044-016; (with Bacitracin Zinc, Clopidol) Coyden 25 ® Medicated Premix; *Rhône-Poulenc.*
049-179; (with Amprolium, Ethopabate) Amprol HI-E ® + Roxarsone; *Merial.*
049-180; (with Amprolium, Bacitracin Methylene Disalicylate; Ethopabate; Roxarsone) Amprol HI-E ® / BMD ® / Roxarsone; *Merial.*
049-464; (with Bacitracin Methylene Disalicylate, Monensin) Monensin & Bacitracin & Roxarsone; *Elanco.*
091-326; (with Bacitracin Zinc, Decoquinate) Deccox ® / 3-Nitro ® / Albac ®; *Alpharma.*
092-953; Roxarsone 10% Type A Medicated Article; Roxarsone 20% Type A Medicated Article; Roxarsone 50% Type A Medicated Article; *Alpharma.*
093-025; 3-Nitro ® W; 3-Nitro ® Soluble; *Alpharma.*
095-546; (with Robenidine Hydrochloride) Robenz ® Plus Roxarsone; *Roche Vitamins.*
095-547; (with Amprolium, Bambermycins, Ethopabate) Amprol HI-E ® + Flavomycin ® + 3-Nitro ®; *Hoechst-Roussel Vet.*
095-548; (with Amprolium, Bambermycins) Amp +3-Nitro ® + Flavomycin ®; *Hoechst-Roussel Vet.*
095-549; (with Amprolium, Bambermycins, Ethopabate) Amprol + 3-Nitro + Flavomycin ®; *Hoechst-Roussel Vet.*
098-341; (with Bambermycins, Monensin) Flavomycin ® + 3-Nitro ® + Coban ®; *Hoechst-Roussel Vet.*
101-628; (with Bambermycins, Zoalene) Flavomycin ® + 3-Nitro + Zoalene; *Hoechst-Roussel Vet.*
102-485; (with Lasalocid Sodium) Avatec ® / 3-Nitro ®; *Roche Vitamins.*
105-758; (with Amprolium, Bacitracin Zinc, Ethopabate) Zinc Bacitracin & Amprol HI-E ®; *Roche Vitamins.*
107-997; (with Lincomycin Hydrochloride, Nicarbazin) Medicated Broiler Feed; *Koffolk.*
108-115; (with Nicarbazin) Nicarbazin Roxarsone Premix; *Koffolk.*
112-661; (with Lasalocid Sodium, Lincomycin Hydrochloride) Avatec ® / Lincomycin / 3-Nitro ®; *Roche Vitamins.*
112-687; (with Bambermycins, Lasalocid Sodium) Avatec ® / 3-Nitro ® / Flavomycin ®; *Roche Vitamins.*
116-082; (with Bacitracin Methylene Disalicylate, Lasalocid Sodium) BMD ® / Avatec ® / 3-Nitro ®; *Alpharma.*
116-088; (with Bacitracin Methylene Disalicylate, Monensin Sodium) BMD ® / Coban ® / 3-Nitro ®; Fortracin / Coban ® / 3-Nitro ®; *Alpharma.*
120-724; (with Monensin Sodium, Virginiamycin) Stafac ® / 3-Nitro ® / Coban ®; *Pfizer.*
123-154; (with Bacitracin Zinc, Monensin Sodium) Coban ® / 3-Nitro ® 10 / Baciferm ® Premix; *Roche Vitamins.*
126-052; (with Bacitracin Zinc, Lasalocid Sodium) Avatec ® / 3-Nitro ® / Baciferm ®; *Roche Vitamins.*
131-894; (with Bacitracin Methylene Disalicylate, Lasalocid Sodium) Avatec ® / Fortracin / 3-Nitro ® Broiler Premix; *Roche Vitamins.*
132-447; (with Salinomycin Sodium) Bio-Cox ® Plus Roxarsone; *Roche Vitamins.*
134-185; (with Bambermycins, Salinomycin Sodium) Biocox ® / 3-Nitro ® / Flavomycin ®; *Roche Vitamins.*
135-321; (with Bacitracin Methylene Disalicylate, Salinomycin Sodium) Biocox ® / 3-Nitro ® / BMD ®; *Roche Vitamins.*
137-536; (with Bacitracin Zinc, Salinomycin Sodium) Bio-Cox ® / 3-Nitro ® Plus Albac ®; *Roche Vitamins.*
138-703; (with Bacitracin Zinc, Monensin Sodium) Albac ® / Coban ® / 3-Nitro ®; *Alpharma.*
138-953; (with Salinomycin Sodium, Virginiamycin) Bio-Cox ® / Stafac ® / 3-Nitro ® Type A Medicated Articles; *Pfizer.*
139-190; (with Bacitracin Zinc, Salinomycin Sodium) Bio-Cox ® / 3-Nitro ® / Baciferm ®; *Roche Vitamins.*
140-445; (with Narasin) Monteban® and Roxarsone; *Elanco.*
140-533; (with Bacitracin Methylene Disalicylate, Halofuginone Hydrobromide) Stenorol ® + 3-Nitro ® + BMD ®; *Hoechst-Roussel Vet.*

140-581; (with Lincomycin Hydrochloride, Salinomycin Sodium) Biocox ® / 3-Nitro ® / Lincomix ®; *Roche Vitamins.*
140-843; (with Bambermycins, Narasin) Monteban ® + Flavomycin ® + 3-Nitro ®; *Hoechst-Roussel Vet.*
140-852; (with Bacitracin Methylene Disalicylate, Narasin) Monteban ® / 3-Nitro ® / BMD ®; *Alpharma.*
140-867; (with Chlortetracycline Calcium Complex, Salinomycin Sodium) Aureomycin ® / Bio-Cox ® / 3-Nitro ®; *Roche Vitamins.*
141-058; (with Bacitracin Methylene Disalicylate, Semduramicin Sodium) Aviax ™ / BMD ® / 3-Nitro ®; *Pfizer.*
141-066; (with Semduramicin Sodium) Aviax ™ / 3-Nitro ®; *Pfizer.*
141-100; (with Bacitracin Methylene Disalicylate, Decoquinate) Deccox ® / BMD ® / 3-Nitro ®; *Alpharma.*
141-112; (with Narasin, Nicarbazin) 3-Nitro ® / BMD ® / Maxiban ®; *Alpharma.*
141-113; (with Narasin, Nicarbazin) 3-Nitro ® / Maxiban ®; *Elanco.*
200-080; (with Bambermycins, Salinomycin Sodium) Sacox ® + 3-Nitro ® + Flavomycin ®; *Hoechst-Roussel Vet.*
200-081; (with Bacitracin Methylene Disalicylate, Salinomycin Sodium) Sacox ® + 3-Nitro ® + BMD ®; *Hoechst-Roussel Vet.*
086; (with Bacitracin Zinc, Salinomycin Sodium) Sacox ® + Albac ® + 3-Nitro ®; *Hoechst-Roussel Vet.*
200-090; (with Lincomycin, Salinomycin Sodium) Sacox ® + Lincomix ® + 3-Nitro ®; *Hoechst-Roussel Vet.*
200-091; (with Chlortetracycline Calcium Complex, Salinomycin Sodium) Sacox ® + 3-Nitro ® + Aureomycin ®; *Hoechst-Roussel Vet.*
200-094; (with Salinomycin Sodium, Virginiamycin) Sacox ® + Stafac ® + 3-Nitro ®; *Hoechst-Roussel Vet.*
200-095; (with Oxytetracycline, Salinomycin Sodium) Sacox ® + Aureomycin ®; *Hoechst-Roussel Vet.*
200-096; (with Oxytetracycline, Salinomycin Sodium) Sacox ® + Terramycin ®; *Hoechst-Roussel Vet.*
200-097; (with Salinomycin Sodium) Sacox ® + 3-Nitro ®; *Hoechst-Roussel Vet.*
200-143; (with Bacitracin Zinc, Salinomycin Sodium) Sacox ® + 3-Nitro ® + Baciferm ®; *Hoechst-Roussel Vet.*
200-170; (with Lincomycin, Nicarbazin) Nicarmix 25 ® / 3-Nitro ® / Lincomix ®; *Planalquimica.*
200-172; (with Nicarbazin) Nicarmix 25 ® / 3-Nitro ®; *Planalquimica.*
200-206; (with Bacitracin Zinc, Decoquinate) Deccox ® / Albac ® / 3-Nitro ®; *Alpharma.*
200-207; (with Bacitracin Methylene Disalicylate, Clopidol) Coyden 25 ® / Albac ® / 3-Nitro ®; *Alpharma.*
200-209; (with Bacitracin Zinc, Salinomycin Sodium) Sacox ® / Albac ® / 3-Nitro ®; *Alpharma.*
200-211; (with Bacitracin Zinc, Monensin) Coban ® / Albac ® / 3-Nitro ®; *Alpharma.*
200-214; (with Amprolium, Bacitracin Zinc, Ethopabate) Amprol Hi-E ® / Albac ® / 3-Nitro ®; *Alpharma.*
200-215; (with Bacitracin Zinc, Salinomycin Sodium) Bio-Cox ® / Albac ® / 3-Nitro ®; *Alpharma.*
200-217; (with Amprolium, Bacitracin Zinc, Ethopabate) Amprol Hi-E ® / Albac ® / 3-Nitro ®; *Alpharma.*
200-259; (with Chlortetracycline, Salinomycin Sodium) ChlorMax ™ / Sacox ® / 3-Nitro ®; ChlorMax ™ / Bio-Cox ® / 3-Nitro ®; *Alpharma.*
200-260; (with Chlortetracycline, Salinomycin Sodium) ChlorMax ™ / Sacox ® / 3-Nitro ®; ChlorMax ™ / Bio-Cox ® / 3-Nitro ®; *Alpharma.*

559 Salinomycin Sodium

55721-31-8 8488

$C_{42}H_{69}NaO_{11}$
Salinomycin sodium salt.
Ovicox; Sacox; Salgain. Anticoccidial. Crystals; mp = 140-142°; $[\alpha]_D^{25}$ = -37° (c = 1, EtOH).

Veterinary products:

128-686; Bio-Cox ® Type A Medicated Article; *Roche Vitamins.*
132-447; (with Roxarsone) Bio-Cox ® Plus Roxarsone; *Roche Vitamins.*
134-185; (with Bambermycins, Roxarsone) Biocox ® / 3-Nitro ® / Flavomycin ®; *Roche Vitamins.*
134-284; (with Bambermycins) Biocox ® / Flavomycin ®; *Roche Vitamins.*
135-321; (with Bacitracin Methylene Disalicylate, Roxarsone) Biocox ® / 3-Nitro ® / BMD ®; *Roche Vitamins.*
135-746; (with Bacitracin Methylene Disalicylate) Biocox ® / BMD ®; *Roche Vitamins.*
137-536; (with Bacitracin Zinc, Roxarsone) Bio-Cox ® / 3-Nitro ® Plus Albac ®; *Roche Vitamins.*
137-537; (with Lincomycin) Bio-Cox ® / Lincomix ®; *Roche Vitamins.*
138-828; (with Virginiamycin) Stafac ® 10,20,50, and 500 Type A Medicated Article/ Bio-Cox ® Type A Medicated Article; *Pfizer.*
138-953; (with Roxarsone, Virginiamycin) Bio-Cox ® / Stafac ® / 3-Nitro ® Type A Medicated Articles; *Pfizer.*
139-190; (with Bacitracin Zinc, Roxarsone) Bio-Cox ® / 3-Nitro ® / Baciferm ®; *Roche Vitamins.*
139-235; (with Bacitracin Zinc) Biocox ® / Baciferm ®; *Roche Vitamins.*
140-448; (with Oxytetracycline Monoalkyl Trimethylammonium Salt) Bio-Cox ® / Terramycin ®; *Pfizer.*
140-581; (with Lincomycin Hydrochloride, Roxarsone) Biocox ® / 3-Nitro ® / Lincomix ®; *Roche Vitamins.*
140-859; (with Chlortetracycline Calcium Complex) Aureomycin ® / Bio-Cox ®; *Roche Vitamins.*
140-867; (with Chlortetracycline Calcium Complex, Roxarsone) Aureomycin ® / Bio-Cox ® / 3-Nitro ®; *Roche Vitamins.*
200-075; Sacox ® Type A Medicated Article; *Hoechst-Roussel Vet.*
200-080; (with Bambermycins, Roxarsone) Sacox ® + 3-Nitro ® + Flavomycin ®; *Hoechst-Roussel Vet.*
200-081; (with Bacitracin Methylene Disalicylate, Roxarsone) Sacox ® + 3-Nitro ® + BMD ®; *Hoechst-Roussel Vet.*
200-082; (with Bacitracin Methylene Disalicylate, Roxarsone) Sacox ® + BMD ®; *Hoechst-Roussel Vet.*
200-083; (with Bambermycins) Sacox ® + Flavomycin ®; *Hoechst-Roussel Vet.*
200-086; (with Bacitracin Zinc, Roxarsone) Sacox ® + Albac ® + 3-Nitro ®; *Hoechst-Roussel Vet.*
200-089; (with Bacitracin Zinc) Sacox ® + Baciferm ®; *Hoechst-Roussel Vet.*
200-090; (with Lincomycin, Roxarsone) Sacox ® + Lincomix ® + 3-Nitro ®; *Hoechst-Roussel Vet.*
200-091; (with Chlortetracycline Calcium Complex, Roxarsone) Sacox ® + 3-Nitro ® + Aureomycin ®; *Hoechst-Roussel Vet.*

200-092; (with Virginiamycin) Sacox ® + Stafac ®; *Hoechst-Roussel Vet.*
200-093; (with Lincomycin) Sacox ® + Lincomix ®; *Hoechst-Roussel Vet.*
200-094; (with Roxarsone, Virginiamycin) Sacox ® + Stafac ® + 3-Nitro ®; *Hoechst-Roussel Vet.*
200-095; (with Oxytetracycline, Roxarsone) Sacox ® + Aureomycin ®; *Hoechst-Roussel Vet.*
200-096; (with Oxytetracycline, Roxarsone) Sacox ® + Terramycin ®; *Hoechst-Roussel Vet.*
200-097; (with Roxarsone) Sacox ® + 3-Nitro ®; *Hoechst-Roussel Vet.*
200-143; (with Bacitracin Zinc, Roxarsone) Sacox ® + 3-Nitro ® + Baciferm ®; *Hoechst-Roussel Vet.*
200-204; (with Bacitracin Zinc) Bio-Cox ® / Albac ®; *Alpharma.*
200-209; (with Bacitracin Zinc, Roxarsone) Sacox ® / Albac ® / 3-Nitro ®; *Alpharma.*
200-210; (with Bacitracin Zinc) Sacox ® / Albac ®; *Alpharma.*
200-215; (with Bacitracin Zinc, Roxarsone) Bio-Cox ® / Albac ® / 3-Nitro ®; *Alpharma.*
200-259; (with Chlortetracycline, Roxarsone) ChlorMax ™ / Sacox ® / 3-Nitro ®; ChlorMax ™ / Bio-Cox ® / 3-Nitro ®; *Alpharma.*
200-260; (with Chlortetracycline, Roxarsone) ChlorMax ™ / Sacox ® / 3-Nitro ®; ChlorMax ™ / Bio-Cox ® / 3-Nitro ®; *Alpharma.*
200-261; (with Chlortetracycline) ChlorMax ™ / Sacox ®; ChlorMax ™ / Bio-Cox ®; *Alpharma.*
200-262; (with Chlortetracycline) ChlorMax ™ / Sacox ®; *Hoechst-Roussel Vet.*

560 Semduramicin Sodium

119068-77-8 8587

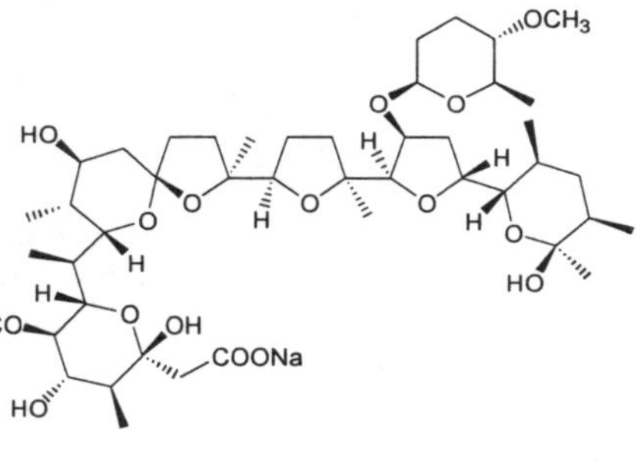

$C_{45}H_{75}NaO_{16}$
[3R,4S,5S,6R,7S,22S(2S,5S,6R)]-23,27-Didemethoxy-2,6,22-tridemethyl-5,11-di-O-demethyl-6-methoxy-22-[(tetrahydro-5-methoxy-6-methyl-2H-pyran-2-yl)oxy]lonomycin A sodium salt.
Aviax; UK-61689-2. Coccidiostat. Crystals; mp = 175-176°; $[\alpha]_D^{25}$ = +25.3° (c = 0.5 MeOH).

Veterinary products:

140-940; Aviax ™ Premix for Broiler Chickens; *Pfizer.*
141-058; (with Bacitracin Methylene Disalicylate, Roxarsone) Aviax ™ / BMD ® / 3-Nitro ®; *Pfizer.*
141-065; (with Bacitracin Methylene Disalicylate) Aviax ™ / BMD ®; *Pfizer.* 141-066; (with Roxarsone) Aviax ™ / 3-Nitro ®; *Pfizer.*
141-114; (with Virginiamycin) Aviax ™ / Stafac ®; *Pfizer.*

561 Sulfanitran

122-16-7 9103

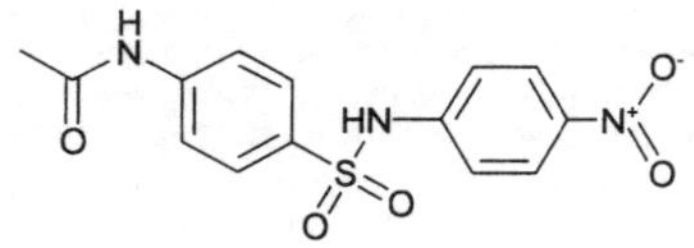

$C_{14}H_{13}N_3O_5S$
4'-[(p-Nitrophenyl)sulfamoyl]acetanilide.
NSC-77120; APNPS. Sulfonamide antibiotic. Also used as a coccidiostat in poultry. mp = 239-240°, 264°; freely soluble in Me_2CO, soluble in hot EtOH, MeOH, sparingly soluble in H_2O, Et_2O.

Veterinary products:

011-141; (with Nitromide) Unistat-2 Type A Medicated Premix; *Fort Dodge Animal Health Divn., Am. Cyanamid.*
014-250; (with Aklomide) Novastat Type A Medicated Premix; *Fort Dodge Animal Health Divn., Am. Cyanamid.*
034-537; (with Aklomide, Roxarsone) Novastat-3 Type A Medicated Article; *Fort Dodge Animal Health Divn., Am. Cyanamid.*
035-388; (with Aklomide) Novastat-W; *Fort Dodge Animal Health Divn., Am. Cyanamid.*
039-666; (with Nitromide, Roxarsone) Unistat-3 Type A Medicated Article; *Fort Dodge Animal Health Divn., Am. Cyanamid.*

562 Sulfaquinoxaline

59-40-5 9109 200-423-2

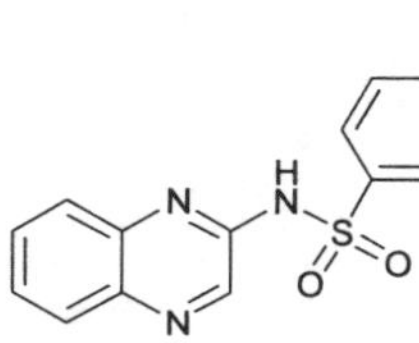

$C_{14}H_{12}N_4O_2S$
N^1-2-Quinoxalinylsulfanilamide.
Compd. 3-120; S.Q.; Sulquin; component of: Sulquin 6-50 Concentrate. Sulfonamide antibiotic. Also used as a coccidiostat in poultry. Crystals; mp = 247-248°; λ_m = 252, 360 nm ($E^{1\%}_{1\,cm}$ 1110, 275, H_2o, pH 6.6); soluble in H_2O at pH 7 (0.00075 g/100 ml), 95% EtOH (0.073 g/100 ml), Me_2CO (0.43 g/100 ml), aqueous $NaHCO_3$, NaOH.

Veterinary products:

006-391; S.Q. 40% Medicated Feed; *Hess & Clark.*
006-677; S.Q. 20% Solution; *Hess & Clark.*
007-087; Sulfaquinoxaline Solubilized; *Hess & Clark.*

563 Sulfaquinoxaline Sodium

9109

$C_{14}H_{11}NaN_4O_2S$
N^1-2-Quinoxalinylsulfanilamide sodium salt.
Aviochina. Sulfonamide antibiotic. Also used as a coccidiostat in poultry. Amorphous, deliquescent solid;

very soluble in H_2O, pH of aqueous solution = 10; concerted to the free acid by atmospheric CO_2.

Veterinary products:

006-707; Sulquin ® 6-50; *Fort Dodge Animal Health Divn., Am. Cyanamid.*
006-891; Liquid Sul-Q-Nox; *Russell.*
007-076; Sulfa-Nox Liquid; *PM Resources.*
008-244; Sulfa-Nox Concentrate; *PM Resources.*

564 Zoalene
148-01-6 3321 205-706-4

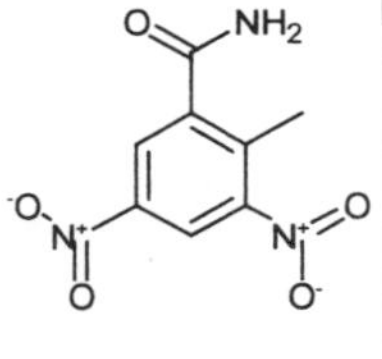

$C_8H_7N_3O_5$
2-Methyl-3,5-dinitrobenzamide.
Zoamix; 3,5-dinitro-o-toluamide; Dinitolmide. Coccidiostat. Crystals; mp = 181°.

Veterinary products:

011-116; Zoamix ® Type A Medicated Article; *Alpharma.*
013-747; Zoalene 90 Medicated Coccidiostat; *Alpharma.*
038-241; (with Arsanilic Acid, Erythromycin Thiocyanate) Erythro ® (High Lev) / Zoalene + Arsanilic Acid; *Merial.*
038-879; (with Carbarsone) Carb-O-Sep ® / Zoamix®; *Alpharma.*
048-954; (with Lincomycin Hydrochloride Monohydrate) Lincomix ® / Zoamix; *Pharmacia & Upjohn.*
101-628; (with Bambermycins, Roxarsone) Flavomycin ® + 3-Nitro + Zoalene; *Hoechst-Roussel Vet.*
101-629; (with Bambermycins) Flavomycin ® + Zoalene; *Hoechst-Roussel Vet.*
141-085; (with Bacitracin Methylene Disalicylate) Zoamix ® / BMD ®; *Alpharma.*

Contraceptives

565 Medroxyprogesterone
520-85-4 5838 208-298-6

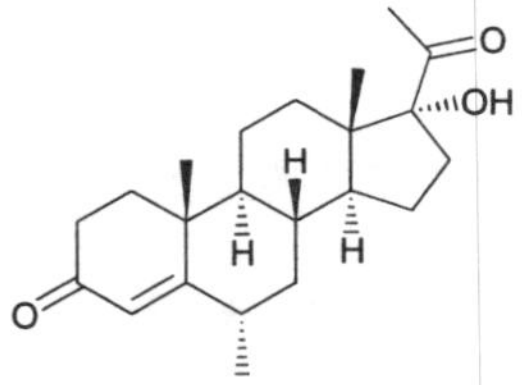

$C_{22}H_{32}O_3$
17-Hydroxy-6α-methylpregn-4-en-3,20-dione.
Progestogen. Used in combination with an estrogen (ethinyl estradiol) as an oral contraceptive. In veterinary medicine, used to manage sexual behavior problems. mp = 220-223.5°; $[\alpha]_D^{25}$ = 75° ($CHCl_3$); λ_m 241 nm (ε 16000 EtOH).

566 Medroxyprogesterone Acetate
71-58-9 5838 200-757-9

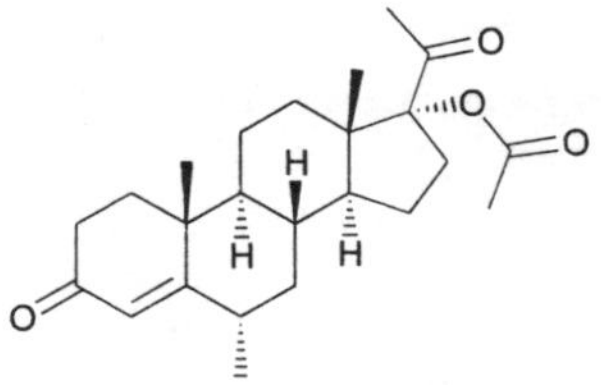

$C_{24}H_{34}O_4$
17-Hydroxy-6α-methylpregn-4-en-3,20-dione acetate.
Curretab; Cycrin; Depo-Provera; Provera; MAP; Amen; Clinovir; Depo-Clinivir; Farlutal; Gestapuran; G-Farlutal; Hysron; Lutoral; Nadigest; Nidaxin; Oragest; Perlutex; Prodasone; Provera; Sodelut G; Veramix. Progestogen. Used in combination with an estrogen (ethinyl estradiol) as an oral contraceptive. In veterinary medicine, used to manage sexual behavior problems. mp = 207-209°; $[\alpha]_D$ = 61° ($CHCl_3$); λ_m = 240 nm (ε 15900 EtOH).

Decongestants

567 d-Ephedrine Sulfate
3645

$C_{20}H_{32}N_2O_6$
D-Erythro-2-(Methylamino)-1-phenylpropan-1-ol sulfate (2:1) salt.
Afrinol. Adrenergic agonist. A stereoisomer of ephedrine. Used to treat incontinence in dogs and cats.

568 Phenylephrine
59-42-7 7440 200-424-8

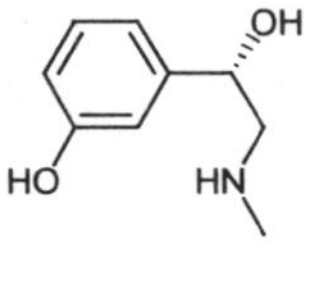

$C_9H_{13}NO_2$
(R)-3-Hydroxy-α-[(methylamino)methyl]-benzenemethanol.
Mydriatic; decongestant. α-Adrenergic agonist. Used to treat hypotension and shock, particularly if cardiac stimulation is not desired. mp = 169-172°.

569 Phenylephrine Hydrochloride
61-76-7 7440 200-517-3

$C_9H_{14}ClNO_2$
(R)-3-Hydroxy-α-[(methylamino)methyl]-benzenemethanol hydrochloride.
metaoxedrin; Adrianol; Ak-Dilate; Ak-Nefrin; Isophrin; m-Sympatol; Mezaton; Neophryn; Neo-Synephrine; Pyracort D; Prefrin; Mydfrin; Alcon Efrin; Biomydrin; Nostril; component of: Cerose DM, Comhist, Cyclomydril, Demazin, Dimetane Decongestant, Dristan Cold, Dristan Nasal Mist, ENTEX; component of Histalet Forte, Hycomine Compound, Isopto Frin, Naldecon,

Novahistine, Op-Isophrin-Z, Phenergan VC, Prefrin-A, PV Tussin Syrup, Relief, Ru-Tuss Liquid, Ru-Tuss Tablets, Vasosulf, Zincfrin. Mydriatic; decongestant. α-Adrenergic agonist. Used to treat hypotension and shock, particularly if cardiac stimulation is not desired. mp = 140-145°; $[\alpha]_D^{25}$ = -46.2° – -47.2°; soluble in H_2O, EtOH; LD_{50} (mus ip) = 17 mg/kg, (mus sc) = 33 mg/kg.

570 Phenylpropanolamine

14838-15-4 7461 238-900-2

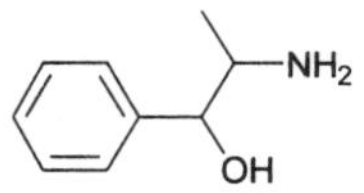

$C_9H_{13}NO$

(1RS,2SR)-2-Amino-1-phenylpropan-1-ol.

(±)-norephedrine; dl-norephedrine. Decongestant; anorexic. Sympathomimetic amine. Used to treat incontinence in dogs and cats. [(dl)-form]: mp = 101-101.5°.

571 Phenylpropanolamine Hydrochloride

154-41-6 7461 205-826-7

$C_9H_{14}ClNO$

(1RS,2SR)-2-Amino-1-phenylpropan-1-ol hydrochloride.

(±)-norephedrine hydrochloride; dl-norephedrine hydrochloride; mydriatin; Kontexin; Monydrin; Obestat; Propadrine; component of: A.R.M., Contac, Contac Severe Cold Formula, Cordicin, Corsym Capsules, Demazin, Diet Gard, Dimetapp, Drize, Endecon, ENTEX, Histabid Duracap, Histalet Forte, Hycomine, Hycomine Pediatric, Naldecon, Naldecon-CX, Naldecon-DX, Naldecon-EX, Naldetuss, Ornade, Tuss-Ornade, Rhinex D-Lay Tablets, Ru-Tuss Caplets, Tablets, Sine-Off Medicine, Sinubid, Trind, Trind-DM. Decongestant; anorexic. Sympathomimetic amine. Used to treat incontinence in dogs and cats. mp = 190-194°; pKa 9.44; soluble in H_2O, EtOH; nearly insoluble in Et_2O, $CHCl_3$, C_6H_6; LD_{50} (rat orl) = 1490 mg/kg.

Diagnostic Agents

572 Diatrizoate Meglumine

131-49-7 5851 205-024-7

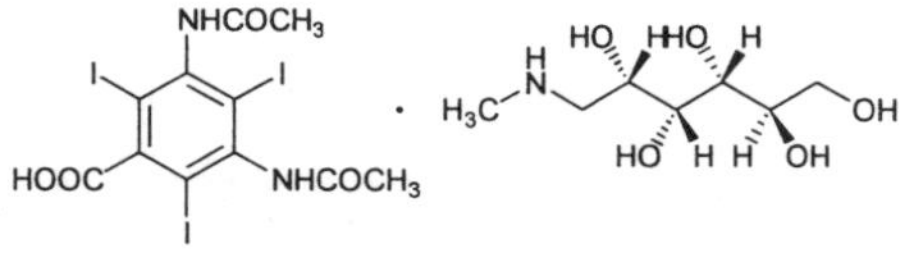

$C_{18}H_{26}I_3N_3O_9$

1-Deoxy-1-(methylamino)-D-glucitol 3,5-bis(acetylamino)-2,4,6-triiodobenzoate salt.

Angiografin; Cardiografin; Cystografin; Hypaque Cysto; Hypaque Meglumine; Renografin; Reno M; Urovist; urografic acid methylglucamine salt; meglumine amidotrizoate. Radiopaque medium; diagnostic aid. Rhombic needles; mp = 189-193°; soluble in H_2O (89 g/100 ml at 20°).

Veterinary products:

091-192; (with Diatrizoate Sodium) Renografin ®-76; *Fort Dodge Animal Health Divn., Am. Cyanamid.*

091-240; (with Diatrizoate Sodium) Renovist; *Fort Dodge Animal Health Divn., Am. Cyanamid.*

091-327; (with Diatrizoate Sodium) Gastrografin ®; *Fort Dodge Animal Health Divn., Am. Cyanamid.*

573 Diatrizoate Sodium

737-31-5 3040 212-004-1

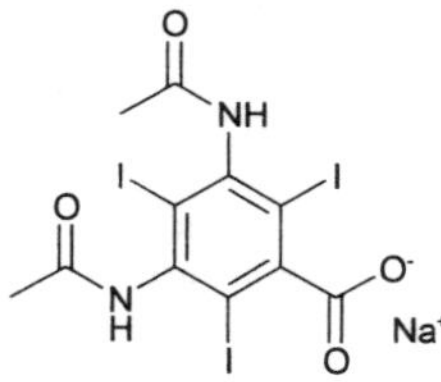

$C_{11}H_8I_3N_2NaO_4$

3,5-Bis(acetylamino)-2,4,6-triiodobenzoic acid sodium salt.

Hypaque; Sodium amidotrizoate; 3,5-Diacetamido-2,4,6-triiodobenzoic acid sodium salt; sodium diatrizoate USP; Triognost; urografic acid sodium salt; sodium diatrizoate. Radiopaque medium used as a diagnostic aid Rhombic needles; mp = 261-262° (dec); soluble in H_2O (60 g/100 ml at 20°); autoclavable; LD_{50} (rat iv) = 14700 mg/kg.

Veterinary products:

091-192; (with Diatrizoate Meglumine) Renografin ®-76; *Fort Dodge Animal Health Divn., Am. Cyanamid.*

091-240; (with Diatrizoate Meglumine) Renovist; *Fort Dodge Animal Health Divn., Am. Cyanamid.*

091-327; (with Diatrizoate Meglumine) Gastrografin ®; *Fort Dodge Animal Health Divn., Am. Cyanamid.*

574 Oleate Sodium

143-19-1 6965 205-591-0

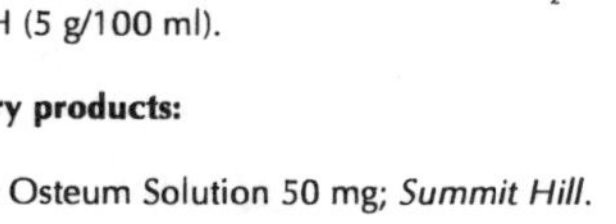

$C_{18}H_{33}NaO_2$

Sodium (Z)-9-octadecenoate.

Eunatrol. Aid in diagnosis of pancreatic disfunction White powder with tallow-like odor; soluble in H_2O (10 g/100 ml), EtOH (5 g/100 ml).

Veterinary products:

099-344; Osteum Solution 50 mg; *Summit Hill.*

Digestive Aids

575 Pancrelipase

53608-75-6 7138 258-659-7

Accelerase; Cotazym; Ilozyme; Ku-Zyme HP; Pancrease; Viokase. Digestive aid. Enzyme concentrate containing primarily lipase, and also amylase and proteases. Used to combat pancreatic enzyme deficiency.

Diuretics

576 Acetazolamide Sodium
1424-27-7 50

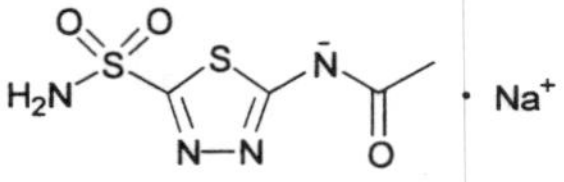

$C_4H_5N_4NaO_3S_2$
N-[5-(Aminosulfonyl)-1,3,4-thiadiazol-2-yl]acetamide, sodium salt.
Diamox Parenteral; Vetamox; component of 011-582, Vetamox Soluble Powder (*American Cyanamid*). Animal diuretic and carbonic anhydrase inhibitor.

Veterinary products:

011-582; Vetamox Soluble Powder; *American Cyanamid Divn., AHP Corp.*

577 Ammonium Chloride
12125-02-9 537 235-186-4
H_4ClN
Hydrochloric acid, ammonium salt.
ammonium muriate; sal ammoniac; salmiac; Amchlor; Darammon. Used as a urinary acidifier. Colorless crystals; d^{25} = 1.5274; soluble in H_2O (22.9 g/100 ml at 0°, 26.0 g/100 ml at 15°, 28.3 g/100 ml at 25°, 39.6 g/100 ml at 80°; pH of 1% aqueous solution at 25° = 5.5, 3% = 5.1, 10% = 5.0; soluble in EtOH, MeOH, insoluble in Me_2CO, Et_2O, EtOAc; LD_{50} (rat im) = 30 mg/kg.

Veterinary products:

009-339; (with Caramiphen Edisylate) Carafen Cough Syrup; Carafen Cough Tablets; *Fort Dodge Animal Health, Divn. AHP.*

578 Chlorothiazide
58-94-6 2221 200-404-9

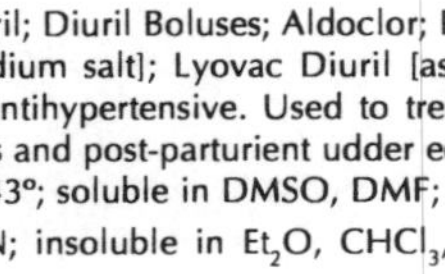

$C_7H_6ClN_3O_4S_2$
6-Chloro-2H-1,2,4-benzothiadiazine-7-sulfonamide 1,1-dioxide.
Chlotride; Diuril; Diuril Boluses; Aldoclor; Diupres; Diuril Lyovac [as sodium salt]; Lyovac Diuril [as sodium salt]. Diuretic and antihypertensive. Used to treat a variety of animal edemas and post-parturient udder edema in cows. mp = 342.5-343°; soluble in DMSO, DMF; less soluble in MeOH, C_5H_5N; insoluble in Et_2O, $CHCl_3$, C_6H_6; poorly soluble in H_O.

Veterinary products:

011-678; Diuril ® Tablet; *Merial.*
012-734; Diuril ®; *Merial.*

579 Ethacrynate Sodium
6500-81-8 3761

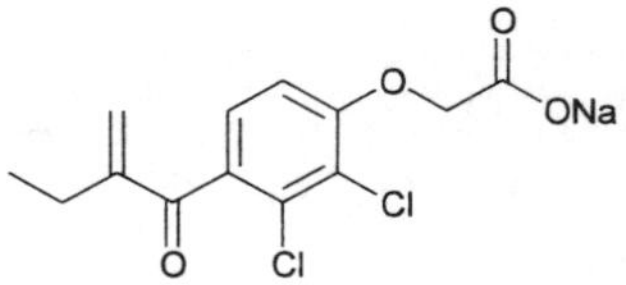

$C_{13}H_{11}Cl_2NaO_4$
Sodium [2,3-dichloro-4-(2-methylenebutyryl)-phenoxy]acetate.
Edecrin sodium; Lyovac Sodium Edecrin. A high-ceiling, or loop diuretic. Has been largely replaced in veterinary medicine by furosemide. λ_m = 225 nm (ε 15287 H_2O); soluble in H_2O (< 9 g/100 ml).

580 Ethacrynic Acid
58-54-8 3761 200-384-1

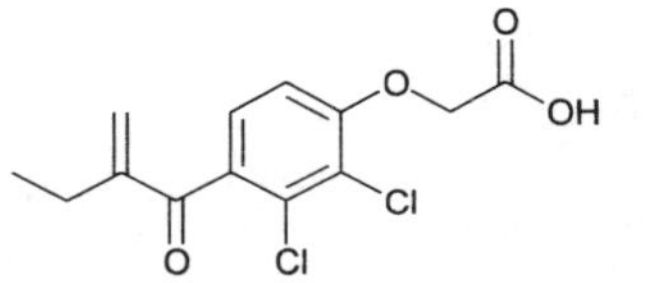

$C_{13}H_{12}Cl_2O_4$
[2,3-Dichloro-4-(2-methylenebutyryl)phenoxy]-acetic acid.
Edecril; Edecrin; MK-595; Crinuril; Endecril; Hydromedin; Reomax; Taladren; Uregit. A high-ceiling, or loop diuretic. Has been largely replaced in veterinary medicine by furosemide. mp = 121-122°; sparingly soluble in H_2O, soluble in $CHCl_3$; LD_{50} (mus iv) = 176 mg/kg, (mus orl) = 627 mg/kg.

581 Furosemide
54-31-9 4830 200-203-6

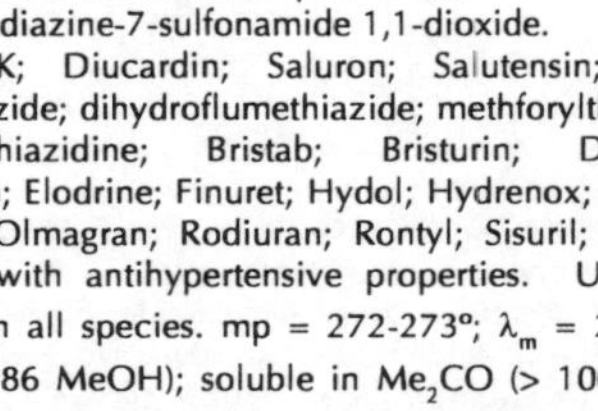

$C_8H_8F_3N_3O_4S_2$
3,4-Dihydro-6-(trifluoromethyl)-2H-1,2,4-benzothiadiazine-7-sulfonamide 1,1-dioxide.
Diumide-K; Diucardin; Saluron; Salutensin; Hydroflumethiazide; dihydroflumethiazide; methforylthiazidine; metflorylthiazidine; Bristab; Bristurin; Di-Ademil; Diucardin; Elodrine; Finuret; Hydol; Hydrenox; Leodrine; NaClex; Olmagran; Rodiuran; Rontyl; Sisuril; Vergonil. Diuretic with antihypertensive properties. Used as a diuretic in all species. mp = 272-273°; λ_m = 272.5 nm (log ε 4.286 MeOH); soluble in Me_2CO (> 100 mg/ml), MeOH (58 mg/ml), CH_3CN (43 mg/ml), H_2O (0.3 mg/ml), Et_2O (0.2 mg/ml), C_6H_6 (< 0.1 mg/ml); LD_{50} (mus orl) > 8000 mg/kg, (mus iv) = 750 mg/kg, (mus ip) = 6280 mg/kg.

Veterinary products:

034-478; Lasix ® Injectable Solution; *Hoechst-Roussel Vet.*
034-621; Lasix ® Tablets 50 mg; *Hoechst-Roussel Vet.*
045-188; Lasix ® Packets; *Hoechst-Roussel Vet.*
102-380; Lasix ® Syrup 1%; *Hoechst-Roussel Vet.*
118-550; Furos-A-Vet; *Anthony Products.*
127-034; Disal ® Injection; *Boehringer Ingelheim Vetmedica.*
129-034; Disal ®; *Boehringer Ingelheim Vetmedica.*
131-538; Disal ®; *Boehringer Ingelheim Vetmedica.*
131-806; Furosemide Tablets; *Teva.*

582 Mannitol

69-65-8 5788 200-711-8

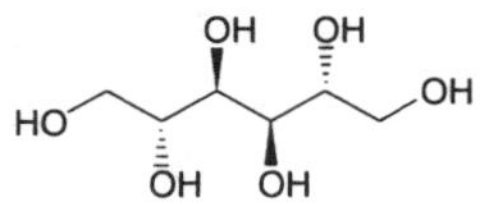

$C_6H_{14}O_6$
D-Mannitol.
Osmitrol; Resectisol; SDM-25. Osmotic diuretic, useful in treating renal failure; also used as a renal function diagnostic aid. mp = 166-168°; $bp_{3.5}$ = 290-295°; d^{20} = 1.52; $[\alpha]_D^{20}$= 23° (borax solution); soluble in H_2O (182 g/l), EtOH (12 g/l).

583 Sodium Polystyrene Sulfonate

9003-59-2 8815
Sulfonated divinylbenzene copolymer with styrene sodium salt.
Amberlite IRP-69 Resin; Kayexalate. Can remove potassium ions selectively and is useful in treatment of hyperkalemia in dogs.

584 Spironolactone

52-01-7 8917 200-133-6

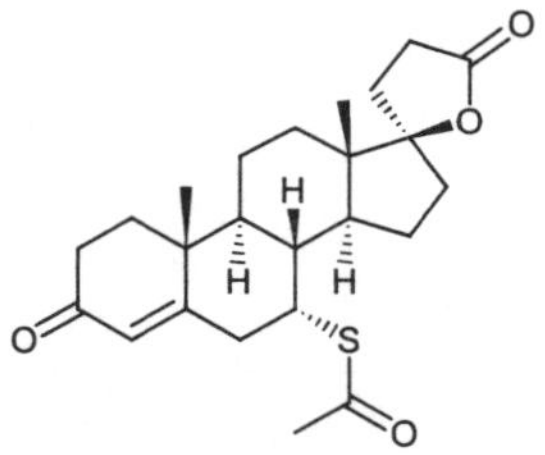

$C_{24}H_{32}O_4S$
17-Hydroxy-7α-mercapto-3-oxo-17α-pregn-4-ene-21-carboxylic acid σ-lactone acetate.
Abbolactone; Aldactone; Aldactazide; SC-9420; Aldace; Aldopur; Almatol; Altex; Aquareduct; Deverol; Diatensec; Dira; Duraspiron; Euteberol; Lacalmin; Lacdene; Laractone; Nefurofan; Osiren; Osyrol; Sagisal; Sincomen; Spiretic; Spiroctan; Spiroderm; Spirolone; Spiro-Tablinen; Supra-Puren; Suracton; Urusonin; Verospiron; Xenalon. Diuretic. Aldosterone antagonist. Potassium sparing. Used for edema in cirrhosis of the liver, nephrotic syndrome, congestive heart failure, potentiation of thiazide and loop diuretics, hypertension and Conn's syndrome. Has found limited veterinary use. mp = 134-135°, 201-202°; $[\alpha]_D^{20}$= -33.5° ($CHCl_3$); λ_m = 238 nm (ε 20200); insoluble in H_2O, soluble in organic solvents.

585 Trichlormethiazide

133-67-5 9754 205-118-8

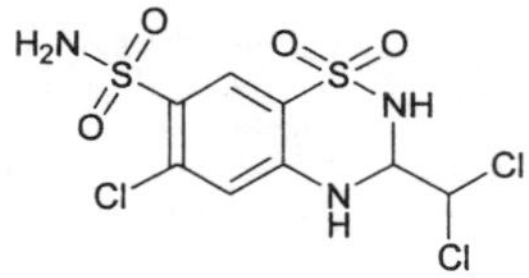

$C_8H_8Cl_3N_3O_4S_2$
6-Chloro-3-(dichloromethyl)-3,4-dihydro-2H-1,2,4-thiadiazine-7-sulfonamide 1,1-dioxide.
Metahydrin; Naqua; Metatensin; Naquasone; Naquival. Diuretic and antihypertensive. Used as a veterinary diuretic. mp= 248-250°, 266-273°; soluble in H_2O (0.8 mg/ml), EtOH (21 mg/ml), MeOH (60 mg/ml); LD_{50} (rat orl) > 20000 mg/kg.

Veterinary products:

030-136; (with Dexamethasone) Naquasone® Bolus; *Schering-Plough Animal Health.*

Ectoparasiticides

586 Amitraz

33089-61-1 510 251-375-4

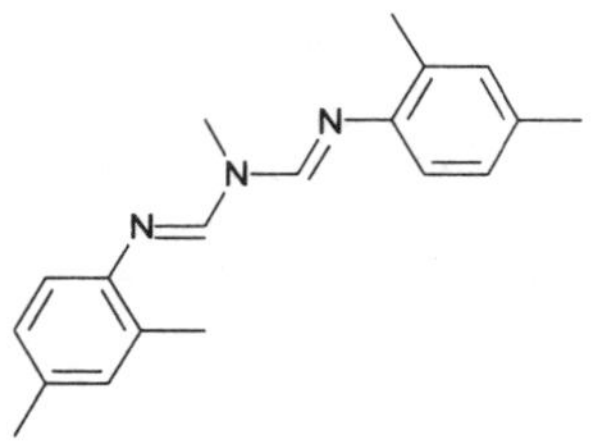

$C_{19}H_{23}N_3$
N-Methyl-N'-2,4-xylyl-N-(N-2,4-xylylformimidoyl)-formamidine.
U-36059; BTS-27419; ENT-27967; BAAM; Mitaban; Mitac; Taktic. Ectoparasiticide and scabicide. mp = 86-87°; insoluble in H_2O, soluble in most organic solvents; LD_{50} (mus orl) = 1600 mg/kg; (rat orl) = 400 mg/kg.

Veterinary products:

120-299; Mitaban® Liquid Concentrate; *Pharmacia & Upjohn.*

587 Cythioate
115-93-5 2854 204-115-9

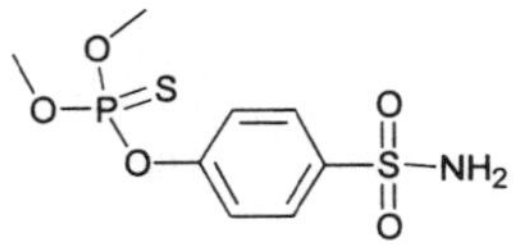

$C_8H_{12}NO_5PS_2$
Dimethyl O-(p-sulfamoylphenyl) phosphorothioate.
Cyflee; Cythioate; O,O-Dimethyl O-p-sulfamoylphenylphosphorothioate; Phosphorothioic acid, O-(4-(aminosulfonyl)phenyl) O,O-dimethyl ester; Proban. Ectoparasiticide. Used to control fleas in dogs. Crystals; mp = 70-71°; LD_{50} (rat orl) = 160 mg/kg.

Veterinary products:

033-342; Proban ® Cythioate Tablets 30 mg; *American Cyanamid Divn., AHP Corp.*
033-606; Proban ® Oral Liquid; *American Cyanamid Divn., AHP Corp.*
132-336; Proban ® Oral Liquid; *Bayer.*
132-337; Proban ® Tablets; *Bayer.*

588 Eprinomectin B_{1b}
123997-26-2 3667

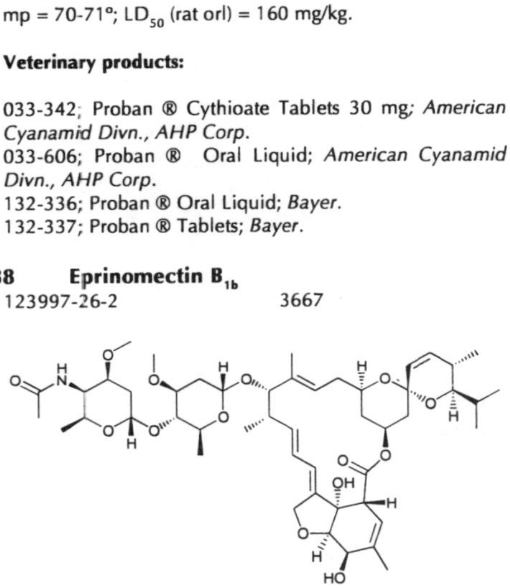

$C_{49}H_{73}NO_{14}$
(4R)-4-(acetylamino)-5-O-demethyl-25-de(1-methylpropyl)-4-deoxy-25-(1-methylethyl)avermectin A_{1a}.
MK-397. Antiparasitic. Used in cattle for various of roundworms and grubs. Crystals; mp = 163.3-165.7°; LD_{50} (mus orl) = 24 mg/kg.

Veterinary products:

141-079; Ivomec® Eprinex™Pour-On for Beef and Dairy Cattle; Ivomec® Eprinex™Pour-On for Cattle; *Merial.*

589 Famphur
52-85-7 3973 200-154-0

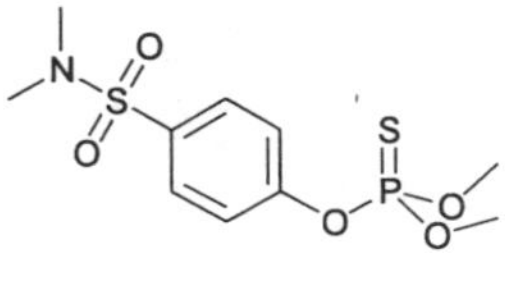

$C_{10}H_{16}NO_5PS_2$
O,O-Dimethyl-O,p-(dimethylsulfamoyl)phenyl phosphorothioate.
Am. Cyanamid 38023; ENT-25644; Warbex. Cholinesterase inhibitor used as a grubicide and insecticide Crystals; mp = 52.5-53.5°; LD_{50} (rat orl) = 35 mg/kg.

Veterinary products:

034-266; Famix Famphur Premix; *Schering-Plough Animal Health.*
034-697; Bo-Ana Famphur Cattle Insect.; Warbex ® Famphur Cattle Pour-on; *Schering-Plough Animal Health.*
043-215; Purina ® Grub-Kill; *PM Resources.*
139-858; (with Levamisole Resinate) Tramisol ®-X-Tra Cattle Anthelmintic and Extoparasitic Paste; *Schering-Plough Animal Health.*

590 Fenthion
55-38-9 4044 200-231-9

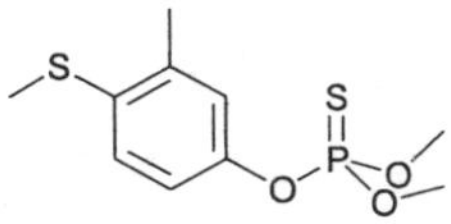

$C_{10}H_{15}O_3PS_2$
O,O-Dimethyl O-[4-(methylthio)-m-tolyl] phosphorothioate.
Bayer 29493; ENT-25540; S-1752; Baycid; Baytex; Entex; Lebaycid; Mercaptophos; Queletox; Talodex; Tiguvon. Ectoparasiticide used for the topical treatment of fleas in dogs, in cattle to control lice and cattle grub and in pigs to control lice. Liquid with garlic odor; $bp_{0.01}$ = 87°; d_4^{20} = 1.250; stable < 210°; soluble in MeOH, EtOH, Et_2O, Me_2CO and other organic solvents, poorly soluble in H_2O (0.0055 g/100 ml); LD_{50} (mrat orl) = 215 mg/kg, (frat orl) = 245 mg/kg.

Veterinary products:

034-641; Tiguvon Pour-On Cattle Insecticide; *Bayer.*
047-138; Spotton 20% Ready-to-Use Cattle Insecticide; *Bayer.*
132-789; Pro-Spot Solution; *Bayer.*

591 Lufenuron
103055-07-8 5624 410-690-9

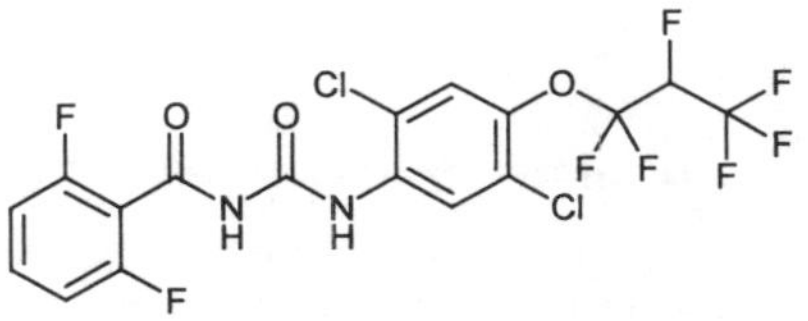

$C_{17}H_8Cl_2F_8N_2O_3$
1-[2,5-Dichloro-4-(1,1,2,3,3,3-hexafluoropropoxy)-phenyl]-3-(2,6-difluorobenzoyl)urea.
fluphenacur; CGA-184699; Program. Ectoparaciticide used in veterinary medicine for control of fleas. mp = 174°; insoluble in water; soluble in organic solvents; LD_{50} (rat orl) > 2000 mg/kg; LC_{50} (rat ihl) > 2350 mg/m^3).

Veterinary products:

141-026; Program ® Suspension; *Novartis Animal Health.*
141-035; Program ®; Program ® Flavor Tabs ™; *Novartis Animal Health.*
141-062; Program ® Cat Flavor Tabs ™; *Novartis Animal Health.*
141-084; (with Milbemycin Oxime) Sentinel ®; Sentinel ® Flavor Tabs ®; *Novartis Animal Health.*
141-105; Program ® 6 Month Injectable for Cats; *Novartis Animal Health.*

592 Phosmet

732-11-6 7492 211-987-4

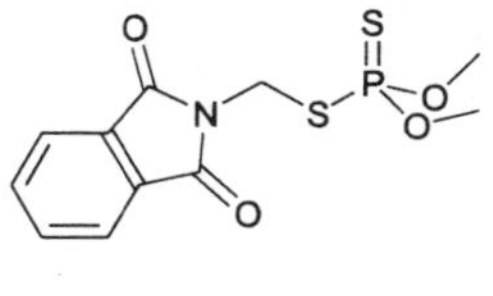

$C_{11}H_{12}NO_4PS_2$
Phosphorodithioic acid S-[(1,3-dihydro-1,3-dioxo-2H-isoindol-2-yl)methyl] O,O-dimethyl ester.
phthalophos; ENT-25705; R-1504; Imidan; Prolate.
Insecticide and acaricide. Off-white crystals; mp = 71.9°; slightly soluble in H_2O (0.0025 g/100 ml); LD_{50} (mrat orl) = 113 mg/kg, (frat orl) = 160 mg/kg.

Veterinary products:

044-757; Prolate I-E; *Schering-Plough Animal Health.*

593 Piperonyl Butoxide

51-03-6 7629 200-076-7

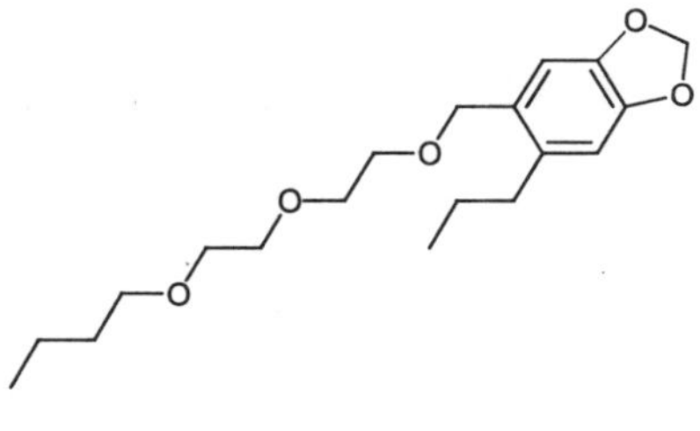

$C_{19}H_{30}O_5$
5-[[2-(2-Butoxyethoxy)ethoxy]methyl]-6-propyl-1,3-benzodioxole.
ENT-14250; Buitacide. Synergist for insecticides, especially pyrethroids. Liquid; $bp_{1.0}$ = 180°; d = 1.04 - 1.07; soluble in MeOH, EtOH, C_6H_6, Freons, Geons and other organic solvents and oils; LD_{50} (mrat orl) = 7500 mg/kg, (frat orl) = 6150 mg/kg.

Veterinary products:

032-984; (with Pyrethrins, Squalene) Cerumite; *Evsco.*

594 Pyrethrin I

121-21-1 8148 204-455-8

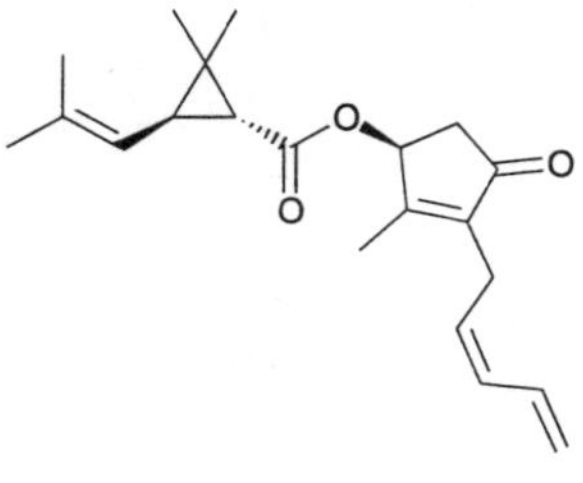

$C_{21}H_{28}O_3$
[1R-[1α[S*(Z)],3β]]-2,2-Dimethyl-3-(2-methyl-1-propenyl)cyclopropanecarboxylic acid 2-methyl-4-oxo-3-(2,4-pentadienyl)-2-cyclopenten-1-yl ester.
chrysanthemum monocarboxylic acid pyrethrolone ester.
Insecticide. Viscous liquid; $bp_{0.0005}$ = 146-150°; $[\alpha]_D^{20}$ = -14° (isooctane); λ_m = 225 nm (ε 36400, 95% EtOH); soluble in EtOH, petroleum ether, kerosene, CCl_4, CH_2Cl_2, CH_3NO_2, insoluble in H_2O.

Veterinary products:

032-984; (with Piperonyl Butoxide, Squalene) Cerumite; *Evsco.*

595 Pyrethrin II

121-29-9 8148 204-462-6

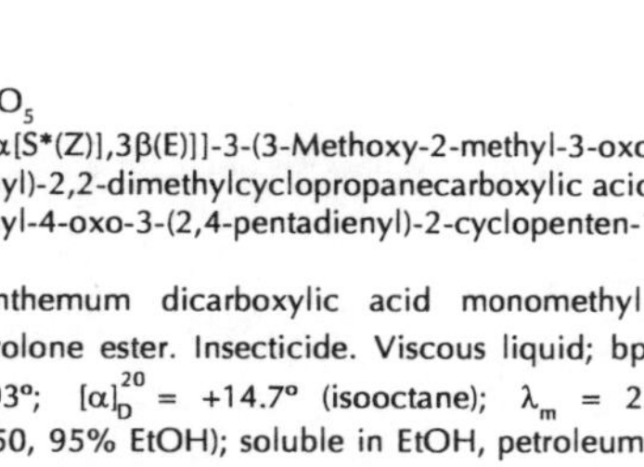

$C_{22}H_{28}O_5$
[1R-[1α[S*(Z)],3β(E)]]-3-(3-Methoxy-2-methyl-3-oxo-1-propenyl)-2,2-dimethylcyclopropanecarboxylic acid 2-methyl-4-oxo-3-(2,4-pentadienyl)-2-cyclopenten-1-yl ester.
chrysanthemum dicarboxylic acid monomethyl ester pyrethrolone ester. Insecticide. Viscous liquid; $bp_{0.007}$ = 192-193°; $[\alpha]_D^{20}$ = +14.7° (isooctane); λ_m = 229 nm (ε 45850, 95% EtOH); soluble in EtOH, petroleum ether, kerosene, CCl_4, CH_2Cl_2, CH_3NO_2, insoluble in H_2O.

Veterinary products:

032-984; (with Piperonyl Butoxide, Squalene) Cerumite; *Evsco.*

596 Trichlorfon

52-68-6 9753 200-149-3

$C_4H_8Cl_3O_4P$

Dimethyl 2,2,2-trichloro-1-hydroxyethylphosphonate.

Ectoparasiticide and anthelmintic targeting nematodes.

Veterinary products:

012-956; Dyrex Bolus; Dyrex Capsules; Dyrex Granules; Dyrex Tablets; *Fort Dodge Animal Health, Divn. AHP.*
013-248; (with Atropine) Freed No. 25; Freed No.10; *Fort Dodge Animal Health, Divn. AHP.*
015-154; (with Phenothiazine, Piperazine Dihydrochloride) Dyrex T.F - 1000; Dyrex T.F - 500; Dyrex T.F. 200; *Fort Dodge Animal Health, Divn. AHP.*
035-650; (with Atropine Sulfate) Dyrex Powder; *Fort Dodge Animal Health, Divn. AHP.*
048-915; Purina ® Bot Control; *Purina Mills.*
091-067; (with Thiabendazole) Equivet-14; Equizole ®-B; *Farnam.*
100-402; (with Mebendazole) Telmin ™ B Equine Wormer; *Schering-Plough Animal Health.*
136-740; (with Oxfendazole) Benzelmin ® Plus Paste; *Fort Dodge Animal Health, Divn. AHP.*
138-954; (with Mebendazole) Telmin ™ B Paste; *Schering-Plough Animal Health.*

Emetics

597 Apomorphine

58-00-4 787 200-360-0

$C_{17}H_{17}NO_2$

5,6,6a,7-Tetrahydro-6-methyl-4H-dibenzo[de,g]quinolin-10,11-diol.

Stimulates dopamine receptors. Used as an emetic in dogs. mp = 195°; soluble in EtOH, Me_2CO, $CHCl_3$; slightly soluble in H_2O, C_6H_6, Et_2O, petroleum ether; λ_m = 336, 399 nm; LD_{50} (mus ip) = 160 mg/kg.

598 Ipecac

8012-96-2 5086 232-385-8

Ipsatol; component of: Dasin. Dried rhizome and roots of *Caphaelis ipecauanha*. Contains 2-2.5% alkaloids, including emetine, cephaeline; emetamine; psychotrine; methyl psychotrine; protoemetine and resin. Emetic. Used to deal with overdoses and accidental poisoning.

Enzyme Cofactors

599 Ascorbic Acid

50-81-7 867 200-066-2

$C_6H_8O_6$

L-Ascorbic acid.

Ascorbicap; Cebione; Cecon; Cenolate; Cetane; Cetane-Caps TC; Cevalin; Cevex; Adenex; Allercorb; Ascorin; Ascorteal; Ascorvit; Cantan; Cantaxin; Catavin C; Cebicure; Cebion; Cecon; Cegiolan; Celaskon; Celin; Cenetone; Cereon; Cergona; Cescorbat; Cetamid; Cetebe; Cetemican; Cevalin; Cevatine; Cevex; Cevimin; Ce-Vi-Sol; Cevitan; Cevitex; Cewin; Ciamin; Cipca; Concemin; C-Vimin; Davitamon C; component of: Chromagen, Freancee, Mediatric, Stuartinic, Tolfrinic, Veliten. Vitamin, vitamin source. Sometimes used as a urinary acidifier, also to treat copper-induced hepatopathy in dogs. mp = 190-192°; d =1.65; $[\alpha]_D^{25}$= 20.5 - 21.5° (c = 1), $[\alpha]_D^{23}$ = 48° (c = 1 MeOH); λ_m = 245 nm (acid solution), 265 nm (neutral solution); soluble in H_2O (33 g/100 ml, 40 g/100 ml at 45°, 80 g/100 ml at 100°), EtOH (3.3 g/100 ml), absolute EtOH (2 g/100 ml); glycerol (1 g/100 ml), propylene glycol (5 g/100 ml), insoluble in Et_2O, $CHCl_3$, C_6H_6, petroleum ether.

600 Phylloquinone

84-80-0 7536 201-564-2

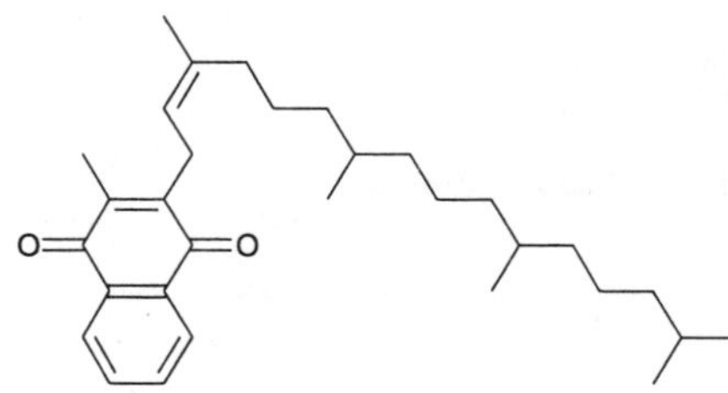

$C_{31}H_{46}O_2$

[R-[R*,R*-(E)]]-2-Methyl-3-(3,7,11,15-tetramethyl-2-hexadecenyl)-1,4-naphthalenedione.

Aquamephyton; Konakion; Mephyton; Mono-kay; Phytonadione; vitamin K_1; 3-phytylmenadione; Veda-K_1; Veta-K_1. Vitamin, vitamin source. Prothrombogenic used as an antidote for anticoagulant rodenticides such as Warfarin. $[\alpha]_D^{25}$= -0.28° (dioxane); λ_m = 242, 248, 260, 269, 325 nm ($E^{1\%}_{1\ cm}$ 396, 4319, 383, 387, 68 petroleum ether); insoluble in H_2O; sparingly soluble in MeOH; soluble in EtOH, Me_2CO, C_6H_6, petroleum ether, C_6H_{14}, dioxane, $CHCl_3$, Et_2O, fats and oils; [dihydro form (phytonadiol; dihydrovitamin K_1)]: insoluble in H_2O, sparingly soluble in petroleum ether, freely soluble in Et_2O; [dihydro form sodium diphosphate (phytonadiol sodium diphosphate; Kayhydrin)]: mp = 138°; soluble in H_2O, MeOH.

601 Thiamine
59-43-8 9430 200-425-3

$C_{12}H_{17}ClN_4OS$
3-[(4-Amino-2-methyl-5-pyrimidinyl)methyl]-5-(2-hydroxyethyl)-4-methylthiazolium chloride.
vitamin B_1, aneurin; thiaminium chloride. Vitamin (enzyme cofactor). Used for prevention and treatment of thiamine deficiency disorders and in the adjunctive treatment of lead and ethylene glycol poisoning.

602 Thiamine Diphosphate
154-87-0 9431 205-836-1

$C_{12}H_{19}ClN_4O_7P_2S$
3-[(4-Amino-2-methyl-5-pyrimidinyl)methyl]-5-(2-hydroxyethyl)-4-methylthiazolium chloride P,P'-dioxide.
Vitamin (enzyme cofactor). Used for prevention and treatment of thiamine deficiency disorders and in the adjunctive treatment of lead and ethylene glycol poisoning. mp = 238-240° (dec), 240-244° (dec); λ_m = 242 nm; soluble in H_2O.

603 Thiamine Disulfide
67-16-3 9432 200-644-4

$C_{24}H_{34}N_8O_4S_2$
N,N'-[Dithiobis[2-(2-hydroxyethyl)-1-methyl-2,1-ethenediyl]]bis[N-[(4-amino-2-methyl-5-pyrimidinyl)methyl]formamide].
aneurin disulfide; Aktivin; Neolamin. Vitamin (enzyme cofactor). Used for prevention and treatment of thiamine deficiency disorders and in the adjunctive treatment of lead and ethylene glycol poisoning. mp = 177°; sparingly soluble in C_6H_6, Me_2CO, Et_2O, EtOH.

604 Thiamine Hydrochloride
67-03-8 9430 200-641-8
$C_{12}H_{18}Cl_2N_4OS$
3-[(4-Amino-2-methyl-5-pyrimidinyl)methyl]-5-(2-hydroxyethyl)-4-methylthiazolium chloride monohydrochloride.
Betalin S; Betaxin; Biamine; Benerva; Betabion; Bewon; Metabolin; Vitaneurin; component of: Spondylonal, Troph-Iron, Trophite, Trophite+Iron. Vitamin (enzyme cofactor). Used for prevention and treatment of thiamine deficiency disorders and in the adjunctive treatment of lead and ethylene glycol poisoning. mp = 248° (dec); soluble in H_2O (100 g/100 ml), glycerol (5.5 g/100 ml), 95% EtOH (1 g/100 ml), EtOH (0.32 g/100 ml); more soluble in MeOH; soluble in propylene glycol; insoluble in Et_2O, C_6H_6, $CHCl_3$, C_6H_{14}; LD_{50} (mus iv) = 89.2 mg/kg, (mus orl) = 8224 mg/kg.

605 Thiamine Mononitrate
532-43-4 9430 208-537-4
$C_{12}H_{17}N_5O_4S$
3-[(4-Amino-2-methyl-5-pyrimidinyl)methyl]-5-(2-hydroxyethyl)-4-methylthiazolium nitrate (salt).
component of: Stuartinic. Vitamin (enzyme cofactor). Used for prevention and treatment of thiamine deficiency disorders and in the adjunctive treatment of lead and ethylene glycol poisoning. mp = 196-200° (dec); soluble in H_2O (2.7 g/100 ml at 25°, 30 g/100 ml at 100°).

606 Vitamin E
59-02-9 10159 200-412-2

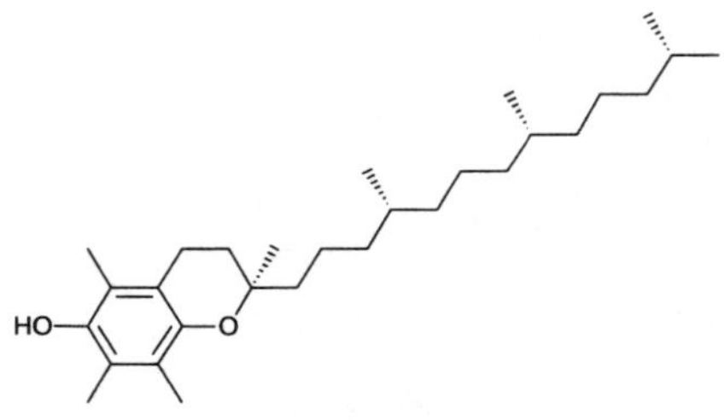

$C_{29}H_{50}O_2$
[2R-2R*(4R*,8R*)]-3,4-Dihydro-2,5,7,8-tetramethyl-2-(4,8,12-trimethyltridecyl)-2H-1-benzopyran-6-ol.
(+)-α-tocopherol; α-tocopherol; antisterility vitamin. Vitamin, vitamin source. Used with Selenium to address Selenium-tocopherol deficiency in sheep and other large animals. $[\alpha]^{25}_{5461} = -3.0°$ (C_6H_6), $[\alpha]^{25}_{5461} = 0.32°$ (EtOH).

Veterinary products:

012-635; (with Sodium Selenite) BO-SE; L-SE; *Schering-Plough Animal Health.*
030-313; (with Sodium Selenite) Seletoc ® Minicaps and Caps; *Schering-Plough Animal Health.*
030-314; (with Sodium Selenite) Mu-Se; *Schering-Plough Animal Health.*
030-315; (with Sodium Selenite) E-SE; *Schering-Plough Animal Health.*
030-316; (with Sodium Selenite) Seletoc ® Injection; *Schering-Plough Animal Health.*
200-109; (with Sodium Selenite) Velenium ™; *Fort Dodge Animal Health, Divn. AHP.*

Enzyme Inhibitors

607 Ketamine
6740-88-1 5306 229-804-1

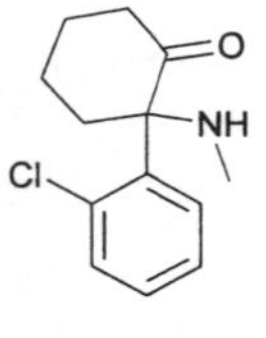

$C_{13}H_{16}ClNO$
(±)-2-(o-Chlorophenyl)-2-(methylamino)cyclohexanone. Intravenous anesthetic. Ion channel blocker. Used as an anesthetic in cats in minor surgery. mp = 92-93°; λ_m = 301, 276, 268, 261 nm ($A^{1\%}_{1\,cm}$ 5.0, 7.0, 9.8, 10.5 0.01N NaOH/95% MeOH).

608 Ketamine Hydrochloride
1867-66-9 5306 217-484-6
$C_{13}H_{17}Cl_2NO$
(±)-2-(o-Chlorophenyl)-2-(methylamino)cyclohexanone hydrochloride.
Ketaject; Ketalar; Ketaset; Ketavet; Ketanarkon; Ketanest; Vetalar; CI-581; CL369; CN-52,372-2; component of: Ketaset Plus. Intravenous anesthetic. Ion channel blocker. Used as an anesthetic in cats in minor surgery. mp = 262-263°; soluble in H_2O (20 g/100 ml); LD_{50} (rat ip) = 229 ± 5 mg/kg, (mus ip) = 224 ± 4 mg/kg.

Veterinary products:

043-304; Ketaset ®; *Fort Dodge Animal Health, Divn. AHP.*
045-290; Vetalar; *Fort Dodge Animal Health, Divn. AHP.*
092-116; (with Aminopentamide Hydrogen Sulfate, Promazine Hydrochloride) Ketaset ® Plus; *Fort Dodge Animal Health, Divn. AHP.*
200-029; Ketamine Hydrochloride Injection; *Boehringer Ingelheim Vetmedica.*
200-042; Ketaject ®; Ketamine Hydrochloride Injection USP; *Phoenix.*
200-055; VetaKet ®; *Lloyd.*
200-073; Ketamine Hydrochloride Injection, USP; *Veterinary Res. Associates.*

Enzymes

609 α-Chymotrypsin
9004-07-3 2320 232-671-2
Avazyme; Catarase; Enzeon; Zolyse; component of: Orenzyme. Proteolytic enzyme. Used topically in veterinary medicine for enzymatic debridement.

Veterinary products:

040-322; (with Hydrocortisone Acetate, Neomycin Palmitate, Trypsin) Kymar Ointment Improved; *Schering-Plough Animal Health.*

610 Pectin
9000-69-5 7194 232-553-0
Citrus pectin.
Genu® HM USP 100; Genu® HM USP L200; Genu® Pectins; Genu® Pectin (citrus) type USP/100; Genu® Pectin (citrus) type USP/200; Genu® Pectin (citrus) type USP-H; Genu® Pectin (citrus) type USP-L/200; Mexpectin LA 100 Range; Mexpectin LC 700 Range; Mexpectin XSS 100 Range. FDA GRAS, Japan, UK approved, Europe listed, FDA approved for dentals and topicals, USP/NF compliance. Used as a protective colloid, thickener, emulsifier, adsorbent, in dentals and topicals. In encapsulated drugs, diarrhea treatment; hemostatic formulations, blood plasma substitutes. Used in detoxication, dietary fiber fortification, cholesterol reduction and glucose metabolism. Used in veterinary medicine as an anti-diarrheal. White powder; soluble in H_2O, insoluble in EtOH or other organic solvents. Non-toxic, non-allergenic.

Veterinary products:

042-548; (with Aminopetamide Hydrogen Sulfate, Attapulgite; Bismuth Subcarbonate, Kanamycin Sulfate) Amforol ® Suspension; *Fort Dodge Animal Health, Divn. AHP.*
042-841; (with Attapulgite, Bismuth Subcarbonate, Kanamycin Sulfate) Amforol ® Veterinary Oral Tablets; *Fort Dodge Animal Health, Divn. AHP.*

611 Trypsin
9002-07-7 9926 232-650-8
Parenzyme; Parenzymol; Typtar; Trypure. Proteolytic enzyme crystallized from an extract of the pancreas gland of the ox.

Veterinary products:

040-322; (with Chymotrypsin, Hydrocortisone Acetate, Neomycin Palmitate) Kymar Ointment Improved; *Schering-Plough Animal Health.*

Estrogens

612 Alfaprostol
74176-31-1 234 277-746-0

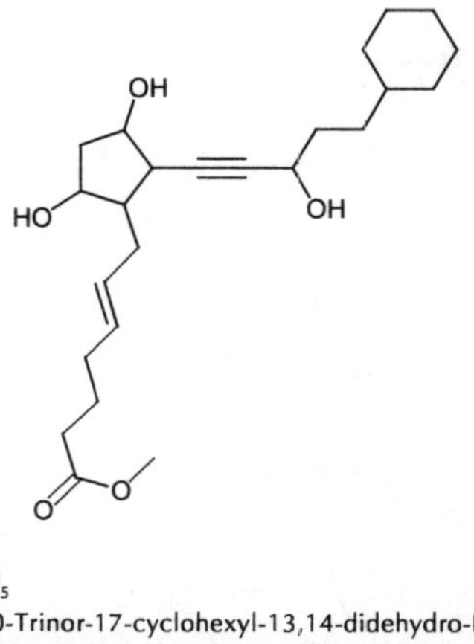

$C_{24}H_{38}O_5$
18,19,20-Trinor-17-cyclohexyl-13,14-didehydro-$PGF_{2}\alpha$.
Ro-22-9000; K-11941; 130-092 Alfavet; Alphacept. Used for estrus control.

Veterinary products:

130-092; Alfavet; *Vetem sPa, Milan, Italy.*

613 Diethylstilbestrol

56-53-1 3177 200-278-5

$C_{18}H_{20}O_2$

(E)-4,4'-(1,2-Diethyl-1,2-ethenediyl)bisphenol.

DES; Antigestil; Bufon; Cyren A; Domestrol; Estrobene; Estrosyn; Oestromensyl; Oestromon; Palestrol; Serral; Sexocretin; Sibol; Stilbetin; Stilboefral; ; Stilboestroform; Stilkap; Synestrin (tablets); Synthoestrin; Vagestrol. Non-steroidal estrogen. Has been used to control estrogen response but is teratogenic and carcinogenic. mp = 169-172°; λ_m = 259 nm ($E^{1\%}_{1\,cm}$ 654 0.1N NaOH); insoluble in H_2O; soluble in EtOH, Et_2O, $CHCl_3$, fatty oils.

614 Diethylstilbestrol Diphosphate

522-40-7 3177 208-328-8

$C_{18}H_{22}O_8P_2$

(E)-4,4'-(1,2-Diethyl-1,2-ethenediyl)bisphenol diphosphate.

Stilphostrol. Non-steroidal estrogen. Has been used to control estrogen response but is teratogenic and carcinogenic. mp = 204-206°; sparingly soluble in cold H_2O.

615 Estradiol

50-28-2 3746 200-023-8

$C_{18}H_{24}O_2$

(17β)-Estra-1,3,5(10)-triene-3,17-diol.

dihydrofollicular hormone; dihydrofolliculin; Dihydroxyestrin; dihydrotheelin; Dimenformon; Diogyn; Estrace; Estraderm; Estroclim; Evorel; Gynoestryl; Macrodiol; Menorest; Oestrogel; Ovocyclin; Ovocylin; Profoliol B; Progynon; Systen; Vagifem; Zumenon; Aquadiol; Climara; Diogyn; Diogynets; Estrace; Estraderm TTS; Estring Vaginal Ring; Progynon; Vivelle; NSC-9895; NSC-20293. Estrogen. Used in mares to induce estrus and as an abortifacient. mp = 173-179°; $[\alpha]_D^{25}$ = 76° to 83° (dioxane); λ_m = 225, 280 nm; insoluble in H_2O; soluble in EtOH, Me_2CO, dioxane.

Veterinary products:

118-123; Compudose 200; Compudose 400; *Elanco.*
140-897; (with Trenbolone Acetate) Revalor ®-G; Revalor ®-S; *Hoechst-Roussel Vet.*
140-992; (with Trenbolone Acetate) Revalor ®-H; *Hoechst-Roussel Vet.*
200-221; (with Trenbolone Acetate, Tylosin Tartrate) Component ™ TE-S; Component ™ TE-S with Tylan ®; *Ivy.*

616 Estradiol Benzoate

50-50-0 3746 200-043-7

$C_{25}H_{28}O_3$

(17β)-Estra-1,3,5(10)-triene-3,17-diol 3-benzoate.

Agofollin; Benzo-Gynoestryl; Benztrone; Pelanin benzoate; Progynon B; Ovahormon Benzoate; NSC-9566. Estrogen. Used in mares to induce estrus and as an abortifacient. mp = 191-196°; $[\alpha]_D^{25}$= 58° to 63° (dioxane); λ_m = 225, 280 nm (c = 2, dioxane); soluble in EtOH, Me_2CO, dioxane; slightly soluble in Et_2O.

Veterinary products:

009-576; (with Progesterone) Synovex ®-C; Synovex ®-S; *Fort Dodge Animal Health, Divn. AHP.*
011-427; (with Testosterone Propionate) Synovex ®-H; *Fort Dodge Animal Health, Divn. AHP.*
110-315; (with Progesterone, Tylosin Tartrate) Calf-Oid; Component ™ E-C with Tylan ®; Component ™ E-S; Component ™ E-S with Tylan ®; Component™ E-C; Implus-C ®; Steer-Oid; *Ivy.*
135-906; (with Testosterone Propionate, Tylosin Tartrate) Component ™ E-H; Component ™ E-H with Tylan ®; Heifer-Oid; *Ivy.*
141-043; (with Trenbolone Acetate) Synovex ® Plus; *Fort Dodge Animal Health, Divn. AHP.*

617 Estradiol Valerate

979-32-8 3746 213-559-2

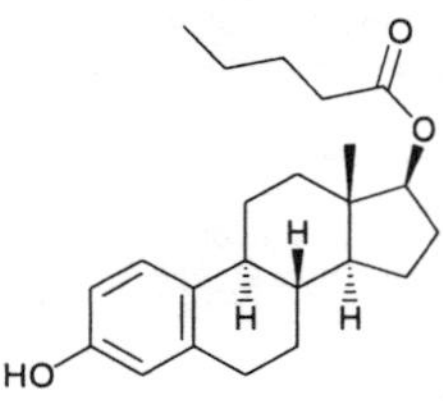

$C_{23}H_{32}O_3$

(17β)-Estra-1,3,5(10)-triene-3,17-diol 17-valerate.

Climaval; Cyclacur; Delestrogen; Gynogen LA; Pelanin Depot; Progynon Depot; Progynova; Primofol; Valergen; Deladumone; Delestrogen; Gynogen L.A. 40; NSC-17590; component of: Deluteval 2X, Ditate. Estrogen. Used in mares to induce estrus and as an abortifacient. mp = 144-145°.

Veterinary products:

097-037; (with Norgestomet) Syncro-Mate-B ®; *Merial.*
134-930; (with Norgestomet) Syncro-Mate-B ®; *Merial.*

618 Melatonin
73-31-4 5857 200-797-7

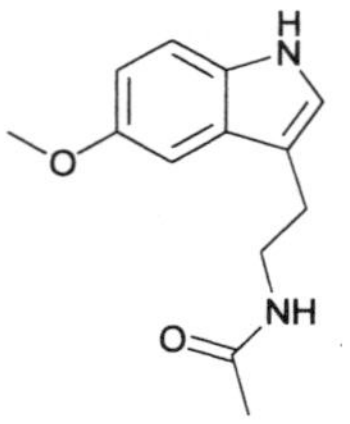

$C_{13}H_{16}N_2O_2$
N-[2-(5-Methoxy-1H-indol-3-yl)ethyl]acetamide.
N-acetyl-5-methoxytryptamine; Regulin. Used in veterinary medicine for control of estrus Pale yellow crystals; mp = 116-118°; λ_m = 223. 2378 nm (ε 27550. 6300).

Veterinary products:

140-846; PrimeX®; *Wildlife.*

Expectorants

619 Acetylcysteine
616-91-1 89 210-498-3

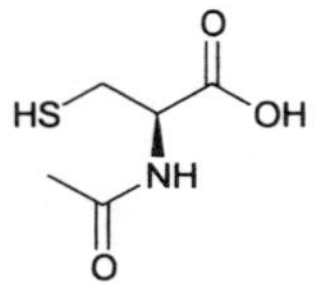

$C_5H_9NO_3S$
N-Acetyl-L-cysteine.
5052; Mucomyst; NSC-111180; Airbron; Broncholysin; Brunac; Fabrol; Fluatox; Fluimucil; Fluimucetin; Fluprowit; Mucocedyl; Mucolator; Mucolyticum; Muco Sanigen; Mucosolvin; Mucret; NAC; Neo-Fluimucil; Parvolex; Respaire; Tixair. Mucolytic, corneal vulnerary and antidote to acetaminophen poisoning. mp = 109-110°; LD_{50} (rat orl) = 5050 mg/kg.

620 Ammonium Chloride
12125-02-9 537 235-186-4
H_4ClN
Hydrochloric acid, ammonium salt.
ammonium muriate; sal ammoniac; salmiac; Amchlor; Darammon. Used as a urinary acidifier. Colorless crystals; d^{25} = 1.5274; soluble in H_2O (22.9 g/100 ml at 0°, 26.0 g/100 ml at 15°, 28.3 g/100 ml at 25°, 39.6 g/100 ml at 80°; pH of 1% aqueous solution at 25° = 5.5, 3% = 5.1, 10% = 5.0; soluble in EtOH, MeOH, insoluble in Me_2CO, Et_2O, EtOAc; LD_{50} (rat im) = 30 mg/kg.

Veterinary products:

009-339; (with Caramiphen Edisylate) Carafen Cough Syrup; Carafen Cough Tablets; *Fort Dodge Animal Health, Divn. AHP.*

621 Guaifenesin
93-14-1 4582 202-222-5

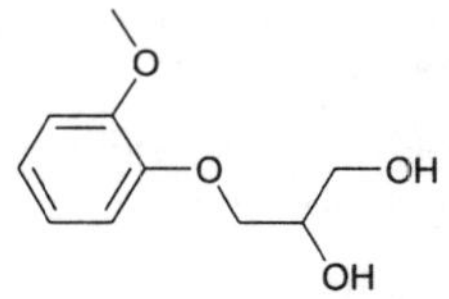

$C_{10}H_{14}O_4$
3-(2-Methoxyphenoxy)-1,2-propanediol.
guaiacyl glyceryl ether; glyceryl guaiacyl ether; glycerol guaiacolate; α-glyceryl guaiacol ether; o-methoxyphenyl glyceryl ether; guaiacol glyceryl ether; guaiphenesin; guaiacuran; MY-301; XL-90; 2-G; Actifed-C; amonidren; Calmipan; Colrex Expectorant; Equicol; Glycodex; Guaiamar; Guayanesin; Miocaina; Myocaine; Myoscain; Oresol; Oreson; Relaxil G; Reorganin; Respenyl; Resyl; Robitussin; Sirotol; Tenntus; Tulyn; component of: Brexin EX, of Brondecon, Congestac, Contac Cough Formula, Contac Cough & Sore Throat Formula, Coricidin, Dilor G, Dimacol, ENTEX, Fedahist, Histalet X Tablets, Hycotuss Tablets, Isoclor Expectorant C, Kwelcof, Lufyllin-GG, Naldecon-CX, Naldecon-DX, Naldecon-EX, Neothylline-GG, Nucofed Expectorant, PV Tussin Tablet, Quibron, Quibron Plus, Ru-Tuss DE Tablets, Ru-Tuss Expectorant, Slo-Phyllin GG, Sudafed Cough Syrup, Tedral Expectorant, Theolair Plus, Triaminic, Tussar-2, Tussar SF. Centrally acting muscle relaxant with expectorant properties. Used as an adjunct to anesthesia. mp = 78.5-79°; bp_{19} = 215°; soluble in H_2O (5 g/100 ml at 25°), EtOH, $CHCl_3$, glycerol, propylene glycol, DMF; moderately soluble in C_6H_6; nearly insoluble in petroleum ether.

Veterinary products:

044-012; Gecolate; *Summit Hill.*
048-854; Gecolate; Glycodex Injection; *Summit Hill.*
136-651; Guailaxin; *Robins.*
200-230; Guaifenesin Injection; *Phoenix.*

Gonad Stimulating Principals

622 Chorionic Gonadotropin
177073-44-8 2273
Gonad-stimulating principal. Used parenterally in cows to suppress frequent or constant heat.

Veterinary products:

006-103; Follutein Veterinary; *Fort Dodge Animal Health Divn., Am. Cyanamid.*
047-055; Chorionic Gonadotropin Injectable; *Steris.*
047-353; Ferti-Cept; *Elanco.*
100-840; Chorionic Gonadotropin For Injection; *Am. Pharm. Partners.*
103-090; Chortropin; *United Vaccines.*
140-856; (with Serum Gonadotropin) P.G. 600 ®; *Intervet.*
140-927; Chorulon ®; *Intervet.*

623 Follicle Stimulating Hormone

9002-68-0 4299 232-662-3

Follicle-stimulating hormone; urofollitrophin; Fertinorm; Follitropin; FSH; Luteoantine; Metrodin. Gonad Stimulating principle. Glycoprotein gonadotropic hormone found in the pituitary tissue of mammals. Molecular weight ≅ 36,000. Directly regulates the metabolic activity of the granulosa cells of the ovaries and the Sertoli cells of the testis. Use in cases of FSH deficiency. Solid, soluble in H_2O.

Veterinary products:

009-505; F.S.H.-P Injectable; *Sioux Biochemical.*
141-014; Super-Ov ®; *Ausa Intl.*

624 Gonadorelin Acetate

52699-48-6 5500

$C_{55}H_{75}N_{17}O_{13}{\cdot}xC_2H_4O_2{\cdot}yH_2O$

Cystorelin; Hypocrine; Lutrelef; Lutrepulse. Gonad Stimulating principle. Used to treat ovarian follicular cysts in cattle.

Veterinary products:

098-379; Cystorelin; *Merial.*
200-134; Fertagyl ®; *Intervet.*

625 Gonadorelin Hydrochloride

51952-41-1 5500

$C_{55}H_{75}N_{17}O_{13}.xC_2H_4O_2.yH_2O$

Gonadorelin acetate.

LH-RH Hydrochloride; AY-240341; HRF; Factrel. Gonad Stimulating principle. Used to treat ovarian follicular cysts in cattle.

Veterinary products:

139-237; Factrel®; *Fort Dodge Animal Health, Divn. AHP.*

626 Pituitary Luteinizing Hormone

9002-67-9 5499 232-661-8

Luteinizing Hormone.

ICSH; LH. A glycoprotein gonadotropic hormone found in the anterior lobe of the pituitary gland. Used as a gonad stimulating principle. White powder, soluble in H_2O.

Veterinary products:

009-167; P.L.H. Injectable; *Schering-Plough Animal Health.*

627 Serum Gonadotropin

9002-70-4 2273 232-663-9

Equine choriogonadotropin.

Equine chorionic gonadotropin; pregnant mare serum gonadotropin; Anteron; Antex; Gestyl; Primantron; PMSG; Seragon; Seragonin; Serotropin. Gonad-stimulating principal. A glycoprotein hormone synthesized by chorionic tissue of the placenta; found in equine blood and urine during pregnancy. Bioactivity similar to that of leutinizing hormone. Soluble in H_2O, aqueous glycerol, glycols; α-subunit is 96 amino acids, β-subunit is 149 amino acids.

Veterinary products:

140-856; (with Chorionic Gonadotropin) P.G. 600®; *Intervet.*

Growth Stimulants

628 Arsanilic Acid

98-50-0 830 202-674-3

$H_2N-C_6H_4-As(=O)(OH)-OH$

$C_6H_8AsNO_3$

4-Aminobenzenearsonic acid.

p-Aminophenylarsenic acid; Arsanilic acid; p-Arsanilic acid; Aminobenzenearsonic. Growth promotant; controls swine dysentery. Needles; slightly soluble in cold H_2O, EtOH, AcOH, moderately soluble in conc. Mineral acids, insoluble in Me_2CO, C_6H_6, $CHCl_3$, Et_2O; LD_{50} (rat orl) > 1000 mg/kg.

Veterinary products:

008-019; Pro-Gen Plus Feed Supplement; *Fleming.*
038-241; (with Erythromycin Thiocyanate, Zoalene) Erythro® (High Lev)/Zoalene + Arsanilic Acid; *Merial.*
038-242; (with Amprolium, Erythromycin Thiocyanate; ethopabate) Erythro® (Low lev)/Amp + Etho; *Merial.*
038-624; (with Erythromycin Thiocyanate) Pro-Gallimycin-10; *Merial.*
049-462; (with Amprolium, Ethopabate, Penicillin G Procaine, Streptomycin) Rainbrook Brolier Premix No. 1; *Stockton Hay & Grain.*

629 Bovine Somatotropin

9002-72-6 8864 232-666-5

$C_{976}H_{1533}N_{265}O_{286}S_8$

Adenohypophyseal growth hormone.

GH; hypophyseal growth hormone; anterior pituitary growth hormone; phyone; pituitary growth hormone; somatotropic hormone; SH; somavubove. Growth stimulant.

Veterinary products:

140-872; Posilac® 1 Step®; *Monsanto.*

630 Efrotomycin
56592-32-6 3567

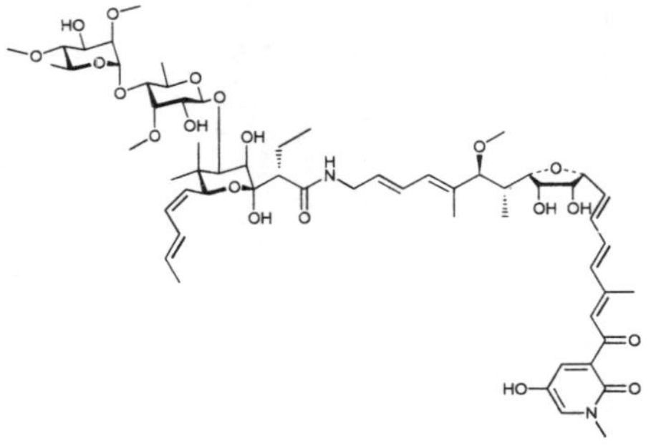

$C_{59}H_{88}N_2O_{20}$
31-O-[6-Deoxy-4-O-(6-deoxy-2,4-di-O-methyl-α-L-mannopyranosyl)-3-O-methylβ-D-allopyranosyl]-1-methylmocimycin.
Producil; MK-621; FR-02A. Veterinary growth stimulant. Yellow solid; λ_m = 232, 327 nm ($E^{1\%}_{1\,cm}$ 464 216 pH 7); LD_{50} (mus orl) > 4000 mg/kg, (mus sc) > 2000 mg/kg.

Veterinary products:

140-818; Producil®; Producil® for Swine; *Merial.*

631 Laidlomycin Propionate Potassium
84799-02-0 5361

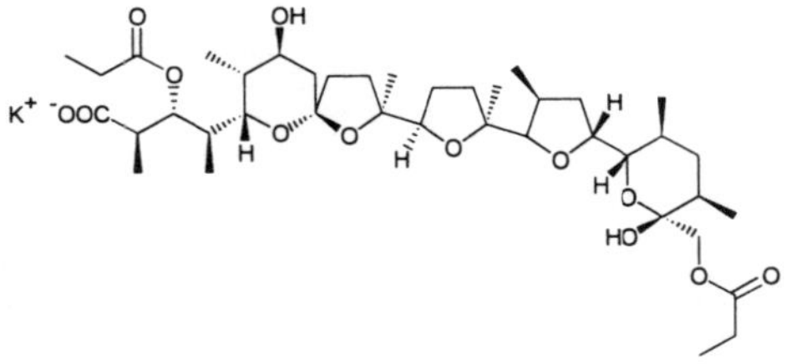

$C_{40}H_{65}KO_{13}$
16-Deethyl-3-O-demethyl-16-methyl-3-O-(1-oxopropyl)monensin 26-propanoate monopotassium salt.
Cattlyst®; RS-11988. Veterinary growth stimulant. mp = 190-192°.

Veterinary products:

141-025; Cattlyst®; *Roche Vitamins.*

632 Narasin
55134-13-9 6506

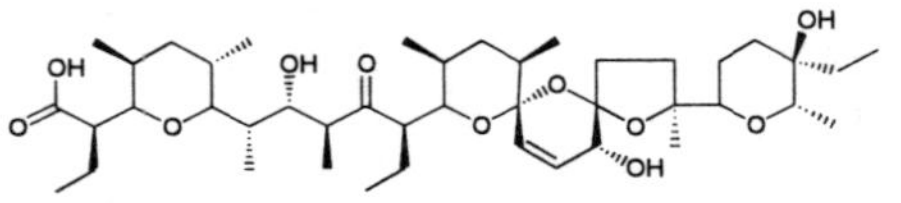

$C_{43}H_{72}O_{11}$
α-Ethyl-6-[5-[2-(5-ethyltetrahydro-5-hydroxy-6-methyl-2H-pyran-2-yl)-15-hydroxy-2,10,12-trimethyl-1,6,8-trioxadispiro[4.1.5.3]pentadec-13-en-9-yl]-2-hydroxy-1,3-dimethyl-4-oxoheptyl]tetrahydro-3,5-dimethyl-2H-pyran-2-acetic acid.
Natacyn; CL-12625; Antibiotic A-5283; Monteban; C-7819B. Veterinary growth stimulant and coccidostat. mp = 98-100°, 198-200°; λ_m 285 nm (ε 58 EtOH); $[\alpha]_D^{25}$ = -54° (c = 0.2 MeOH); insoluble in H_2O, soluble in organic solvents; LD_{50} (mus ip) = 7.15 mg/kg.

Veterinary products:

118-980; Monteban ®; *Elanco.*
138-952; (with Nicarbazin) Maxiban ® 72 / Narasin / Nicarbazin; *Elanco.*
140-445; (with Roxarsone) Monteban ® and Roxarsone; *Elanco.*
140-843; (with Bambermycins, Roxarsone) Monteban ® + Flavomycin ® + 3-Nitro ®; *Hoechst-Roussel Vet.*
140-845; (with Bambermycins) Flavomycin ® + Monteban ®; *Hoechst-Roussel Vet.*
140-852; (with Bacitracin Methylene Disalicylate, Roxarsone) Monteban ® / 3-Nitro ® / BMD ®; *Alpharma.*
140-853; (with Bacitracin Methylene Disalicylate) Monteban ® / BMD ®; *Alpharma.*
140-926; (with Bacitracin Methylene Disalicylate, Nicarbazin) BMD ® / Maxiban ®; *Elanco.*
140-947; (with Lincomycin, Nicarbazin) Maxiban ® / Lincomix ®; *Elanco.*
141-112; (with Nicarbazin, Roxarsone) 3-Nitro ® / BMD ® / Maxiban ®; *Alpharma.*
141-113; (with Nicarbazin, Roxarsone) 3-Nitro ® / Maxiban ®; *Elanco.*

Hematinics

633 Epoetin-α
113427-24-0 3729
$C_{809}H_{1301}N_{229}O_{240}S_5$
1-165 Erythropoietin (human clone λHEPOFL13 protein moiety), glycoform α.
Epogen; Procrit; Epoade; Eprex; Erypo; Espo; A 165-mer glycoprotein. Anti-anemic and hematinic.

634 Epoetin-β
122312-54-3 3729
$C_{809}H_{1301}N_{229}O_{240}S_5$
1-165 Erythropoietin (human clone λHEPOFL13 protein moiety), glycoform β.
Marogen; EPOCH; BM-06.019; Epogin; Recormon; A 165-mer glycoprotein. Anti-anemic;hematinic.

635 Ferric Oxide, Colloidal
1309-37-1 4072 215-168-2
Fe_2O_3
Ferric sesquioxide.
Jeweler's rouge. Used as a stain for polysaccharides.

636 Ferrous Sulfate, Dried
13463-43-9 4105 231-753-5
$FeSO_4.xH_2O$
Iron(2+) sulfate (1:1) hydrate.
Feromax; Feroritard; Ferro-Gradumet; Fespan; Tetucur; component of: Cytoferin, Mediatric. Anti-anemic and hematinic used to treat iron deficiency. Soluble in H_2O.

637 Ferrous Sulfate Heptahydrate
7782-63-0 4105 231-753-5

$FeSO_4 \cdot 7H_2O$
Iron(2+) sulfate (1:1) heptahydrate.
Feosol; Fero-Gradumet; Mol-Iron; Natabec; Slow-Fe; copperas green; green vitriol; iron vitriol; Feospan; Fesotyme; Fer-in-Sol; Haemofort; Ironate; Presfersul; Sulferrous; component of: Plastulen-N. Anti-anemic and hematinic used to treat iron deficiency. d = 1.897; soluble in H_2O, insoluble in EtOH; LD_{50} (mus iv) = 65 mg/kg, (mus orl) = 1520 mg/kg.

638 Gleptoferron
57680-55-4

$(FeO_2H)_m(C_6H_{10}O_5)_x \cdot (C_7H_{13}O_7)_n$
Complex of ferric hydroxide and dextran glucoheptonic acid. Used to treat anemia.
Gleptosil; Heptomer; iron heptonate. Hematinic used as a parenteral nutritional factor in iron deficiency anemia, mainly in pigs .

Veterinary products:

110-399; Gleptosil; Heptomer Injection; *Fisons.*

639 Iron-Dextran Complex
9004-66-4 2991

A complex of trivalent iron and dextran.
Fenate; Imferon; Ironorm. Hematinic. Used in veterinary medicine and as an anti-anemic factor. LD_{50} (mus iv) = 2240 mg/kg.

Veterinary products:

010-793; Nonemic; *Schering-Plough Animal Health.*
010-865; Ferrextran 100; *Fort Dodge Animal Health, Divn. AHP.*
011-879; Rubrafer; Rubrafer S-100; *Fort Dodge Animal Health Divn., Am. Cyanamid.*
015-207; Hyferdex Injection; *Elanco.*
099-667; Imposil; *Alstoe.*
106-772; Iron Dextran Complex; *Boehringer Ingelheim Vetmedica.*
113-748; Purina ® Oral Pigemia; *Purina Mills.*
119-142; PVL Iron Dextran Injectable; *Merial.*
134-708; Iron Dextran Complex Inj.; *Boehringer Ingelheim Vetmedica.*
138-255; Iron Hydrogenated Dextran Injection; *Veterinary Labs.*
200-254; Iron Dextran Injection; Iron Dextran Injection - 100mg; *Phoenix.*
200-256; Iron Dextran Injection - 200 mg; *Phoenix.*

Hematopoietics

640 Erythropoietin
11096-26-7 3729 234-317-2

erythropoiesis stimulating factor; hempoietine; ESF; Ep; Epo; Has two components, Epo-α and Epo-β, both produced by recombinant technology. Epo-α (Epoetin alfa; Epoade; Epogen; Eprex; Erypo; Espo; Procrit), has 165 residues, molecular weight 30,400. Epo-β (Epoetin beta; Epogin; Marogen; Recormon), has 165 residues, molecular weight 30,000. The α and β forms differ only in their carbohydrate moieties. Hematopoietic. Glycoprotein hormone that stimulates red blood cell formation. Found in plasma and urine. Used in treatment of anemia.

Hemostatics

641 ε-Aminocaproic Acid
60-32-2 451 200-469-3

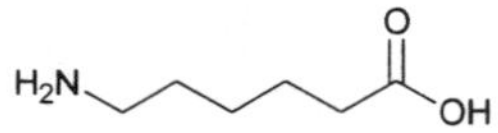

$C_6H_{13}NO_2$
6-Aminohexanoic acid.
Amicar; CL-10304; CY-116; EACA; 177 J.D.;NSC-26154; Ipsilon; Hemocaprol; Caprocid; Capramol; Afibrin; Epsikapron; Hepin. Hemostatic and antifibrinolytic. Used as a treatment for degenerative myelopathy (mainly in German shepherds). mp = 204-206°; freely soluble in H_2O, sparingly soluble in MeOH, insoluble in EtOH; LD_{50} (rat ip) → 7000 mg/kg, (rat iv) → 3300 mg/kg; [hydrobromide ($C_6H_{14}BrNO_2$)]: mp = 105°; [hydrochloride ($C_6H_{14}ClNO_2$)]: mp = 128-129°.

Hepatoprotectants

642 Methionine, DL-form
59-51-8 6053 200-432-1

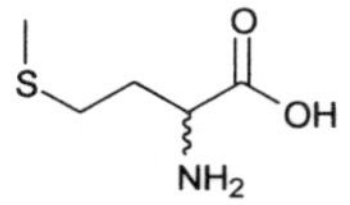

$C_5H_{11}NO_2S$
DL-2-Amino-4-(methylthio)butyric acid.
racemethionine; Amurex; Banthionine; Dyprin; Lobamine; Metione; Pedameth; Urimeth. Hepatoprotectant. Used as a nutritional supplement and a urine acidifying agent. d = 1.340; soluble in H_2O (1.82 g/100 ml at 0°, 3.38 g/100 ml at 25°, 6.07 g/100 ml at 50°, 10.52 g/100 ml at 75°, 17.60 g/100 ml at 100°); soluble in dilute acids, alkalies; slightly soluble in 95% EtOH; insoluble in Et_2O.

643 Methionine, D-form
348-67-4 6053 206-483-6

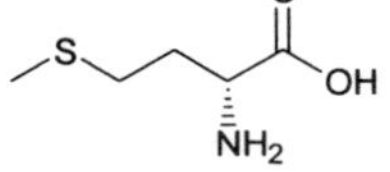

$C_5H_{11}NO_2S$
D-2-Amino-4-(methylthio)butyric acid.
Hepatoprotectant. Used as a nutritional supplement and as a urine acidifying agent. $[\alpha]_D^{25}$ = 8.12° (c = 0.8), $[\alpha]_D^{25}$ = -21.18° (c = 0.8 0.2N HCl).

644 Methionine, L-form
63-68-3 6053 200-562-9

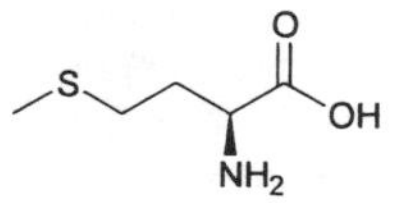

$C_5H_{11}NO_2S$
L-2-Amino-4-(methylthio)butyric acid.
Acimethin. Hepatoprotectant. Used as a nutritional supplement and a urine acidifying agent. mp = 280-282° (dec); $[\alpha]_D^{25}$ = -8.11° (c = 0.8), $[\alpha]_D^{20}$ = 23.40° (c = 5.0 3N HCl); soluble in H_2O, warm, dilute EtOH; insoluble in EtOH, Et_2O, Me_2CO, C_6H_6, petroleum ether.

645 Tiopronin
1953-02-2 9597 217-778-4

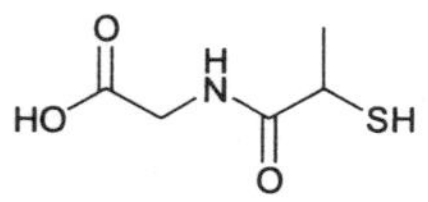

$C_5H_9NO_3S$
N-(2-Mercapto-1-oxopropyl)glycine.
Acadione; Capen; Epatiol; Mucolysin; Thiola; Thiosol. Hepatoprotectant. Used to counter cystine urolithiasis. mp = 95-97°; LD_{50} (mus iv) = 2100 mg/kg.

Immunomodulators

646 Levamisole
14769-73-4 5486 238-836-5

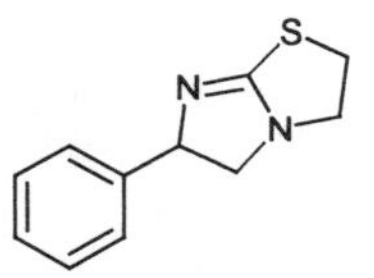

$C_{11}H_{12}N_2S$
(-)-2,3,5,6-Tetrahydro-6-phenylimidazo[2,1-b]thiazole.
Levovermax; Totalon. Immunomodulator with anthelmintic activity against nematodes. mp = 60-61.5°; $[\alpha]_D^{25}$ = -85.1° (c = 10 $CHCl_3$); [DL-form (tetramisole, Tetramizole)]: mp = 87-89°; [D-(+)-form (dexamisole)]: mp = 60-61.5°; $[\alpha]_D^{25}$ = 85.1° (c = 10 $CHCl_3$).

Veterinary products:

139-877; Totalon ™ Topical Cattle Anthelmintic; *Schering-Plough Animal Health.*
140-844; Tramisol ® Pour-On; *American Cyanamid Divn., AHP Corp.*

647 Levamisole Hydrochloride
16595-80-5 5486 240-654-6
$C_{11}H_{13}ClN_2S$
(-)-2,3,5,6-Tetrahydro-6-phenylimidazo[2,1-b]thiazole monohydrochloride.
Ergamisole; R-12654; Ascaridil; Decaris; Ergamisol; Levacide; Levadin; Levasole; Meglum; Nemicide; Nilverm; Ripercol; Solaskil; Spartakon; Tramisol. Immunomodulator with anthelmintic activity against nematodes. mp = 227-229°; $[\alpha]_D^{20}$ = -124° ± 2° (c = 0.9 H_2O); soluble in H_2O; [DL-form (Bayer 9051, McN-JR-8299, R-8299)]: mp = 264-265°; soluble in H_2O (21 g/100 ml), MeOH, propylene glycol; sparingly soluble in EtOH; slightly soluble in $CHCl_3$, Me_2CO, C_6H_{14}; LD_{50} (mus iv) = 22 mg/kg, (mus sc) = 84 mg/kg, (mus orl) = 210 mg/kg, (rat iv) = 24 mg/kg, (rat sc) = 130 mg/kg, (rat orl) = 480 mg/kg; [D-(+)-form]: mp = 227-227.5°; $[\alpha]_D^{20}$ = 125° (c = 0.7 H_2O).

Veterinary products:
039-356; Tramisol ® Cattle Wormer Bolus; Ripercol L Bolus; *Cyanamid Agri., Puerto Rico.*
039-357; Ripercol L Soluble Drench Powder; *Cyanamid Agri., Puerto Rico.*
042-740; Tramisol ® Soluble Drench Powder for Sheep; Ripercol L; *Cyanamid Agri., Puerto Rico.*
042-837; Tramisol ® Sheep Wormer Oblets; Ripercol L Wormer Oblets; *Cyanamid Agri., Puerto Rico.*
044-015; Tramisol ® Type A Medicated Article; *Cyanamid Agri., Puerto Rico.*
045-455; Tramisol ® Type A Medicated Article; *Cyanamid Agri., Puerto Rico.*
045-513; Ripercol L; *Cyanamid Agri., Puerto Rico.*
091-826; Levasole ® Cattle Wormer Bolus; Tramisol Cattle Wormer Oblets Tablets; *Schering-Plough Animal Health.*
092-237; (with Piperazine Dihydrochloride) Ripercol L-Piperazine Soluble Dihydrochloride; *Cyanamid Agri., Puerto Rico.*
093-688; (with Piperazine Dihydrochloride) Ripercol L-Piperazine Dihydrochloride; *Cyanamid Agri., Puerto Rico.*
107-085; Tramisol ®; *Cyanamid Agri., Puerto Rico.*
112-049; Levasole ® Soluble Pig Wormer; *Schering-Plough Animal Health.*
112-050; Levasole ® Soluble Drench Powder; Tramisol Soluble Drench Powder; *Schering-Plough Animal Health.*
112-051; Levamisole Soluble Drench Powder; Levasole Soluble Drench Powder; Tramisol ® Soluble Drench Powder; *Schering-Plough Animal Health.*
112-052; Levasole ® Sheep Wormer Bolus; Tramisole Sheep Wormer Oblets Tablets; *Schering-Plough Animal Health.*
126-237; Tramisol ® Gel; *Cyanamid Agri., Puerto Rico.*
200-225; Prohibit ™ Soluble Drench Powder; *Agri Labs.*

648 Levamisole Phosphate
5486
Immunomodulator with anthelmintic activity against nematodes.

Veterinary products:

049-553; Ripercol L; *Cyanamid Agri., Puerto Rico.*
102-437; Tramisol ® Injectable Solution; *Cyanamid Agri., Puerto Rico.*
126-742; Levasole ® Injectable Solution; *Schering-Plough Animal Health.*

649 Levamisole Resinate
5486
Immunomodulator with anthelmintic activity against nematodes.

Veterinary products:

101-079; Tramisol ® 10% Pig Wormer; Tramisol ® Hog Wormer; *Cyanamid Agri., Puerto Rico.*
139-858; Tramisol ®-X-Tra Cattle Anthelmintic and Extoparasitic Paste; *Schering-Plough Animal Health.*

Immunosuppressants

650 Cyclosporin A

59865-13-3 2821

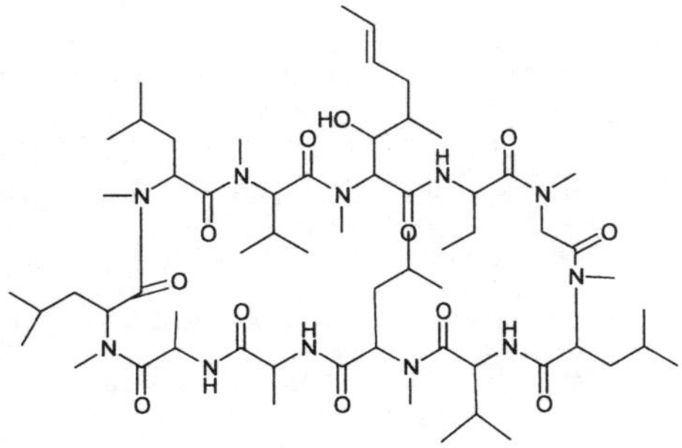

$C_{62}H71_{11}N_{11}O_{12}$
Cyclosporine.
ciclosporin; 27-400; Sandimmun; Sandimmune; Neoral. Immunosuppressant. mp = 148-151°; $[\alpha]_D^{20}$ = -244° (c = 0.6 $CHCl_3$), -189° (c = 0.5 MeOH), soluble in MeOH, EtOH, Me_2CO, Et_2O, $CHCl_3$; slightly soluble in H_2O and saturated hydrocarbons; LD_{50} (mus iv) = 107 mg/kg, (mus orl) = 2329 mg/kg, (rat i.

Veterinary products:

141-052; Optimmune®; *Schering-Plough Animal Health.*

Laxatives and Cathartics

651 Bisacodyl

603-50-9 1282 210-044-4

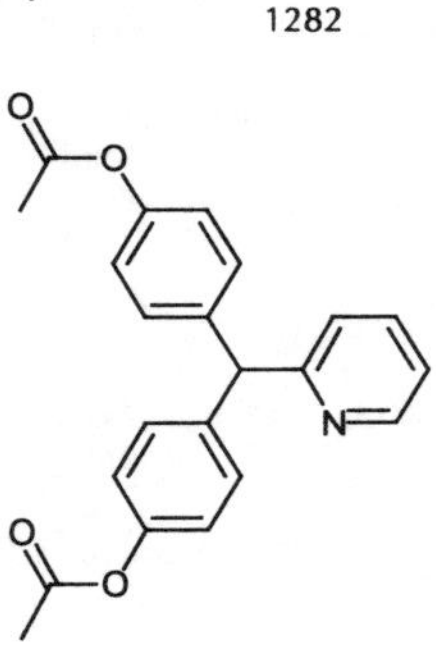

$C_{22}H_{19}NO_4$
4,4'-(2-Pyridylmethylene)diphenol diacetate.
Correctol Tablets; Correctol Caplets; Dulcolax; Feen-a-Mint Tablets; SK-Bisacodyl; Theralax; Bicol; Broxalax; Contralax; DAMP; Dulcolan; Durolax; Endokolat; Eulaxan; Godalax; Laxadin; Laxanin N; Laxorex; Nigalax; Perilax; Pyrilax; Stadalax; Telemin; Ulcol. Laxative-cathartic. Used orally or rectally in dogs and cats as a stimulant-cathartic. mp = 138°; insoluble in H_2O, alkaline solutions; soluble in acids, EtOH, Me_2CO, propylene glycol; LD_{50} (rat orl) > 3000 mg/kg.

652 Docusate Calcium

128-49-4 3459 204-889-8

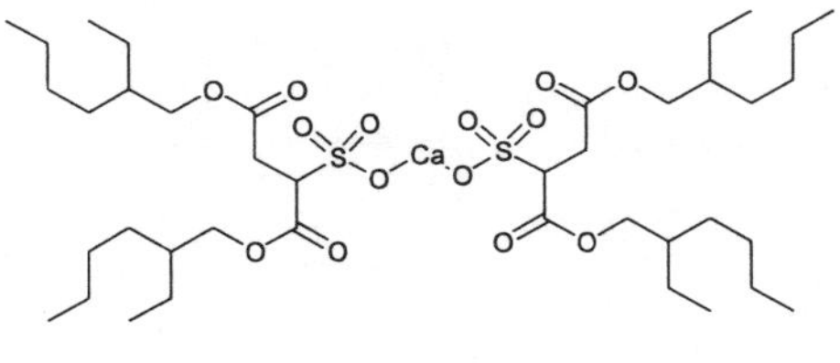

$C_{40}H_{74}CaO_{14}S_2$
1,4-Bis(2-ethylhexyl)sulfosuccinate calcium salt.
Surfak; component of: Doxidan. Laxative-cathartic. Used as a stool softener. Soluble in mineral and vegetable oils, polyethylene glycol; insoluble in glycerol.

653 Docusate Potassium

7491-09-0 3460 231-308-5

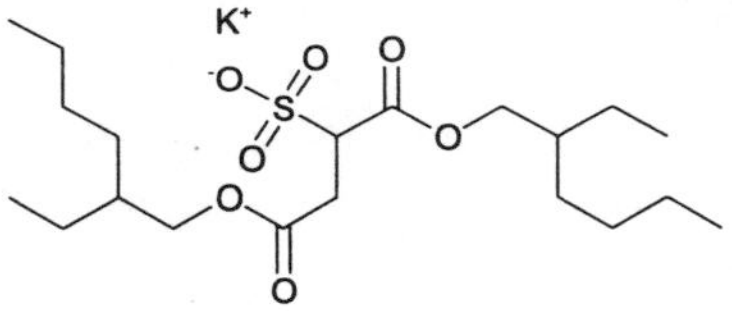

$C_{20}H_{37}KO_7S$
1,4-Bis(2-ethylhexyl)sulfosuccinate potassium salt.
Rectalad Enema. Laxative-cathartic. Used as a stool softener.

654 Docusate Sodium

577-11-7 3460 209-406-4

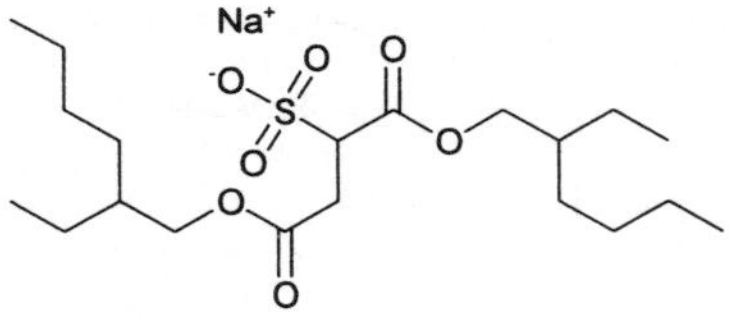

$C_{20}H_{37}NaO_7S$
1,4-Bis(2-ethylhexyl)sulfosuccinate sodium salt.
Colace; Correctol Stool Softener Laxative; Dialose; Doxinate; D.S.S.; Modane Soft; Molofac; Aerosol OT; Comfolax; Coprola; Dioctylal; Dioctyl; Diotilan; Disonate; Disposaject; Doxol; Dulcivac; Jamylène; Molatoc; Molcer; Nevax; Regutol; Soliwax; Velmol; Waxsol; Yal; component of: Correctol Caplets, Correctol Tablets, Dialose Plus, Dorbantil, Doxan, D.S.S. Plus, Feen-a-Mint Pills, Geriplex, Laxcaps, Modane Plus, Peri-Colace, Phillips Gelcaps, Senokap DSS, Senokot S, Unilax. Laxative-cathartic. Used as a stool softener.

Soluble in H_2O (1.5 g/100 ml at 25°, 2.3 g/100 ml at 40°, 3.0 g/100 ml at 50°, 5.5 g/100 ml at 70°), CCl_4, petroleum ether, naphtha, xylene, dibutyl phthalate, liquid petrolatum; Me_2CO, EtOH, vegetable oils; very soluble in H_2O + EtOH or H_2O-miscible solvents.

655 Lactulose

4618-18-2 5360 225-027-7

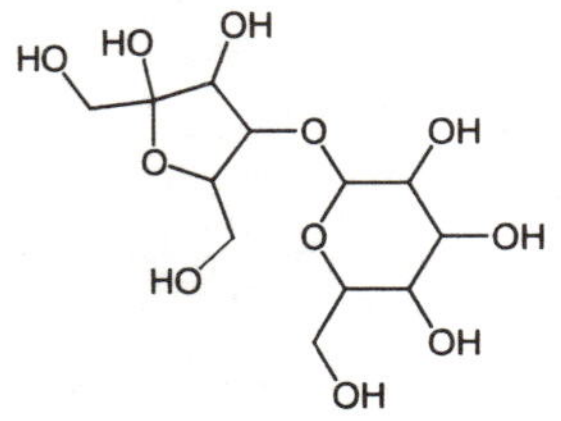

$C_{12}H_{22}O_{11}$
4-O-β-D-Galactopyranosyl-D-fructofuranose.
Bifiteral; Cephulac; Duphalac; Generlac; Lactuflor; Laevilac; Normase. Laxative-cathartic. Use to reduce ammonia levels in the blood. mp = 169°; soluble in H_2O (76.4 g/100 ml at 30°, 81.0 g/100 ml at 60°, >86 g/100 ml at 90°).

656 Magnesium Sulfate

7487-88-9 5731 231-298-2

MgO_4S
Sulfuric acid magnesium salt (1:1).
Laxative-cathartic.

657 Magnesium Sulfate Heptahydrate

10034-99-8 5731

$MgO_4S.7H_2O$
Sulfuric acid magnesium salt (1:1) heptahydrate.
bitter salts, Epsom salts. Laxative-cathartic. Anticonvulsant, electrolyte replenisher. Laxative-cathartic. d = 1.67; soluble in H_2O (71 g/100 ml at 20°, 91 g/100 ml at 40°), slightly soluble in EtOH.

Veterinary products:

046-789; (with Chloral Hydrate, Pentobarbital) Chloropent; *Fort Dodge Animal Health, Divn. AHP.*

658 Magnesium Sulfate Trihydrate

15320-30-6 5731

$MgO_4S{\cdot}3H_2O$
Sulfuric acid magnesium salt (1:1) trihydrate.
Laxative-cathartic. Laxative-cathartic. Colorless crystalline material.

659 Mineral Oil

8012-95-1 7327 232-384-2

905 (mineral hydrocarbons); adepsine oil; Alboline®; Bayol f; Bayol 55; Benol®; blandlube; blandol white mineral oil; Blandol®; Britol®; Britol® 6NF; Britol® 7NF; Britol® 9NF; Britol® 20USP; Britol® 35USP; Britol® 50USP; Cable oil; Carnation®; Carnea 21; clearteck; crystol 325; Crystosol; Crytosol USP 200; Crytosol USP 240; Crytosol USP 350; Drakeol®; Drakeol®; Drakeol® 5; ; Drakeol® 6; Drakeol® 7; Drakeol® 9; Drakeol® 13; Drakeol® 19; Drakeol® 21; Drakeol® 32; Drakeol® 34; Drakeol® 35; Electrical Insulating Oil; Ervol®; Filtrawhite®; Fonoline®; Frigol®; Gloria®; Glymol®; Heat-treating oil; heavy mineral oil; Hevyteck; Hydraulic oil; Hydrobrite 200PO; Hydrobrite 300PO; Hydrobrite 380PO; Hydrobrite 550PO; hydrocarbon oils; jute batching oil; ; paraffin oil (class); Paraffin oil; Paraffin oils; paroleine; PD-23; peneteck; Penreco; perfecta; petrogalar; petrolatum, liquid; Petroleum hydrocarbons; primol; primol d; primol 355; Protol®; protopet; Saxol; Superla® No. 5; Superla® No. 6; Superla® No. 7; Superla® No. 9; Superla® No. 10; Superla® No. 13; Superla® No. 18; Superla® No. 21; Superla® No. 31; Superla® No. 35; tech pet f; triona b; Uvasol; white oil; white mineral oil; Emollient, solvent, lubricant, oleaginous vehicle. FDA limit 0.008% in wash water for foods, 150 ppm in yeast, FDA approved for ophthalmics, orals, topicals, USP/NF, BP, Ph.Eur. compliance. Used as an oleaginous vehicle, filler, solvent, lubricant and laxative. Light mineral oil as a tablet/capsule lubricant, used in ophthalmic ointments, suspensions, oral tablets and capsules, topical creams, lotions, ointments and suppositories. Used in veterinary medicine as a laxative. Colorless oil; insoluble in H_2O, EtOH, soluble in C_6H_6, $CHCl_3$, Et_2O, petroleum ether; d = 0.83 - 0.86 (light), 0.875 - 0.905 (heavy); Eye irritant. Non-toxic, LD_{50} (mus orl) = 22000 mg/kg.

660 Sodium Sulfate [anhydrous]

7757-82-6 8829 231-820-9

Na_2SO_4
Sulfuric acid disodium salt.
component of: Colyte. Calcium regulator Laxative-cathartic. Used in food animals as a saline cathartic. mp = 800°; d = 2.7; soluble in H_2O (50 g/100 ml at 33°; solubility decreases with temperature to 41.6 g/100 ml at 100°); insoluble in EtOH.

661 Sodium Sulfate Decahydrate

7727-73-3 8829

$Na_2SO_4.10H_2O$
Sulfuric acid disodium salt decahydrate.
Glauber's salt; component of: Colyte. Calcium regulator Laxative-cathartic. Used in food animals as a saline cathartic. mp =32.4°; d = 1.46; loses all H_2O at 100°; soluble in H_2O (66 g/100 ml at 25°, 30 g/100 ml at 15°); soluble in glycerol; insoluble in EtOH.

Veterinary products:

036-361; (with Amprolium, Chlortetracycline Calcium Complex, Ethopabate) Amp Ethopabate CTC® Sodium Sulfate; *Roche Vitamins.*

LH-RH Agonists

662 Deslorelin Acetate

2968

6-D-Tryptopan-9-(N-ethyl-prolinamide)-10-deglycin-amide acetate luteinizing hormone-releasing factor (pig).
LH-RH antagonist, used in treatment of precocious puberty.

Veterinary products:

141-044; Ovuplant™; *Peptech Animal Health.*

Mineralocorticoids

663 Betamethasone Acetate

987-24-6 1226 213-578-6

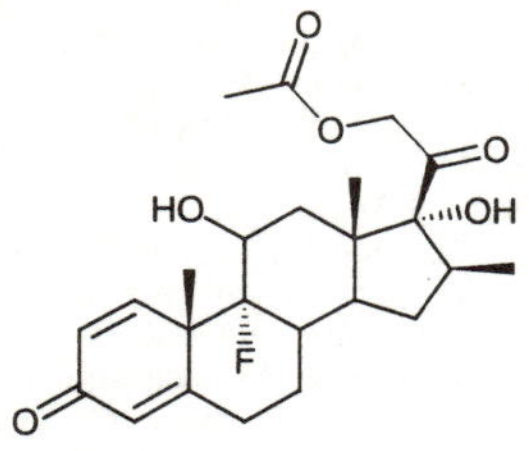

$C_{24}H_{31}FO_6$
(11β,16β)-9-Fluoro-11,17,21-trihydroxy-16-methylpregna-1,4-diene-3,20-dione 21-acetate.
Betafluorene; Celestovet; component of: Betavet Soluspan, Celestone Soluspan. Glucocorticoid used in dogs as an antipruritic. mp = 205-208° [also reported as 196-201°]; $[\alpha]_D$ = +140° (in chloroform); λ_m = 238 nm (ε 14800).

Veterinary products:

034-010; (with Betamethasone Sodium Phosphate) Betavet Soluspan Suspension; *Schering-Plough Animal Health.*
034-267; (with Gentamicin Sulfate) Gentocin ® Durafilm Ophthalmic Solution; *Schering-Plough Animal Health.*

664 Betamethasone Dipropionate

5593-20-4 1226 227-005-2

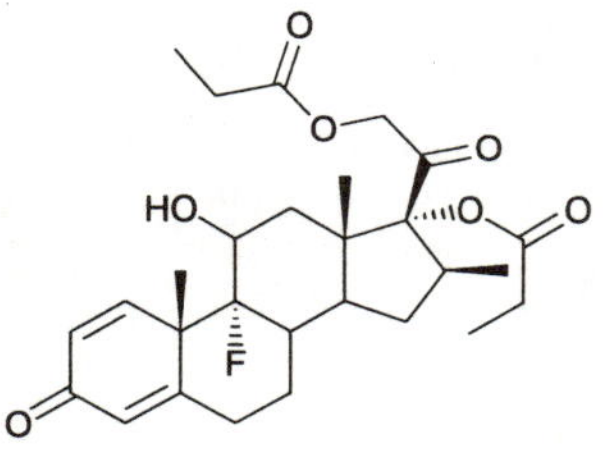

$C_{28}H_{37}FO_7$
(11β,16β)-9-Fluoro-11,17,21-trihydroxy-16-methylpregna-1,4-diene-3,20-dione 17,21-dipropionate.
betamethasone dipropionate; Sch-11460; Diprolene; Diproderm; Diprophos; Diprosis; Diprosone; Maxivate; Psorion; Rinderon-DP; component of: Alphatrex, Betasone, Lotrisone. Glucocorticoid used in dogs as an antipruritic. mp = 170-179° (decomposes); $[\alpha]_D^{26}$ = +65.7° (in dioxane); λ_m = 238 nm (ε 15700 in methanol).

Veterinary products:
049-185; (with Betamethasone Sodium Phosphate) Betasone Aqueous Suspension; *Schering-Plough Animal Health.*

665 Betamethasone Sodium Phosphate

151-73-5 1226 205-797-0

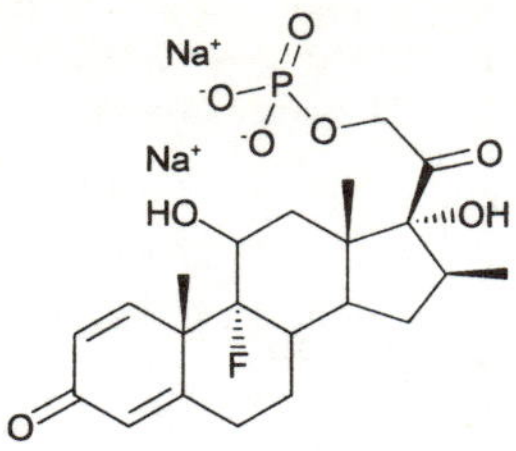

$C_{22}H_{28}FNa_2O_8P$
(11β,16β)-9-Fluoro-11,17,21-trihydroxy-16-methylpregna-1,4-diene-3,20-dione 21-phosphate disodium salt.
betamethasone 21-(dihydrogen phospate) disodium salt; Bentelan; Betnesol; Celestan; Durabetason; Vista-Methasone; component of: Betasone, Betavet Soluspan, Celestone Soluspan. Glucocorticoid used in dogs as an antipruritic.

Veterinary products:
034-010; (with Betamethasone Acetate) Betavet Soluspan Suspension; *Schering-Plough Animal Health.*
049-185; (with Betamethasone Dipropionate) Betasone Aqueous Suspension; *Schering-Plough Animal Health.*

666 Betamethasone Valerate

2152-44-5 1226 218-439-3

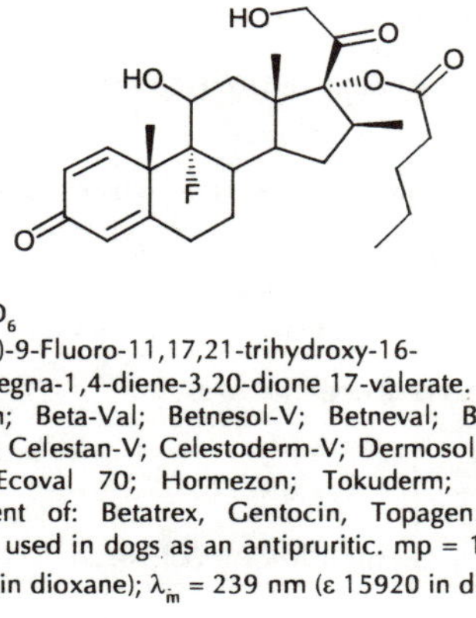

$C_{27}H_{37}FO_6$
(11β,16β)-9-Fluoro-11,17,21-trihydroxy-16-methylpregna-1,4-diene-3,20-dione 17-valerate.
Bedermin; Beta-Val; Betnesol-V; Betneval; Betnovate; Bextasol; Celestan-V; Celestoderm-V; Dermosol; Dermovaleas; Ecoval 70; Hormezon; Tokuderm; Valisone; component of: Betatrex, Gentocin, Topagen. Glucocorticoid used in dogs as an antipruritic. mp = 183-184°; $[\alpha]_D$= ${}_7$° (in dioxane); λ_m = 239 nm (ε 15920 in dioxane).

Veterinary products:
046-821; (with Gentamicin Sulfate) Gentocin ® Otic Solution; *Schering-Plough Animal Health.*
113-231; (with Gentamicin Sulfate) Topagen ® Ointment; *Schering-Plough Animal Health.*
132-338; (with Gentamicin Sulfate) Gentocin ® Topical Spray; *Schering-Plough Animal Health.*
140-896; (with Clotrimazole, Gentamicin Sulfate) Otomax ®; *Schering-Plough Animal Health.*
200-183; (with Gentamicin Sulfate) Gentavet ® Otic Solution; *Med-Pharmex.*
200-188; (with Gentamicin Sulfate) Betagen ™ Topical Spray; *Med-Pharmex.*
200-229; (with Clotrimazole, Gentamicin Sulfate) Tri-Otic Ointment; *Med-Pharmex.*

667 Deoxycorticosterone Pivalate

808-48-0 2947 212-366-0

$C_{26}H_{38}O_4$

21-(2,2-Dimethyl-1-oxopropoxy)pregn-4-ene-3,20-dione.

desoxycorticosterone pivalate; desoxycortone pivalate; Percorten Pivalate. Mineralocorticoid.

Veterinary products:

141-029; Percorten®-V; *Novartis Animal Health.*

668 Fludrocortisone

127-31-1 4166 204-833-2

$C_{21}H_{29}FO_5$

(11β)-9-Fluoro-11,17,21-trihydroxypregn-4-ene-3,20-dione.

9α-fluorcortisol; fluodrocortisone; fluohydrisone; fluohydrocortisone; Astonin H. Mineralocorticoid. Used to treat Addison's disease. mp = 260-262°; $[\alpha]_D^{23}$ = +139° (c = 0.55 95% EtOH); λ_m = 239 nm (ε 17600 MeOH); soluble in H_2O (0.14 mg/ml).

669 Fludrocortisone Acetate

514-36-3 4166 208-180-4

$C_{23}H_{31}FO_6$

(11β)-9-Fluoro-11,17,21-trihydroxypregn-4-ene-3,20-dione 21-acetate.

fludrocortisone 21-acetate; Alflorone; F-Cortef; Florinef. Mineralocorticoid. Used to treat Addison's disease. mp = 233-234°; $[\alpha]_D^{23}$ = +123° (c = 0.64 $CHCl_3$); λ_m = 238 nm (ε 16800 MeOH); very slightly soluble in H_2O (0.04 mg/ml); more soluble in Me_2CO (56 mg/ml), $CHCl_3$ (20 mg/ml), Et_2O (4 mg/ml).

670 Fluocinolone Acetonide

67-73-2 4185 200-668-5

$C_{24}H_{30}72O_6$

(6α,11β,16α)-6,9-Difluoro-11,21-dihydroxy-16,17-[(1-methylethylidene)bis(oxy)]pregna-1,4-diene-3,20-dione.

Coriphate; Cortiplastol; Dermalar; Fluonid; Fluovitef; Fluvean; Fluzon; Jellin; Localyn; Synalar; Synamol; Synandone; Synemol; Synotic; Synsac. Glucocorticoid; anti-inflammatory. mp = 265-266°; $[\alpha]_D$ = +95° (in chloroform); λ_m = 238 nm (log ε 4.21).

Veterinary products:

015-151; (with Neomycin Sulfate) Neo-synalar Cream; *Medicis Dermatologics.*

015-152; Synalar ® Cream Veterinary; *Medicis Dermatologics.*

015-298; Synalar ® Solution Veterinary; *Medicis Dermatologics.*

045-512; (with Dimethyl Sulfoxide) Synotic ® Otic Solution; *Fort Dodge Animal Health, Divn. AHP.*

047-334; Synsac Solution; *Fort Dodge Animal Health, Divn. AHP.*

671 Hydrocortisone

50-23-7 4828 200-020-1

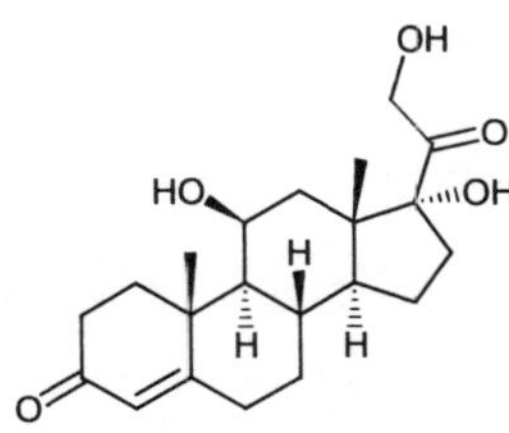

$C_{21}H_{30}O_5$

(11β)-11,17,21-Trihydroxypregna-4-ene-3,20-dione.

cortisol; 17-hydroxycorticosterone; anti-inflammatory hormone; Kendall's compound F; Reichstein's substance M; Aeroseb-HC; Ala-Cort; Anflam; Cetacort; Cort-Dome; Cortef; Cortenema; Cortril; Dermacort; Dermocortal; Dermolate; Dioderm; Efcortelan; Evacort; Ficortril; Hydracort; Hydro-Adreson; Hydrocort; Hydrocortisyl; Hydrocortone; Hytone; Lacticare-HC; Medicort; Mildison; Nutracort; Penecort; Proctocort; Scheroson F; Synacort; Texacort; Timocort; Zenoxone; Acticort; Alphaderm; CaldeCort Spray; Eldecort; component of: Cor-Tar-Quin, Cort-Quin Cortisporin, Drotic, Neo-Cort-Dome, Nystaform-HC, Otalgine, Otic-Neo-Cort-Dome, Otobiotic, Otocort, Pediotic Suspension, Racet, Vioform-Hydrocortisone, VoSol HC, Vytone. Principal glucocorticoid hormone produced by the adrenal cortex. Used in veterinary medicine as an anti-inflammatory. mp = 217-220°; $[\alpha]_D^{22}$ = +167° (in absolute ethanol); λ_m =

242 nm ($E^{1\%}_{1cm}$ 445); slightly soluble in water (0.28 mg/ml); more soluble in ethanol, methanol, acetone, chloroform, propylene glycol, ether; soluble in concen.

Veterinary products:

013-293; (with Oxytetracycline Hydrochloride) Liqua-Cortril Spray; Terra-Cortril Spray; *Pfizer*.

672 Hydrocortisone Acetate

50-03-3 4828 200-004-4

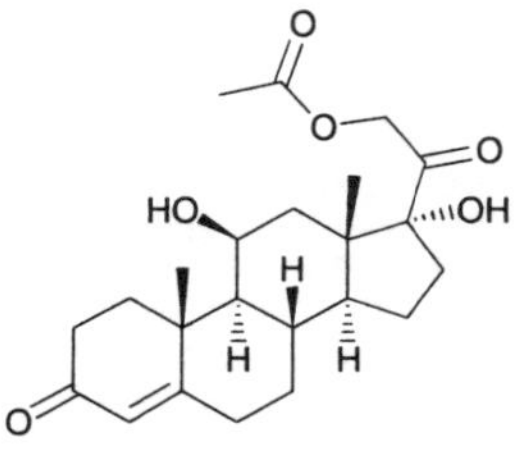

$C_{23}H_{32}O_6$

(11β)-21-(Acetoxy)-11,17,21-trihydroxypregna-1,4-diene-3,20-dione.

Anusol-HC; CaldeCort; Colifoam; Colofoam; Cortaid; Cordes; Cortef; Cortifoam; Cortril Acetate-AS; Efcolin; Hc45; Hydrin-2; Hydrocal; Hydrocortistab; Hydrocortone Acetate; Lanacort; Lenirit; Sigmacort; Sintotrat; Velopural; component of: Chloromycetin Hydrocortisone Ophthalmic, Clear-Aid, Coly-MycinS Otic, Cirticaine Cream, Cortisporin Cream, Epifoam, Lidaform-HC, Lidamantle-HC, Mantadil, Neo-Cortef, Ophthocort, ProctoFoam, Protef. Glucocorticoid. Used in veterinary medicine as an anti-inflammatory. Decomposes at 223°; d_4^{20}= 1.289; $[\alpha]_D^{25}$ = +166° (c = 0.4 in dioxane); $[\alpha]_D^{25}$ = +150.7° (c = 0.4 in acetone); λ_m = 242 nm ($E^{1\%}_{1cm}$ 390 in methanol); slightly soluble in water (1 mg/100 ml).

Veterinary products:

010-524; (with Neomycin Sulfate, Tetracaine Hydrochloride) Neo-Cortef ® with Tetracaine; *Pharmacia & Upjohn*.

040-322; (with Chymotrypsin, Neomycin Palmitate, Trypsin) Kymar Ointment Improved; *Schering-Plough Animal Health*.

043-784; (with Amphomycin Calcium, Kanamycin Sulfate) Kanfosone Ointment; *Fort Dodge Animal Health, Divn. AHP*.

047-997; (with Amphomycin Calcium, Kanamycin Sulfate) Amphoderm Ointment; *Fort Dodge Animal Health, Divn. AHP*.

065-015; (with Bacitracin Zinc, Neomycin Sulfate, Polymyxin B Sulfate) Bacitracin-Neomycin-Polymyxin With Hydrocortisone Acetate ; *Altana*.

065-119; (with Neomycin Sulfate, Penicillin G Procaine, Polymyxin B Sulfate) Forte Topical ® Ointment; *Pharmacia & Upjohn*.

065-476; (with Bacitracin Zinc, Neomycin Sulfate, Polymyxin B Sulfate) Cortisporin Veterinary Ophthalmic Ointment; *Schering-Plough Animal Health*.

093-514; (with Neomycin Sulfate) Neo-Cortef ® Sterile Ointment; *Pharmacia & Upjohn*.

673 Methylprednisolone

83-43-2 6189 201-476-4

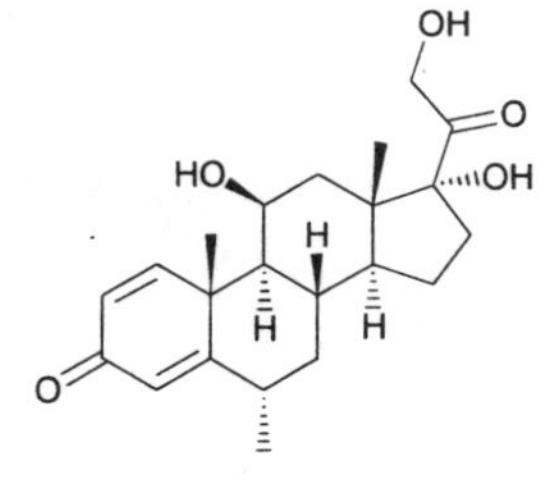

$C_{22}H_{30}O_5$

(6α,11β)-11,17,21-Trihydroxy-6-methylpregna-1,4-diene-3,20-dione.

NSC-19987; Medrate; Medrol; Medrone; Metastab; Metrisone; Promacortine; Suprametil; Urbason; component with acetylsalicylic acid of 011-700, Cortaba® Tablets (Pharmacia & Upjohn). Glucocorticoid used as an analgesic and, in veterinary medicine, as an anti-inflammatory. mp = 228-237°; $[\alpha]_D^{20}$ = 83° (in dioxane); λ_m = 243 nm (α_M 14875 in 95% ethanol).

Veterinary products:

011-403; Medrol ® Tablets; *Pharmacia & Upjohn*.

011-700; (with Acetylsalicylic Acid) Cortaba ® Tablets; *Pharmacia & Upjohn*.

135-771; Methylprednisolone Tablets; *Boehringer Ingelheim Vetmedica*.

674 Methylprednisolone Acetate

53-36-1 6189 200-171-3

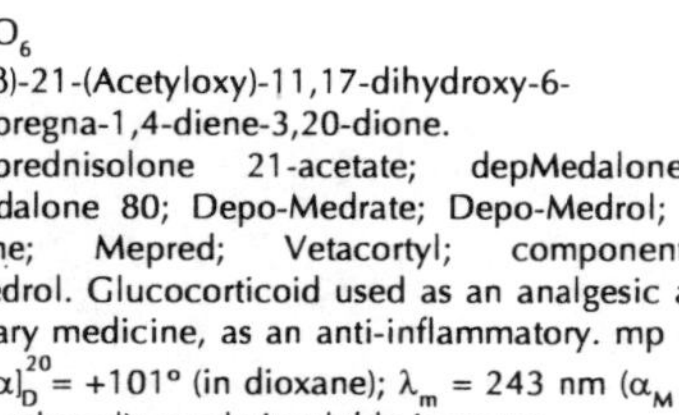

$C_{24}H_{32}O_6$

(6α,11β)-21-(Acetyloxy)-11,17-dihydroxy-6-methylpregna-1,4-diene-3,20-dione.

methylprednisolone 21-acetate; depMedalone 40; depMedalone 80; Depo-Medrate; Depo-Medrol; Depo-Medrone; Mepred; Vetacortyl; component of: NeoMedrol. Glucocorticoid used as an analgesic and, in veterinary medicine, as an anti-inflammatory. mp = 205-208°; $[\alpha]_D^{20}$= +101° (in dioxane); λ_m = 243 nm (α_M 14825 in 95% ethanol); nearly insoluble in water.

Veterinary products:

012-204; Depo-Medrol ® (20 mg) Sterile Aqueous Suspension; *Pharmacia & Upjohn*.

136-212; Methylprednisolone Acetate Inj.; *Boehringer Ingelheim Vetmedica*.

675 Prednisolone

50-24-8 7901 200-021-7

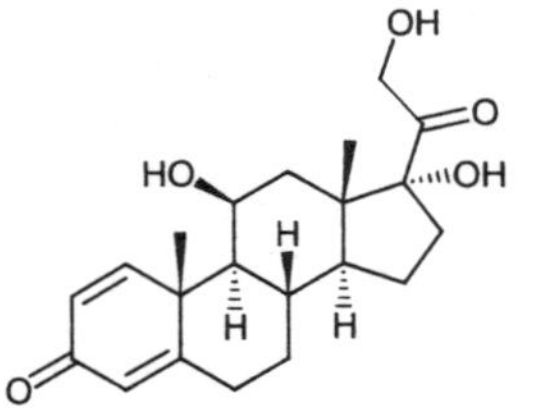

$C_{21}H_{28}O_5$

(11β)-11,17,21-Trihydroxypregna-1,4-diene-3,20-dione.

metacortandralone; delta F; δ^1-dehydrocortisol; δ^1-hydrocortisone; δ^1-dehydrohydrocortisone; hydroretrocortine; NSC-9120; Codelcortone; Cortalone; Decaprednil; Decortin H; Delta-Cortef; Deltacortril Enteric; Deltastab; Deltasolone; Flamasone; Hydeltra; Hydrodeltalone; Klismacort; Meticortelone; Meti-Derm; Paracortol; Precortancyl; Precortilon; Precortisyl; Prednelan; Prednicen; Predni-Dome; Predniretard; Predonine; Solone; Sterolone; component of: K-Predne-Dome. Synthetic corticosteroid; metabolically interconvertible with prednisone. Used to treat septic shock and in adjunct treatment of neoplasms. mp = 240-241° (decomposes); $[\alpha]_D^{25}$ = +102° (in dioxane); λ_m = 242 nm (ε 15,000 in methanol); slightly soluble in water; soluble in alcohol, chloroform, acetone, methanol, dioxane.

Veterinary products:

012-437; (with Trimeprazine Tartrate) Temaril-P-Tablets; *Pfizer.*

032-322; (with Hexamethyltetracosane, Neomycin Sulfate, Tetracaine) Liquisone F With Cerumene; *Evsco.*

035-161; (with Trimeprazine Tartrate) Temaril-P Spansule Capsule No.1; Temaril-P Spansule Capsule No. 2; *Pfizer.*

055-005; (with Chloramphenicol, Squalene, Tetracaine) Liquichlor With Cerumene; *Evsco.*

065-090; (with Novobiocin Sodium, Tetracycline Hydrochloride) Delta Albaplex ® 3x Tablets; Delta Albaplex ® Tablets; *Pharmacia & Upjohn.*

140-921; PrednisTab ®; *Lloyd.*

676 Prednisolone Acetate

52-21-1 7901 200-134-1

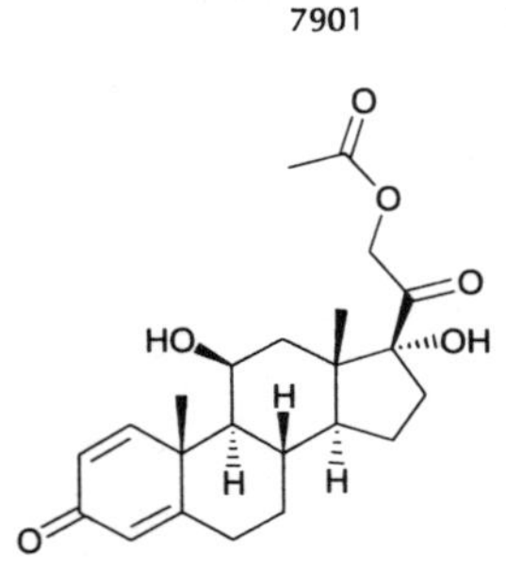

$C_{23}H_{30}O_6$

(11β)-21-(Acetyloxy)-11,17-dihydroxypregna-1,4-diene-3,20-dione.

Ak-Tate; Econopred; Hostacortin H; Inflanefran; Meticotelone Acetate; Predalone 50; Pred Forte; Pred Mild; Scherisolon; Sterane; component of: Blephamide Liquifilm, Blephamide SOP, Cetapred Ointment, Isopto Cetapred, Metimyd, Mydapred, Neo-Delta-Cortef, Poly-Pred, Pred-G Liquifilm, Pred-G SOP, Vasocidin Ointment. Glucocorticoid. Used to treat septic shock and in adjunct treatment of neoplasms. Decomposes at 237-239°; $[\alpha]_D^{25}$ = +116° (in dioxane).

Veterinary products:

010-312; Meticortelone Acetate; *Schering-Plough Animal Health.*

011-703; (with Neomycin Sulfate, Tetracaine Hydrochloride) Neo-Delta-Cortef ® With Tetracaine Ointment; *Pharmacia & Upjohn.*

045-288; (with Neomycin Sulfate) Optisone; *Evsco.*

065-259; (with Chloramphenicol) Chlorasone Ophthalmic Ointment; *Evsco.*

091-534; (with Neomycin Sulfate) Neo-Delta Cortef ® Sterile Solution; *Pharmacia & Upjohn.*

677 Prednisolone 21-tert-Butylacetate

7681-14-3 7901 231-661-5

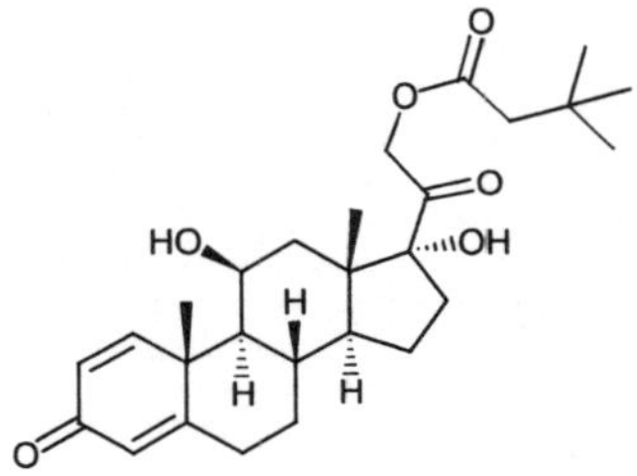

$C_{27}H_{38}O_6$

(11β)-21-(3,3-Dimethyl-1-oxobutoxy)-11,17-dihydroxypregna-1,4-diene-3,20-dione.

prednisolone tebutate; Codelcortone-T.B.A.; Hydeltra-T.B.A.; Predalone-T.B.A. Glucocorticoid. Used to treat septic shock and in adjunct treatment of neoplasms. Crystals; mp = 266-273°.

Veterinary products:

011-080; Hydeltrone-TBA Suspension; *Merial.*

678 Prednisolone Sodium Phosphate

125-02-0 7903 204-722-9

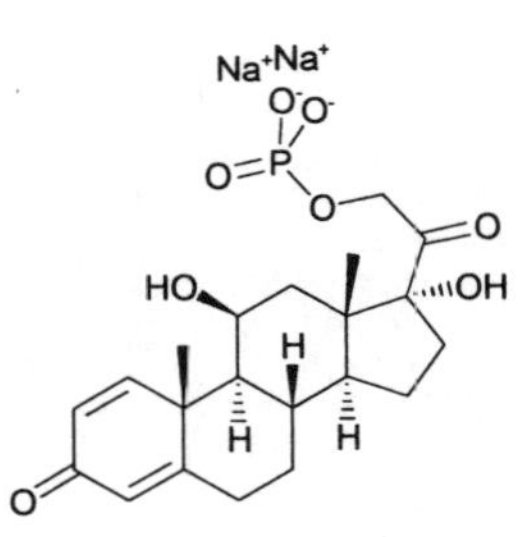

$C_{21}H_{27}Na_2O_8P$

(11β)-11,17-Dihydroxy-21-(phosphonooxy)pregna-1,4-diene-3,20-dione disodium salt.

21-prednisolonephosphoric acid disodium salt; prednisolone phosphate disodium; disodium prednisolone 21-phosphate; Ak-pred; Codelsol; Hefasolon; Hydeltrasol; Inflamase; Metreton; Pediapred; Prednesol; Predsol; Solucort; Solu-Predalone; component of: Optimyd, Vasocidin Solution. Glucocorticoid. Used to treat septic shock and in adjunct treatment of neoplasms. $[\alpha]_D^{25} = +102.5°$; λ_m = 243 nm ($A_{1cm}^{1\%}$ 308 in methanol); soluble in water, methanol, ethanol; pH (1% aqueous solution) = 7.5-8.5.

Veterinary products:

011-437; (with Neomycin Sulfate) Hydeltrone Ointment; *Merial.*
098-288; Prednis-A-Vet Injection; *Anthony Products.*

679 Prednisolone Sodium Succinate

1715-33-9 7901 216-995-1

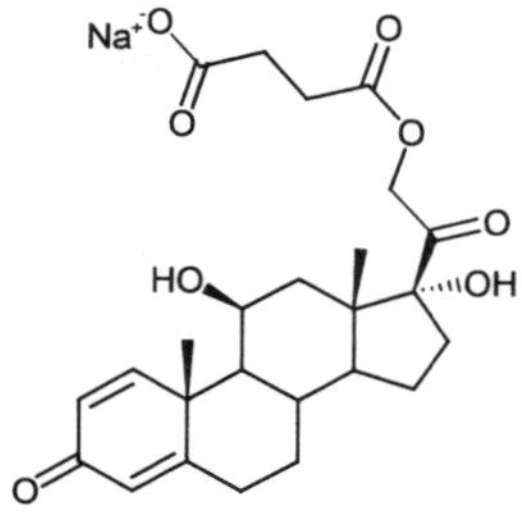

$C_{25}H_{31}NaO_8$
(11β)-21-(3-Carboxy-1-oxopropoxy)-11,17-dihydroxy-pregna-1,4-diene-3,20-dione monosodium salt.
prednisolone 21-succinate sodium salt; Di-Adreson-F; Meticotelone Soluble; Solu-Decortin-H. Glucocorticoid. Used to treat septic shock and in adjunct treatment of neoplasms.

Veterinary products:

011-593; Solu-Delta Cortef ® Sterile Powder; *Pharmacia & Upjohn.*
117-973; Prednisolone Sodium Succinate; *Steris.*

680 Prednisone

53-03-2 7904 200-160-3

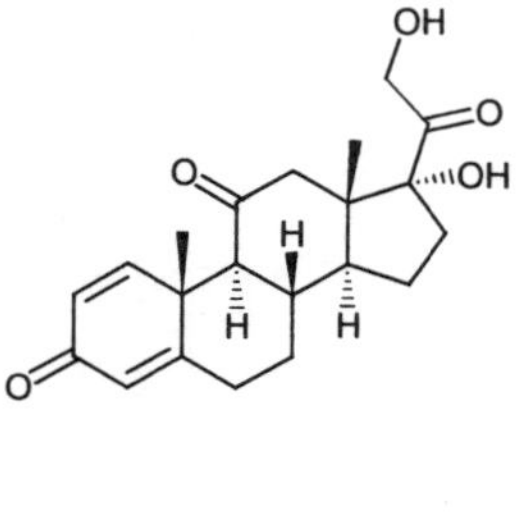

$C_{21}H_{26}O_5$
17,21-dihydroxypregna-1,4-diene-3,11,20-trione.
Δ^1-cortisone; Δ^1-dehydrocortisone; deltacortisone; delta E; metacortandracin; retrocortine; NSC-10023; Ancortone; Colisone; Cortancyl; Dacortin; Decortancyl; Decortin; Delcortin; Deltacortone; Deltasone; Deltison; Di-Adreson; Encorton; Meticorten; Nurison; Orasone; Paracort; Prednilonga; Pronison; Rectodelt; Sone; Ultracorten. Glucocorticoid. Used to treat septic shock and in adjunct treatment of neoplasms. Crystals; mp = 233-235°; $[\alpha]_D^{25} = +172°$ (dioxane); λ_m = 238 nm (ε 15500, MeOH); very slightly solulbe in H_2O, soluble in EtOH (0.66 g/100 ml), $CHCl_3$ (0.5 g/100 ml), slightly soluble in MeOH, dioxane.

Veterinary products:

009-963; Meticorten ®; *Schering-Plough Animal Health.*

Muscle Relaxants

681 Aminophylline

317-34-0 485 206-264-5

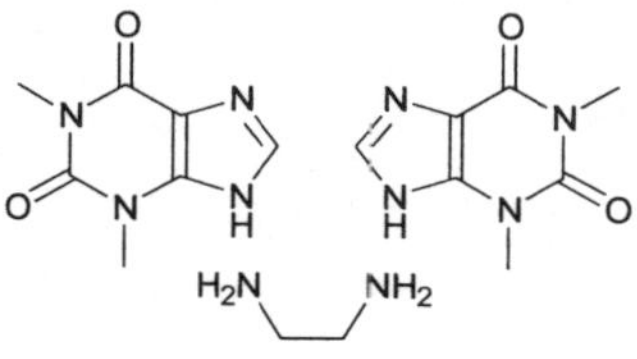

$C_{16}H_{24}N_{10}O_4$
3,7-Dihydro-1,3-dimethyl-1H-purine-2,6-dione compound with ethylenediamine (2:1).
Aminophyllin; Phyllocontin; Rectalad Aminophylline; Somophyllin; component of: Mudrane Tablets, Mudrane 2 Tablets, Mudrane GG Tablets, Mudrane GG-2 Tablets; theophyllamine; Carena; Inophylline; Metaphyllin; Theophyldine; Aminocardol; Aminodur; Ammophyllin; Cardiofilina; Cardophylin; Phylcardin; Tefamin; Cardiomin; Grifomin; Minaphil; Pecram; Peterphyllin; Phyllocontin; Somophyllin; Stenovasan; Theodrox; Cardophyllin; Diophyllin; Euphyllin CR; Genophyllin; Phyllindon; Theolamine; Theomin. Smooth muscle relaxant, used in animals as a bronchodilator. Soluble in H_2O (20 g/100 ml); insoluble in EtOH, Et_2O; LD_{50} (mus orl) = 540 mg/kg.

682 Atracurium Besylate

64228-81-5 900 264-743-4

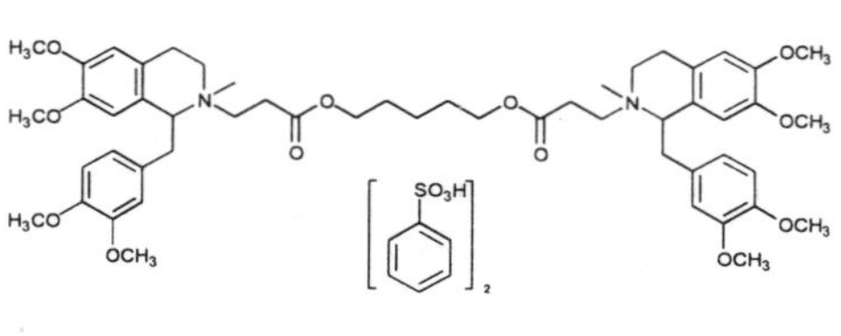

$C_{65}H_{82}N_2O_{18}S_2$
2-(2-Carboxyethyl)-1,2,3,4-tetrahydro-6,7-dimethoxy-2-methyl-1-veratrylisoquinolinium benzenesulfonate pentamethylene ester.
Tracrium; BW-33A; Wellcome 33-A-74. Skeletal muscle relaxant. Used as an adjunct to general anesthesia. mp = 85-90°.

683 Atracurium Besylate, 1R-cis, 1R'-cis form
96946-42-8 900

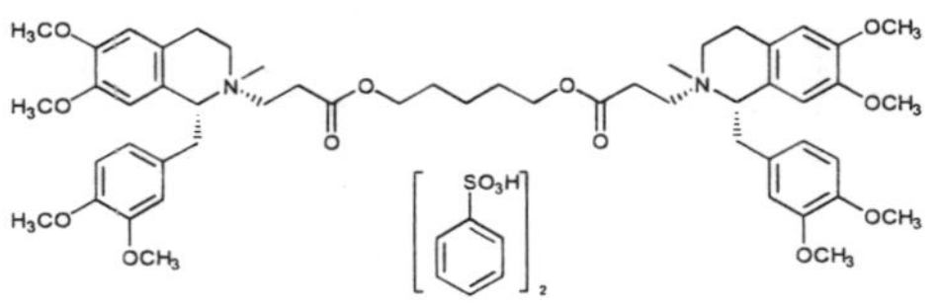

$C_{65}H_{82}N_2O_{18}S_2$
(1R-cis,1R'-cis)-2-(2-Carboxyethyl)-1,2,3,4-tetrahydro-6,7-dimethoxy-2-methyl-1-veratrylisoquinolinium benzenesulfonate pentamethylene ester.
Tracrium; BW-33A; 51W89; Nimbex; cisatracurium besylate. Skeletal muscle relaxant. Used as an adjunct to general anesthesia. White solid.

684 Chlorphenesin Carbamate
886-74-8 2231 212-954-7

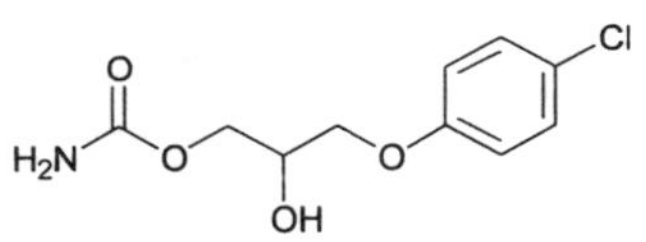

$C_{10}H_{12}ClNO_4$
3-(p-Chlorophenoxy)-1,2-propanediol 1-carbamate.
Maolate; U-19646; Rinlaxer. Skeletal muscle relaxant. mp = 89-91°; insoluble in H_2O, C_6H_6, C_6H_{12}, soluble in EtOH, Me_2CO, EtOAc, dioxane; LD_{50} (rat orl) = 748 mg/kg, (mus iv) = 239 mg/kg.

Veterinary products:

038-160; Maolate ® Tablets; Maolate Veterinary; *Pharmacia & Upjohn.*

685 Dantrolene
7261-97-4 2879 230-684-8

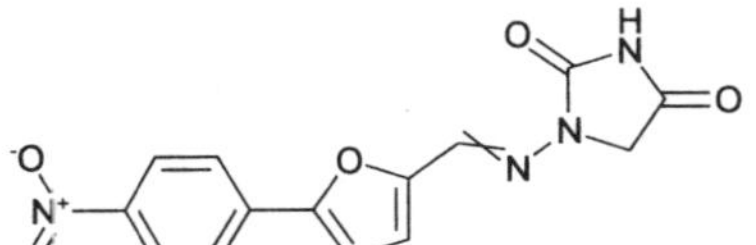

$C_{14}H_{10}N_4O_5$
1-[[5-(p-Nitrophenyl)furfurylidene]amino]hydantoin.
F-368. Skeletal muscle relaxant. May have value for treatment of malignant hyperthermia syndrome and urethral obstruction in dogs and cats. mp = 279-280°.

686 Dantrolene Sodium Salt
24868-20-0 2879

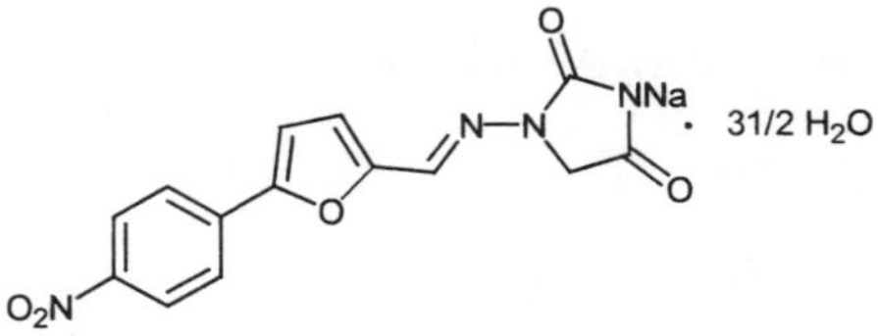

$C_{14}H_9N_4NaO_5 \cdot 3\ 1/2\ H_2O$
1-[[5-(p-Nitrophenyl)furfurylidene]amino]hydantoin sodium salt hydrate.
Dantrium [as hemiheptahydrate]; Dantamacrin; Dantrix [as tetrahydrate]. Skeletal muscle relaxant. May have value for treatment of malignant hyperthermia syndrome and urethral obstruction in dogs and cats. Slightly soluble in H_2O.

687 Methocarbamol
532-03-6 6060 208-524-3

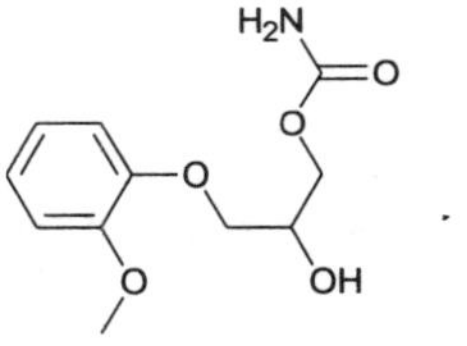

$C_{11}H_{15}NO_5$
3-(o-Methoxyphenoxy)-1,2-propanediol 1-carbamate.
AHR-85; Neuraxin; Miolaxene; Lumirelax; Etroflex; Delaxin; Robamol; Traumacut; Tresortil; Relestrid; Robaxin. Skeletal muscle relaxant. Used in animals to reduce muscular spasms. mp = 92-94°; soluble in H_2O (2.5 g/100 ml at 20°), soluble in EtOH, propylene glycol.

Veterinary products:

038-838; Robaxin ®-V Injectable; *Robins.*
045-715; Robaxin ®-V Tablets; *Robins.*

688 Pancuronium Bromide
15500-66-0 7139 239-532-5

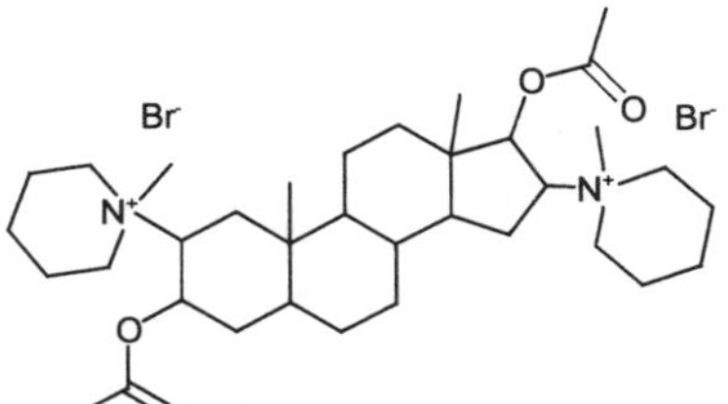

$C_{35}H_{60}Br_2N_2O_4$
1,1'-(3α,17β-Dihydroxy-5α-androstan-2β,16β-ylene)bis[1-methylpiperidinium] dibromide diacetate.
Pavulon; Org NA 97; NA 97; Mioblock; Poncuronium bromide. Skeletal muscle relaxant. Used as an adjunct in general anesthesia. Soluble in H_2O (100 g/100 ml), $CHCl_3$

(5 g/100 ml); LD_{50} (mus iv) = 0.047 mg/kg, (mus ip) = 0.152 mg/kg, (mus sc) = 0.167 mg/kg, (mus orl) = 21.9 mg/kg, (rat iv) = 0.153 mg/kg, (rbt iv) = 0.016 mg/kg.

689 Succinylcholine Bromide

55-94-7 9043 200-248-1

$C_{14}H_{30}Br_2N_2O_4$

Choline bromide succinate.

Suxamethonium bromide; IS-370; compd. 48/268; LT-1; M&B-2207; Brevidil M. Neuromuscular blocking agent. Skeletal muscle relaxant. In animals provides short term muscle relaxation useful for diagnosis or minor surgery. mp = 225°; freely soluble in H_2O.

690 Succinylcholine Chloride

71-27-2 9044 200-747-4

$C_{14}H_{30}Cl_2N_2O_4$

Choline chloride succinate.

Suxamethonium chloride; Anectine; Querlicin; Sucostrin. Neuromuscular blocking agent. Skeletal muscle relaxant. In animals provides short term muscle relaxation useful for diagnosis or minor surgery. mp = 156-163°, 190°; soluble in H_2O (100 g/100 ml), EtOH (0.42 g/100 ml); sparingly soluble in C_6H_6, $CHCl_3$; insoluble in Et_2O; LD_{50} (mus iv) = 0.45 mg/kg.

691 Succinylcholine Iodide

541-19-5 9045 208-770-1

$C_{14}H_{30}I_2N_2O_4$

Choline iodide succinate.

Suxamethonium iodide; Celocurine. Neuromuscular blocking agent. Skeletal muscle relaxant. In animals provides short term muscle relaxation useful for diagnosis or minor surgery. mp = 243-245°; freely soluble in H_2O.

692 Vecuronium Bromide

50700-72-6 10075 256-723-9

$C_{34}H_{57}BrN_2O_4$

1-(3α,17β-Dihydroxy-2β-piperidino-5α-androstan-16β,5α-yl)-1-methylpiperidinium bromide diacetate.

Norcuron; Org-NC45; NC45; Musculax. Neuromuscular blocker; skeletal muscle relaxant; adjunct to general anesthesia. mp = 227-229°; LD_{50} (mus iv) = 0.061 mg/kg.

Narcotic Antagonists

693 Diprenorphine Hydrochloride

16808-86-9 3403

$C_{26}H_{36}ClNO_4$

(5α,7α)-17-(Cyclopropylmethyl)-4,5-epoxy-18,19-dihydro-3-hydroxy-6-methoxy-α,α-dimethyl-6,14-ethenomoprhinan-7-methanol hydrochloride.

Revivon. Narcotic antagonist. Crystals; LD_{50} (mus sc) = 316 ± 20 mg/kg.

Veterinary products:

047-870; (with Etorphine Hydrochloride) M50-50 Diprenorphine; M99 Etorphine; *Wildlife*.

694 Nalorphine Hydrochloride

57-29-4 6448 200-321-8

$C_{19}H_{22}ClNO_3$

17-Allyl-7,8-didehydro-4,5α-epoxymorphinan-3,6α-diol hydrochloride.

Nalline. Narcotic antagonist. A Federally controlled substance. mp = 260-263°; soluble in H_2O; somewhat soluble in EtOH; λ_m = 285 nm (H_2O); LD_{50} (rat sc) = 1460 mg/kg.

Veterinary products:

010-424; Nalline Hydrochloride; *Merial.*

695 Naloxone Hydrochloride

357-08-4 6449 206-611-0

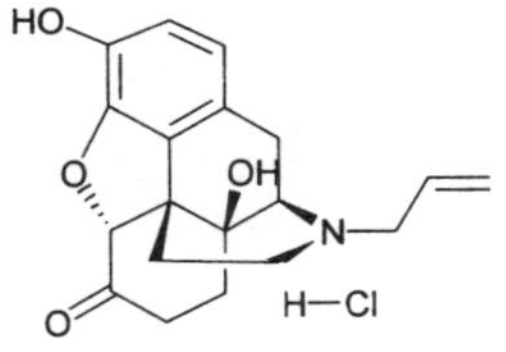

$C_{19}H_{22}ClNO_4$
17-Allyl-4,5α-epoxy-3,14-dihydroxymorphinan-6-one hydrochloride.
N-allylnoroxymorphone hydrochloride; Narcan; EN-15304; Nalone; Narcanti. Narcotic antagonist. Used for its opiate reversal effects. mp = 200-205°; soluble in H_2O, EtOH, insoluble in Et_2O; [free base]: mp = 184°, 178°; $[\alpha]_D^{20}$ = -194.5° (c = 0.93 in $CHCl_3$); soluble in $CHCl_3$; nearly insoluble in petroleum ether.

Veterinary products:

035-825; Narcan Injection; *Endo.*

696 Naltrexone Hydrochloride

16676-29-2 6450 240-723-0

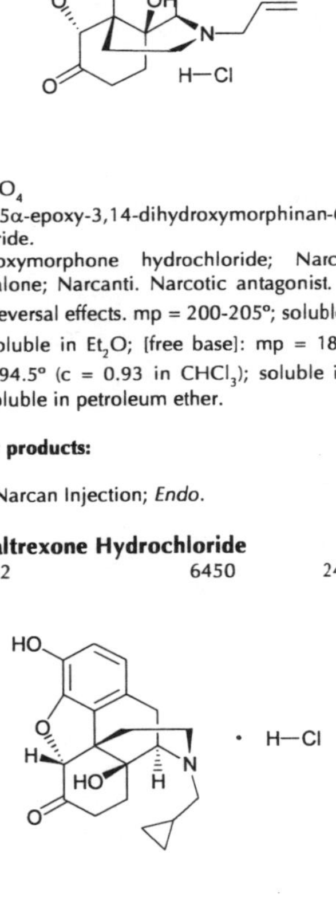

$C_{20}H_{23}NO_4 \cdot HCl$
17-(Cyclopropylmethyl)-4,5α-epoxy-3,14-dihydroxymorphinan-6-one hydrochloride.
EN-1639A; Antaxone; Celupan; Nalorex; Trexan. Narcotic antagonist, related to cyprenorphine. Used to counter tail-chasing behavior in dogs and cats. mp = 274-276°; [free base]: mp = 168-170°; LD_{50} (mus sc) = 586 mg/kg.

Veterinary products:

141-074; Trexonil ®; *Wildlife.*

Oxytocics

697 Oxytocin

50-56-6 7114 200-048-4

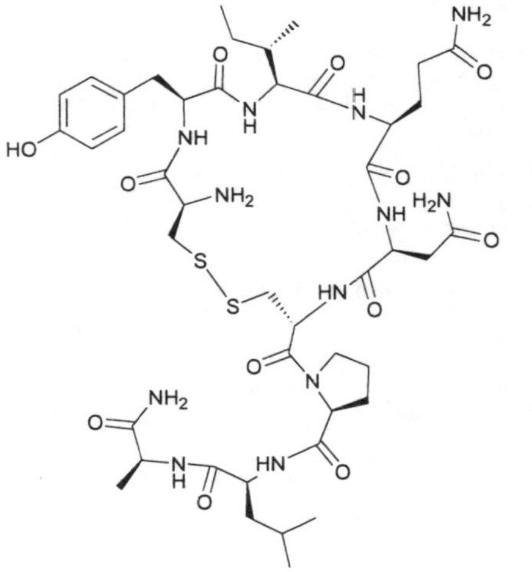

$C_{44}H_{66}N_{12}O_{12}S_2$
Oxytocin.
Pitocin; Syntocinon; Uteracon; Alpha-hypophamine; ocytocin; Intertocine-S; Perlacton; Orasthin; Oxystin; Partocon; Synpitan; H-Cys-Tyr-Ile-Glu-Asp-Cys-Pro-Leu-Gly-NH_2 cyclic (1→6)-disulfide. Oxytocic. Used to induce uterine contractions. $[\alpha]_D^{22}$ = -26.2° (c = 0.53); soluble in H_2O, 1-BuOH, 2-BuOH.

Veterinary products:

042-889; Oxytocin; *Veterinary Labs.*
044-585; Oxytocin Injection; *Steris.*
046-788; Oxytocin Injection; *Fort Dodge Animal Health, Divn. AHP.*
046-822; Vetocin Injection; *United Vaccines.*
099-169; Oxytocin (Label Under Distributors); *Phoenix.*
109-305; Oxytocin Injectable; *Boehringer Ingelheim Vetmedica.*
124-241; PVL Oxytocin Injectable; *Merial.*
130-136; Oxytocin Injection; *Anthony Products.*

698 Oxytocin Citrate

7563-62-4 7114

$C_{50}H_{74}N_{12}O_{19}S_2$
Oxytocin citrate (salt).
Pitocin-Buccal. Oxytocic. Used to induce uterine contractions.

Pharmaceutical Aids

699 Hyaluronate Sodium

9067-32-7 4793
Hyaluronic acid, sodium salt.
hyaluronic acid FCH; Kelisema Sodium Hyaluronate Bio; Pronova™; ARTZ; Connettivina; Equron; Healon; Healonid; Hyacid; Hyalgan; Hyalovet; Hyonate' lal;

Opegan; Provisc; Synacid. Used as a gelling agent and as an adjunct in treatment of non-infectious synovitis. White powder; $[\alpha]_D^{25}$ = -74° (c = 0.25 H_2O); moderately toxic if ingested.

Veterinary products:

112-048; Hylartin ® V Injection; *Pharmacia & Upjohn.*
122-578; Hyvisc; *Anika.*
139-913; Equron ®; *Fort Dodge Animal Health, Divn. AHP.*
140-474; Synacid ™; *Schering-Plough Animal Health.*
140-806; Hyalovet ®; *Fort Dodge Animal Health, Divn. AHP.*
140-883; Legend ™ Injectable Solution; *Bayer.*

700 Hexamethyltetracosane

111-01-3 8923 203-825-6

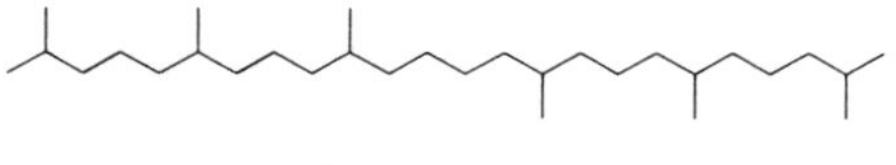

$C_{30}H_{62}$
Dodecahydrosqualene.
squalane; spinacane; 2,6,10,15,19,23-hexamethyltetracosane; perhydrosqualene; Cosbiol; perhydrosqualene; Robane; Robane®; Saturated branched chain hydrocarbon obtained by hydrogenation of shark liver oil or other natural oils; high-grade lubricating oil, perfume fixative, gas chromatographic analysis, transformer oil; in cosmetics and pharmaceut. FDA approved for topicals, USP/NF, BP compliance. Used as an oleaginous vehicle and bactericide. Lubricant in topical pharmaceuticals, carrier of lipid-soluble drugs in suppositories, ointment base. mp = -38°; bp_{10} = 263°; d_4^{15} = 0.8115; n_D^{15} = 1.4530; insoluble in H_2O, soluble in organic solvents. No known toxicity.

Veterinary products:

032-322; (with Neomycin Sulfate, Prednisolone, Tetracaine) Liquisone F With Cerumene; *Evsco.*

Plasma Volume Expanders

701 Dextran 70

9004-54-0 2989 232-677-5
Polysaccharide produced by the action of *Leuconostoc mesenteroides* on sucrose.
Hyskon; Macrodex; Aquasite; component of: Biontears, Estivin II, Tears Naturale, Tears Naturale II, Tears Naturale Free. Plasma volume extender. Used in treatment of shock resulting from decreased volume of circulating plasma. Average molecular weight 70,000.

702 Hetastarch

9004-62-0 4707
Starch 2-hydroxyethyl ether.
Hydroxyethyl starch; HES; 6-H.E.S.; Hespan; Hespander; Hestar; Hestat; Hestsol; Plasmasteril; Volex. An amylopectin derivative used as a cryoprotective agent for erythrocytes and in small animals as a plasma volume extender.

Progestogens

703 Altrenogest

850-52-2 327 212-703-1

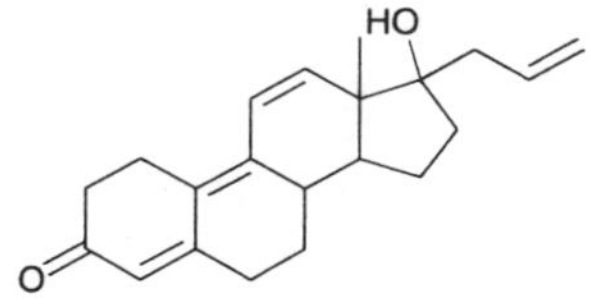

$C_{21}H_{26}O_2$
17α-Allyl-17β-hydroxyestra-4,9,11-trien-3-one.
A-35957; RU-2267; Regumate; 131-310 Regu-Mate® (*Hoechst-Roussel Vet.*). Progestogen used to suppress and control estrus in mares. mp = 120°; $[\alpha]_D^{20}$ = -72° (c = 0.5 EtOH).

Veterinary products:

131-310; Regu-Mate®; *Hoechst-Roussel Vet.*

704 Flurogestone Acetate

2529-45-5 4235 219-776-9

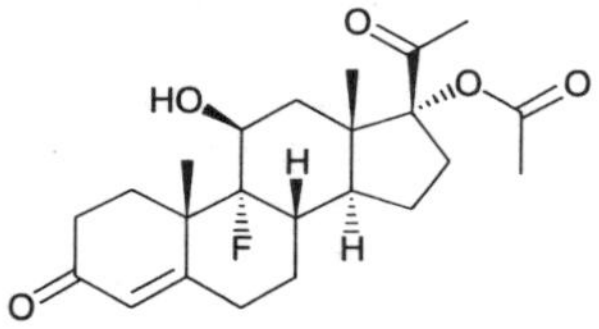

$C_{23}H_{31}FO_5$
9-Fluoro-11β,17-dihydroxypregn-4-ene-3,20-dione 17-acetate.
SC-9880; NSC-65411; Chronogest; Cronolone; Synchronate; component of: Syncro-Mate. Progestogen used in veterinary medicine for estrus regulation. mp = 266-269°; $[\alpha]_D$ = 77.6° ($CHCl_3$); λ_m = 238 nm (ε 17500 MeOH).

Veterinary products:

034-601; Synchro-Mate ®; *Searle.*

705 Megestrol Acetate

595-33-5 5849 209-864-5

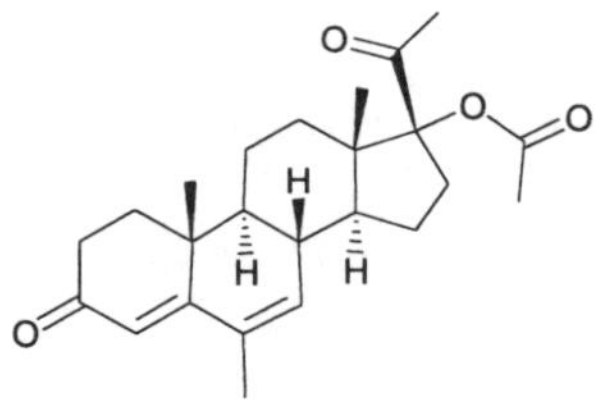

$C_{24}H_{32}O_4$
17-Hydroxy-6-methylpregna-4,6-diene-3,20-dione acetate.

Maygace; Megace; Megestat; Megestil; Nia; Niagestin; Ovaban; component of: Co-Ervonum, Kombiquens, Noval, Nuvacon, Planovin, Triu-Ervonum, Volidan, Weradys, Delpregnin. Progestogen. Used for palliative treatment of breast cancer and in veterinary medicine as an estrus regulator. mp = 214-216°; $[\alpha]_D^{24}$= 5° ($CHCl_3$); λ_m = 287 nm (log ε 4.40); soluble in H_2O (2 σg/ml), plasma (24 σg/ml).

Veterinary products:

091-603; Ovaban ® Tablets 20 mg; Ovaban ® Tablets 5 mg; *Schering-Plough Animal Health.*

706 Melengestrol Acetate

2919-66-6 5859 220-859-7

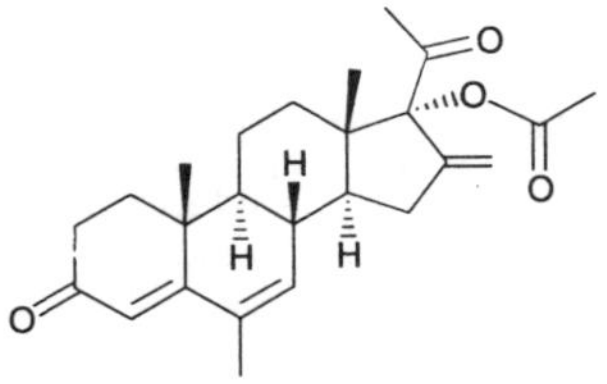

$C_{25}H_{32}O_4$
17α-Hydroxy-6-methyl-16-methylenepregna-4,6-diene-3,20-dione acetate.
MGA. Used as a progestogen (in veterinary medicine) and as an antineoplastic agent. Crystals; mp = 224-226°; $[\alpha]_D^{23}$= -127° (c = 0.31 $CHCl_3$); λ_m = 287 nm (log ε = 4.35 EtOH).

Veterinary products:

034-254; MGA ® 100 Liquid Premix; MGA ® 200 Liquid Premix; MGA ® 500 Liquid Premix; *Pharmacia & Upjohn.*
039-402; MGA ® 500 Liquid Premix; *Pharmacia & Upjohn.*
046-718; (with Oxytetracycline) MGA ® (liquid) / Terramycin ®; *Pharmacia & Upjohn.*
046-719; (with Oxytetracycline) MGA ® (dry) / Terramycin ®; *Pharmacia & Upjohn.*
124-309; (with Monensin Sodium) MGA ® 100-200 / Rumensin ®; *Pharmacia & Upjohn.*
125-476; (with Monensin Sodium) MGA ® 500 / Rumensin ®; *Pharmacia & Upjohn.*
138-792; (with Monensin Sodium, Tylosin Phosphate) MGA ® 100-200 / Rumensin ® / Tylan ®; *Pharmacia & Upjohn.*
138-870; (with Monensin Sodium, Tylosin Phosphate) MGA ® (liquid) / Rumensin ® / Tylan; MGA ® 100-200 Premix / MGA 500 Liquid Premix / Rumensin ® / Tylan ®; *Pharmacia & Upjohn.*
138-904; (with Lasalocid Sodium, Tylosin Phosphate) MGA ® 100-200 Premix / MGA ® 500 Liquid Premix / Bovatec ® / Tylan ®; MGA ® 100 / Bovatec ® / Tylan ®; *Pharmacia & Upjohn.*
138-992; (with Lasalocid Sodium, Tylosin Phosphate) MGA ® (liquid) / Bovatec ® / Tylan ®; *Pharmacia & Upjohn.*
138-995; (with Tylosin Phosphate) MGA ® 100 / Tylan ®; MGA ® 200 / Tylan ®; *Pharmacia & Upjohn.*
139-192; (with Tylosin Phosphate) MGA ® 500 (liquid) / Tylan ®; *Pharmacia & Upjohn.*
139-876; (with Lasalocid Sodium) MGA ® 100 Bovatec ®; MGA ® 200 Bovatec ®; *Pharmacia & Upjohn.*
140-288; (with Lasalocid Sodium) MGA ® 500 / Bovatec ®; *Pharmacia & Upjohn.*

707 Norgestomet

25092-41-5 246-611-8

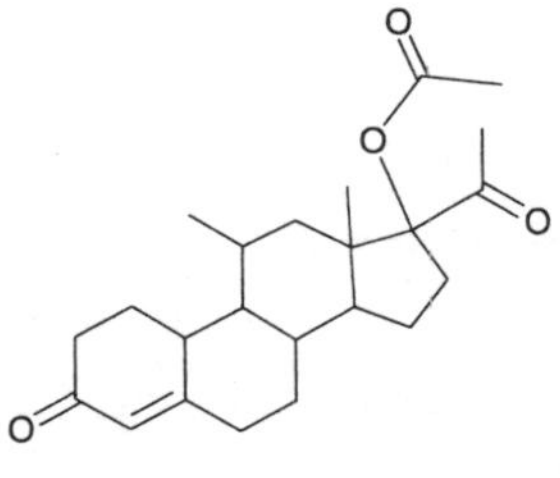

$C_{23}H_{32}O_4$
17-Hydroxy-11β-methyl-19-norpregn-4-ene-3,20-dione acetate.
SC-21009. Progestogen.

Veterinary products:

097-037; (with Estradiol Valerate) Syncro-Mate-B ®; *Merial.*
134-930; (with Estradiol Valerate) Syncro-Mate-B ®; *Merial.*

708 Progesterone

57-83-0 7956 200-350-6

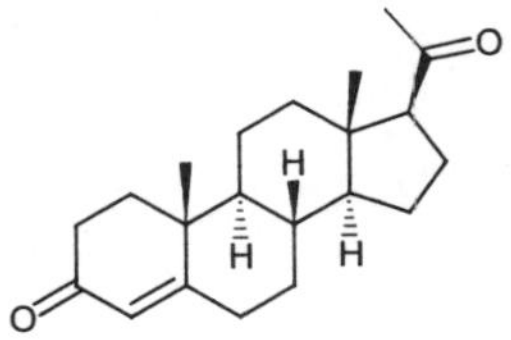

$C_{21}H_{30}O_2$
Pregn-4-en-3,20-dione.
corpus luteum hormone; luteohormone; Corlutina; Corluvite; Cyclogest; Gestiron; Gestone; Lipo-Lutin; Lutocuclin M; Lutogyl; Lutromone; Progestasert; Progestin; Progestogel; Progestol; Progeston; Prolidon; Proluton; Syngesterone; Utrogestan. Progestogen. [α form]; mp = 127-131°; d^{23} = 1.166; [β form]; mp = 121°; d^{20} = 1.171; $[\alpha]_D^{20}$= 172° to 182° (c = 2 dioxane)λ_m = 240 nm; soluble in EtOH, Me_2CO, dioxane.

Veterinary products:

009-576; (with Estradiol Benzoate) Synovex ®-C; Synovex ®-S; *Fort Dodge Animal Health, Divn. AHP.*
110-315; (with Estradiol Benzoate, Tylosin Tartrate) ; *Ivy.*

Prostaglandins

709 Cloprostenol Sodium

55028-72-3 2461 259-439-3

$C_{22}H_{28}ClNaO_6$

Sodium (±)-(Z)-7-[(1R*,2R*,3R*,5S*)2-[(E)-(3R*)-4-(m-chlorophenoxy)-3-hydroxy-1-butenyl]-3,5-dihydroxycyclopentyl]-5-heptenoate.

Estrumate; ICI-80996; Planate. Prostaglandin used in treatment of infertility and synchronization of estrus in farm animals.

Veterinary products:

113-645; Estrumate ®; *Bayer.*

710 Dinoprost Tromethamine

38562-01-5 8065 254-002-3

$C_{24}H_{45}NO_8$

(E,Z)-(1R,2R,3R,5S)-7-[3,5-Dihydroxy-2-[(3S)-(3-hydroxy-1-octenyl)]cyclopentyl]-5-heptenoic acid compound with 2-amino-2-(hydroxyethyl)-1,3-propanediol.

Prostaglandin $F_{2\alpha}$ tromethamine; $PGF_{2\alpha}$ THAM; U-14583E. Prostaglandin used as an oxytocic agent. Used in cattle to synchronize estrus and as an abortifacient. mp = 25-35°; $[\alpha]_D^{25}$ = 23.5° (c = 1 THF); freely soluble in MeOH, EtOH, EtOAc, $CHCl_3$, slightly soluble in H_2O; LD_{50} (rbt iv, im) = 2.5-5.0 mg/kg; [tromethamine salt ($C_{24}H_{45}NO_8$)]: mp = 100-101°, soluble in H_2O (> 20 g/ml).

Veterinary products:

100-202; Prostin F2 Alpha ® Sterile Solution; *Pharmacia & Upjohn.*
108-901; Lutalyse Sterile Solution; *Pharmacia & Upjohn.*
200-253; Dinoprost Tromethamine Injection; Prosta-Mate ™; ProstaMate ™ Sterile Solution; *Phoenix.*

711 Fenprostalene

69381-94-8 4037 273-982-3

$C_{23}H_{30}O_6$

(±)-9α,11α,15α-Trihydroxy-16-phenoxy-17,18,19,20-tetranorprosta-4,5,13-trans-trienoic acid methyl ester.

RS-84043; Bovilene; Porcilene; Synchrocept B. Prostaglandin used as a luteolytic agent. Used as an estrus regulator and abortifacient in cows. λ_m = 220, 265, 271, 278 nm (log ε 3.99, 3.11, 3.23, 3.16, MeOH).

Veterinary products:

128-549; Bovilene; *Fort Dodge Animal Health, Divn. AHP.*
138-903; Porcilene ® Solution; *Fort Dodge Animal Health, Divn. AHP.*

712 Fluprostenol Sodium

55028-71-2 4231 259-438-8

$C_{23}H_{28}F_3NaO_6$

Sodium (±)-(zZ)-7-[(1R*,2R*,3R*,5S*)-3,5-dihydroxy-2-[(E)-(3R*)-3-hydroxy-4-[(α,α,α-trifluoro-m-tolyl)oxy]-1-butenyl]cyclopentyl]-5-heptenoate.

ICI-80008. Prostaglandin used in treatment of infertility in mares.

Veterinary products:
111-529; Equimate ®; *Bayer.*

713 Luprostiol

67110-79-6 5638

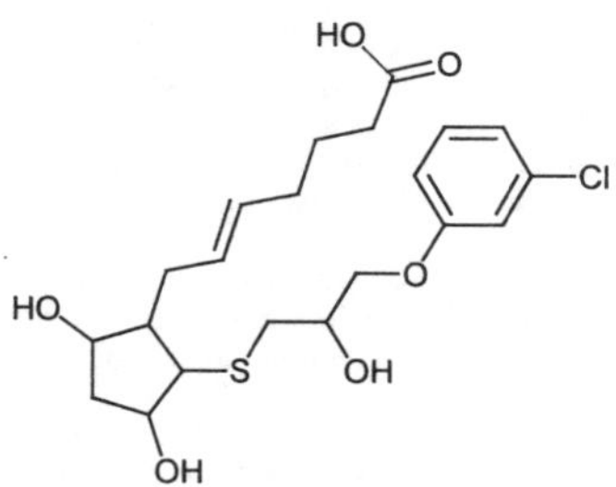

$C_{21}H_{29}ClO_6S$

(±)-(Z)-7-[(1R*,2S*,3S*,5R*)-2-[[(2R*)-3-(m-Chlorophenoxy)-2-hydroxypropyl]thio]-3,5-dihydroxycyclopentyl]-5-heptenoic acid.

prostianol;EMD-34946; Pronilin; Prosolvin; Reprodin. Synthetic prostaglandin; luteolytic (veterinary). Analog of prostaglandin $F_{2\alpha}$ used in veterinary medicine for induction of luteolysis.

Veterinary products:

140-857; Equestrolin ™; Prosolvin ®; *Intervet.*

714 Prostalene

54120-61-5 8066 258-984-4

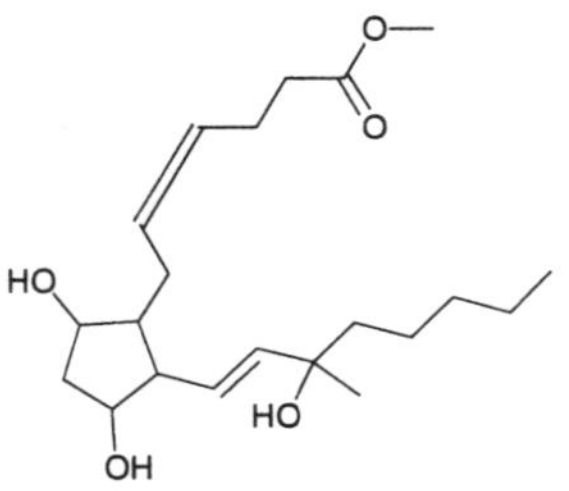

$C_{22}H_{36}O_5$
(±)-Methyl 7-[(1R*,2R*,3R*,5S*)-3,5-dihydroxy-2-[((E)-3-hydroxy-3-methyl-1-octenyl]cyclopentyl]-4,5-heptadienoate.
Synchrocept; RS-9390. Prostaglandin. Used as a veterinary luteolytic.

Veterinary products:

100-254; Synchrocept Solution; *Fort Dodge Animal Health, Divn. AHP.*

Replenishers

715 Magnesium Sulfate Heptahydrate

10034-99-8 5731

$MgSO_4 \cdot 7H_2O$
Sulfuric acid magnesium salt (1:1) heptahydrate.
bitter salts; Epsom salts. Magnesium replenisher. Also anticonvulsant and laxative. d = 1.67; soluble in H_2O (71 g/100 ml at 20°, 91 g/100 ml at 40°), slightly soluble in EtOH.

Veterinary products:

046-789; (with Chloral Hydrate, Pentobarbital) Chloropent; *Fort Dodge Animal Health, Divn. AHP.*

716 Sodium Chloride

7647-14-5 8742 231-598-3

ClNa
Hydrochloric acid, sodium salt.
chlorure de sodium; cloreto de sódio; common salt; dendritis; extra fine 200 salt; extra fine 325 salt; h.g. blending; halite; NaCl; natural halite; Purex; rock salt; saline; Saline solution; salt; sea salt; Stat trak plus; sterling; table salt; top flake; white crystal ; Inorganic salt; occurs in nature as the mineral halite; source of chlorine and sodium; preservative, seasoning for foods; manufacture of soaps, dyes; in freezing mixtures; dyeing and printing fabrics, glazing pottery. Curing hides, metallurgy, herbicide, electrolyte replenisher, emetic, topical anti-inflammatory. FDA listed, BP, Ph. Eur compliance. Used as a flavoring agent and preservative and as a tonicity agent (in injectables, nasals, dentals, parenterals, inhalants, ophthalmics, orals, rectals and topicals). Essential nutrient factor in veterinary medicine. mp = 804°; bp = 1439°; d = 0.93; slightly soluble in EtOH, soluble in 95% EtOH (0.4 g/100 ml), H_2O (35.7 g/100 ml at 25°; 38.5 g/100 ml at 100°), glycerin (10 g/100 ml), insoluble in organic solvents. LD_{50} (rat orl) = 3000 mg/kg, (mus sc) = 3000 mg/kg, (mus orl) = 4000 mg/kg, (mus iv) = 650 mg/kg, (mus ip) = 6610 mg/kg.

Veterinary products:

006-281; (with Gelatin) Intragel; *Fort Dodge Animal Health, Divn. AHP.*
125-961; (with Citric Acid; Dextrose; Glycine; Potassium Citrate; Potassium Phosphate) Sodium Re-Sorb; Vy'trate; *Pfizer.*

Respiratory Stimulants

717 Doxapram Hydrochloride Monohydrate

7081-53-0 3488

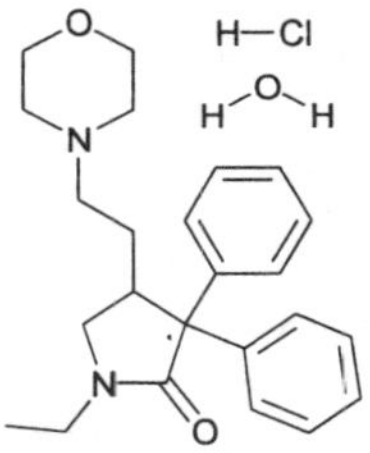

$C_{24}H_{31}ClN_2O_2,H_2O$
1-Ethyl-4-(2-morpholinoethyl)-3,3-diphenyl-2-pyrrolidinone monohydrochloride monohydrate.
Dopram; AHR-619; Doxapril; Stimulexin. Respiratory stimulant. Used in dogs, cats and horses to stimulate respiration during and after general anesthesia. mp = 217-219°; soluble in H_2O, slightly soluble in EtOH, $CHCl_3$; LD_{50} (rat orl) = 261 mg/kg.

Veterinary products:

034-879; Dopram®-V Injectable; *Robins.*

Sedatives and Hypnotics

718 Azaperone

1649-18-9 931 216-715-8

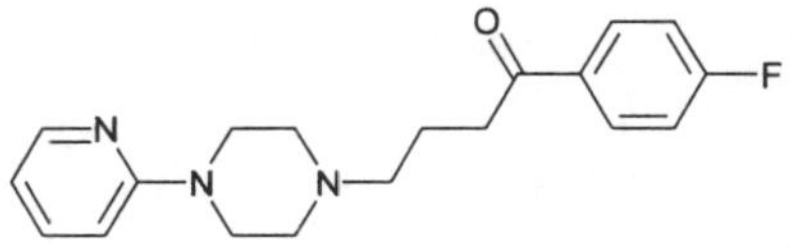

$C_{19}H_{22}FN_3O$
4'-Fluoro-4-[4-(2-pyridinyl)-1-piperazinyl]butyrophenone.
R-1929; Stresnil; Suicalm. Sedative (veterinary); tranquilizer (veterinary). mp = 73-75°.

Veterinary products:

115-732; Stresnil® Injection; *Schering-Plough Animal Health.*

719 Chloral Hydrate

302-17-0 2113 206-117-5

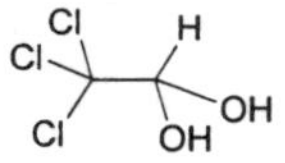

$C_2H_3Cl_3O_2$
2,2,2-Trichloro-1,1-ethanediol.
Noctec; SK Chloral Hydrate; Escre; Nycton; Somnos; Chloraldurat. Used in veterinary medicine as an anesthetic, sedative and hypnotic. mp = 57°; d = 1.91; bp = 98°; freely soluble in H_2O (240 g/100 ml at 0°, 830 g/100 ml at 25°, 1430 g at 40°), EtOH (76.9 g/100 ml), $CHCl_3$ (50 g/100 ml), Et_2O (66.6 g/100 ml), glycerol (200 g/100 ml), CS_2 (1.47 g/100 ml); sparingly solution.

Veterinary products:

046-789; (with Magnesium Sulfate, Pentobarbital) Chloropent; *Fort Dodge Animal Health, Divn. AHP.*

720 Chlorobutanol

57-15-8 2180 200-317-6

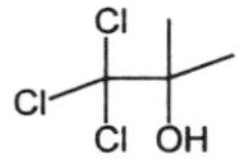

$C_4H_7Cl_3O$
1,1,1-Trichloro-2-methyl-2-propanol.
trichlorobutanol; β,β,β-Trichloro-tert-butyl alcohol; acetone chloroform; chlorbutol; Chloretone; Coliquifilm; Methaform; Sedaform. Analgesic, dental. Has been used in veterinary medicine as a mild sedative, antipruritic and antiseptic. [anhydrous]: mp = 97°; [hemihydrate]: mp = 78°; bp_{760} = 167°; bp_{246} = 135°; soluble in hot H_2O, alc, glycerol, $CHCl_3$, Et_2O, Me_2CO, petroleum ether, glacial acetic acid, oils; MLD (rbt orl) = 213 mg/kg.

Veterinary products:

006-983; (with Doxylamine Succinate) A-H Injection; *Schering-Plough Animal Health.*

721 Detomidine Hydrochloride

76631-46-4 2981

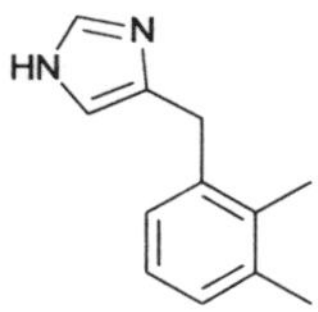

$C_{12}H_{15}ClN_2$
4-(2,3-Dimethylbenzyl)imidazole hydrochloride.
MPV-253 AII. An α_2 adrenoceptor agonist with sedative properties. Used as a sedative and analgesic in horses. mp = 114-116°; LD_{50} (mus iv) = 35 mg/kg; [hydrochloride ($C_{12}H_{15}ClN_2$, Domosedan)]: mp = 160°.

Veterinary products:

140-862; Dormosedan ™; *Orion-Farmos.*

722 Metoserpate Hydrochloride

1178-29-6 6237

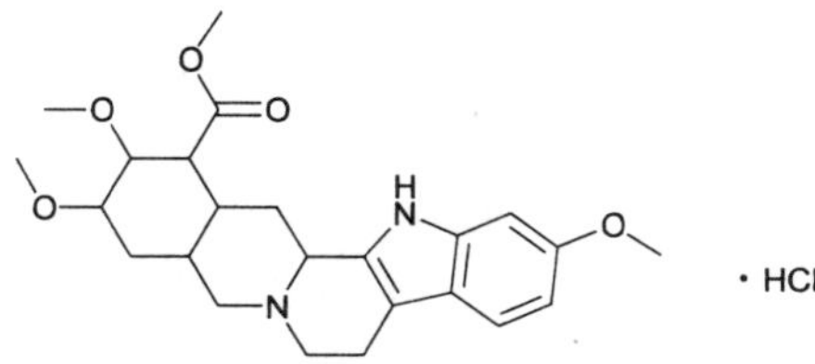

$C_{24}H_{33}ClN_2O_5$
(3β,16β,17α,18α,2-α)-11,17,18-Trimethoxyyohimban-16-carboxylic acid methyl ester hydrochloride.
Su-9064; Avicalm; Pacitran. Used as a sedative Solid; mp = 240-242° (dec); soluble in H_2O.

Veterinary products:

032-738; Pacitran; *Fort Dodge Animal Health, Divn. AHP.*

723 Pentobarbital

76-74-4 7272 200-983-8

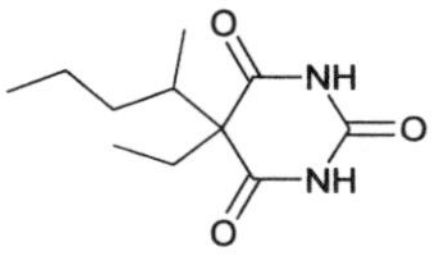

$C_{11}H_{18}N_2O_3$
5-Ethyl-5-(1-methylbutyl)-2,4,6(1H,3H,5H)-pyrimidinetrione.
Neodorm; pentobarbitone; mebubarbital. Sedative/hypnotic. Originally a widely used anesthetic agent but has been supplanted by inhalation anesthetics. mp = 129-130°.

Veterinary products:

046-789; (with Chloral Hydrate, Magnesium Sulfate) Chloropent; *Fort Dodge Animal Health, Divn. AHP.*

724 Pentobarbital Sodium

57-33-0 7272 200-323-9

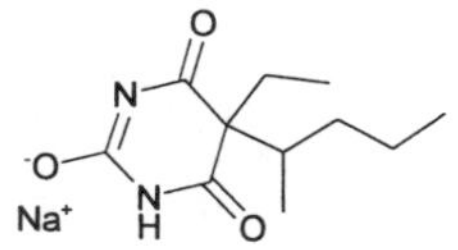

$C_{22}H_{34}CaN_4O_6$
5-Ethyl-5-(1-methylbutyl)-2,4,6(1H,3H,5H)-pyrimidinetrione calcium salt.

Nembutal; pentobarbital sodium; soluble pentobarbital; Carbrital; Narcoren; Pentone; Praecicalm; Somnopentyl; Sopental; component of: Beuthanasia-D, Carbrital, Sedalixir. Sedative/hypnotic. Originally a widely used anesthetic agent but has been supplanted by inhalation anesthetics. mp ≅ 127° (dec); freely soluble in H_2O, EtOH, insoluble in C_6H_6, Et_2O; LD_{50} (rat orl) = 118 mg/kg.

Veterinary products:

004-536; Somnopentyl Injection; *Schering-Plough Animal Health.*
010-346; (with Thiopental Sodium) Combuthal Powder; *Merial.*
045-737; Sodium Pentobarbital Injection; *Steris.*
119-807; (with Phenytoin Sodium) Beuthanasia-D-Special; *Schering-Plough Animal Health.*
200-071; (with Phenytoin Sodium) Euthasol ®; *Delmarva.*

725 Secobarbital Sodium

309-43-3 8563 206-218-4

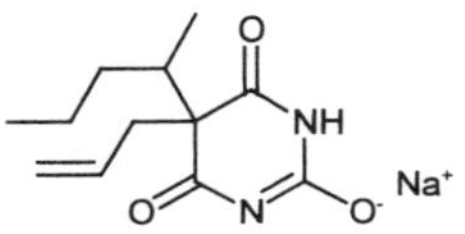

$C_{12}H_{17}N_2NaO_3$
5-Allyl-5-(1-methylbutyl)-2,4,6(1H,3H,5H)-pyrimidinetrione sodium salt.
meballymal sodium; qinalbarbitone sodium; Barbosec; Immenoctal; Pramil; Quinalspan; Sedutain; Seconal Sodium; seotalnatrium. Short acting sedative/hypnotic. Very soluble in H_2O, soluble in EtOH, insoluble in Et_2O; LD_{50} (rat orl) = 125 mg/kg.

Veterinary products:

128-967; (with Dibucaine Hydrochloride) Repose Euthanasia Solution; *Fort Dodge Animal Health, Divn. AHP.*

726 Zolazepam Hydrochloride

33754-49-3 251-668-7

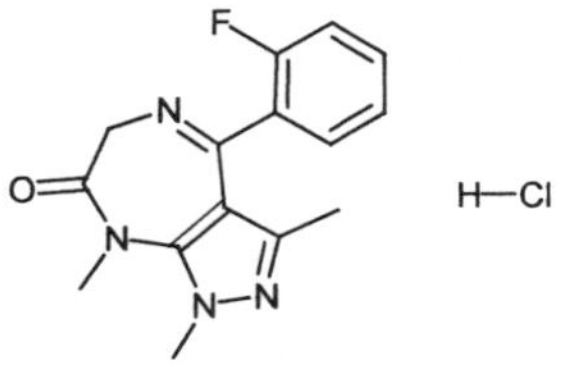

$C_{15}H_{16}ClFN_4O$
4-(o-Fluorophenyl)-6,8-dihydro-1,3,8-trimethylpyrazolo[3,4-e][1,4]diazepin-7(1H)-one hydrochloride.
CI-716; component of: Telazol. Sedative/hypnotic.

Veterinary products:

106-111; (with Tiletamine Hydrochloride) Telazol ®; *Robins.*

Sequestrants

727 Potassium Citrate

6100-05-6 7785 212-755-5

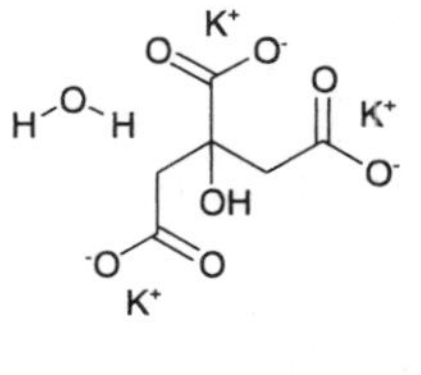

$C_6H_5K_3O_7 \cdot H_2O$
1,2,3-Propanetricarboxylic acid, 2-hydroxy-, tripotassium salt.
citrate of potash; citric acid potassium salt; E332; 2-hydroxy-1,2,3-propanetricarboxylic acid, tripotassium salt monohydrate; potassium citrate tertiary; tripotassium citrate monohydrate; tripotassium citrate. Alkalizing and buffering agent, sequestering agent. FDA GRAS; Japan, UK approved, Europe listed; FDA approved for orals, USP/NF, BP, Ph.Eur. compliant. In the FDA list of inactive ingredients, (oral solutions). Buffering agent, antiurolithic; antacids, used in orals and extended release tablets. White, hygroscopic crystals; mp = 230° (loses H_2O of crystallization at 180°); d = 1.98; soluble in H_2O (154 g/100 ml), slowly soluble in glycerol (40 g/100 ml), insoluble in 95% EtOH; pH of saturated aqueous solution = 8.5. LD_{50} (dog iv) = 167 mg/kg.

Veterinary products:

125-961; (with Citric Acid, Dextrose, Glycine, Potassium Phosphate, Sodium Chloride) Re-Sorb; Vy'trate; *Pfizer.*

728 Potassium Phosphate

7778-77-0 7829 231-913-4

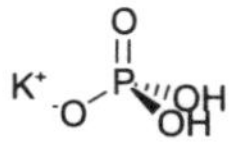

H_2KO_4P
Phosphoric acid, monopotassium salt.
Dipotassium dihydrogenphosphate; Potassium dihydrogen phosphate; Monopotassium phosphate; potassium dihydrogenorthophosphate; Sörensen's potassium phosphate. Japan, UK approved, Europe listed, FDA approved for injectables, parenterals, intravenous, ophthalmics, orals, otics; USP/NF, BP, Ph.Eur. compliant. Buffering agent and sequestrant. Used in injectables, parenterals, intravenous, ophthalmics, orals, otics; used as a urinary acidifier. dec 400°; d = 2.34; soluble in H_2O (22.2 g/100 ml)insoluble in EtOH, pH of aqueous solution is 4.4 - 4.7. Nuisance dust.

Veterinary products:

125-961; (with Citric Acid, Dextrose, Glycine, Potassium Citrate, Sodium Chloride) Re-Sorb; Vy'trate; *Pfizer.*

Serotonin Uptake Inhibitors

729 Fluoxetine

54910-89-3 4222

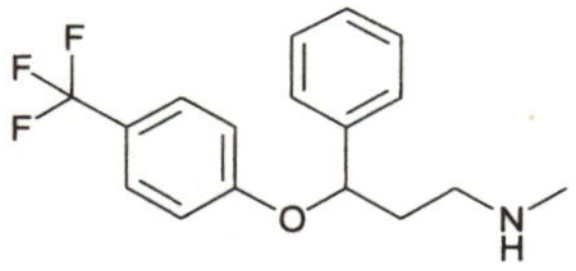

$C_{17}H_{18}F_3NO$

(±)-N-Methyl-3-phenyl-3-[(α,α,α-trifluoro-p-tolyl)oxy]-propylamine.

Serotonin uptake inhibitor. Antidepressant used in management of behavior disorders.

730 Fluoxetine Hydrochloride

59333-67-4 4222

$C_{17}H_{19}ClF_3NO$

(±)-N-Methyl-3-phenyl-3-[(α,α,α-trifluoro-p-tolyl)oxy]-propylamine hydrochloride.

LY-110140; Adofen; Fluctin; Fluoxeren; Fontex; Foxetin; Prozac, Reneuron. Serotonin uptake inhibitor. Antidepressant used in management of behavior disorders. mp = 158-159°; slightly soluble in H_2O (<14 mg/ml), more soluble in organic solvents; λ_m = 227, 264, 268, 275 nm ($E^{1\%}_{1\ cm}$ 372, 29,29,22 MeOH); LD_{50} (rat orl) = 452 mg/kg.

Thyrotropic Hormones

731 Thyroid Stimulating Hormone

9002-71-5 9931 232-664-4

Thyroid-stimulating hormone.

TSH; thyrotropic hormone; thyreotrophic hormone; TTH; Dermathycin; Thytropar. A glycoprotein hormone. Thyrotropic hormone. Also used as a diagnostic aid.

Veterinary products:

011-893; Dermathycin Injectable; *Schering-Plough Animal Health.*

Tocolytics

732 Albuterol

18559-94-9 217 242-424-0

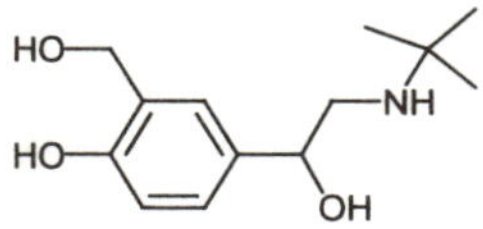

$C_{13}H_{21}NO_3$

2-(tert-Butylamino)-1-(4-hydroxy-3-hydroxymethyl-phenyl)ethanol.

salbutamol; Proventil Inhaler; Ventalin Inhaler. Bronchodilator. Ephedrine derivative used as a bronchodilator in dogs and cats. mp = 151° (also as 157-158°); soluble in most organic solvents.

733 Albuterol Sulfate

51022-70-9 217 256-916-8

$C_{26}H_{44}N_2O_{10}S$

2-(tert-Butylamino)-1-(4-hydroxy-3-hydroxymethyl-phenyl)ethanol sulfate (2:1).

Sch-13949W Sulfate; Aerolin; Asmaven; Broncovaleas; Cetsim; Cobutolin; Ecovent; Loftan; Proventil; Salbumol; Salbutard; Salbutine; Salbuvent; Sultanol; Ventelin; Ventodiscks; Ventolin; Volma. Bronchodilator and tocolytic. Ephedrine derivative used as a bronchodilator in dogs and cats.

Topical Protectants

734 Bismuth Subcarbonate

5892-10-4 1324 227-567-9

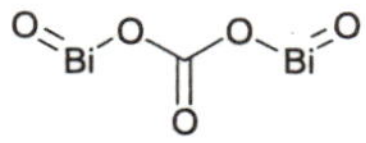

CBi_2O_5

1,3,5-Trioxo-2,4-dioxa-1,5-dibismapentane.

Basic bismuth carbonate; Bismuth oxycarbonate; Bismuth; component with Aminopentamide Hydrogen Sulfate and Attapulgite of 042-548, Amforol® Suspension (Fort Dodge Animal Health). FDA permanently listed. Used as a skin protectant and as a color additive for externally applied pharmaceuticals. Odorless, tasteless powder; insoluble in H_2O, EtO; d = 6.860; [hemihydrate]: soluble in mineral acids, concentrated acetic acid.

Veterinary products:

042-548; (with Aminopentamide Hydrogen Sulfate, Kanamycin Sulfate, Pectin) AttapulgiteAmforol ® Suspension; *Fort Dodge Animal Health, Divn. AHP.*

042-841; (with Kanamycin Sulfate, Pectin) Amforol ® Veterinary Oral Tablets; *Fort Dodge Animal Health, Divn. AHP.*

735 Bismuth Subsalicylate

14882-18-9 1327 238-953-1

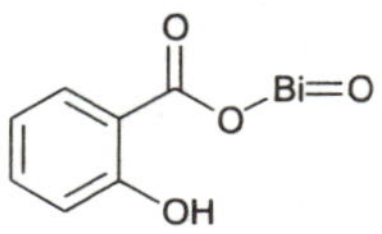

$C_7H_5BiO_4$

2-Hydroxybenzoic acid bismuth (3+) salt.

basic bismuth salicylate; oxo(salicylato)bismuth; Bismogenol Tosse Inj.; Stabisol. Antidiarrheal, antacid and antiulcerative. Also used as a lupus erythematosus suppressant. Used in animals to manage diarrhea. Insoluble in H_2O, EtOH.

Vasodilators, Coronary

736 Diltiazem
42399-41-7 3247 255-796-4

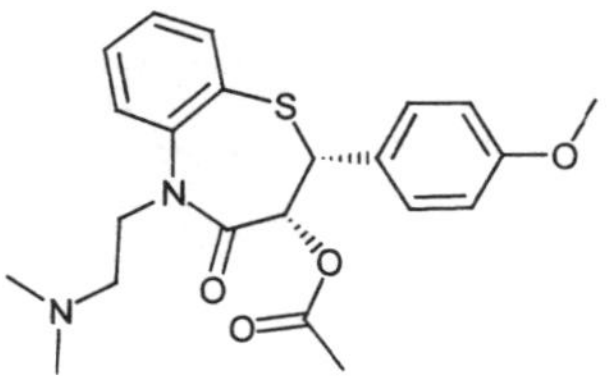

$C_{22}H_{26}N_2O_4S$
(+)-5-[2-(Dimethylamino)ethyl]-cis-2,3-dihydro-3-hydroxy-2-(p-methoxyphenyl)-1,5-benzothiazepin-4(5H)-one acetate (ester).
Calcium channel blocker with coronary vasodilating properties. Antianginal; antihypertensive; antiarrhythmic (class IV). Used to treat tachycardia in small animals.

737 Diltiazem Hydrochloride, d-cis form
33286-22-5 3247 251-443-3

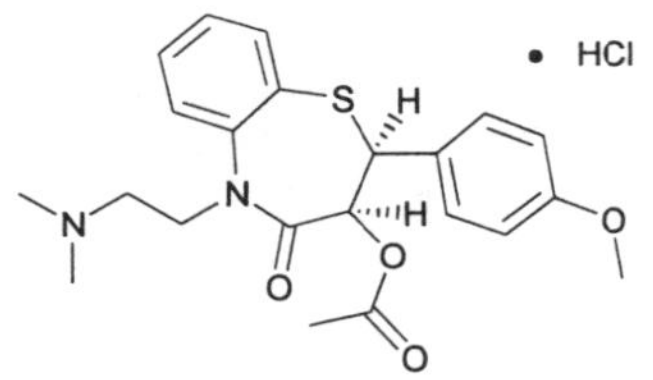

$C_{22}H_{27}ClN_2O_4S$
(+)-5-[2-(Dimethylamino)ethyl]-cis-2,3-dihydro-3-hydroxy-2-(p-methoxyphenyl)-1,5-benzothiazepin-4(5H)-one acetate (ester) monohydrochloride.
CRD-401; RG-83606; Adizem; Altiazem; Anginyl; Angizem; Britiazim; Bruzem; Calcicard; Cardizem; Citizem; Cormax; Deltazen; Diladel; Dilpral; Dilrene; Dilzem; Dilzene; Herbesser; Masdil; Tildiem. Calcium channel blocker with coronary vasodilating properties. Antianginal; antihypertensive; antiarrhythmic (class IV). Used to treat tachycardia in small animals. mp = 212°; $[\alpha]_D^{24}$ = +98.3 ± 1.4° (c = 1.002 in MeOH); soluble in H_2O, MeOH, $CHCl_3$; slightly soluble in absolute EtOH; practically insoluble in C_6H_6; LD_{50} (mmus iv) = 61 mg/kg, (mmus sc) = 260 mg/kg, (mmus orl) = 740 mg/kg.

738 Nitroglycerin
55-63-0 6704 200-240-8

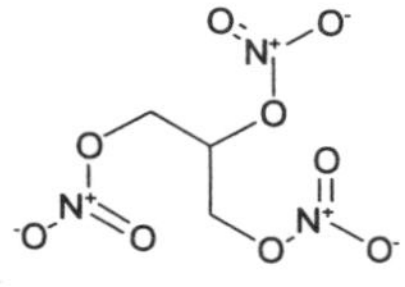

$C_3H_5N_3O_9$
1,2,3-Propanetriol trinitrate.
glyceryl trinitrate; glycerol nitric acid triester; nitroglycerol; trinitroglycerol; glonoin; trinitrin; blasting gelatin; blasting oil; SNG; Adesitrin; Angibid; Angiolingual; Anginine; Angorin; Aquo-Trinitrosan; Cardamist; Cordipatch; Coro-Nitro; Corditrine; Deponit; Diafusor; Discotrine; Gilucor; GTN; Klavikordal; Lenitral; Lentonitrina; Millisrol; Minitran; Myoglycerin; Niong; Nitradisc; Nit; Percutol; Perlinganit; Reminitrol; Susadrin; Suscard; Sustac; Sutonit; Transderm-Nitro; Transiderm-Nitro; Tridil; Trinalgon; Trinitrosan; Vasoglyn; component of: SDM No. 27, No. 37. Antianginal; vasodilator (coronary). Used in manufacture of dynamite. CAUTION: Accute poisoning can cause nausea, vomiting abdominal cramps, headache, mental confusion, delirium, bradypnea, bradycardia, paralysis, convulsions, methemoglobinemia, cyanosis, circulatory collapse, death. Chronic poisoning can cause severe headaches, hallucinations, skin rashes. Alcohol aggravates symptoms. Toxic effects may occur by ingestion, inhalation, absorption. Also as a vasodilator in heart failure. [labile form]:mp = 2.8°; [stable form]: mp = 13.5°; begins to decompose at ≅ 50°; d_{15}^{15} = 1.5991; n_D^{15} = 1.474; heat of combustion = 1580 cal/g; slightly soluble in H_2O (0.125 g/100 ml), EtOH (0.5 g/100 ml); more soluble in MeOH (2.25 g/100 ml), CS_2 (15 g/100 ml); miscible with Et_2O, Me_2CO, glacial AcOH, EtOAc, C_6H_6, nitrobenzene, C_5H_5N, $CHCl_3$, ethylene bromide, dichloroethylene; sparingly soluble in petroleum Et_2O, liquid petrolatum, glycerol.

Vasodilators, Peripheral

739 Isoxsuprine
395-28-8 5259 206-898-2

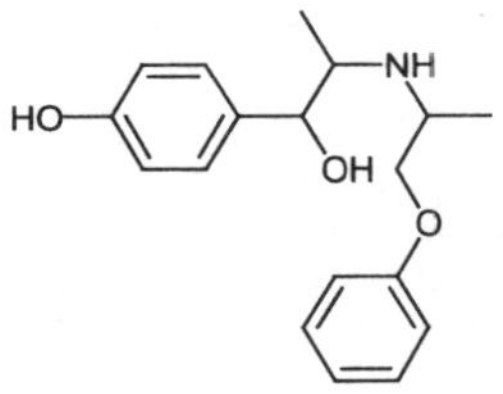

$C_{18}H_{23}NO_3$
4-Hydroxy-α-[1-[(1-methyl-2-phenoxyethyl)-amino]ethyl]benzyl alcohol.
Peripheral vasodilator. Used in horses. mp = 102.5-103.5°.

740 Isoxsuprine Hydrochloride
579-56-6 5259 209-443-6

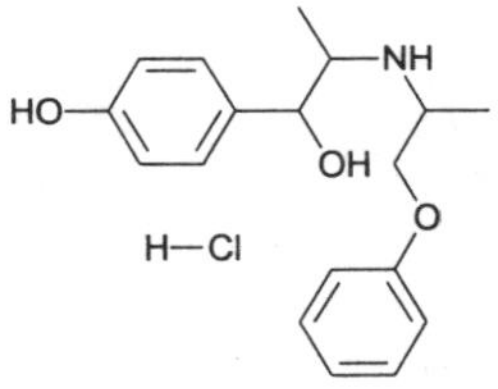

$C_{18}H_{24}ClNO_3$
4-Hydroxy-α-[1-[(1-methyl-2-phenoxyethyl)amino]-ethyl]benzyl alcohol hydrochloride.
Dilavase; Divadilan; Duviculine; Isolait; Navilox;

Suprilent; Vadosilan; Vasodilan; Vasoplex; Vasotran. Peripheral vasodilator. Used in horses. mp = 203-204°; slightly soluble in H_2O (2% at 25°); soluble in EtOH; LD_{50} (rat orl) = 1750 mg/kg, (rat ip) = 164 mg/kg.

741 Pentoxifylline

6493-05-6 7278 229-374-5

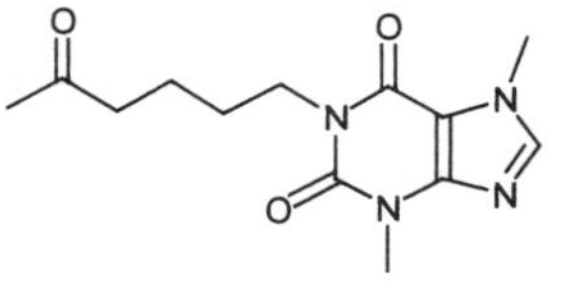

$C_{13}H_{18}N_4O_3$
3,7-Dihydro-3,7-dimethyl-1-(5-oxohexyl)-1H-purine-2,6-dione.
oxpentifylline; vazofirin; BL-191; Azupentat; Durapental; Rentylin; Torental; Trental. Vasodilator. Used to enhance microcirculation, alleviating dermatosis in dogs. mp = 105°; λ_m = 273, 208 nm ($E^{1\%}_{1\ cm}$ 365, 935); soluble in H_2O (0.0077 g/100 l), more soluble in organic solvents; LD_{50} (mus orl) = 1385 mg/kg.

742 Sodium Nitroprusside

14402-89-2 8794 238-373-9

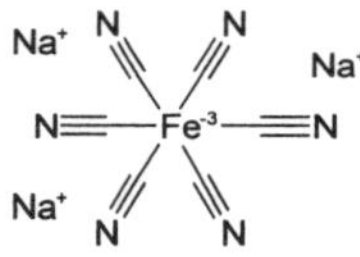

$C_5FeN_6Na_2O$
Disodium pentacyanonitrosyl ferrate(2-) dihydrate.
sodium nitroferricyanide; sodium nitroprussiate; Nipruss; Nipride [dihydrate]; Nitropress [dihydrate]. Antihypertensive. Also used as a reagent in the detection of many organic compounds and alkali sulfides. Used in veterinary medicine in management of hypertensive crises. Soluble in H_2O; slightly soluble in EtOH.

743 Tolazoline Hydrochloride

59-97-2 9645 200-447-3

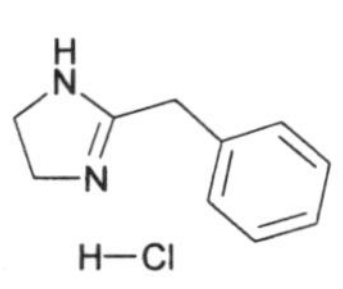

$C_{10}H_{13}ClN_2$
2-Benzyl-2-imidazoline monohydrochloride.
Lambral; Priscol; Priscoline; Vaso-Dilatan. α-adrenergic blocker used as a peripheryl vasodilator. mp = 174°; soluble in H_2O, EtOH, $CHCl_3$; slightly soluble in Et_2O, EtOAc; pH (2.5%) = 4.9-5.3.

Veterinary products:

140-994; Tolazine ™ Injection; *Lloyd.*

Vitamins and Vitamin Sources

744 Ascorbic Acid

50-81-7 867 200-066-2

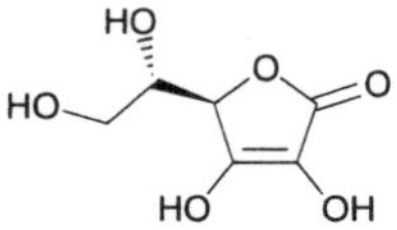

$C_6H_8O_6$
L-Ascorbic acid.
Ascorbicap; Cebione; Cecon; Cenolate; Cetane; Cetane-Caps TC; Cevalin; Cevex; Adenex; Allercorb; Ascorin; Ascorteal; Ascorvit; Cantan; Cantaxin; Catavin C; Cebicure; Cebion; Cecon; Cegiolan; Celaskon; Celin; Cenetone; Cereon; Cergona; Cescorbat; Cetamid; Cetebe; Cetemican; Cevalin; Cevatine; Cevex; Cevimin; Ce-Vi-Sol; Cevitan; Cevitex; Cewin; Ciamin; Cipca; Concemin; C-Vimin; Davitamon C; component of: Chromagen, Freancee, Mediatric, Stuartinic, Tolfrinic, Veliten. Vitamin, vitamin source. A urinary acidifier; also to treat copper-induced hepatopathy in dogs. mp = 190-192°; d = 1.65; $[\alpha]^{25}_D$ = 20.5 - 21.5° (c = 1), $[\alpha]^{23}_D$ =48° (c = 1 MeOH); λ_m = 245 nm (acid), 265 nm (neutral); soluble in H_2O (33 g/100 ml, 40 g/100 ml at 45°, 80 g/100 ml at 100°), EtOH (3.3 g/100 ml), absolute EtOH (2 g/100 ml); glycerol (1 g/100 ml), propylene glycol (5 g/100 ml), insoluble in Et_2O, $CHCl_3$, C_6H_6, petroleum ether.

745 Phylloquinone

84-80-0 7536 201-564-2

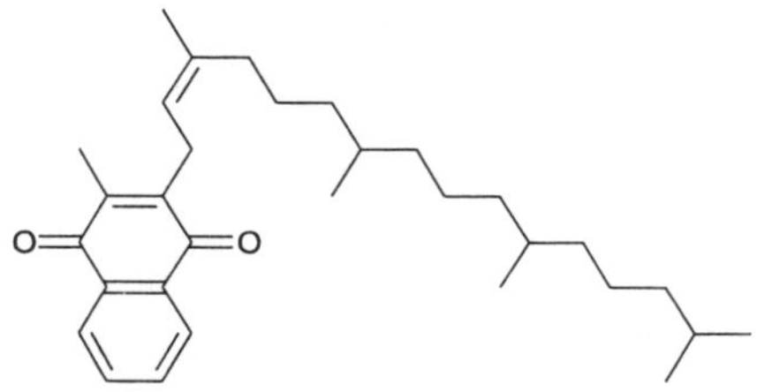

$C_{31}H_{46}O_2$
[R-[R*,R*-(E)]]-2-Methyl-3-(3,7,11,15-tetramethyl-2-hexadecenyl)-1,4-naphthalenedione.
Aquamephyton; Konakion; Mephyton; Mono-kay; Phytonadione; vitamin K_1; 3-phytylmenadione; Veda-K_1; Veta-K_1. Vitamin, vitamin source. Prothrombogenic used as an antidote for anticoagulant rodenticides such as Warfarin. $[\alpha]^{25}_D$ = -0.28° (dioxane); λ_m = 242, 248, 260, 269, 325 nm ($E^{1\%}_{1\ cm}$ 396, 4319, 383, 387, 68 petroleum ether); insoluble in H_2O; sparingly soluble in MeOH; soluble in EtOH, Me_2CO, C_6H_6, petroleum ether, C_6H_{14}, dioxane, $CHCl_3$, Et_2O, fats and oils; [dihydro form (phytonadiol; dihydrovitamin K_1)]: insoluble in H_2O, sparingly soluble in petroleum ether, freely soluble in Et_2O; [dihydro form sodium diphosphate ($C_{31}H_{48}Na_2O_8P_2$; phytonadiol sodium diphosphate; Kayhydrin)]: mp = 138°; soluble in H_2O, MeOH.

746 Vitamin E
59-02-9 10159 200-412-2

$C_{29}H_{50}O_2$
[2R-2R*(4R*,8R*)]-3,4-Dihydro-2,5,7,8-tetramethyl-2-(4,8,12-trimethyltridecyl)-2H-1-benzopyran-6-ol.
(+)-α-tocopherol; α-tocopherol; antisterility vitamin. Vitamin, vitamin source. Used with Selenium to address selenium-tocopherol deficiency in sheep and other large animals. $[\alpha]^{25}_{5461}$ = -3.0° (C_6H_6), $[\alpha]^{25}_{5461}$ = 0.32° (EtOH).

Veterinary products:

012-635; (with Sodium Selenite) BO-SE; L-SE; *Schering-Plough Animal Health.*
030-313; (with Sodium Selenite) Seletoc ® Minicaps and Caps; *Schering-Plough Animal Health.*
030-314; (with Sodium Selenite) Mu-Se; *Schering-Plough Animal Health.*
030-315; (with Sodium Selenite) E-SE; *Schering-Plough Animal Health.*
030-316; (with Sodium Selenite) Seletoc ® Injection; *Schering-Plough Animal Health.*
200-109; (with Sodium Selenite) Velenium ™; *Fort Dodge Animal Health, Divn. AHP.*

Vulnerary Agents

747 Balsam Peru
8007-00-9 976 232-352-8
Balsam of Peru; Balsamum peruvianim; Black balsam; China oil; Honduras balsam; Indian balsam; Peruvian balsam; Surinam balsam; Balsams, Peru; Balsam Peru oil; Oil balsam peru; Peru balsam; Myroxylon pereirae klotzsch resin; Myrosperum pereira balsam; Toluifera pereira balsam; Peru balsam oil; Myroxylon pereirae klotzsch oil. FDA, FEMA GRAS, Japan approved; BP, Ph.Eur. compliance. A flavor ingredient, mild antiseptic, scabicide, for treatment of skin ulcers. Dark brown liquid with aromatic odor; d = 1.150 - 1.170; insoluble in H_2O, olive oil; soluble in EtOH, $CHCl_3$, AcOH; slightly soluble in Et_2O, petroleum ether. Mildly allergenic.

Veterinary products:

031-555; (with Castor Oil, Trypsin) Trypzyme Aerosol; *Farnam.*
039-583; (with Castor Oil, Trypsin) Granulex Aerosol Spray; *Hickam.*

Miscellaneous Agents

748 β-Aminopropionitrile
151-18-8 489 205-786-0
$C_3H_6N_2$
3-Aminopropanenitrile.
3-aminopropionitrile. Liquid; bp = 185°, bp_{20} = 87-89°, bp_5 = 66-69°, bp_3 = 50-55°.

749 Citric Acid
77-92-9 2387 201-069-1

$C_6H_8O_7$
2-Hydroxy-1,2,3-propanetricarboxylic acid.
β-hydroxytricarballylic acid; aciletten; citretten; Citro; hydrocerol a; 2-Hydroxytricarballylic acid; Hydroxytricarballylic acid; citrate ion; Citralite; ; Descote® Citric Acid; Preparation of citrates, flavoring extracts, confections, soft drinks; antioxidant in foods; sequestering agent; detergent builder; metal cleaner. FDA GRAS, approved for injectables, buccals, inhalants, nasals, ophthalmics, topicals, orals, otics. Acidifier, buffering agent, pH adjuster, flavorant; anticoagulant. Used primarily as an alkalizer and bicarbonate source. mp = 153°; d = 1.665; pK_1 = 3.128, pK_2 = 4.761, pK_3 = 6.396; soluble in H_2O (59.2% w/w 20°), moderately soluble in organic solvents. LD_{50} (rat orl) = 6730 mg/kg, LD_{50} (rat ip) = 975 mg/kg.

Veterinary products:

125-961; (with Dextrose, Glycine, Potassium Citrate, Potassium Phosphate, Sodium Chloride) Re-Sorb; Vy'trate; *Pfizer.*

750 Copper Naphthenate
1338-02-9 215-657-0
Naphthenic acid, copper salts
Fungicide.

Veterinary products:

012-991; Kopertox ™; *Fort Dodge Animal Health, Divn. AHP.*
100-616; Thrush-XX; *Farnam.*

751 Cupric Glycinate
13479-54-4 2707 236-783-2

$C_4H_8CuN_2O_4$
Copper(II) glycinate.
Nutritional factor used to address copper deficiency in ruminants Deep blue, rhombic needles; loes H_2O at 123°; dec = 213-228°; soluble in H_2O.

Veterinary products:

031-971; Cuprate; *Walco.*

752 Iodinated Casein
9005-97-4 9554
Used as a feed supplement in cases of thyroid insufficiency.

Veterinary products:

005-633; Protomone Thyroactive Casein; *Agri-Tech.*
013-502; Protamone-D; *Agri-Tech.*

753 Poloxalene

9003-11-6 7721

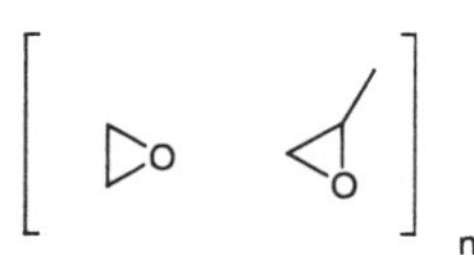

$C_{86}H_{174}O_{37}$
Polyoxypropylene, polyoxyethylene block polymer.
Meroxapol. FDA listed. Used as a non-ionic surfactant in creams. Used to prevent bloat in cattle.

Veterinary products:

032-704; Bloat Guard Top Dressing; *Pfizer.*
033-760; Bloat Guard Drench Concentrate; *Pfizer.*
033-773; Sweetlix Bloat Guard Block; *PM Ag Products.*
038-281; Bloat Guard Liquid Premix 2.2; Bloat Guard Liquid Premix 1.65; *Pfizer.*
039-729; Therabloat ®; *Pfizer.*
140-869; Purina ® Saf-T-Block Bg; Purina Bloat Block; *Purina Mills.*

754 Sodium Selenite

10102-18-8 8822 233-267-9

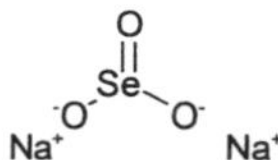

Na_2O_3Se
Selenious acid disodium salt.
Tetragonal prisms; soluble in H_2O, insoluble in EtOH; LD_{50} (rat orl) = 7 mg/kg.

Veterinary products:

012-635; (with Vitamin E) BO-SE; L-SE; *Schering-Plough Animal Health.*
030-313; (with Vitamin E) Seletoc ® Minicaps and Caps; *Schering-Plough Animal Health.*
030-314; (with Vitamin E) Mu-Se; *Schering-Plough Animal Health.*
030-315; (with Vitamin E) E-SE; *Schering-Plough Animal Health.*
030-316; (with Vitamin E) Seletoc ® Injection; *Schering-Plough Animal Health.*
200-109; (with Vitamin E) Velenium ™; *Fort Dodge Animal Health, Divn. AHP.*

755 Toluene

108-88-3 9667 203-625-9

C_7H_8
Methylbenzene.
toluol; phenylmethane; Methacide. Used as a solvent, for example, for Dichlorophen. CAUTION: Direct contact can cause dermatitis; chronic overexposure may ve hepatotoxic and nephrotoxic. Colorless, flammable liquid; mp = -95°; bp = 110.6°; d_4^{20} = 0.866; slightly soluble in H_2O (0.067 g/100 ml at 23.5°), freely soluble in EtOH, HCl_3, Et_2O, Me_2CO, AcOH, CS_2; LD_{50} (rat orl) = 7530 mg/kg.

Veterinary products:

006-691; (with Dichlorophene) Vermiplex; *Schering-Plough Animal Health.*
048-010; (with Dichlorophene) Anaplex Caps; Canine and Feline Wormer Caps; *Boehringer Ingelheim Vetmedica.*
095-333; (with Dichlorophene) Difolin No. 1 Capsules; Difolin No. 2.5 Capsules; Difolin No. 5 Capsules; Difolin No. 25 Capsules; Difolin No. 40 Capsules; *Fort Dodge Animal Health, Divn. AHP.*
101-497; (with Dichlorophene) Tiny Tiger ™ Worming Capsules For Cats; Worming Capsules For Puppies, Small Dogs And Cats; *Lambert-Kay.*
101-498; (with Dichlorophene) L.K. Worming Capsules For Dogs; *Lambert-Kay.*
102-020; (with Dichlorophene) Tri-Plex Worm Capsules; *Schering-Plough Animal Health.*
102-942; (with Dichlorophene) Pulvex Multipurpose Worm Caps; *Zema.*
103-953; (with Dichlorophene) Trivermicide Worm Capsules; *Happy Jack.*
107-189; (with Dichlorophene) Worm Capsules; *Farnam.*
117-688; (with Dichlorophene) Dichlorophene & Toluene Caps; *Texas Vitamin.*
120-671; (with Dichlorophene) Pet-Worm-Caps; *K. C. Pharmacal.*
121-557; (with Dichlorophene) THR Worm Capsules; *Natchez Animal Supply.*
126-676; (with Dichlorophene) D & T Worm Capsules; *Boehringer Ingelheim Vetmedica.*
138-900; (with Dichlorophene) Dichlorophene / Toluene; *Global.*
140-850; (with Dichlorophene) Elite Dog & Cat Wormer; *RSR Labs.*

PART II

INDEXES

CAS Registry Number Index

EINECS Number Index

NADA Number Index

Name and Synonym Index

PART III

MANUFACTURER AND SUPPLIER DIRECTORY

Manufacturer and Supplier Directory

Abbott Laboratories, Inc.
1401 Sheridan Rd.
North Chicago, IL 60064
USA
Tel: (888) 299-7416
(847) 935-4849
Fax: (847) 938-6659
www.abbott.com

ADM Animal Health & Nutrition
P.O. Box 1470
Decatur, IL 62525
USA
Tel: (217) 424-4544
www.adm-nutrition.ca

Ag-Mark, Inc.
see: Iowa Veterinary Supply

Agri Laboratories, Ltd.
P.O. Box 3103
St. Joseph, MO 64503
USA
Tel: (800) 542-8916
(816) 233-9533
Fax: (816) 233-1858
www.agrilabs.com

Agri-Tech Inc.
Address Unknown

Akey, Inc.
6531 St. Rt. 503
CS 5002
Lewisburg, OH 45338
USA
Tel: (800) 251-2339
(937) 962-2661
Fax: (937) 962-4753
www.akey.com

Alpharma, Inc.
P. O. Box 1399
1 Executive Drive
Fort Lee, NJ 07024
USA
Tel: (201) 947-7774
Fax: (201) 947-4879
www.alpharma.com

Alstoe Limited Animal Health
Granary Chambers
37-39 Burton St.
Melton, Mowbray
Leicestershire
LE13 1AF
United Kingdom
Tel: +44 0 1664411663
Fax: +44 0 1664 481527

Altana, Inc.
60 Bayliss Road
Melville, NY 11747
USA
Tel: (631) 454-7677
Fax: (631) 756-5119
www.altanainc.com

American Cyanamid Division of AHP Corp.
5 Giralda Farms
Madison, NJ 07940
USA
Tel: (973) 660-5000
www.ahp.com

American Pharmaceutical Partners Inc.
1101 Perimeter Dr.
Suite 300
Schaumburg, IL 60173-5837
USA
Tel: (888) 386-1300
(847) 330-1287
Fax: (800) 743-7082

Anika Therapeutics
236 West Cummings Park
Woburn, MA 01801
USA
Tel: (781) 932-6616
Fax: (781) 935-4120

Anthony Products Co.
Address Unknown

Argent Laboratories
8702 152nd Ave. NE
Redmond, WA 98052
USA
Tel: (425) 885-3777
Fax: (425) 885-2112
www.argent-labs.com

Ausa International Inc.
Address Unknown

Aventis Animal Nutrition
3480 Preston Ridge Rd.
Suite 650
Alpharetta, GA 30005-8891
USA
Tel: (800) 727-4433
(678) 339-1500
Fax: (678) 727-4433
www.an.aventis.com

BASF Agricultural Products
26 Davis Dr.
Research Triangle Park, NC
27709-1957
USA
Tel: (919) 547-2000
www.basf.com

BASF Aktiengesellschaft
Carl-Bosch-Strasse 38
67056 Ludwigshafen
Germany
Tel: +49 621 600
Fax: +49 621 6042525
www.basf.de/en

Baxter Pharmaceutical Products, Inc.
95 Spring St.
New Providence, NJ 07974
USA
Tel: (800) 667-0959
(908) 286-7000
Fax: (877) 702-3580
www.baxter.com

Bayer Corporation Agriculture
8400 Hawthorn Rd.
Kansas City, MO 64120-2306
USA
Tel: (816) 242-2000
www.bayer-animal-health.com

Biocraft Laboratories, Inc.
see: TEVA Pharmaceuticals

Bioproducts, Inc.
Address Unknown

Boehringer-Ingelheim Vetmedica, Inc.
2621 N. Belt Highway
St. Joseph, MO 64506-2002
USA
Tel: (800) 821-7467
Fax: (816) 233-4767
www.bi-vetmedica.com

Cargill Nutrena Feed Division
P.O. Box 9300
Minneapolis, MN 55440
USA
Tel: (952) 742-7575
www.cargill.com

Chanelle Pharmaceuticals Manufacturing Ltd.
Dublin Rd.
Main St.
Loughrea County, Galway
Ireland
Tel: +353 91841788
Fax: +353 91841303

Combe, Inc.
1101 Westchester Ave.
White Plains, NY 10604
USA
Tel: (800) 873-7400
www.combe.com

ConAgra Pet Products
see: Sergeant's Pet Products

Contemporary Products, Inc.
Address Unknown

Cooperative Research Farms
Address Unknown

Cross Vetpharm Group Ltd.
Broomhill Road, Tallaght
Dublin 24
Ireland
Tel: +3531 451 5011
Fax: +3531 451 5803
www.bimeda.com

Custom Feed Blenders Corp.
Address Unknown

Custom Feed Services Corp.
2100 N. 13th St.
Norfolk, NE 68701-2262
USA
Tel: (402) 371-6111

Cyanamid Agriculture de Puerto Rico, Inc.
see: BASF Agricultural Products

Delmarva Laboratories, Inc.
1500 Huguenot Rd.
#106
Midlothian, VA 23113-2478
USA
Tel: (804) 794-7064

Elanco Animal Health
500 East 96th St.
Suite 125 - IN
Indianapolis, IN 46240
USA
Tel: (800) 428-4441
www.elanco.com

Endo Pharmaceuticals, Inc.
220 Lake Dr.
Newark, DE 19702
USA
Tel: (800) 462-3636
Fax: (877) 329-3636
www.endoinc.com

Eon Laboratories Manufacturing, Inc.
22715 N. Conduit Blvd.
Jamaica, NY 11413-3134
USA
Tel: (718) 276-8600

Equi Aid Products, Inc.
1517 West Knudsen Drive
Phoenix, AZ 85027
USA
Tel: (800) 735-3399
(623) 492-9190
Fax: (623) 492-9385
www.equiaid.com

Eridania Beghin-Say America
Suite 1500
National City Center
110 W. Berry St.
Fort Wayne, IN 46802
USA
Tel: (219) 425-5100

EVSCO Pharmaceuticals
P.O. Box 209
Harding Hwy.
Buena, NJ 08310
USA
Tel: (800) 225-0270
(856) 691-2411
Fax: (856) 697-9711
www.evscopharm.com

Farmland Industries, Inc.
P.O. Box 20111
Kansas City, MO 64195
USA
Tel: (816) 713-7000
www.farmland.com

Farnam Companies, Inc.
P.O. Box 34820
301 West Osborn
Phoenix, AZ 85067-4820
USA
Tel: (800) 234-2269
www.farnam.com

Feed Service Co., Inc.
P.O. Box 698
303 Lundin Blvd.
Mankato, MN 56001-2719
USA
Tel: (507) 387-4464

Ferrante, John J.
Address Unknown

Fisons plc,
Pharmaceutical Division
see: Alstoe Ltd. Animal Health

Fleming Laboratories
P. O. Box 34384
2215 Thrift Rd.
Charlotte, NC 28234
USA
Tel: (704) 372-5613

Fort Dodge Animal Health
Division of American Home
Products
9401 Indian Creek Pkwy
Overland Park, KS 66210
USA
Tel: (913) 664-7000
www.ahp.com

Fort Dodge Animal Health
Division of American
Cyanamid
see: BASF Aktiengesellschaft

Furst-McNess Co.
120 E. Clark St.
Freeport, IL 61032-3300
USA
Tel: (800) 435-5100
Fax: (815) 232-9724
www.mcness.com

Global Pharmaceutical Corp.
Division of IMPAX Labs
Castor and Kensington Ave.
Philadelphia, PA 19124
USA
Tel: (800) 296-5227
(215) 289-2220
Fax: (215) 289-2223

Golden Sun Feeds
628 E. Fairchild
Danville, IL 61832
USA

Gossett Nutrition, Inc.
1676 Cascade Dr.
Marion, OH 43302
USA
Tel: (614) 383-5265

Halocarbon Products Corp.
P.O. Box 661
887 Kinderkermack Rd.
River Edge, NJ 07661
USA
Tel: (201) 262-8899
Fax: (201) 262-0019
www.halocarbon.com

Hanford Pharmaceuticals
304 Oneida Street
P.O. Box 1017
Syracuse, NY 13201-1017
USA
Tel: (800) 234-4263
(315) 476-7418
Fax: (315) 476-7434
www.hanford.com

Happy Jack, Inc.
P.O. Box 475
Snow Hill, NC 28580
USA
Tel: (800) 326-5225
www.happyjackinc.com

Hartz Mountain Products
400 Plaza Dr.
Secaucus, NJ 07094
USA
Tel: (800) 275-1414
www.hartz.com

Heinold Feeds, Inc.
P.O. Box 639
Kouts, IN 46347-0639
USA
Tel: (800) 331-8673

Heska Corporation
1613 Prospect Pkwy
Fort Collins, CO 80525
USA
Tel: (970) 493-7272
Fax: (970) 484-9505
www.heska.com

Hess and Clark, Inc.
Seventh and Orange Streets
Ashland, OH 44805-1799
USA
Tel: (419) 289-9129

Hickam, Dow B., Inc.
see: Mylan Pharmaceuticals

Hoechst-Roussel Vet.
see: Intervet International B.V.

Honeggers and Co., Inc.
Address Unknown

Hubbard Feeds, Inc.
424 N. Riverfront Dr.
P.O. Box 8500
Mankato, MN 56001
USA
Tel: (507) 388-9400
www.hubbardfeeds.com

I.D. Russell Co. Laboratories
1301 Iowa Ave.
Longmont, CO 80501
USA
Tel: (800) 821-5811
Fax: (303) 678-8983
www.alpharma.com

IMS, Inc.
Address Unknown

Inhalon Pharmaceuticals, Inc.
see: Medeva Pharmaceuticals

International Nutrition, Inc.
P.O. Box 27540
7706 I Plaza
Omaha , NE 68127-1841
USA
Tel: (402) 331-0123
Fax: (402) 331-0169
www.ini-agworld.com

Intervet International B.V.
P.O. Box 318
405 State St.
Millsboro, DE 19966
USA
Tel: (302) 934-8051
Fax: (302) 934-6087
www.intervet.com

Iowa Veterinary Supply
124 Country Club Rd.
Iowa Falls, IA 50126
USA
Tel: (800) 392-5636
(641) 648-2529
Fax: (641) 648-5994
www.iowavet.com

Ivy Animal Health
8857 Bond St.
Overland Park, KS 66214
USA
Tel: (800) 828-2192
(913) 888-2192
Fax: (913) 888-3007
www.ivyanimalhealth.com

J & R Specialty Supply Co.
Address Unknown

J. C. Feed Mills
P.O. Box 224
1050 Sheffield Ave.
Waterloo, IA 50704
USA
Tel: (319) 232-0000

Jorgensen Laboratories, Inc.
1450 N. Van Buren Ave.
Loveland, CO 80538
USA
Tel: (800) 525-5614
(970) 669-2500
Fax: (970) 663-5042

K. C. Pharmacal
8235 Melrose Dr.
Lenexa, KS 66214
USA
Tel: (800) 821-5569
(913) 888-0945
Fax: (800) 888-4940
www.kcpharmacal.com

Kerber Milling Co.
P.O. Box 96
1817 E. Main St.
Emmetsburg, IA 50536-0096
Tel: (712) 852-2712

Koffolk
P.O. Box 1098
Tel Aviv 61010
Israel
Tel: +972-3-9273100
Fax: +972-3-9230341
www.koffold.com

Lambert Kay
P.O. Box 1418
Cranbury, NJ 08512-0187
USA
Tel: (609) 951-4700
Fax: (609) 951-4742
www.lambertkay.com

Lloyd, Inc.
P.O. Box 130
Shenandoah, IA 51601
USA
Tel: (800) 831-0004
www.lloydinc.com

Luitpold Pharmaceuticals, Inc.
26 Precision Drive
Shirley, NY 11967
USA
Tel: (800) 458-0163
(516) 924-4000
www.luitpold.com

M&M Livestock Products Co./ Biotechnologies
310 E. Broadway
Eagle Grove, IA 50533
USA
Tel: (800) 247-4820
(515) 448-5371
Fax: (515) 448-3407
www.mmbiotech.com

Macleod Pharmaceuticals, Inc.
2600 Canton Court
Fort Collins, CO 80525
USA
Tel: (970) 482-7254
Fax: (970) 482-7454

Marsam Pharmaceuticals, Inc.
24 Olney Ave.
Cherry Hill, NJ 08034
USA
Tel: (856) 424-5600
Fax: (856) 751-8784
www.schein-rx.com

Medeva Pharmaceuticals, Inc.
755 Jefferson Rd.
Rochester, NY 14623
USA

Medicis Dermatologics / Medicis Pharmaceutical Corp.
8125 N. Hayden Rd.
Scottsdale, AZ 85258-2463
USA
Tel: (480) 808-8800
Fax: (480) 808-0822
www.medicis.com

Med-Pharmex, Inc.
2727 Thompson Creek Rd.
Pomona, CA 91767-1861
USA
Tel: (909) 593-7875
www.med-pharmex.com

Merial Ltd.
2100 Ronson Rd.
Iselin, NJ 08830-3077
USA
Tel: (888) 637 4251
www.merial.com

Micro Beef Technologies
P.O. Box 9262
Amarillo, TX 79105
USA
Tel: (800) 858-4330

Micro Chemical, Inc.
see: Micro Beef Technologies

Mid-Continent Agrimarketing, Inc.
1465 North Winchester
Olathe, KS 66061-5881
USA
Tel: (800) 547-1392
(913) 768-8967
Fax: (913) 768-8968
www.mid-conagri.com

Monsanto Co.
800 N. Lindbergh Blvd.
St. Louis, MO 63167
USA
Tel: (314) 694-1000
www.monsanto.com

MoorMan Manufacturing Co.
1000 N. 30th. St.
Quincy, IL 62301-3400
USA
Tel: (217) 222-7100
www.moormans.com

Mylan Pharmaceuticals
781 Chestnut Ridge Rd.
P.O. Box 4310
Morgantown, WV 26505
USA
Tel: www.mylan.com

Natchez Animal Supply Co.
201 John Junkin Dr.
Natchez, MS 39120
USA
Tel: (800) 647-6760

Neogen Corp.
620 Lesher Pl.
Lansing, MI 48912
USA
Tel: (517) 372-9200
Fax: (517) 372-0108
www.neogen.com

Norbrook Laboratories, Ltd.
Station Works
Camlough Road
Newry, Co Down
BT35 6JP
Northern Ireland
Tel: +44 0 28 3026 4435
Fax: +44 0 28 3026 1721
www.norbrook.co.uk

Norco Mills of Norfolk, Inc.
P.O. Box 77
Norfolk, NE 68701
USA
Tel: (402) 371-6897
Fax: (402) 371-0415

Novartis Animal Health, Inc.
3200 Northline Ave.
Suite 300
Greensboro, NC 27408
USA
www.ah.novartis.com

Nutra-Blend Corp.
3200 2nd St.
Neosho, MO 64850
USA
Tel: (417) 451-6111

Nutribasics Co.
3801 N. Hawthorne St.
Chattanooga, TN 37406-1310
USA
Tel: (423) 629-1469

Nylos Trading Co., Inc.
49 S. Ridge Rd.
Pomona, NY 10970
USA
Tel: (845) 354-7077

Orion-Farmos/Orion Pharma
P.O. Box 65
FIN-02101
Espoo
Finland
Tel: +358 10 4291
Fax: +358 10 429 2801
www.orion.fi

Orphan Medical, Inc.
13911 Ridgedale Dr.
Suite 250
Minnetonka, MN 55305
USA
Tel: (888) 867-7426
www.orphan.com

Oxis International, Inc.
6040 N. Cutter Circle
Suite 317
Portland, OR 97217-3935
USA
Tel: (503) 283-3911
Fax: (503) 283-4058
www.oxis.com

Peavey Co.
Address Unknown

Pegasus Laboratories, Inc.
8809 Ely St.
Pensacola, FL 32514-7064
USA
Tel: (850) 478-2770
www.pegasusanimalcare.com

Pennfield Oil Co.
14040 Industrial Rd.
Omaha, NE 68144-3351
USA
Tel: (402) 330-6000

Peptech Animal Health Pty
Locked Bag 2053
North Ryde, NSW
Australia 1670
Tel: 612 98708788
Fax: 612 98708786
www.peptech.com

Pfizer Animal Health
812 Springdale Dr.
Exton, PA 19341-3100
USA
Tel: (610) 363-3100
www.pfizer.com

Pharmacia Corp.
100 Route 206 North
Peapack, NJ 07977
USA
Tel: (888) 768-5501
(908) 901-8000
Fax: (908) 901-8379
www.pharmacia.com

Phoenix Pharmaceutical
P.O. Box 6457
St. Joseph, MO 64506
USA
Tel: (800) 759-3644
Fax: (816) 232-2539

Planalquimica Industrial Ltda.
Rua des Magnolias nr. 2405,
Jardim das Bandeiras
Campinas, Sao Paulo
Brazil
Tel: CEP 13053-120

PLIVA USA
150 E. 58th St.
New York, NY 10155
USA
Tel : (212) 832-8970
Fax: (212) 832-8971
www.pliva.hr

PM Ag Products, Inc.
7600 JW Peavy
Houston, TX 77011
USA
Tel:(713) 967-5145

PM Resources, Inc.
13001 St. Charles Rock Rd.
Bridgeton, MO 63044
USA

Premo Pharmaceutical Labs.
Purina Mills, Inc.
1401 S. Hanley Rd.
St. Louis, MO 63144-2987
USA
Tel: (800) 227-8941
(314) 768-4100
Fax: (314) 768-4894
www.purinamills.com

Quali-Tech Products, Inc.
318 Lake Hazeltine Dr.
Chaska, MN 55318
USA
Tel: (952) 448-5151
Fax: (952) 448-3603
www.qualitechco.com

Rath Packing Co.
POB 330
Waterloo, IA 50704
USA

R.P. Scherer, Inc.
645 Martinsville Rd.
Suite 200
Basking Ridge, NJ 07920
USA
Tel: (888) 636-1919
Fax: (908) 580-1500

Rhodia Ltd.
259 Prospect Plains Rd.
CN 7500
Cranbury, NJ 08512-7500
USA
Tel: (609) 860-4000
www.rhodia.com

Rhone-Poulenc
see: Aventis Animal Nutrition

Ridley, Inc.
17 Speers Rd.
Winnipeg, Manitoba
R2J 1M1
Canada
Tel: (204) 956-1717
Fax: (204) 956-1687
www.ridleyinc.com

Robins, A. H., Co.
Roche Vitamins, Inc.
45 Waterview Blvd.
Parsippany, NJ 07054-1298
USA
Tel: (800) 526-0189
www.roche.com/vitamins

RSR Laboratories, Inc.
Address Unknown

Schering-Plough Animal Health Corp.
10409 I St.
Omaha, NE 68127
USA
Tel: (800) 521-5767
Fax: (800) 462-3720
usa.spah.com

Searle World Headquarters
5200 Old Orchard Rd.
Skokie, IL 60077
USA
Tel: (847) 982-7000
Fax: (847) 470-1480
www.searlehealthnet.com

Seeco, Inc.
Address Unknown

Sergeant's Pet Products
4242 BF Goodrich Rd.
Memphis, TN 38118
USA
Tel: (901) 362-1950
Fax: (901) 795-9525

Sioux Pharm, Inc.
206 3rd St. NW
Sioux Center, IA 51250
USA
Tel: (712) 722-4697
Fax: (712) 722-4698

South St. Paul Feeds, Inc.
Address Unknown

Southern Micro-Blenders, Inc.
Address Unknown

Springfield Milling Corp.
see: Ridley Inc.

Squire Laboratories, Inc.
see: Neogen Corp.

Steris Laboratories, Inc.
620 N. 51st Ave.
Phoenix, AZ 85043-4705
USA
Tel: (800) 692-9995
(602) 278-1400
Fax: (602) 447-3537
www.schein-rx.com

Stockton Hay & Grain Co.
see: Cargill Nutrena Feed Division

Summit Hill Laboratories
Tinton Falls Business Ctr.
One Sheila Dr.
Tinton Falls, NJ 07724-2658
USA
Tel: (732) 933-0800

Swisher Feed Division
117 2nd St. SE
Swisher, IA 52338-9602
USA
Tel: (319) 857-4111

TEVA Pharmaceuticals, USA
1090 Horsham Rd.
P.O. Box 1090
North Wales, PA 19454-1090
USA
Tel: (215) 591-3000
www.tevapharmusa.com

Texas Vitamin Co.
Address Unknown

Triple "F", Inc
10104 Douglas Ave.
Des Moines, IA 50322-7983
USA
Tel: (515) 254-1200
www.fff.com

United Vaccines
2826 Latham Dr.
Madison, WI 53713-4616
USA
Tel: (608) 277-2030
Fax: (608) 277-2120

Vetem SpA
Cento direzionale Colleoni
Palazzo Orione 2, 20041
Agrate Brianza
Milano
Italy

Veterinary Laboratories, Inc.
12340 Santa Fe Dr.
Lenexa, KS 66215
USA
Tel: (913) 888-7500

Veterinary Research Associates, Inc.
113 Commerce Dr.
Fort Collins, CO 80524
Tel: (970) 498-0363
Fax: (970) 498-0394

Veterinary Services, Inc.
Address Unknown

Veterinary Specialties, Inc.
785 Oakwood Rd.
S-108
Lake Zurich, IL 60047
USA
Tel: (847) 726-1179
Fax: (847) 726-1469

Walco International, Inc.
520 S. Main St.
Grapevine, TX 76051
USA
Tel: (817) 601-6000
Fax: (817) 601-3098

Waterloo Mills Co.
P.O. Box 1227
Waterloo, IA 50704
USA
Tel: (319) 234-7756

Wayne Feed Division
see: Hubbard Feeds Inc.

Webel Feeds, Inc.
see: Eridania Beghin-Say America

Wendt Laboratories, Inc.
100 Nancy Dr.
Belle Plain, MN 56011
USA
Tel:(800) 328-5890

West Agro, Inc.
11100 N. Congress Ave.
Kansas City, MO 64153-1296
USA
Tel: (816) 891-1600
Fax: (816) 891-1595
www.westagro.com

Western Chemical, Inc.
1269 Lattimore Rd.
Ferndale, WA 98248
USA
Tel: (800) 283-5292
Fax: (360) 384-5898

Wildlife Laboratories, Inc.
1401 Duff Drive
Suite 600
Fort Collins, CO 80524-2739
USA
Tel: (877) 883-9283
www.wildpharm.com

Wyeth-Ayerst Pharmaceuticals
555 Lancaster Ave.
St. Davids, PA 19087
USA
Tel: (610)902-1200
www.wyeth.com

Yoder Feed
Address Unknown

Youngs, Inc.
Address Unknown

Zema Corp.
6514 Chapel Hill Rd.
Raleigh, NC 27607
USA
Tel: (800) 334-5530
(919) 851-9494
Fax: (919) 859-2170